Sustainable Agriculture Reviews

Volume 48

Series Editor

Eric Lichtfouse, Aix-Marseille University, CNRS, IRD, INRAE, Coll France, CEREGE, Aix-en-Provence, France

Other Publications by Dr. Eric Lichtfouse

Books
Scientific Writing for Impact Factor Journals
https://www.novapublishers.com/catalog/product_info.php?products_id=42242

Environmental Chemistry
http://www.springer.com/978-3-540-22860-8

Sustainable Agriculture
Volume 1: http://www.springer.com/978-90-481-2665-1
Volume 2: http://www.springer.com/978-94-007-0393-3

Book series
Environmental Chemistry for a Sustainable World
http://www.springer.com/series/11480

Sustainable Agriculture Reviews
http://www.springer.com/series/8380

Journals
Environmental Chemistry Letters
http://www.springer.com/10311

Sustainable agriculture is a rapidly growing field aiming at producing food and energy in a sustainable way for humans and their children. Sustainable agriculture is a discipline that addresses current issues such as climate change, increasing food and fuel prices, poor-nation starvation, rich-nation obesity, water pollution, soil erosion, fertility loss, pest control, and biodiversity depletion.

Novel, environmentally-friendly solutions are proposed based on integrated knowledge from sciences as diverse as agronomy, soil science, molecular biology, chemistry, toxicology, ecology, economy, and social sciences. Indeed, sustainable agriculture decipher mechanisms of processes that occur from the molecular level to the farming system to the global level at time scales ranging from seconds to centuries. For that, scientists use the system approach that involves studying components and interactions of a whole system to address scientific, economic and social issues. In that respect, sustainable agriculture is not a classical, narrow science. Instead of solving problems using the classical painkiller approach that treats only negative impacts, sustainable agriculture treats problem sources.

Because most actual society issues are now intertwined, global, and fast-developing, sustainable agriculture will bring solutions to build a safer world. This book series gathers review articles that analyze current agricultural issues and knowledge, then propose alternative solutions. It will therefore help all scientists, decision-makers, professors, farmers and politicians who wish to build a safe agriculture, energy and food system for future generations.

More information about this series at http://www.springer.com/series/8380

Inamuddin • Mohd Imran Ahamed
Eric Lichtfouse
Editors

Sustainable Agriculture Reviews 48

Pesticide Occurrence, Analysis and Remediation Vol. 2 Analysis

Springer

Editors
Inamuddin
Department of Applied Chemistry
Zakir Husain College of Engineering and Technology
Faculty of Engineering and Technology
Aligarh Muslim University
Aligarh, India

Mohd Imran Ahamed
Department of Chemistry
Faculty of Science
Aligarh Muslim University
Aligarh, India

Eric Lichtfouse
Aix-Marseille University
CNRS, IRD, INRAE, Coll France, CEREGE
Aix-en-Provence, France

ISSN 2210-4410 ISSN 2210-4429 (electronic)
Sustainable Agriculture Reviews
ISBN 978-3-030-54721-9 ISBN 978-3-030-54719-6 (eBook)
https://doi.org/10.1007/978-3-030-54719-6

This Springer imprint is published by the registered company Springer Nature Switzerland AG
The registered company address is: Gewerbestrasse 11, 6330 Cham, Switzerland

Preface

Pesticides are a group of complex chemical compounds that are used to kill pests. They are widely used in agriculture to increase crop yield and minimize post-harvest losses in addition to pest control. Pesticides enter into the biological system and their residues in different body fluids like blood, urine, serum, breast milk, and semen are analyzed. This book overviews the pesticide residues analysis using electroanalytical, chromatographic, and spectral analytical techniques. This book is an archival reference guide for faculty, students, researchers, and professionals who are working in environmental science, geography, toxicology, agriscience, and analytical chemistry.

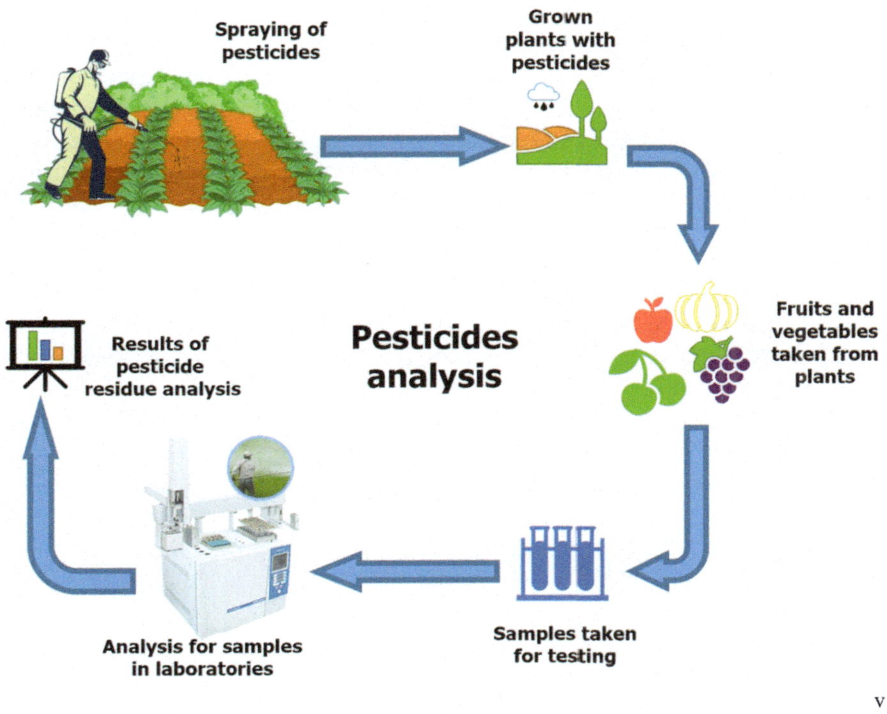

Chapter 1 by De Souza et al. presents the description of the main electroanalytical techniques used in pesticide analysis, which include potentiometry, amperometry, voltammetry, and impedance electrochemical spectroscopy. Some discussions about the specificity of each technique, the components of instrumentation and appropriate choice regarding the electroanalytical technique, electrode types, and the experimental and electroanalytical parameters are presented. Chapter 2 by de Córdova and Martínez describes in detail the main analytical methodologies available today for the treatment of food and environmental samples for pesticide analysis. Particular emphasis is placed on the techniques developed to miniaturize and simplify the extraction and purification stages and to meet the requirements of green chemistry. Their advantages, disadvantages, and prospects are also discussed. Chapter 3 by Torres et al. describes several methods to analyze pesticide residues in food and environmental samples by large volume injection or by on-line coupled liquid chromatography-gas chromatography using the through oven transfer adsorption–desorption interface after presenting a description of the through oven transfer adsorption–desorption interface and its operation mode. Chapter 4 by Patel discusses in detail the current analytical techniques available to detect the pesticides that are being used heavily in recent times. The routinely used quantification procedures are described in general. Additionally, it details the specific and other detection methods employed for water, air, soil, and food products. Chapter 5 by Saha et al. discusses the use of bioindicator using a living organism or its part or group of the organism for monitoring environmental pesticide pollution. Pesticides contamination in air, soil, and water and their route of entry into the environmental matrices is discussed and a general idea is given on well-known natural species used for bioindication of environmental pesticide pollution. Chapter 6 by Anuradha and Singh discusses the characteristics of pesticides in detail. Their presence in the water system have adverse effects on ecosystems. Their removal becomes necessary from our environment, and very innovative and effective technologies are used for their removal. Chapter 7 by Rani et al. describes the recent status of types of hazardous pesticides and their removal from water by engineered nanomaterials like TiO_2 and ZnO. Engineered nanomaterials are highly used for the treatment of organochlorine and organophosphorus pesticides. Major gaps in implementing hybrid technologies in water treatment and their future scope are also discussed. Chapter 8 by Ali et al. discusses the impacts of injudicious use of pesticides on human health and the environment. Tons of synthetic chemicals are deposited into the agroecosystems which influence the non-target organism directly or indirectly. This exposes these organisms to hazards and health problems through pesticide residues in water, soil, and air. Chapter 9 by Kumar and Joshiba explains the pollution of water due to pesticides and the different analyses available for the measurement of pesticides which include chromatography, electrophoresis, spectrometry, biological assay, and biosensors. Different types of pesticides and their impact on the environment are presented in this chapter. The scope for further research is explained in a simple manner to readers so they can begin their research on this topic.

Aligarh, India Inamuddin

Aligarh, India Mohd Imran Ahamed

Aix-en-Provence, France Eric Lichtfouse

Contents

About the Editors

Inamuddin is an assistant professor in the Department of Applied Chemistry, Aligarh Muslim University, Aligarh, India. He has extensive research experience in multidisciplinary fields of analytical chemistry, materials chemistry, electrochemistry, renewable energy and environmental science. He has published about 175 research articles in various international scientific journals, 18 book chapters, and 115 edited books with multiple well-known publishers. His current research interests include ion exchange materials, a sensor for heavy metal ions, biofuel cells, supercapacitors and bending actuators.

Mohd Imran Ahamed received his Ph.D degree on the topic "Synthesis and characterization of inorganic-organic composite heavy metals selective cation-exchangers and their analytical applications", from Aligarh Muslim University, Aligarh, India in 2019. He has published several research and review articles in the journals of international recognition. He has also edited various books which are published by Springer, CRC Press Taylor & Francis Asia Pacific and Materials Science Forum LLC, U.S.A. He has completed his B.Sc. (Hons) Chemistry from Aligarh Muslim University, Aligarh, India, and M.Sc. (Organic Chemistry) from Dr. Bhimrao Ambedkar University, Agra, India. His research work includes ion-exchange chromatography, wastewater treatment, and analysis, bending actuator and electrospinning.

Eric Lichtfouse is a biogeochemist at Aix Marseille University who has invented carbon-13 dating, a molecular-level method allowing to study the dynamics of organic compounds in temporal pools of complex environmental media. He is Chief Editor of the journal Environmental Chemistry Letters, and the book series Sustainable Agriculture Reviews and Environmental Chemistry for a Sustainable World. He is the author of the book Scientific Writing for Impact Factor Journals, which includes an innovative writing tool: the Micro-Article.

Contributors

Sajjad Ali Department of Entomology, Faculty of Agriculture & Environmental Sciences, The Islamia University of Bahawalpur, Bahawalpur, Pakistan

Jesús Villén Altamirano Escuela Técnica Superior de Ingenieros Agrónomos y de Montes, Departamento de Ciencia y Tecnología Agroforestal y Genética, Universidad de Castilla-La Mancha, Albacete, Spain

Anuradha Department of Zoology, Raghuveer Singh Government P.G. College, Lalitpur, Uttar Pradesh, India

Partha P. Choudhury ICAR-Indian Institute of Horticultural Research, Bengaluru, Karnataka, India

Djenaine De Souza Laboratory of Electroanalytical Applied to Biotechnology and Food Engineering (LEABE), Multidisciplinary Group of Research, Science and Technology (RMPCT) Chemistry Institute, Federal Uberlândia University, Patos de Minas, MG, Brazil

María Luisa Fernández de Córdova Department of Physical and Analytical Chemistry, Faculty of Experimental Sciences, University of Jaén, Jaén, Spain

Diego L. Franco Laboratory of Electroanalytical Applied to Biotechnology and Food Engineering (LEABE), Multidisciplinary Group of Research, Science and Technology (RMPCT) Chemistry Institute, Federal Uberlândia University, Patos de Minas, MG, Brazil

Rakesh Kumar Ghosh ICAR-National Institute of Natural Fibre Engineering and Technology, Kolkata, West Bengal, India

Danielle Gonçalves-Filho Laboratory of Electroanalytical Applied to Biotechnology and Food Engineering (LEABE), Multidisciplinary Group of Research, Science and Technology (RMPCT) Chemistry Institute, Federal Uberlândia University, Patos de Minas, MG, Brazil

Azhar Hussain Department of Soil Science, Faculty of Agriculture & Environmental Sciences, The Islamia University of Bahawalpur, Bahawalpur, Pakistan

G. Janet Joshiba Department of Chemical Engineering, Sri Sivasubramaniya Nadar College of Engineering, Chennai, Tamil Nadu, India

P. -K. Jesna Research Centre of ICAR-Central Inland Fisheries Research Institute, Bengaluru, Karnataka, India

Keshu Dr B R Ambedkar National Institute of Technology, Jalandhar, Punjab, India

Eulogio José Llorent Martínez Department of Physical and Analytical Chemistry, Faculty of Experimental Sciences, University of Jaén, Jaén, Spain

Hitika Patel Food Chemist and Microbiologist, Analytical and Environmental Services, Vadodara, Gujarat, India

Manviri Rani Malaviya National Institute of Technology, Jaipur, Rajasthan, India

Ajoy Saha Research Centre of ICAR-Central Inland Fisheries Research Institute, Bengaluru, Karnataka, India

Asif Sajjad Department of Entomology, Faculty of Agriculture & Environmental Sciences, The Islamia University of Bahawalpur, Bahawalpur, Pakistan

P. Senthil Kumar Department of Chemical Engineering, Sri Sivasubramaniya Nadar College of Engineering, Chennai, Tamil Nadu, India

Qaiser Shakeel Department of Plant Pathology, Faculty of Agriculture & Environmental Sciences, The Islamia University of Bahawalpur, Bahawalpur, Pakistan

Uma Shanker Dr B R Ambedkar National Institute of Technology, Jalandhar, Punjab, India

Jagvir Singh Department of Chemistry, ARSD College (University of Delhi), New Delhi, India

Rosa M. Toledano Torres Facultad de Educación de Albacete, Departamento de Química-Física, Universidad de Castilla-La Mancha, Albacete, Spain

Muhammad Irfan Ullah Department of Entomology, College of Agriculture, University of Sargodha, Sargodha, Pakistan

Ana M. Vázquez Moliní Facultad de Educación de Albacete, Departamento de Química-Física, Universidad de Castilla-La Mancha, Albacete, Spain

Jyoti Yadav Malaviya National Institute of Technology, Jaipur, Rajasthan, India

Chapter 1
Pesticide Residues Analysis by Electroanalytical Techniques

Djenaine De Souza, Danielle Gonçalves-Filho, and Diego L. Franco

Abstract Despite the historical importance of pesticide in crops to control pests, in the last decades its indiscriminate use has been responsible by harmful to human health resulting in diseases as cancer and neurological disorders. This allowed the emergence of novel perspectives in pesticide analysis, in which the electroanalytical techniques have been widely used as fast, reliable, sensitive, low-cost methods to identify and quantify these compounds with few or none sample pretreatments, reduced waste generation, possibility of detection of pesticides from different chemical class and in situ detection. In the electroanalytical techniques are measured some electrical properties such as potential, current, resistance, charge among others, which can be related to the concentration and identification of the pesticide of under study. The resulting electroanalytical signals are ever related to the physicochemical properties of pesticides such as the presence of functional groups that enable reduction and or oxidation reactions, presence ionizable hydrogens and the ability of promote a specific enzyme inhibition.

This chapter presents the most important advances in pesticide electroanalysis obtained from the evaluation in the scientific databases over the last 10 years (2009–2019). It was observed the extensive use of potentiometry, amperometry, voltammetry, stripping voltammetry, and impedance electrochemical spectroscopy. We reviewed the published works and noticed the predominance in the use of voltammetry, mainly square wave voltammetry, and a considerable increase in the employ of the electrochemical impedance spectroscopy in pesticide analysis. This occurs because the square wave voltammetry permits the acquisition of high analytical sensitivity, while the electrochemical impedance spectroscopy permits the analysis of pesticides without electronic transference. It was observed that the success in the identification and quantification of pesticides, from different chemical class, is related to the adequate choice of working electrode material, solvent,

D. De Souza (✉) · D. Gonçalves-Filho · D. L. Franco
Laboratory of Electroanalytical Applied to Biotechnology and Food Engineering (LEABE), Multidisciplinary Group of Research, Science and Technology (RMPCT) Chemistry Institute, Federal Uberlândia University, Patos de Minas, MG, Brazil
e-mail: djenaine@ufu.br; diegoleoni@ufu.br

Inamuddin et al. (eds.), *Sustainable Agriculture Reviews 48*, Sustainable Agriculture Reviews 48, https://doi.org/10.1007/978-3-030-54719-6_1

supporting electrolyte type and pH values, and also the proper selection of electroanalytical technique and their respective parameters, which will define the sensitivity and selectivity in the pesticides analysis. In this way, the use of electroanalytical techniques in pesticide analysis requires previous knowledge about the specificity of each electroanalytical technique, and the main components of the instrumentation, such as the equipment type and the electrochemical cell, which is constituted by ionic and electric conductors. Furthermore, the analysis of experimental information obtained after an adequate evaluation in the experimental and electroanalytical parameters provides the understanding of the interface electrode/solution, allows the evaluation of the redox potential of functional groups, which leads to a suitable identification and quantification of pesticides, even in complex systems like natural waters, soil, and *in natura* foods.

Keywords Pesticides · Electroanalytical techniques · Potentiometry · Amperometry- Voltammetry · Electrochemical impedance spectroscopy · Carbon-based electrodes · Mercury-based electrodes · Chemically modified electrodes · Biosensors

1.1 Introduction – Pesticides Analysis

In the last three decades, the indiscriminate and abusive use of pesticides in agricultural practices, with few guidelines or restrictions, has promoted undesirable transport and contamination in the drinking water, air, soil, and foods (Nowell et al. 2014; Uwizeyimana et al. 2017). Many pesticides, from different chemical class, present intense mobility through the environment, persistent effects and considerable toxicity. If employed in inadequate quantities and without the respective regard between the time of application and food consumption can promote health problems such as cancer, neurological disorders, and reproductive disorders (Mostafalou and Abdollahi 2013; Kim et al. 2017; Kalliora et al. 2018; Requena et al. 2018; Sabarwal et al. 2018; Ma et al. 2019; Rohlman et al. 2019). Besides, the lixiviation process by rain and wind promote the dislodgment of the pesticides to natural water and soil to other regions without pesticide use, resulting in the environmental and public health problems (Ochoa and Maestroni 2018).

Therefore, different science areas have studied the effects of the inappropriate use of pesticides and evaluated the specificity of remediation process. Development of the analytical methodologies for identification and quantification of these compounds in environmental, biological and industrial samples has also been a key area of research, as shown in other chapters of this book. The toxics effects, the remediation and analysis techniques are dependent of the chemical nature of pesticides, which are differentiated by specific functional groups that characterize them structurally, producing pesticides physicochemical properties, such as partition coefficients, solubility, and chemical stability (Eddleston and Bateman 2012; Samsidar et al. 2018). Besides, the mechanisms of action, degradation and elimination processes of pesticides are also related to chemical classes of pesticides (Hornsby et al. 1996; Crossley 2004).

In most countries, the health and environment regulatory agencies indicate that the analysis of pesticides be performed by chromatographic techniques, that permit the separation, identification, and quantification of its residues in different samples (Ochoa and Maestroni 2018). However, the adequate choices of the analytical technique is based on physicochemical characteristics of pesticide with the availability instrumentation (Gaweł et al. 2019). For pesticides that are volatile and thermally stable, gas chromatography is the recommended technique, whereas the use of high-performance liquid chromatography depends only that the solubility of the pesticide in the mobile phase choice, which can be changed according to necessity of the analysis (Miller 2005; Francesquett et al. 2019).

Chromatographic techniques require the adequate choices of an adequate stationary and a mobile phase, and its selection also depends on physicochemical properties of pesticides. After this, the resultant data, known as chromatograms, are obtained using specific chromatographic detectors, where the retention time and peak area data are employed in the identification and quantification of the pesticide, respectively, with suitable sensitivity and selectivity (Christian et al. 2014). This technique requires that the analysis of the sample such as natural water, soil, in nature foods, foodstuffs, biological fluids, pharmaceutical or industrial samples, among them, occurs only after preliminary separation and preparation steps. The use of chromatographic techniques produce toxic waste generation from the use of organic solvents, excessive cost and can be time consuming (Primel et al. 2017; Madej et al. 2018; Sajid and Alhooshani 2018; Narenderan et al. 2019).

The evolution in the chemical instrumentations, the use of modern technologies eco-friendly, the necessity the obtain suitable selectivity and sensitivity in the pesticide's analysis in reduced time, and multiresidues detection necessity, led to development of two-dimensional liquid chromatography and or two-dimensional gas chromatography. In these techniques, the samples are subjected to combination of two independent separation mechanisms, that improve the analytical detection, separation and resolution of the pesticides from different chemical classes (Muscalu and Górecki 2018; Bahaghighat et al. 2019; Brandão et al. 2019; López-Ruiz et al. 2019). Moreover, use of ultra-high-performance liquid chromatography permit fast analyses, without compromising efficiency in the separation and quantification (Kharbouche et al. 2019).

Although the instrumentation and technologies are available with different mobile and stationary phases that permit obtain the best separation performance, the use of chromatographic techniques in the pesticides analysis is still a great challenge in the chemical analysis, because pesticides present high variability in the chemical structures, isomerization process, low stability and lack of commercial standards for the identification of compounds from complex samples. Besides, the excessive cost of the instrumentation, maintenance of the equipment, and the need for highly skilled-labor personal are the main disadvantages of these techniques (Sajid and Alhooshani 2018). For this, electroanalytical techniques have been presented as an alternative technique in the pesticide analysis from different chemical class in natural water, soil, foods and compounds of biological interests (Bakirhan et al. 2018).

This chapter shows the main electroanalytical techniques that are employed in the pesticides analysis that are potentiometry, amperometry, voltammetry, stripping voltammetry and electrochemical impedance spectroscopy, indicating the characteristics of each technique and some specific application in pesticides analysis. A section provides a solid theoretical background on the main steps in the electroanalysis of pesticides, from the right choices of adequate electroanalytical technique, solvents, types of the supporting electrolyte, reference electrode, auxiliary electrode, working electrodes from diverse disponible materials, and the voltammetric optimization, in order to gain information about the electrochemical redox reaction and an adequate identification and quantification of pesticides from different chemical class.

1.2 Electroanalytical Techniques in Pesticides Analysis

The pioneering work about the electroanalytical determination of the pesticides was published in 1970 by Hence, which was used as working electrode a dropping mercury electrode allied to the derivative polarography (Hance 1970). In this work, were evaluated 38 pesticides in different supporting electrolytes, producing sensitivity around 10^{-5} mol L^{-1}. Since this work, the innovations in the electroanalytical instrumentation from the development of modern electronic components such as integration, sampling and storage elements, current-load, current-voltage and pulse-polarity converters, permitted the development and commercialization of modern instruments, the potentiostat/galvanostat. These instruments control the voltage between the working and auxiliary electrodes to maintain the potential difference between the working and reference electrodes, according to previously defined by a function generator, and measure the current in both electrodes (Bockris et al. 2000).

Allied to the modern software for acquisition and treatment of the analytical information, the use of electroanalytical techniques has allowed a suitable improvement in the signal-to-noise ratio even further, thus reducing the detectable concentration of pesticides in different media. Besides, the development of materials to prepare diverse types of working electrodes and or sensors has promoted a suitable increase in the analytical selectivity, extending the applicability of electroanalytical techniques to pesticides analysis from different chemical classes. In the last decade, the potentiometry, the amperometry, the electrochemical impedance spectroscopy and voltammetry are the most employed electroanalytical techniques for pesticides analysis in different samples type, as can be observed by analysis of data presented in Table 1.1.

Electroanalytical techniques are based on measurements of some electrical properties that arise when two separate electrodes are maintained in contact with an electrolyte containing the compounds of interest. The analytical responses obtained are related to electrochemical process that occurs in the electrode-solution interface,

Table 1.1 Main electroanalytical techniques employed in the pesticide analysis from different chemical class, showing some examples of specific pesticide with respective electroanalytical applications

Electroanalytical technique	Pesticide chemical classes	Pesticide/applications	References
Amperometry	Carbamate	Mancozeb/commercial formulations	Simões et al. (2007)
	Bipyridinium	Paraquat/NR	Rocha et al. (2018)
	Benzoylurea	Diflubenzuron/forestry matrices	Rodríguez et al. (1999)
	Organophosphorus	Fenitrothion, ethyl-parathion, methyl- parathion, paraoxon and guthion/NR	Martinez et al. (1993)
	Phenylurea	Fenuron, neburon, chlorotoluron and linuron/NR	Kunert et al. (2002)
Potentiometry	Organophosphorus	Malathion, parathion-methyl and methamidophos/NR	Timur and Telefoncu (2004)
	Organophosphorus	Trichlorfon/NR	Reybier et al. (2002)
	Pyrethroid	Permethrin/treated wood	Arip et al. (2013)
	Glycine derivative	Glyphosate/NR	Vaghela et al. (2018)
	Bipyridinium	Diquat dibromide/NR	Abu Shawish et al. (2012)
	Organophosphorus	Paraoxon/NR	Schöning et al. (2003)
Cyclic voltammetry	Carbamate	Pirimicarb and propoxur/NR	Selva et al. (2017)
	Nitropesticide	Pendimethalin/baby food	Galli et al. (2011)
	Organophosphorus	Methyl parathion and dichlorvos/ natural waters	De Souza and Machado (2006)
	Triazines	Atrazine and ametrine/natural waters	De Souza et al. (2006)
	Organophosphorus	Chlorpyrifos, fenthion and Methyl parathion/NR	Zheng et al. (2017)
Differential Pulse Voltammetry	Chlorophenoxy acetic acid	2,4-dichlorophenoxyacetic acid/ water samples	Skrzypczyńska et al. (2016)
	Carbamate	Diethofencarb/NR	Sinha et al. (2019)
	Imidazolinone	Imazaquin/	Imidazolinone
	Phenylureas	Diuron/soil solutions	Phenylureas
	Triazine	Atrazine/NR	Vaz et al. (2008)

(continued)

Table 1.1 (continued)

Electroanalytical technique	Pesticide chemical classes	Pesticide/applications	References
Square wave voltammetry	Organophosphorus	Methyl parathion and dichlorvos/NR	De Souza and Machado (2006)
	Pyrimidines	Diazinon/NR	Akyüz and Koca (2019)
	Organophosphorus	Methyl parathion and dichlorvos/NR	De Souza and Machado (2006)
	Nitropesticide	Trifluralin/human plasma	Jafari et al. (2019)
	Urea	Metobromuron/soil samples	Sipa et al. (2018)
	Organophosphorus	Fenitrothionin/fruit samples	Govindasamy et al. (2018)
	Triazolone	Sulfentrazone/soymilk and groundwater samples	Silva et al. (2019)
Stripping voltammetry	Organophosphorus	Fenitrothion in natural water and orange juice samples	Itkes et al. (2019)
	Organophosphorus	Chlorpyriphos/irrigation agricultural water	Melo et al. (2018)
	Carbamates	Mancozeb, maneb, propineb, nabam, zineb, ziram, ferbam and thiram/NR	Amorello and Orecchio (2013)
	Carbamate	Thiodicarb/in food samples	Lucca et al. (2017)
	Pyrethroid	Cypermethrin/NR	Nurdin et al. (2019a)
	Nitrophenyl	Aclonifen/NR	Shetti et al. (2019)
	Organophosphorus	Methyl parathion, diazinon and chlorpyrifos/water and apple juice	Yola (2019)
Electrochemical impedance spectroscopy	Organophosphorus	Paraoxon-ethyl, fenitrothion and chlorpyrifos/NR	Sgobbi and Machado (2018)
	Organophosphorus	Malathion/NR	Bao et al. (2019)
	Bipyridinium	Paraquat/milk and tomato samples	Farahi et al. (2014)
	Organophosphorus	Carbaryl and dichlorvos/tap water and lettuce samples	Malvano et al. (2017)
	Triazine	Atrazine/NR	Madianos et al. (2018)

NR no reported

or interfacial techniques, and are classified as potentiometric or potentiostatic techniques. If an electric potential is established across a membrane or a specific sensor, this potential is proportional to the concentration of the compound of interest, characterizing the potentiometric techniques, such as occurs in pH measurements, a most popular between electroanalytical techniques. In the analysis of pesticides, potentiometric techniques present applicability, because some chemical structures of pesticides permit the development of some specific indicator or selective membranes electrodes. Electric potential established in the interface of the potentiometric sensor is measured and related to quantity of pesticides under study (Wang 2006; Scholz 2010).

In the potentiostatic techniques or controlled potential, a potential is held constant or is varied with time in a predetermined manner. This way can occur transfer electrons reaction, which a compound of interest receives and or loss electrons in reduction and or oxidation reaction, respectively, or occurs interaction with other medium components. Therefore, electrical currents produced are measured as a time function or applied potential, and present values directly proportional to the species concentration (Dahmen 1986). In general, the applied potential presents enough energy to promote electron-transfer reaction, the resultant current allows quantification and the potential value where the electron transfer takes place can be used in the identification of the compound under study. A wide variety of potentiostatic electroanalytical techniques have been used to identify and quantify pesticides in different samples such as natural water, soil, *in nature* foods and foodstuffs (Díaz-González et al. 2016; Zhao et al. 2018).

1.3 An Overview of Pesticides Electroanalysis

A detailed evaluation in the scientific databases Web of Sciences®, Scopus® and ScienceDirect® over the last 10 years from 2009 to 2019 using the keywords pesticides, potentiometry, amperometry, cyclic voltammetry, differential pulse voltammetry, square wave voltammetry and electrochemical impedance spectroscopy, considering the employ of carbon-based, mercury-based, chemically modified electrodes or biosensors, has shown the pesticides analysis from different chemical classes, as previously shown in Table 1.1. Additionally, these evaluation in the scientific databases has shown the predominance in the use of voltammetry and considerable increase in the employ of the electrochemical impedance spectroscopy in pesticides analysis, as shown in Fig. 1.1 considering the last 5 year old. The electrochemical impedance spectroscopy permits the analysis of pesticides without electroactivity using several electrode surfaces as square wave voltammetry permits the acquisition of a high analytical sensitivity like the ones obtained by traditional chromatographic techniques, as will be discussed in the following sections.

In potentiometry is measured the electrical potential that circulates in an electrochemical cell, where no current or only a negligible current flow occurs. This potential is related to pesticide concentration using a specific sensor, called

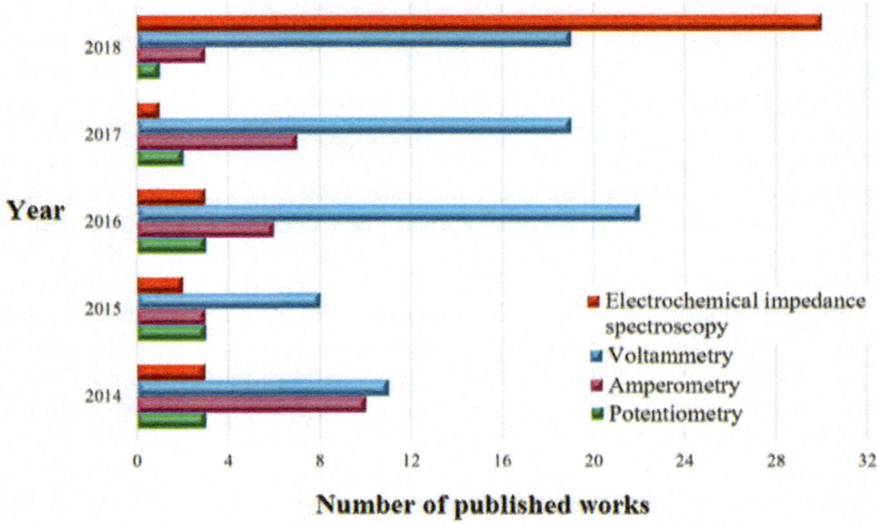

Fig. 1.1 Number of scientific works published which performed the pesticides analysis using potentiometry, amperometry, voltammetry and electrochemical impedance spectroscopy. These data considerate literature reports from scientific databases Web of Sciences®, Scopus® and ScienceDirect® in the years from 2009 to 2019, and were searched using the keywords pesticides, potentiometry, amperometry, voltammetry (cyclic voltammetry, differential pulse voltammetry and square wave voltammetry) and electrochemical impedance spectroscopy along with keywords carbon-based, mercury-based, chemically modified electrodes or biosensors. It is possible to observe the predominance of voltammetric techniques in all years and an intense increase in the application of electrochemical impedance spectroscopy in 2018

potentiometric sensor, which a potentiometer indicates the potential difference between this sensor, while responds to the activity of pesticide, and the reference electrode that present a known and constant potential. In amperometry are measured the resulting current after the application of a constant potential to the working electrode promoting a gain or loss of electrons, known as reduction or oxidation reaction, respectively, which is related to the compound concentration under study.

In the electrochemical impedance spectroscopy, the potentiostat apply a fixed sinusoidal voltage across a three electrodes cell, which contain the solution of electrolyte plus the pesticide of interest. After the interaction of pesticides with working electrode surface, all changes in the interfacial properties at working electrode are measured. In voltammetry, a potential difference, which present enough energy to promote the electrons transfer, is applied in the electrochemical cell producing an electrical current directly proportional to pesticide concentration, among other factors inherent to the chosen voltammetric technique. The mode of the potential vary will define the voltammetric profile, producing differences in the analytical results, considering sensitivity and selectivity.

In all electroanalytical techniques, the reactions occur at the electrode/solution interface from functional groups present in the pesticide molecule, where the phosphates, chlorides, nitro and nitro-derivatives, carbonyl, aromatic rings, double-bonds, triple-bonds, and sulphur-bonds groups, produce a measurable electrical signal. The profile, position and intensity of this electrical signal will determine the analytical sensitivity and selectivity. Moreover, the adequate choice of a working electrode material allied to the mode of potential difference application and the specific potential interval applied will be determinant in the success of the identification and quantification of pesticide of interest.

In general, the use of electroanalytical techniques, mainly voltammetric techniques, permits the study of the nature of redox reactions or mechanism reactions and measure the concentration of species in solution after a simplified pre-treatment in the sample, reducing the time and cost of the analysis. The results are obtained in an abbreviated time, and the instrumentation has a low cost and operates easily compared to chromatographic techniques. Besides, its use makes possible to identify and quantify organic and inorganic compounds at below of the parts per billion level, keeping the concentration analyte practically constant, making the technique like non-destructive.

Electroanalytical techniques permit the evaluation of the different forms of the same element such as Fe(II) or Fe(III), Cr(IV) or Cr(VI), Ru(II) or Ru(III), As(III) or As(V) in compounds containing these metals or its complexed forms in complex mixtures (Dahmen 1986). They also enable analysis to be performed in colored solutions or containing suspended particles, using unusual media such as non-polar solvents, supercritical fluids, gases and even solids. Finally, electroanalytical techniques permit the study of nature of redox reactions or mechanism reactions and measure concentration of chemical species in solution (Christian et al. 2014).

All advantages described above, allied to improvements in the analytical instrumentation have led to a considerable increase in the use of electroanalytical techniques (potentiometry, electrochemical impedance spectroscopy, amperometry and voltammetry) in pesticides analysis. The automation, miniaturization, and easy handling of different working electrodes used as sensors, biosensors or lab-on-a-chip devices or electrochemical transducers, promoted a suitable increment in the analytical performance, mainly in the analytical selectivity. A wide variety of pesticides from various chemical class, have been analyzed in environmental, pharmaceutical, and industrial samples, with sensitivity and selectivity comparable to other well-established analytical techniques such as chromatography and spectroscopy.

Various books and literature reports show the main mathematical treatments that can be used in interpreting and predicting the experimental responses from steps, sweeps or the combination of both potential changes, considering of a simple electrochemical reaction and or complex electrochemical systems, such as coupled chemical reactions, which can be consulted to obtain information about electrochemical mechanism study of the pesticides (Bard and Faulkner 1944; Wang 2006; Mirceski et al. 2007). For this, it will be presented only a brief discussion about the mode of pulse potential application and resulting currents, main equations for determination of peak current, peak potential and half-height width, indicating the advantage and disadvantage of each voltammetric technique. A brief description of the potentiometry, amperometry, and electrochemical impedance spectroscopy in pesticides analysis and some case studies for each technique also will be presented.

1.3.1 Potentiometry

Potentiometry can be considerate the simplest, with a wide linear range and fast between all electroanalytical technique. It's related to potential measures in the absence of appreciable currents, for this is called as static technique, and has the output potential readings *vs*. time, where the potential value is determinate by Nernst equation, that relate the analyte concentrations:

$$E = E^0 + \frac{RT}{nF} \, ln \, \frac{C_{ox}}{C_{red}} \tag{1.1}$$

where E^0 is the standard potential given in volts, C_{red} and C_{ox} are the concentration of chemical species in the reduced and oxidized forms given in mol L^{-1}, respectively, n is the number of electrons involved in the half-reaction given in equivalents per mole, *R* is the gas constant and present of value 8.3143 V C K^{-1} mol^{-1}, *T* is the absolute temperature in Kelvin, and *F* is the Faraday constant and present the 96,487 C eq^{-1}. Substituting the known constant values, the Nernst equation produces the most basic relation to evaluate the relationships between potential values and concentration of the redox compounds in all electroanalytical techniques, including potentiometry, Eq. 1.2:

$$E = E^0 + \frac{0.05916}{n} \, log \, \frac{C_{ox}}{C_{red}} \tag{1.2}$$

In potentiometric analysis, a pair of electrodes are inserted in the cell containing the sample solution of the compounds under study. After this, is measured the electrical potential difference between pair of electrodes, and the data are used to quantify the compound of interest present in the sample. So, in a potentiometric analysis are used a reference electrode, an indicator electrode and a potential measuring device, as indicated in Fig. 1.2. The reference electrode gives references for potential measurements, it is an electrode independent of compounds potential, and independent of temperature. Its design relies on an aqueous bridge electrolyte, typically high concentration of a potassium chloride solution in contact with the sample solution, as will be explained in the next section.

Indicator electrode is where the compounds of interest is evaluated due to the measured potential difference, which depends mainly on the concentration of these compounds. When the pair of electrodes are taken in contact, occurs some change in the free energy due to the chemical phenomena proceed until the equilibrium condition has been satisfied. The potential measuring device is constituted by a potentiometer with a high-input impedance, that indicate the cell voltage, providing

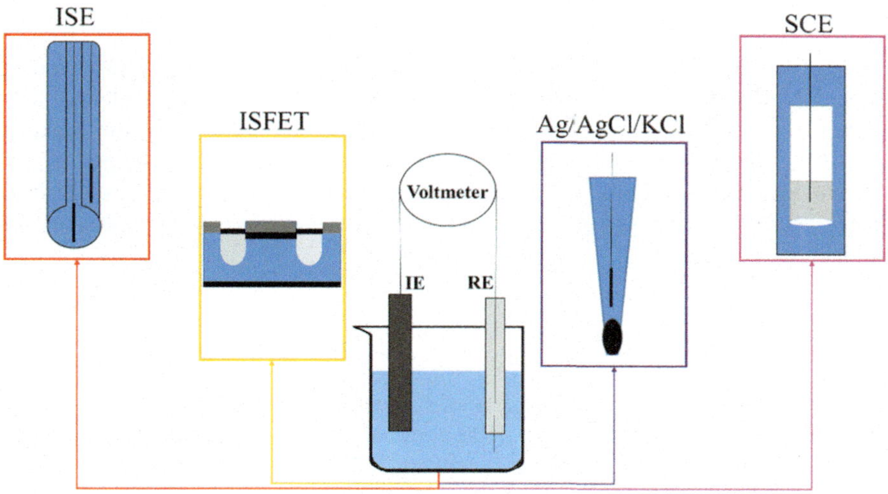

Fig. 1.2 Potentiometric cell, containing indicator (IE) that can be an ion-selective electrode (ISE) or an ion-sensitive field-effect transistor (ISFET) and reference (RE) electrodes, that can be a saturated calomel electrode (SCE) or a silver-silver chloride in potassium chloride solution electrode (Ag/AgCl/KCl). IE and RE are connected in a voltmeter to measure the potential in the electrochemical cell.

a digital potential lecture, either in terms of voltage or pH, without noticeable currents (Christian et al. 2014; Fatibello-Filho et al. 2019).

In practice, potentiometric cell can be represented by:

$$E_{ref}|solution|E_{ind} \tag{1.3}$$

where E_{ind} is the potential of the indicator electrode, E_{ref} is the potential from reference electrode and the single lines represent a boundary between either an electrode phase and a solution phase, where a liquid-junction potential, E_j, can arise (Néher-Neumann 2009). The potentiometer measures the potential established between the indicator and reference electrode, but when a salt bridge is employed, the liquid-junction potential must be included, according to:

$$E_{cell} = \left(E_{ind} - E_{ref}\right) + E_j \tag{1.4}$$

As the reference electrode potential is constant, any alterations in potential of the indicator electrode will be reflected by an equal change in the cell potential. As the liquid-junction potential in the indicator electrode does not differ significantly from one solution to the next, and can't be evaluated under most circumstances, and for this can accept that is related to constant k:

$$k = E^{0}_{ind} - E_{ref} + E_{j} \tag{1.5}$$

where k is a constant that related the liquid-junction potential and E^{0}_{ind} is standard potential of indicator electrode. In both electrodes, the electrical potential will dependent of the concentrations of the species and will varies from the standard potential from reference electrode. This potential dependence is described by the Nernst Equation, Eq. 1.2, and permits to make them practical by establishing quantitative relationships between potential and concentrations of chemical species under study:

$$E_{cell} = k' - \frac{2.303RT}{nF} \ log \frac{C_{red}}{C_{ox}} \tag{1.6}$$

Temperature is a fundamental parameter, and at 25 °C or 298.16 K, the value of 2.3026RT/F is 0.05916 V, for this, in the commercial potentiometers there is a meter for the solution temperature (Néher-Neumann 2009). Advances in the development of specific indicator electrodes, known as ion selective electrodes, has promoted an increase in the potentiometry application for pesticides analysis in the last years. The ion selective electrodes are indicator electrodes capable of realize selective measures of concentration or activity of some chemical species. They are constituted by an ion-selective membrane that is normally placed between two aqueous phases. The first one is the sample and the second is the inner solutions that contain the compounds of interest in a known concentration, as shown in Fig. 1.2. This way arises a potential difference across the membrane, that may be a glass, a crystalline solid, or a liquid, that is measured using two reference electrodes.

The most known ion selective electrode is the combined glass electrode or pH electrode, where a permeable membrane allows the quantification only of hydrogen ions. However, in the last decades, some enzymatic potentiometric electrodes have been developed to pesticides analysis, which a normal pH electrode is coated with an impregnated gel containing a specific enzyme, that if placed in contact with a sample containing the pesticide, promote an enzymatic reaction resulting in an electric potential that is related to the amount of the pesticide that permeate the gel. These enzymatic potentiometric electrodes can be developed for any permeable species in which a selective reaction promotes alterations in pH values (Ghindilis et al. 1996; Cho et al. 1999; Ding and Qin 2009; Justino et al. 2015; Songa and Okonkwo 2016; Trojanowicz 2016; Kudr et al. 2017; Li et al. 2019).

In this context, some works reported the immobilization of enzyme under pH sensor surface to organophosphate pesticides detection, where a change in the output potential corresponds to change in concentration of the protons as a result of the enzymatic reaction (Patel et al. 2019). Timur and Telefoncur used chitosan membrane to the immobilization of acetylcholinesterase enzyme, which was inhibited by malathion, parathion-methyl and methamidophos, producing alterations in the pH values measured (Timur and Telefoncu 2004). Reybier et al. employed glutaraldehyde on an electropolymerized polyethyleneimine film to the immobilization of butyrylcholinesterase under pH electrode surface to quantify trichlorfon (Reybier et al. 2002).

Other modifications was also proposed, such as the immobilization of cells of the some fungus types to produce a potentiometric biosensor for the permethrin determination, a pyrethroid pesticide, in treated wood (Arip et al. 2013). Others ion selective electrodes also can be modified by specific enzymes to pesticides detection. Vaghela et al. used an ammonium ion selective electrodes covered with agarose-guar gum remained bio-nanoconjugate of urease with gold nanoparticles for glyphosate detection, and the potential of ammonium ion selective electrodes modified was measured and related to inhibition of urease activity by glyphosate, which was related to glyphosate concentration (Vaghela et al. 2018).

Ion-sensitive field-effect transistor has also been used as sensor in potentiometric detection pesticides due to the advantages over ion selective electrodes, such as very fast response, high sensitivity, batch processing capability, potential for on-chip circuit integration, small size, low cost and robustness, which makes portable analytical applications possible (Sajini et al. 2019). Ion-sensitive field-effect transistor is indicator electrode where the chemically sensitive membrane is substituted by ordinary metal oxide silicon field effect transistor with the gate electrode plus the, solution and a reference electrode. Ion-sensitive field-effect transistor use presents a sensitive area and incorporates the means of transduction from an ion concentration to a voltage, Fig. 1.2. When the ion-sensitive field-effect transistor sensor comes into contact with the solution containing pesticide, interaction or ion exchange with the measurement system occurs (Yuqing et al. 2003; Dzyadevych et al. 2006; Nehra and Pal Singh 2015).

The changes in the nature of the chemically sensitive membrane or insulating layer has been tested permitting the development of different ion-sensitive field-effect transistor sensor (Cichosz et al. 2018; Xu et al. 2019). For some metals and metallic oxides, polymer and carbon-based materials act as conductive components or a matrix of an electrically conductive system. Among carbon-based materials, Shawish et al. prepared an ion-sensitive field-effect transistor using carbon paste electrode for diquat dibromide, a bipyridinium pesticide, and the ion pair diquat-phospho-tungstate dissolved in 2-nitrophenyloctyl ether as pasting liquid with 1.0% sodium triphosphate buffer as an additive (Abu Shawish et al. 2012). Anirudhan and Alexander employed ion-sensitive field-effect transistor based on the use of carbon nanotube with multiwalled modified by molecularly imprinted polymer in the determination of the pesticide 2,4-dichlorophenoxyacetic acid, a substituted derivative from benzene pesticide, in natural water samples (Anirudhan and Alexander 2014).

However, in most of the research already done, the ion-sensitive field-effect transistors are based on the enzyme or antibodies modifier to improve the selectivity and sensitivity in the pesticide potentiometric analysis (Dzyadevych et al. 2006; Jiang et al. 2008; Gubanova et al. 2017). The enzyme organophosphorus hydrolase was used in the modification of ion-sensitive field-effect transistor pH sensitive by Schöning et al. to analysis of paraoxon, an organophosphorus pesticide (Schöning et al. 2003). Thanh et al. recovered an ion-sensitive field-effect transistor modified by graphene films with urease enzyme to carbaryl enzymatic inhibition analysis towards urease (Thanh et al. 2018). An immunosensor was developed by Starobud et al. based on ion-sensitive field-effect transistor gate modified by polyclonal antibodies against simazine attached to the via staphylococcal protein A, which was employed in the determination of the simazine, a triazine pesticide, by analysis of the competition between the peroxidase-labelled simazine molecules and the antibodies on the ion-sensitive field-effect transistor surface (Starodub et al. 2000). Additionally, the chlorpyrifos, an organophosphorus pesticide, was evaluated using a surface modified by Chlorella sp., based on the detection principle of its inhibition by chlorpyrifos, where the enzyme was dephosphorylate the phosphate monoester of substrate 2-phospho-L-ascorbic acid to release L-ascorbic acid, which was monitored using the ion-sensitive field-effect transistor as potentiometric tool (Pabbi et al. 2018).

1.3.2 *Amperometry*

In the amperometry and voltammetry, that are controlled potential techniques, some electrochemical events occur to produce currents signals related to the concentration of pesticide of interest, as indicated in Fig. 1.3. Initially, it is applied a difference between the electric potential of a cell under the passage of a current and the

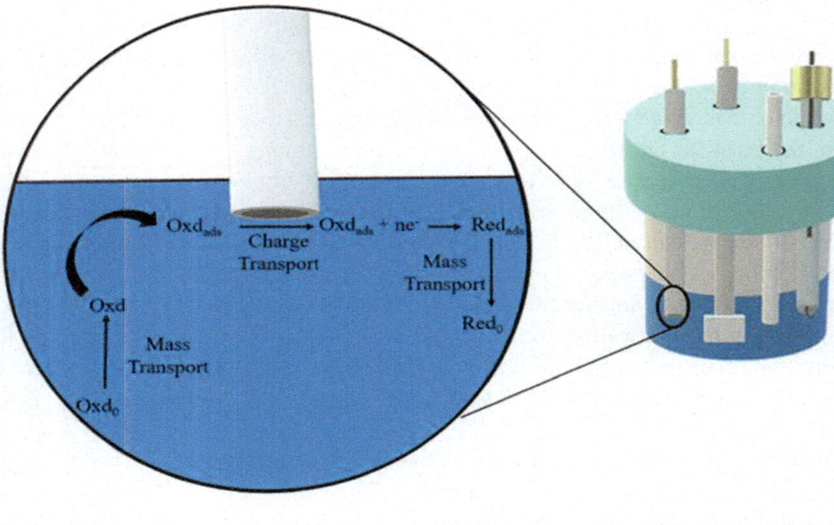

Fig. 1.3 Main electrochemical events that occurs in the interface electrode/solution (mass transport, reactant and product adsorption and charge transport), where Ox_0 and Red_0 represents the oxidized and reduced forms in the bulk solution, Ox and Red represents the oxidized and reduced forms in any stage, Ox_{ads} Red_{ads} indicate the oxidized and reduced forms adsorbed in the electrode surface and ne^- is the number of electrons transferred. Modified after (Escarpa et al. 2015)

thermodynamic value of the cell potential in the absence of electrolysis in the electrochemical cell, or high potential (Amine and Mohammadi 2018). This potential is enough to ensure that the electroactive chemical species are transferred by mass transport from the bulk solution, (Ox) containing supporting electrolyte unstirred, to interface of working electrode (Ox_{ads}), promoting a rapid decrease of the electroactive species under the electrode surface, known as the electron transfer ($Ox + ne^- \rightarrow Red$).

Therefore, current signals are produced according to Cottrell Equation, Eq. 1.7, derived from Fick's second law of diffusion, (Bard and Faulkner 1944), where the potential step is applied under conditions of high overpotential variation, producing the limit current related to time by:

$$i_l = \frac{nFAD_{ox}^{1/2}C_{ox}^*}{(\pi\, t)^{1/2}} \tag{1.7}$$

where i_l is the limit current, D_{ox} is the diffusion coefficient for the species O given in $cm^2\ s^{-1}$, t is the time in seconds, A is the electrode area given in cm^2 and other terms were previously defined. This equation is the fundamental base in all voltammetry techniques and in amperometry.

Amperometry is a type of electroanalytical technique that uses a potential step at stationary system, without agitation in the solution and after this, the behavior of an electrochemical system is evaluated from analysis of the current-potential curves, Fig. 1.4. Before the potential application, the chemical species are electro inactive,

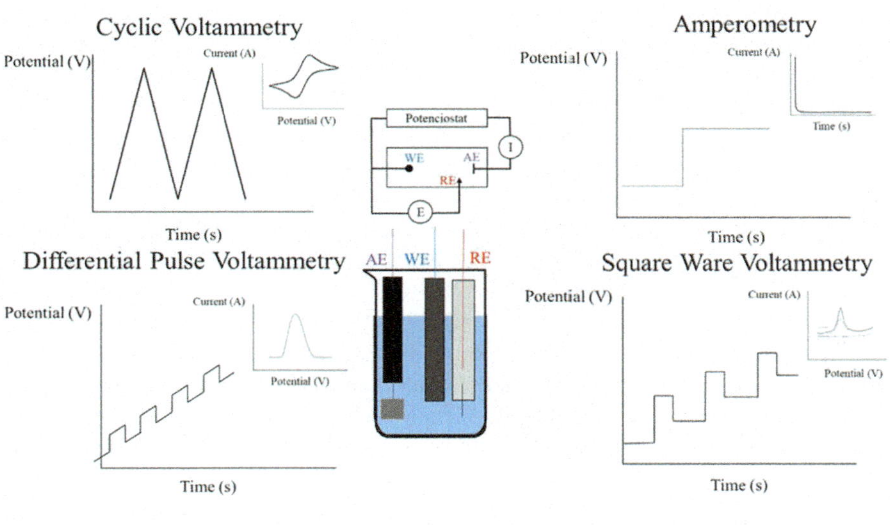

Fig. 1.4 The potential application mode and the voltammetric profile obtained from electroanalytical analysis using amperometry, cyclic voltammetry, differential pulse voltammetry and square wave voltammetry. In the center is presented a single electrochemical cell containing auxiliary (AE), working (WE) and reference (RE) electrodes, and its respective electric circuit representation. (Modified after Zoski 2007)

and by applying an adequate potential step, electrons are transferred by the redox reaction, producing an intense current signal, that is measured according to Cottrell Equation, Eq. 1.7. This current becomes close to zero when the electroactive species are consumed around electrode surface (Bakirhan et al. 2018). The current resultant, considering a fixed potential, is proportional to the compound of interest concentration, and it can be used for direct measurement of pesticides. Double-step amperometry are employed in the study of mechanisms of electrode processes, which the second step is used to probe the fate of a species generated in the first one (Galus 1994; Bockris et al. 2000; Zoski 2007).

Amperometry permits the distinction between several electroactive pesticides in solution by application of an adequate potential and or by proper chooses of the electrode material. Moreover, amperometry is the detection method most widely employed in many electrochemical enzyme-based biosensors, immunosensors and aptasensors for pesticides detection, which will be described in the Sect. 1.4. In these systems, the success in pesticides detection is related to the immobilization steps of the biological material and strategies of signal amplification, commonly realized by amperometry detection (Songa and Okonkwo 2016; Amine and Mohammadi 2018; Bakirhan et al. 2018; Samsidar et al. 2018; Liu et al. 2019; Patel et al. 2019). Considering the employ of enzymatic sensors, the amperometry detection can be performed by pesticide direct detection and or using an enzymatic reaction previously to the pesticide quantification, as shown by Llorent-Martínez et al. (Llorent-Martínez et al. 2011; Stanković and Kalcher 2016). In the last two decades, the electrochemical detection of pesticides using amperometry allied to sensor and biosensors has also been largely used in flow-based analytical methods, (Miró and Frenzel 2018), as will described at final of this section.

The simplicity of the amperometry also permit its use in batch-injection analysis. In this application, a microvolume of sample is dripped close to the working electrode surface and after this, is compounds of interesting present in the sample is instantly reduced or oxidized producing a transient signal that vanish away as the convective transport dilutes the analyte in the bulk of the solution in the electrochemical cell. The batch-injection analysis use only an electronic micropipette, which permits fast and high-precise sample partition, a simple cell and not necessity of the system with pumps and valves (Tormin et al. 2014). So, batch-injection analysis with multiple-pulse amperometry detection was used in the determination of picoxystrobin, a β-methoxyacrilic acid pesticide, in natural waters (Dornellas et al. 2015), in the determination of mancozeb, a carbamate pesticide, in commercial formulations (Simões et al. 2007), in paraquat, a bipyridinium pesticide, determinations, among others class of pesticides (Rocha et al. 2018). Finally, the amperometry is the main electroanalytical technique employed in the electrochemical chromatographic detection, as will be explained in the final of this section (Rivoira et al. 2015).

1.3.3 Voltammetry

In the voltammetric techniques, the potential application modes will determine the profile of the analytical signal, known as voltammogram and shown in Fig. 1.4. Besides, the current intensities and peak potential values are responsible by quantification and identification, respectively, of pesticide of interest. The voltammetric techniques most frequently used are related to application of the potential steps such as amperometry, described above, and potential sweeps such as linear and cyclic voltammetry (Wang 2006). The combination between steps and sweep potential techniques, known as pulse voltammetry, promote a suitable improve in the analytical sensitivity and selectivity, due to a substantially increase in the ratio between faradaic current, from electrons transfer, and no faradaic responses, from other process that occurs in electrochemical cell such as double electric layer formation, known as capacitive current (Zoski 2007).

In pulse voltammetry, the potential is held at some value, E_o where no electrons transfer occurs and then is stepped instantaneously to a second value, E_l, where electrons are transferred. After pulse potential application, the capacitive current flow decreases with time much faster than the faradaic current flow. Current is measured sometime after the pulse application, in a time where occurs a considerable increase in the ratio between faradaic and non-faradaic currents, allowing the obtention of sensitivity in a level like obtained using traditional chromatographic techniques. Pulse techniques present different excitation waveforms and current sampling, characterizing the normal pulse voltammetry, differential pulse voltammetry, staircase voltammetry, and square wave voltammetry.

In these techniques, occurs the application of a potential difference in typical waveforms producing analytical signals, or voltammograms, from redox reactions producing electron transfers, reduction and or oxidation reactions. They are often far more sensitive than spectroscopic techniques or most approaches performed by chromatographic techniques (Escarpa et al. 2015). In addition, the use of pulse voltammetry can provide information about the chemical form in which an compound of interest appears, is possible define the oxidation states, detect complexation reactions and characterize acid-base chemistry, that are information frequently overlooked by others analytical techniques (Bockris et al. 2000; Wang 2006).

In the normal pulse voltammetry, the potential is maintained at an initial value, where no current flows occur, and then a series of pulses with increasing amplitude are applied, followed by currents sampled, at shortly after each pulse application to eliminate the capacitive component and promote improvements in the analytical sensitivity. The resultant voltammograms present a sigmoidal shape, where the position and the slope depend on the reversibility of the redox reaction and current signal is related to concentration of pesticide of interest. However, the results of current present an intense influence of capacitive/background components, and for this, the sensitivity of the normal pulse voltammetry is lower than obtained using others pulse potential techniques. So, its use in quantification of pesticides and inorganic compound is limited to electrochemical process with a slow electron transfer (Wang 2006; Escarpa et al. 2015).

In the staircase voltammetry, potential steps of 10 mV height and about 50 milliseconds duration are applied. After this time, the current is measured producing a peak-shaped current response, like obtained by linear scan experiments. However, staircase voltammetry requires three parameters, the step height, step width and fraction of the step width at which current is sampled, while linear scan voltammetry is specified by one parameter, the scan rate. This way, staircase voltammetry present a considerable discrimination against the capacitive current and consequently a higher sensitivity (Christie and Lingane 1965; Seralathan et al. 1987; Batchelor-McAuley et al. 2015).

Considering modern electrochemical instrumentation commercially available, pesticides electroanalyses has been performed using cyclic voltammetry, differential pulse voltammetry and square wave voltammetry due to inherent properties of each technique, and for this, these techniques will be reviewed next, indicating the main characteristics and some important applications considering some pesticides types analysis from different chemical classes.

1.3.3.1 Cyclic Voltammetry

Cyclic voltammetry is a potential sweep technique, where the current is measured when the electrode potential is varied linearly with time, begin from a potential where no reaction occurs, E_1, and driving to potentials where occurs reduction or oxidation reactions, E_2. In linear sweep voltammetry, the scan concludes at E_2, already in cyclic voltammetry, when the potential reaches E_2 the sweep direction is reversed and the potential returned to E_1, Fig. 1.4, producing one cycle of the cyclic voltammogram. The number of cycles may be evaluated in a determinate time, producing signals of the current and potentials, in the cathodic and anodic directions, the half-peak potential and the half-wave potential values provide useful information about redox mechanisms of pesticides. These information include the electrochemical reversibility, kinetic of electron transfer, determination of the energy levels, intermediary formation, quantification of the electron transfer number across the electrode/solution interface as well as identification of adsorption process and following chemical reactions that are coupled to electron transfer (Wang 2006; Compton and Banks 2007; Zoski 2007; Forster and Cumba 2018).

Like other voltammetric techniques, the current signals (I) measured by linear and cyclic voltammetry correspond to capacitive (I_c) and faradaic (I_f) current contributions, which occurs as interfacial charging and the electron transfer events that take place at the electrode surface Eq. 1.8:

$$I = I_c + I_f \tag{1.8}$$

where the capacitive contribution can be eliminated using the subtraction of the blank response procedure, where the responses correspond to the only electrolytic solution without electroactive species. For this, detection limits for linear scan voltammetry and cyclic voltammetry are considered high, with values around 10^{-5} M, and for this, no adequate to analytical purposes.

Some mathematical procedure has already been employed to improve the sensitivity, as demonstrated by Chen and Chen, where a mathematical method was applied in the analysis of dates from the analytical determination of the carbofuran, a carbamate pesticide. This mathematical method employed calculous of specific area from the cyclic voltammetry curves as the dependent variables and the carbofuran concentrations as the independent variables (Chen and Chen 2013). Even with the use of modern equipment, or advanced mathematical treatments, or modified electrodes the sensitivity is not adequate for environmental analyses, and the pulse techniques are the most adequate for this purpose, as will be explained next.

However, as reaction of pesticides in the electrochemical cell can be like reactions in the environment, soil and natural waters, the use of cyclic voltammetry makes possible the study of the mechanism reaction and estimate eventually intermediary compounds and or products of the redox reactions. So, information about intermediary reactions, toxicity and persistence of these compounds can be obtained. In practice, cyclic voltammetry is the most appropriate tool to obtain information about the kinetic, thermodynamic and mechanisms of many chemical reactions, such as electron and proton transfers, isomerization, and ejection of leaving chemical groups (Wang 2006). Additionally, cyclic voltammetry is the first technique used to obtain basic information about the redox process and preliminary experimental and voltammetric conditions (Compton and Banks 2007). Commonly, are evaluated the peak height, i_p in the cathodic and anodic scan, which for a planar diffusion-controlled reaction can be related to the Randles-Sevčik equation, presented in Eq. 1.9:

$$i_p = 0.4463 \frac{(nF)^{3/2}}{(RT)^{1/2}} A D_O^{1/2} C_O^* \nu^{1/2} \tag{1.9}$$

where i_p is the peak current, ν is the scan rate given in V s^{-1} and other terms were previously described. This equation indicates the proportionality between the peak current and the square root of the scan rate and is the key feature in cyclic voltammetry. Additionally, in specialized literature can be found specific equations that are used in diagnostic criteria to characterize a reversible, irreversible, chemical coupled reactions and reactants or products adsorption process in cyclic voltammograms (Zoski 2007).

So, Selva et al. studied the electrochemical behavior of pesticide pirimicarb and propoxur, carbamate pesticides, using cyclic voltammetry allied to boron doped diamond as working electrode. To pirimicarb it was observed the presence of three independent irreversible oxidation peaks related to the transfer of one electron for each peak, across a pH range from 2.0 to 8.0, and the analysis of voltammograms permitted a reaction proposal for the electrochemical reaction of pirimicarb, supported by mass spectrometry experiments. To propoxur, results indicated a single and irreversible electron transfer by oxidation process, that was observed without any derivatization processes, which are commonly used for the electrochemical detection of this pesticide class (Selva et al. 2017).

Galli et al. described the electroanalytical behavior of pendimethalin, a nitro pesticide, based on cyclic voltammetry responses from its electro-reduction using a hanging mercury drop electrode, where they observed that two well-defined voltammetric peaks appeared during the direct scan, while the reverse scan presented no peaks, indicating a totally irreversible redox process (Galli et al. 2011). De Souza and Machado used cyclic voltammetry in the redox behavior study of methyl parathion and dichlorvos that are organophosphorus pesticides, and paraquat and diquat that are bipyridinium pesticides, using gold microelectrodes (De Souza and Machado 2006). Additionally, was also evaluated the electrochemical behavior of atrazine and ametrine, triazines pesticides, using solid amalgam electrode allied to diagnosis criterions of cyclic voltammetry to electrochemical reactions (De Souza et al. 2006). Zheng et al. used cyclic voltammetry to characterize the redox properties of pralidoxime chloride, an antidote to organophosphate pesticides, as an electrochemical probe in the development of a non-enzymatic inhibition method for the determination of organophosphorus pesticide chlorpyrifos, fenthion, and methyl parathion (Zheng et al. 2017).

Some reports have demonstrated the use of cyclic voltammetry in the electropolymerization process, in which occurs the formation of a thin polymeric film under material that presents conductivity, typical of the chemically modified electrodes. These modifications alter the properties of the interface producing adequate sensors and biosensors to use in the pesticides analysis (Akyüz et al. 2017). Based on the electric prospects from specific monomers and the knowledge of the most appropriate supporting electrolyte and solvent, the immersion of the electrode in a monomer solution allied to cyclic voltammetry technique is enough for polymer formation, because the parameters of the techniques can be easily changed to shape the polymer over the electrode surface regarding its thickness, conductivity and doping properties.

The presence of a polymer over an electrode surface is useful in biosensor development to quantify specific pesticides. The formation of a conductive polymer can increase the electronic transference throughout the transducer and the presence of non-polymerized functional groups can be used to attach the biomolecules (Edwards et al. 2007). Alves et al. used cyclic voltammetry in the study the electropolymerization of the monomer 2-hydroxybenzylamide over a carbon graphite electrode with further modification with the enzyme tyrosinase through physical adsorption to detection of fenitrothion, an organophosphorus pesticide, based on its enzymatic inhibition, evaluated through chronoamperometry and flow injection analysis (Alves et al. 2018).

Yildiz et al. also used cyclic voltammetry in the electropolymerization of monomer 4-(2,5-di(thiophen-2-yl)-1H-pyrrol-1-yl)benzenamine under carbon graphite electrode modified by multi-walled carbon nanotube (Yildiz et al. 2008). This electropolymerization promoted the electrode activation through carbodiimide reaction and the enzyme acetylcholinesterase, that was posteriorly used in the detection of paraoxon, parathion and chlorfenvinphos, organophosphorus pesticides, through chronoamperometry based on enzymatic inhibition (Kesik et al. 2014).

1.3.3.2 Differential Pulse Voltammetry

In differential pulse voltammetry the potential waveform is characterized by application of a series of potential pulses with constant amplitude, superimposed in the linear ramp. The generate current is sampled at the beginning and the at the end of each pulse. The difference between these two current values is recorded and shown as function of the applied potential, as shown in Fig. 1.4. The resulting voltammogram is characterized by peak-shaped, that present a considerable decrease in the capacitive/background current, resulting in the analytical signals with superior sensitivity, compared to cyclic voltammetry and normal pulse voltammetry (Escarpa et al. 2015). Like normal pulse voltammetry, considering a reversible system, are also employed mathematical expressions that relate peak current, I_p, with concentration Eq. 1.10, and peak potential, E_p, with diffusion of electroactive species Eq. 1.11 based on Nernst Equation, according to:

$$I_p = \frac{nFAD_O^{\frac{1}{2}}C_O^*}{(\pi t_p)^{\frac{1}{2}}}\left(\frac{1-\sigma}{1+\sigma}\right) \tag{1.10}$$

$$E_p = E^0 + \frac{RT}{nF}\,ln\left(\frac{D_{red}}{D_{ox}}\right)^{1/2} - \frac{\Delta E_p}{2} \tag{1.11}$$

where $\sigma = \exp(nFa/2RT)$ and a is the pulse amplitude. Considering small a value, peak width at half height, $w_{1/2}$, can be employed in the calculous of the number of electrons transferred by Eq. 1.12:

$$w_{1/2} = 3.52\frac{RT}{nF} \tag{1.12}$$

So, Selva et al. employed differential pulse voltammetry in the carbamate pesticide pirimicarb quantification using diamond boron doped electrode, producing a analytical procedure very simple, whit low-cost, accurate and that possibility the portable in natural waters electroanalysis (Selva et al. 2017). Skrzypczyńska et al. used differential pulse voltammetry coupled to different carbonaceous materials as working electrodes, in the determination of 2,4-dichlorophenoxyacetic acid, usually called as 2,4-dichlorophenoxyacetic acid, in water samples (Skrzypczyńska et al. 2016). Hashemi et al. employed differential pulse voltammetry in the determination of carbaryl and fenamiphos, a carbamate and an organophosphorus pesticides, respectively, using a non-enzymatic sensor prepared with reduced graphene oxide as working electrode, generating an electroanalytical procedure very simple and an efficient sensor for use in the detection of low levels of these pesticides in fruit and vegetable samples (Hashemi et al. 2019).

Duarte et al. developed used the differential pulse voltammetry allied to a boron-doped diamond electrode associated to solid phase extraction in the simultaneous determination of multiresidues of diuron, 2,4-dichlorophenoxyacetic acid and tebuthiuron and posterior analysis in lake and well water samples, where the results attested the feasibility of method for pesticides determination in water samples (Duarte et al. 2018). Sinha et al. used differential pulse voltammetry in the development of a rapid and selective screening of carbamate pesticides diethofencarb applying a nanosensor platform based that permitted analysis from of 1 ng μL^{-1} to 35 ng μL^{-1} in environmental monitoring (Sinha et al. 2019).

Soil contaminated by pesticides from different chemical class has also been successfully analyzed using differential pulse voltammetry. Some examples include the work developed by Moreno et al. that evaluated the dynamics of triazine pesticides in soils and the formation of complexes with cupper ions, indicating that in crops treated with triazines plus with cupper salts occurs a decrease in persistence and effectiveness of pesticides (Moreno et al. 2018). Vaz et al. used differential pulse voltammetry for evaluation of the degradation products of imazaquin, methyl-parathion, bentazon and atrazine pesticides, generated from the ultraviolet irradiation under pesticides solutions, They performed this study due to the degradation products of these pesticides can be even more toxic then the parent products and such studies should be encouraged (Vaz et al. 2008).

Castanho et al. determined sorption coefficients of the pesticide imazaquin using the differential pulse voltammetry in some soil classes, contributing to the predictions of pesticide fate and transport mechanisms in the environment as well as its exposure and interactions with different soil materials (Castanho et al. 2016). Finally, Soares et al. evaluated the diuron pesticide in soil solutions using differential pulse voltammetry and a graphite-polyurethane electrode and observed that the organic matter and clay content influence directly in the differential pulse voltammetry responses (Soares et al. 2012).

1.3.3.3 Square Wave Voltammetry

Square wave voltammetry, sometimes called as Osteryoung square wave voltammetry, is the most used electroanalytical technique to electrochemical behavior study and to detection of pesticides from different chemical class, using a varied of materials as voltammetric sensors (De Souza et al. 2003; De Souza et al. 2004). In square wave voltammetry, a series of equal amplitude pulses are applied under a staircase potential, considering the use of a cathodic pulse and an anodic pulse, as shown in Fig. 1.4. During each forward pulse, the analyte diffuses to the electrode surface and it is immediately reduced or oxidized. During the backward pulse, analyte that was just oxidized or reduced returns to the initial condition in a reversible reaction, or no reaction occurs if the system is irreversible (Mirceski et al. 2007; Mirčeski et al. 2007). The current values are measured just before, called as forward current (I_f) and at the end of each pulse, called as backward current (I_b). The net or resulting currents (ΔI) are plotted as a function of the corresponding potential of the staircase waveform, producing a voltammogram with gaussian profile:

$$\Delta I = I_f - I_b \tag{1.13}$$

The peak current (I_p) and the peak potential (E_p) in the maximum signal can be used to identify and quantify pesticides, respectively, from different chemical class. Since in the square wave voltammetry the resultant current is higher than forward or backward currents, the analytical signals are higher than signals observed by use other electroanalytical techniques, producing very low detection limits, from 1.0×10^{-6} mol L^{-1} until 1.0×10^{-12} mol L^{-1}. Besides, the peak current and peak potential values are strongly dependents of the type of working electrode employed, due to electrons transfer properties depend on the kinetics and thermodynamics of the electrode materials, and also depends of the use of previous stripping voltammetric steps (Mirčeski et al. 2007). Additionally, square wave voltammetry makes possible the potential to be scanned faster than the obtained by other electroanalytical techniques and produce a well-defined peak that are more easily interpreted.

The theoretical description and several application of square wave voltammetry are described in several reviews, and the mathematical models were previously developed considering electrochemically reversible, quasi reversible and irreversible electrode reactions and coupled chemical reactions, with or without reactants and products adsorption process at working electrode (Mirčeski et al. 2007). The adequate use of some mathematical models makes possible to estimate pesticides electrochemical mechanisms and to evaluate its behavior in the environment, since reactions studied in electrochemical cell are like electrons transfer in natural water, soil and food samples. For this, it is necessary to observe the behavior of I_p, E_p and half-height width ($w_{1/2}$) when are individually changes the height of potential pulse (a), the height between each potential step (ΔE_s) and the duration of each pulse potential (f).

The linear relationships between I_p and f square root are typical of the redox reaction controlled by diffusion rate of the pesticide from the bulk solution to working electrode interface, Eq. 1.14. The linear relationship between f and I_p values are typical of redox reactions controlled by adsorption kinetic, followed by product adsorption, which are coupled by intense peak enlargement Eq. 1.15. Besides, the I_p are also related to a and ΔE_s values, according to Eq. 1.16 to reversible redox reaction with reactant and products adsorbed or Eq. 1.17 to reversible redox reaction with only reactant adsorbed or and Eq. 1.18 to irreversible redox reaction with only reactant adsorbed or with reactant and products adsorbed).

$$I_p = kf^{1/2} \quad (1.14)$$

$$I_p = kf \quad (1.15)$$

$$I_p = \text{const.} n^x f^{1/2} a^y t_0^{1/2} A^* \quad \text{for } 1 < x < 2; y < 1 \quad (1.16)$$

$$I_p = const.n^x \Delta E_s f a^y t_0^{1/2} A^* \quad \text{for } 1 < x < 2; y < 1 \quad (1.17)$$

$$I_p = \text{const.} \alpha n^2 \Delta E_s f a^x t_0^{1/2} A^* \quad \text{for } x < 1 \quad (1.18)$$

where n is the number of electrons transferred, k is the kinetic constant, α is the electron transfer coefficient, t_0 is delay time, A is the area of working electrode and other terms were previously defined.

In the square wave voltammetry signals from redox reaction of pesticides, can be observed a shift towards less or more negative values in the E_p, which varied linearly with the logarithmic value of the f according to the Eq. 1.19 to reversible redox reaction without products adsorption and Eq. 1.20 to irreversible redox reactions, known as Lovric's Equations.

$$\frac{\Delta E_p}{\Delta \log(f)} = \frac{-2.3RT}{2nF} \quad (1.19)$$

$$\frac{\Delta E_p}{\Delta \log(f)} = \frac{-2.3RT}{\alpha nF} \quad (1.20)$$

Finally, the $w_{1/2}$ during the variation in the voltammetric parameters can be related to number of electron transfers, according to Eq. 1.21 considering reversible redox reaction with only reactant adsorbed, Eq. 1.22 considering reversible redox reaction with reactant and product adsorbed and Eq. 1.23 considering irreversible redox reaction with only reactant adsorbed or with reactant and product adsorbed.

$$w_{1/2} = \frac{101}{n} \quad (1.21)$$

$$w_{1/2} = \frac{97}{n} \quad (1.22)$$

$$w_{1/2} = \frac{(63,5 \pm 0,5)}{\alpha n} \quad (1.23)$$

Based on the above-mentioned advantages, the square wave voltammetry has been largely employed in the electrochemical behavior study and in the analysis of pesticides from different chemical classes and various samples, including natural water, soil, foods and biological fluids samples. In this context, De Souza and Machado have proposed the use of square wave voltammetry diagnostic parameter to evaluate the electrochemical mechanisms of electrons transfer from bipyridiniums pesticides paraquat and diquat, and organophosphorus pesticides methyl-parathion and dichlorvos and quantified its residues in natural water samples using a simple gold microelectrode (De Souza and Machado 2006).

Akyüz and Koca used square wave voltammetry in the characterization of an electrochemical sensor enzyme less based on the modifications by manganese phthalocyanine polyaniline, which was employed in the detection of the fenitrothion, eserine, and diazinon. This pesticides were determined with sensitivity around 0.049 μmol L^{-1} for fenitrothion, 0.088 μmol L^{-1} for eserine and 0.062 μmol L^{-1} for diazinon, presented wide linear ranges, and especially higher selectivity (Akyüz and Koca 2019). Furthermore, Jafari et al. used square wave voltammetry and a modified electrode by silver nanoparticles and reduced graphene oxide in the sensitive detection of trifluralin, a nitro pesticide, in human plasma samples, obtaining a reliable, low cost, rapid, and user-friendly analytical procedure (Jafari et al. 2019).

Sipa et al. employed square wave voltammetry and a pencil graphite electrode modified by carbon nanotubes in the determination of metobromuron, a urea pesticide, in soil samples with good selectivity and recovery values around 99.1% (Sipa et al. 2018). Govindasamy et al. used a niobium carbide supported on molybdenum nanoparticles in the construction of a voltammetric sensor that allied to square wave voltammetry technique permitted the fenitrothion, a organophosphorus pesticide, detection in a linear range from 0.01 μmol L^{-1} to 1889 μmolL^{-1} and the with detection limit of 0.15 nmol L^{-1} in fruit samples (Govindasamy et al. 2018).

Diquat and paraquat, bipyridiniums pesticides, were evaluated in natural water samples using square wave voltammetry allied to bismuth film electrode, and the results presented results similar to results obtained employing the high-performance liquid chromatography, without the use of special mobile phase or preparations steps (de Figueiredo-Filho et al. 2017). The sulfentrazone, a triazolone pesticide, was evaluated in soymilk and groundwater samples using square wave voltammetry and multi-walled carbon nanotube screen-printed electrode (Silva et al. 2019).

Square wave voltammetry allied to the use of enzyme modified electrodes has also been used to improve the selectivity in electroanalysis of pesticides. So, Qiu et al. used a nanoenzyme material based amino acids attached titanium oxide nanoparticles (TiO_2) in order to construct a voltammetric sensor that allied to square wave voltammetry permitted the quantification of organophosphorus pesticides methyl paraoxon, methyl parathion and ethyl paraoxon in lettuces samples (Qiu et al. 2019). Bao et al. constructed an acetylcholinesterase biosensor using graphene-copper oxide nanoflowers nanocomposites for malathion, an organophosphorus pesticides, which exhibited a sensitivity around 0.31 parts by trillion (Bao et al. 2019). Da Silva et al. quantified carbaryl, a carbamate pesticide, in tomato samples using square wave voltammetry allied to a biosensor based on reduced graphene oxide and acetylcholinesterase enzyme (da Silva et al. 2018). A pencil graphite electrode modified by diphenylalanine peptide nanotubes was considerate as very simple and economically viable sensor that allied to square wave voltammetry permitted the determination of fenitrothion, an organophosphorus pesticide, in natural water samples (Bolat et al. 2018).

Simple electrodes, without modifications in its surface has also been used with success in the pesticide analysis by square wave voltammetry technique, which carbons forms, mainly diamond electrode and noble metals are the most commons materials used as voltammetric sensor. Ribeiro et al. used square wave voltammetry allied to boron-diamond doped electrode to evaluate the electrochemical redox behavior of formetanate, a carbamate pesticide. They observed that the reaction mechanism was diffusion-controlled, involved the participation of one electron and was influenced by the electrolyte pH, indicating no protonation participating in the rate-determining step in the redox process. Square wave voltammetry allowed the development of a simple and accurate methodology for formetanate in fruits using the forward current component (Ribeiro et al. 2018).

Costa et al. evaluated the electrochemical behavior and developed an electroanalytical procedure to methomyl, a carbamate pesticide, in commercial formulation, river and tap water using square wave voltammetry allied to boron-diamond doped electrode (Costa et al. 2017). In a similar work, square wave voltammetry has employed in the pethoxamid determination, a chloroacetamide pesticide, in river water and commercial formulations samples (Jevtić et al. 2018). Andrade et al. used square wave voltammetry and graphite-polyurethane composite electrode to direct determination of the 2,4-dichlorophenoxyacetic acid in soil solutions without necessity of the steps of extraction or pretreatment of the samples, which was observed that the effect of components of the soil solution on the electroanalytical responses was minimal (Andrade et al. 2014).

Shrives et al. used square wave voltammetry and an electrode of the silver nanoparticles in the development of a simple, rapid, selective and low cost method for detection of diazinon, an organophosphorus pesticide, in fruit and vegetable samples (Shrivas et al. 2019). Geto et al. employed square wave voltammetry in the detection of bentazone, a thiadiazine pesticide, in ground water and lake water with high precision by use of simple screen-printed carbon electrode (Geto et al. 2019). Oliveira et al. used square wave voltammetry and carbon-fiber microelectrode to development of a fast, simple, low cost and eco-friendly electroanalytical procedure for the direct analysis of dimethomorph, a morpholine pesticide, in natural water samples (Oliveira et al. 2013). Square wave voltammetry allied to a gold microelectrode also was employed in the determination of dichlorvos, an organophosphorus pesticide, in natural water and cow milk samples, without pretreatments steps in the samples (De Souza and Machado 2005). Deroco et al. analyzed mesotrione, a pyrazolium pesticide, in natural lake, tap water and sugarcane juice samples using square wave voltammetry and a carbon black-modified glassy carbon electrode (Deroco et al. 2017).

Square wave voltammetry allied to hanging mercury drop electrode has also been used in the determination of the nitro pesticide pendimethalin in soil samples, the organophosphorus pesticide chlorpyriphos in natural water samples, the dithiocarbamate pesticide ziram in vegetables, the triazine atrazine in river waters and formulation samples, sulfonylurea pesticide triasulfuron in river water, among other pesticides (Dos Santos et al. 2004; Qiu and Ni 2008; Galli et al. 2011; Bandžuchová et al. 2013; Melo et al. 2018).

However, the electroanalytical techniques, similar other techniques, has used procedures according to the principles of green analytical chemistry by use of a nontoxic electrode and methodologies that make it possible to reduce the amount of waste produced, promoting the use of more environmentally friendly electroanalysis. So, in the last decades, solid amalgam electrodes have been frequently used as suitable voltammetric sensor for reducible pesticides, considering electrochemical behavior study and analytical determination from various chemical class using the square wave voltammetry, the most sensitivity electroanalytical technique.

Silva and De Souza studied the behavior of carbamate pesticide methomyl over silver solid amalgam electrode in media containing a cationic surfactant to improve the sensitivity obtained by square wave voltammetry experiments and decrease the necessary energy in the electron transfer, also reducing the interferents effects in voltammetric signals in natural waters samples (Silva and De Souza 2018). Besides, square wave voltammetry signals from ziram, a dithiocarbamate pesticide, reduction reaction using the silver solid amalgam were used to study the electrochemical mechanism for ziram reduction, indicating that initially the pesticide was totally reversible reduces producing an anion radical and metallic cation, in a process reaction winding two electrons transfer. The radical anion was immediately converted in a highly reactive dianion, that undergo an electrophilic substituent by H^+ ions in a following chemical reaction, plus the reduction reaction of metallic cation to metallic Zn (Silva and De Souza 2017).

Amalgam electrode was also used by de Jesus et al. in the development of the electroanalytical method to dithiocarbamate pesticide molinate in river water and rice samples, obtaining good accuracy and sensitivity comparable to the use of high-performance liquid chromatography (de Jesus et al. 2017). A dental amalgam electrode, prepared similar to in dental clinics, was used by Melo et al. to evaluate the electrochemical behavior and determinate diquat herbicide in natural water and potato samples, which permitted a complete elimination of toxic residues due to the use of liquid mercury and minimized eventually adsorption process observed by use of other solid surfaces (Melo et al. 2009).

1.3.4 Stripping Voltammetry

Stripping voltammetry is a complementary procedure which the compound of interest is initially preconcentrate from the solution into or onto the working electrode, followed by an electroanalytical measurement of the preconcentrated compound, that can be using cyclic voltammetry, differential pulse voltammetry or square wave voltammetry. The adequate combination between preconcentration steps with measurements procedures of redox currents can produce extremely sensitive and selective signals (Wang 1985). Stripping voltammetry is largely applied in quantitative determination and speciation of metals and metallic complexes. However, many pesticides, from different chemical class can also be previously preconcentrate in working electrodes surfaces, called as accumulation step, followed by a potential sweep to strip the preconcentrated species of interest, called as stripping analysis, and in this way, to be adequately identified and quantified (Zoski 2007).

The accumulation steps generally involve very fast and reversible or irreversible reactions, in which the equilibrium between the concentration of accumulated compound and the concentration of compound at the electrode surface is established, producing a concentration profile near the electrode and the posterior diffusion of compound under study towards its surface. During the redissolution step, or measurement step, occurs the dissolution of deposited layer in the electrode surface and the production of an electric current that is proportional to the quantity of previously deposited compounds, which in turn is proportional to the concentration in the solution under study.

Stripping voltammetry can be classified according to the nature of deposition or preconcentration and measurements steps as: anodic stripping voltammetry, cathodic stripping voltammetry, adsorptive stripping voltammetry and potentiometric stripping analysis (Abollino et al. 2018). The difference among those techniques is related to the stripping step, more specifically to the manner that the compounds of interest are stripped from the electrode and to the resulting signals are registered (Esteban and Casassas 1994). In all stripping voltammetry modes, the amount accumulated over the electrode surface depends of some factors such as: solvent, electrode material, pH, ionic strength, mass-transport, temperature and potential values.

Stripping voltammetry has limitations to higher pesticides concentrations due to the limited number of adsorption sites at the electrode surfaces, related to electroactive area, that are dependent of the materials and pretreatments steps in the electrodes. Besides, the sensitivity for pesticides analysis depends strongly on the chemical structure of pesticide, characterized by specific functional groups, the interactions with electrode surface used, the electroanalytical techniques to measure dissolution of the deposit, stripping currents, and the interferers compounds presents during the preconcentration steps. Some specific chemical functional groups, in some pesticides structure such as nitro, phosphorous, sulphur, among others, are easily adsorbed on the electrodes surface, and for this, the use of stripping voltammetry, mainly adsorptive stripping voltammetry, is considerate as the most suitable electroanalytical method for the determination of these pesticides (Wang 1985).

In anodic stripping voltammetry and cathodic stripping voltammetry occurs stripping (oxidation) of the reduced compound and the stripping (reduction) of the oxidized compound, followed by the application of an anodic or cathodic potential scan, respectively. Some examples of applications of these technique in pesticides analysis include pesticides from different chemical class in a wide variety of samples. The concentration intervals analyzed correspond to trace or ultratrace of pesticides, with suitable accuracy by use of differential pulse and square wave voltammetry as the technique of measurement. Paraquat, a bipyridinium pesticides was evaluated in olives and olive oil using square wave anodic stripping voltammetry with chitin-modified carbon paste electrode in the range from

5.0×10^{-9} to 1.0×10^{-5} mol L^{-1} and the relative standard deviation was 5.2% (El Harmoudi et al. 2013). Picoxystrobin, a triazole pesticide, was determinate in urine and water samples employing differential pulse anodic stripping voltammetry on a glassy carbon electrode modified by bismuth film (Dornellas et al. 2013).

Fenitrothion, a organophosphorus pesticide, was evaluated using differential pulse anodic stripping voltammetry under glassy carbon electrode modified with prolamin, a type-protein find in maize, with a good repeatability and reproducibility in the voltammetric experiments performed in natural water and orange juice samples (Itkes et al. 2019). Carbendazim and isoproturon, benzimidazole and phenylurea pesticides respectively, were quantified in water, soil and vegetable samples using square wave anodic stripping voltammetry allied to a graphene-based electrochemical sensor, and the results were statistically like those obtained from the standard high performance liquid chromatography technique (Noyrod et al. 2014). Some organophosphorus pesticides, such as azinphos-methyl, parathion, parathion-methyl and fenitrothion were determined by use of cathodic stripping voltammetry under a hanging mercury dropping electrode, due to previously preconcentration from sulphur at mercury surface followed by stripping in cathodic scan (Tsiafoulis and Nanos 2010).

Considering the adsorptive stripping voltammetry, the pesticides that present hydrophobic and adsorptive properties, can form metal complex at the working electrode surface, such as noble metals, chemically modified, carbon and or mercury electrodes, which are initially adsorbed and posteriorly reduced at the electrode surface. Some chemical class of pesticides has been evaluated using hanging mercury dropping electrode or static mercury drop electrode allied to square wave adsorptive voltammetry. Therefore, fenoxanil, a propionamide pesticide, was determined after optimization of the supporting electrolyte, pH of the medium, potential and time accumulation, frequency, amplitude and step potential, which were used in trace level determination of in natural water spiked rice samples (Brycht et al. 2015). Similar research was developed to determine chlorpyriphos, an organophosphorus pesticide, in irrigation agricultural water (Melo et al. 2018). Additionally, the pesticides ziram, zineb, cyclosulfamuron, imidacloprid, fenthion and fenthion-sulfoxide also were evaluated using square wave adsorptive voltammetry at hanging mercury dropping electrode (Shan Lin et al. 1999; Guiberteau et al. 2001; Galeano Díaz et al. 2008; Qiu and Ni 2008; Sarigül and Inam 2009).

The use of differential pulse adsorptive stripping voltammetry has also been reported to pesticides analysis at hanging mercury dropping electrode. Amorelo and Orechio used differential pulse adsorptive stripping voltammetry in the determination of mancozeb, maneb, propineb, nabam, zineb, ziram, ferbam and thiram, that are dithiocarbamates pesticides. The recovery efficiency in spiked commercial formulate samples was calculated, with recovery percentage ranging between 85% and 97% and precision ranging from 1.3% to 6.1% (Amorello and Orecchio 2013). Qiu et al. used differential pulse adsorptive stripping voltammetry allied to artificial neural network at static mercury drop electrode in the analysis of a mixture containing the pesticides captafol, thiram and phoxim, without any previous chemical separation steps (Qiu et al. 2013).

Mercury film electrodes has also been used in the adsorptive voltammetry of pesticides. Brychet et al. used mercury film under silver amalgam surfaces in the analysis of chlornitrofen and clothianidin, nitro and neonicotinoid pesticide, respectively, in natural waters at micromolar concentrations (Brycht et al. 2013; Brycht et al. 2016). A similar work was developed by Fisher et al. that evaluated the voltammetric behavior of chlorpyrifos, an organophosphorus pesticide, based on its catalytic action after adsorption on a mercury film under amalgam electrode (Fischer et al. 2016). Maleki et al. used mercury film electrode in the determination of atrazine, a triazine pesticide, in soil and water samples (Maleki et al. 2007).

However, the mercury toxicity has promoted its substitution by solid amalgam and other materials as voltammetric sensors. Among the use of solid amalgam electrodes, it has proved be an eco-friendly alternative material to pesticides analysis with high analytical sensitivity promoted by the use of adsorptive voltammetry. Therefore, Lucca et al. described an electroanalytical procedure using square wave adsorptive voltammetry allied to silver solid amalgam electrode to investigate the electrochemical behavior of thiodicarb, a carbamate pesticide, and carried out its electroanalytical determination in food samples (Lucca et al. 2017).

Carbon-based materials such as carbon nanoparticles, diamond, graphene, carbon nanotubes and carbon paste have also been used in the working electrodes preparations to adsorptive voltammetry of pesticides from various chemical class (Asadian et al. 2019). Carbon paste nanocomposite electrode modified by titanium dioxide allied to adsorptive voltammetry was used by Nurdin et al. in the voltammetric detection of the cypermethrin and fipronil, pyrethroid and pyrazole pesticides respectively (Nurdin et al. 2019a, b). Shetti et al. determined aclonifen, a nitrophenyl pesticide in water and soil sample by use of adsorptive voltammetry on the glassy carbon electrode modified with carbon nitride (Shetti et al. 2019). Carbon paste electrode was used by De Lima et al. and Wu et al. to parathion-methyl and linuron, organophosphorus and sulphonylurea pesticides, respectively, determination in vegetables samples (De Lima et al. 2011; Wu et al. 2011).

Finally, in potentiometric stripping analysis is measured the potential variation that occurs during the stripping process, which can be performed either by applying a cathodic or anodic constant current or by the action of an oxidant agent in solution. The analytical signal is produced during the monitoring the potential of the working electrode as a function of time (Jaya and Rao 1982; Estela et al. 1995). In the use of biosensors, the working electrode is maintained in contact with the solution of analysis for a time, in after which, the potential is measured, indicating the enzyme inhibition, that is related to pesticide concentration. Sanghavi et al. employed the potentiometric stripping analysis coupled to carbon paste electrode modified with carbon nanoparticles in the analytical determination of methyl-parathion and ethyl-parathion, organophosphorus pesticides, in fruits, vegetables, water and soil samples, obtaining analytical signals with high sensitivity and consequently with very low detection limits and also excellent reproducibility (Sanghavi et al. 2012).

Khaled et al. used potentiometric stripping analysis and a fabricated screen-printed sensor based on multi-walled carbon nanotube in development of a potentiometric sensors for determination of the ethion, an organophosphorus pesticide, and its degradation residues in soybean, natural water and ethion commercial samples. The proposed procedure presented excellent sensitivity and selectivity towards butyrylcholine analysis, producing spontaneous response time and long shelf-lifetime (Khaled et al. 2014). This disposable potentiometric sensor also was employed in the determination of malathion, an organophosphorus pesticide (Khaled et al. 2010).

1.3.5 Electrochemical Impedance Spectroscopy

In recent years, the use of the aptamers, defined as short single and stranded oligonucleotides that are selectively linked to targets with high affinity characteristics, the immunosensors that present selectivity antibody and antigen interactions, the biosensors that are electrodes modified by antibodies or enzymes, and electrode chemically modified by polymeric film, or oxide layer, or carbon forms, has been used as alternative sensors in pesticides analysis (Liu et al. 2019; Skládal 1997; Pundir et al. 2019; Rhouati et al. 2018; Rana et al. 2019). If allied to electrochemical impedance spectroscopy, a stationary, non-destructive and interfacial technique, these sensors provide useful information about the electric differences between they and the solution containing the compounds of interest. So, electrochemical impedance spectroscopy coupled to these sensor employ has been an excellent technique for pesticide detection producing excellent sensitivity, high specificity and easiness due to the possibility of label-free systems.

In electrochemical impedance spectroscopy a fixed sinusoidal voltage is applied in an electrochemical cell generating a current signal, which can be related to some impedimetric changes in the electrode interface when a biological or a chemical group reacts with its surface modified by a bioreceptor, typical of use of enzyme and other biological substrates. The potentiostat software then converts the potential/current responses into an impedance value, and for this, electrochemical impedance spectroscopy is considerate an excellent alternative in the transduction technology, permitting the direct detection of pesticides from the evaluation of electrical properties in the interface of the sensors and biosensors. So, in the electrochemical impedance spectroscopy experiments are measure the resistive and capacitive properties of the electrode surface in solution containing the pesticides of interest, after a small amplitude perturbation in a system from steady state, which are related to pesticides concentration in the bulk solution or in a complex sample (Wang 2006).

The electrical resistance (*R*), shown in Eq. 1.24, for a direct current (*DC*) with zero frequency circuit is the opposition of an electric circuit to the current flow in a system when a voltage is applied. The impedance (*Z*), shown in Eq. 1.25 considering a frequency different of zero, has the same meaning as resistance with the same unit, ohm or Ω, but is used in an alternating current (*AC*) circuit instead (Randviir and Banks 2013).

$$R(Ohm) = \frac{E_{DC}(Volts)}{I_{DC}(Ampere)} \tag{1.24}$$

$$Z(Ohm) = \frac{E_{AC}(Volts)}{I_{AC}(Ampere)} \tag{1.25}$$

The electrochemical impedance spectroscopy experiment is based in a sinusoidal voltage applied with amplitude usually ranging between 1 and 10 mV changing the frequency of oscillation in scales that ranges from a few mHz to kHz (de Carvalho et al. 2006). The current response present also a sinusoidal shape but with a phase shifting with time. The resulting signals, known as impedance spectrums, can be shown in two main ways, the Bode plot in which the x axis is a frequency logarithmic scale and the y axis is the logarithm of the impedance (Z) while the second y axis is the phase shift Φ, or in the Nyquist plot way, in which are plotted the negative imaginary impedance (−Z'') versus the real part of the impedance Z', as shown in Fig. 1.5. The Nyquist plot way is very sensitive to changes in the electrode/solution interface, and for this, is largely used in pesticides analysis.

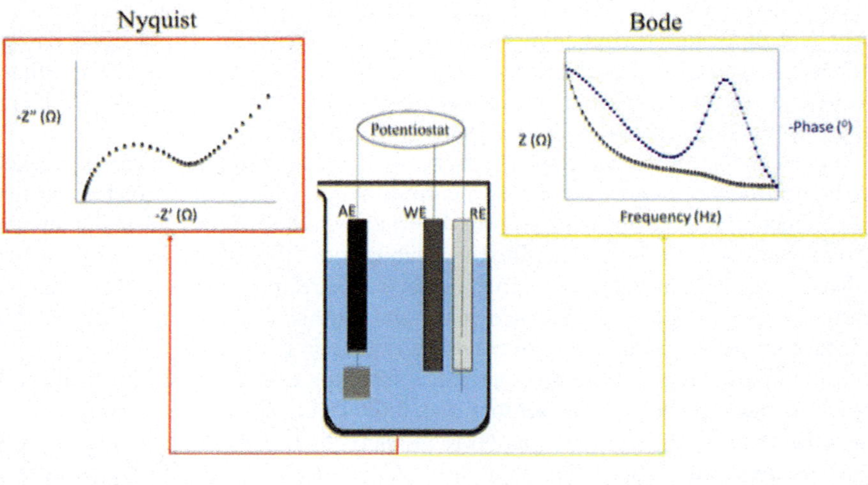

Fig. 1.5 Representation of an electrochemical cell used in impedance spectroscopy experiments, containing auxiliary (AE), working (WE), and reference (RE) electrodes, all connected to a potentiostat. The Nyquist and Bode plots, used concomitantly to check the electrical responses using specific equivalent circuits, are shown adjacent to the cell. The Nyquist plots contain the real and the imaginary impedance components (Z' and Z'') and give information about the electron transfer resistance and solution resistance. The Bode plots contain the phase angle and impedance (Z) both plotted against the frequency and give information about capacitive or inductive effects of electrochemical systems.

The impedance between the electrode and the solution is measured regarding the differences in values during the analysis, Eq. 1.26:

$$Z = E = \frac{Esin(\omega t)}{Isin(\omega t + \varnothing)} \tag{1.26}$$

where ω is the radial frequency, t is time and ø is the phase shift. With some complex mathematic considerations, the impedance data can be analyzed in function of a real (Z') and an imaginary component (Z'') in a graphic known as the Nyquist plot (Tertiş et al. 2017), Fig. 1.6, Eq. 1.27:

$$Z(\omega) = R_s + \frac{R_{ct}}{1 + \omega^2 R_{ct}^2 C_{dl}^2} - \frac{jR_{ct}^2 C_{dl}(\omega)}{1 + \omega^2 R_{ct}^2 C_{dl}^2} = Z'(\omega) - jZ''(\omega) \tag{1.27}$$

where $j = (-1)^{1/2}$, R_s is the solution resistance, R_{ct} is the resistance of charge transfer and C_{dl} is the double layer capacitance (Chang and Park 2010). As both axis (Z' and $-Z''$) has the same unit (Ω), it is necessary an approach to understand and to evaluate the data and its multiple components.

An equivalent circuit is the most suitable option and it is the main choice of several researchers, it is an appropriate choice because to fit with the experimental data, the right settings must be done. Two or more equivalent circuits can fit perfectly to the experimental data, but it doesn't mean that they are all accurate. It is necessary a deep knowledge of the electrochemical system under evaluation and its components to simulate correctly with proper circuit components (Halim et al. 2017; Tertiş et al. 2017; Yoon et al. 2017).

A common typical equivalent circuit for an electrochemical experiment known as Randles circuit, consists in the uncompensated solution resistance (R_s), which occurs at highest frequency it is presented in the right side of the circuit, shown in Fig. 1.6, following the description of the electrochemical cell, in a typical interface Fig. 1.6 part A (Randles 1947). However, it is usually drawn backwards because it is how the data are normally analyzed, from the left to right, from high to low frequencies, Fig. 1.6, part C. The introduction of a new component at a low-frequency range, Z_W is referred to a Warburg resistance regarding the mass transport of the solution in the diffusion layer. The resistance of charge transfer represents the faradaic response (alongside R_s) from an electronic transference between the interfaces in the double layer region. The non-faradaic component arises from the charging of the double layer, generating a capacitor (C_{dl}). By counting all of these parameters, the Randles circuit can fit the experimental data similar to the presented in the Nyquist plot, Fig. 1.6 part C.

Any modification in the electrode configuration whether by a polymeric film, an oxide layer, an immobilized biomolecule or else, will certainly change the interface and thus, the equivalent circuit will need a revaluation. Once the proper circuit is found, the data can be interpreted correctly in several environmental applications, such as chemical characterization of sensor and biosensor to pesticides analysis (Skládal 1997).

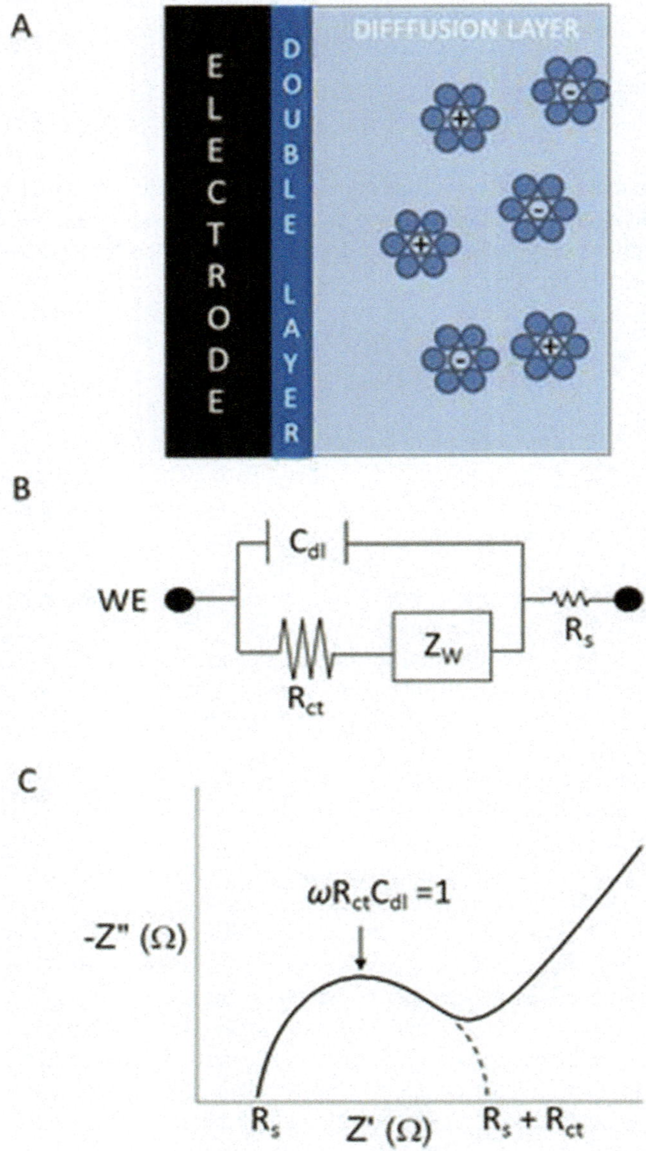

Fig. 1.6 (**a**) Interface electrode solution where is formed the double layer electric when the electrode is polarized. (**b**) The most common equivalent circuit (Randles) used in the Nyquist plots to obtain information about solution resistance and electron transfer resistance. (**c**) The Nyquist diagram referred to the Randles circuit. The WE and AE are the working and auxiliary electrodes, respectively, R_s is the solution resistance, R_{ct} is the resistance of charge transfer, C_{dl} is the double layer capacitance, Z_W is the Warburg resistance, ω is the radial frequency, Z' and Z'' are the real and imaginary impedance. (**c**) was modified from (Randviir and Banks 2013)

In this context, Yola used electrochemical impedance spectroscopy in the voltammetric characterization of the electrochemical sensor prepared from the graphene oxide modification by boron nitride quantum dots, which the results were from the synergistic interactions between modifying agent and graphene oxide used as substrate. The resultant sensor presented a suitable specificity and sensitivity in the simultaneous determination of methyl parathion, diazinon and chlorpyrifos, organophosphorus pesticides, in some samples such as water and apple juice (Yola 2019). Sgobbi and Machado used electrochemical impedance spectroscopy in the evaluation of the electrode surface, prepared from functionalized polyacrylamide to mimics the performance of the acetylcholinesterase enzyme, posteriorly used in the amperometric detection of paraoxon-ethyl, fenitrothion and chlorpyrifos, organophosphorus pesticides (Sgobbi and Machado 2018).

Borah et al. employed electrochemical impedance spectroscopy in the characterization of an amperometric biosensor prepared from glutathione-S-transferase immobilized on platinum electrode, that was used in the analysis of benzimidazoles, organochlorines, thiophosphates, carbamates, polyphenols and pyrethroids pesticides (Borah et al. 2018). Bao et al. evaluated the performance of the biosensor prepared by acetylcholinesterase immobilization under graphene-copper oxide for use in the malathion, an organophosphorus pesticide, analysis with sensitivity around 0.31 parts by trillion, with good selectivity and ideal stability (Bao et al. 2019). Additionally, electrochemical impedance spectroscopy also was used in the study of the reduction kinetic of vinclozolin, a chloroaniline pesticide, in acetonitrile medium (Pospíšil et al. 2000).

Considering the chemically modified electrodes, the interaction between the modifying agent with the pesticide of interest promote changes in the sensor interface, which can be used with success in the pesticide's quantification, mainly organophosphorus, carbamates and dithiocarbamates classes. In these cases, the use of an adequate equivalent circuit allied with electrochemical knowledge, once is done, results in satisfactory data and can be the solution in the pesticides detection. In this context Farahi et al. used the Nyquist plots, modeled with a Randles equivalent circuit from the electrochemical impedance spectroscopy measurements, in the identification of the charge transfer resistance between a silver electrode and paraquat pesticide in different concentrations range. This procedure was applied with suitable sensitivity in the determination of paraquat residues in milk and tomato samples (Farahi et al. 2014).

Malvano et al. performed a comparison between amperometric and impedimetric responses from carbamates and organophosphates pesticides using an acetylcholinesterase-modified gold electrode, in which the detection was performed by electrochemical impedance spectroscopy, based on the enzymatic inhibition by pesticides, without any substrate, in a label-free direct detection. Regarding to the overall toxicity levels in the samples, even with both systems showing no specificity, the responses of electrochemical impedance spectroscopy to impedimetric proved to be better than amperometric system, producing a linear range and limit of detection suitable to measure carbaryl and dichlorvos spiked in tap water and lettuce samples (Malvano et al. 2017).

Facure et al. developed a free-enzyme sensor based on the reduced graphene oxide that allied to electrochemical impedance spectroscopy permitted the optimization of the an alternative method to detect organophosphorus pesticides with speed, simplicity, low cost and sensitivity adequate to analysis in the concentration range permitted by legislation of several countries (Facure et al. 2017). Madianos et al. used electrochemical impedance spectroscopy and an impedimetric aptasensor, allowing the highly sensitive and selective detection of acetamiprid and atrazine down to 6.0×10^{-10} mol L^{-1} and 4.0×10^{-9} mol L^{-1} respectively (Madianos et al. 2018). Bolat et al. used electrochemical impedance spectroscopy allied to a disposable pencil graphite electrode modified with self-assembled peptide-nanotubes in the analytical determination of fenitrothion, an organophosphorus pesticide, in tap water with recovery efficiencies between 94.9% and 103.6% (Bolat et al. 2018).

The use of immunosensors in pesticides analysis produce suitable specificity when are realized through impedimetric detection, or electrochemical impedance spectroscopy, and this way, considering the recent advances in the immunologic field, it is possible to design and to develop specific anti-pesticides antibodies and use them as a recognition biomolecule for biosensors. In this aspect, Mehta et al. used electrochemical impedance spectroscopy and a sensor prepared from the carbon screen-printed electrode modified by an anti-parathion antibody in the parathion detection, an organophosphorus pesticide. The limit of detection found in the optimized system was 0.052 ng L^{-1} with a concentration range from 0.1 ng L^{-1} to 1000 ng L^{-1}, which was satisfactorily applied in direct analysis of parathion in tomato and carrot samples (Mehta et al. 2016). Tran et al. used electrochemical impedance spectroscopy allied to a label-free electrochemical immunosensor containing hydroxyatrazine as bioreceptor element to determine atrazine in environmental monitors (Tran et al. 2012).

1.3.6 Electrochemical Detection Systems

Pesticides with redox properties can be chromatographically separated, and subsequently detected by electroanalytical techniques, which can operate as an appropriate electrochemical detection system in the high-performance liquid chromatography. Furthermore, electrochemical detection can be used in flow injection analysis, capillary electrophoresis or methodologies based in the use of microfluidics systems, providing improvements in the sensitivity and selectivity to pesticides analysis from different chemical class (Llorent-Martínez et al. 2011; Trojanowicz 2011). In electrochemical detection devices, the electrodes are connected to an electronic circuit containing an adequate low noise amplifier, that measure the picoampere or microampere current in a potential range of ± 1.5 Volt, and consequently, the data are collected and stored in data acquisition software. In the resulting signals, the intensities of current are used to quantification and the retention time identifies the compound of interest (Llorent-Martínez et al. 2011).

Electrochemical detection devices coupled to high performance liquid chromatography, compared with traditionally ultraviolet detection, present better sensitivity and selectivity (Zhu et al. 2019). Basically, after the pretreatments or preconcentration steps the samples are introduced in the high performance liquid chromatography system, following a separation in an adequate chromatographic column, that is connected to an electrochemical detection cell containing an electrochemical sensor. Electrochemically active compounds elute from the column and when reach the cell occurs an electrochemical reaction resulting in an electrical current. In electrochemical detection devices, the usual detection modes involve the potentiometry and amperometry as electroanalytical technique of the identification and quantification, both described above (Nagels and Poels 2000; Mahé et al. 2015).

Electrochemical detection devices coupled to high performance liquid chromatography exhibit a wide linear range of concentration, comprising more than 3 orders of magnitude, from nano mol L^{-1} to micro mol L^{-1} or more, and for this provides a highly sensitive and selective detection device. Besides, in these analyses, low sample volume is required, occurs a significative elimination of matrix interferences and main the excellent selective detection (Mahé et al. 2015). Therefore, Rodríguez et al. used electrochemical detection devices coupled to high performance liquid chromatography in the determination of the diflubenzuron, benzoylurea pesticide, and its metabolites in forestry matrices, where the reverse phase separation was followed by amperometric detection using a glassy carbon working electrode at +1.35 V vs Ag/AgClsat., allowing detection limits from 0.07 μg L^{-1} to 1.8 μg L^{-1} (Rodríguez et al. 1999), similar to evaluated by Kunert et al. (Kunert et al. 2002).

Additionally, fenitrothion, ethyl-parathion, methyl-parathion, paraoxon and guthion, organophosphorus pesticides, were simultaneously separated during 25 min using electrochemical detection device coupled to high performance liquid chromatography allowing detection in interval concentration below 4 μg L^{-1} (Martinez et al. 1993). Kunert et al. determined fenuron, neburon, chlorotoluron and linuron, phenylurea pesticides, using electrochemical detection device coupled to high performance liquid chromatography and amperometric detection, with sensitivity lower than picomole range, and in a manner rapid, simple and highly selective (Kunert et al. 2002). Similar detection of thiamethoxam and imidacloprid were performed in samples that were previously treated using a photochemical reactor, where the electrochemical detection device coupled to high performance liquid chromatography permitted identify and quantify phenylurea pesticide in the samples evaluated containing different interference contribution (Rancan et al. 2006a, b).

Flow injection analysis coupled to electrochemical detection device can be used in determination of compound that are either oxidizable or reducible producing this way, an electron flow caused by the electrons transfer reaction that takes place in the electrodes surfaces. The use of the flow injection analysis coupled to electrochemical detection device in the pesticides detection allow analysis with excellent accuracy and precision in the results, very low consumption of reactants and suitable robustness. Additionally, in the flow injection analysis coupled to electrochemical detection device the flowing liquid continuously retain the clean in the working electrode surface, due to the removal of the reaction products and impurities leached from the electrode, allow improvements in the detection limit due to the convective transport of compounds of interest and mainly, improve the analytical selectivity (Llorent-Martínez et al. 2011). It can be classified into direct detection of the pesticide, which the pesticide is directly measured by means of its electrochemical behavior without requiring enzymatic reaction, and into the enzymatic reaction followed by its determination, which is used an enzymatic reaction previously to the analyte quantification. However, the use of the enzymatic reaction is the most common flow injection analysis coupled to electrochemical detection device method employed in the pesticides analysis (Llorent-Martínez et al. 2011).

Shi et al. used flow injection analysis coupled to electrochemical detection device allied to an amperometric biosensor to determine organophosphorus pesticides, with a wide linear for the responses from the inhibition by dichlorvos, in the range from 0.1 μmol L^{-1} to 80 μmol L^{-1}, corresponding from 7.91% to 84.94% inhibition for acetylcholinesterase enzyme, and detection limit of 10 nmol L^{-1} in the simulated seawater, considering 15 min for inhibition time (Shi et al. 2006). Wei et al. also used an acetylcholinesterase enzyme biosensor allied to flow injection analysis coupled to electrochemical detection device in order to obtain flow system fast, sensitive, and with stable responses to trichlorfon, an organophosphorus pesticide, with the detection in a concentration range from 0.01 μmol L^{-1} to 20 μmol L^{-1} and detection limit of 1.0×10^{-8} mol L^{-1} (Wei et al. 2009).

Nikolelis et al. used flow injection analysis coupled to electrochemical detection device in the development of a fast, selective and sensitive procedure for carbofuran a carbamate pesticide, analysis in foods, using an sensor based on the acetylcholinesterase enzyme incorporated in air stable lipid films (Nikolelis et al. 2005). The highly sensitive method using flow injection analysis coupled to electrochemical detection device with an amperometric biosensor, prepared from the modification of the glassy carbon electrode with carbon nanotube and self-assembled acetylcholinesterase, was used by Liu and Lin in the online monitoring of organophosphate pesticides and nerve agents (Liu and Lin 2006). Valdés-Ramírez et al. quantified the pesticides dichlorvos and methylparaoxon, organophosphorus pesticides, in water samples using flow injection analysis coupled to electrochemical detection device with amperometric detection based on screen-printed amperometric biosensors of acetylcholinesterase enzyme, without pretreatments steps to eliminate interfering species (Valdés-Ramírez et al. 2009).

The use of capillary electrophoresis coupled to electrochemical detection device present some advantages in comparison with capillary electrophoresis and other detection modes, such as: high speed, high efficiency, ultra-small sample volume, low consumption of solvent, simplicity, selectivity, large separation capacity, and relatively low cost. In practices, in the capillary electrophoresis separations is applied of electrical field to a solution of charged molecules that promote their movement. Each charged molecule present different mobility, which are dependent of the ratio between its charge and size, that also are dependent of the molecular weight, the degree of solvation and the three-dimensional structures. At the end, the electroactive species can be detected by electrochemical detection device, mainly by the use of amperometry technique (Malik and Faubel 2001; Picó et al. 2003; Shin et al. 2003).

Chicarro et al used capillary electrophoresis coupled to electrochemical detection device and the mixture of the 0.020 mol L^{-1} phosphoric acid at pH 7.0 with 0.020 mol L^{-1} of sodium dodecylsulfate in the development of an analytical procedure to determine the phenylureas fenuron, chlorotoluron, monuron, monolinuron, isoproturon, diuron, and linuron, and the triazines atrazine and simazine in environmental water samples (Chicharro et al. 2005). These pesticides were separated in less than 30 min with detection limit lower than 5.0 μmol L^{-1}. In other work, these authors used the mixture of the 0.020 mol L^{-1} of boric acid at pH 8.3 with 0.025 mol L^{-1} of sodium dodecylsulfate in the micellar electrokinetic capillary chromatography with electrochemical detection device to obtain selectivity in the analysis of urea, triazines and triazinones pesticides in environmental waters, using separations at less than 14 min (Chicharro et al. 2007).

Moreno et al. employed capillary electrophoresis coupled to electrochemical detection device allied to extraction steps using matrix solid-phase dispersion in the amitrole, a triazole pesticides, and its metabolites urazole in apple samples with recovery efficiencies around 85% and detection limits of the 0.4 μg g^{-1} (Moreno et al. 2008). Organophosphorus pesticides paraoxon, methyl parathion, fenitrothion, and ethyl parathion were determined by Wang et al. using capillary electrophoresis coupled to electrochemical detection device with amperometric detection at sub micromolar detection limits and good precision (Wang et al. 2001). Additionally, Cheng et al. determined the carbamates pesticides fenobucarb, isoprocarb, metolcarb and carbaryl with recovery efficiencies of the 105%, 104%, 110% and 98%, respectively, by use of capillary electrophoresis coupled to electrochemical detection device (Cheng et al. 2007).

Despite the applicability of all the techniques shown above, there is still a need to develop pesticide detection platforms that are fast, portable, inexpensive and easy to handle. For this, the microfluidics system has been used as these specific platforms that miniaturize all systems controlling fluid flow in microchannels producing this way, rapid detection and identification of compounds, using a small amount of samples and reagent, and the potential for monitoring of pesticides from different chemical class (Wang 2002; Sia and Whitesides 2003). Microfluidics devices with amperometric detection permits the use of inexpensive test devices, such as paper-based indicators, dipstick test assays, and even paper chromatograph microfluidic paper-based separation. However, the use of microelectrodes and biosensors has become commonplace (Jokerst et al. 2012; Mross et al. 2015).

Therefore, Arduini et al. determined paraoxon, 2,4-dichlorophenoxyacetic acid and atrazine using a paper-based platform modified by enzymes due to these pesticides can inhibit the enzymatic activity of the butyrylcholinesterase, alkaline phosphatase, and tyrosinase enzymes, respectively. The degree of inhibition was evaluated by monitoring the enzymatic activity in the absence and in the presence of pesticides using the amperometry and the proposed sensor coupled to electrochemical detection device (Arduini et al. 2019). A phenol sensor, an important metabolite in pesticides degradation, coupled to electrochemical detection device, was evaluated using a microfluidic device based on commercial textile threads and modified by carbon nanotubes and gold nanoparticles anchored through covalent bond with the tyrosinase enzyme (Caetano et al. 2018).

Carbaryl and chlorpyrifos pesticides were evaluated using electrochemical detection device and a droplet-based microfluidic system, which consisted of polydimethylsiloxane microfluidic chip integrated with microelectrodes modified by acetylcholinesterase enzyme (Gu et al. 2013). Medina-Sánchez et al. prepared a lab-on-a-chip platform from the use of a boron doped diamond electrode for use in the electrochemical detection of the pesticide atrazine by electrochemical detection device in amperometric mode, which produced 3.50 pmol L^{-1} of the detection limit (Medina-Sánchez et al. 2016).

In every work demonstrated in this subsection, the procedure is markedly subjected to the interfering effects caused by adsorption of pesticides on the electrode surfaces or irreversible enzymatic inhibition, impairing the sensitivity and selectivity in the pesticide detection. Besides, in high performance liquid chromatography coupled to electrochemical detections device the excessive cost of the instrumentation, the necessity of pre-treatments steps in the samples and the demand on highly skilled labor, make the use of electrochemical detections device less popular in pesticides analysis. Considering flow injection analysis coupled to electrochemical detections device, special attention must be given to suitable flow-through cells construction, where will occur an efficient and repeatable mass transport to the electrode surface with complete clean and renovation by use of small dead volumes, reducing the waste generations and consequently the cost of the analysis. Microfluidics devices coupled to electrochemical detection device has been presented as a suitable alternative because the use of disposable sensors could solve the problem of the working electrode poisoning.

1.4 Application of Electroanalytical Techniques in Pesticides Determination

Pesticide contains in its chemical structures electroactive groups that can be electrochemically reduced and or oxidized. So, many pesticides have well-defined electroactivity on some known electrode surface, and for this, they can be identified and quantified using some electroanalytical technique, as described above. In some cases, the pesticide can inhibit an enzymatic activity, proportionally to its concentration, permitting the use of biosensor prepared from specific enzymes. In some cases, the chemical structure of pesticides permits its complexation reaction with metals producing metallic complex easily reduced and or oxidized electrochemically, characterizing an indirect analysis. In other cases, the pesticide can react with a specific compound, and the generated products can be quantified, in a typical indirect analysis of pesticides.

In all the cases cited above, the success in the pesticides electroanalysis is related to adequate choice of experimental parameters such as solvent, supporting electrolyte and pH values. Additionally, the proper selection of electroanalytical technique and the analysis of influence of voltammetric parameters, typical of each technique employed, in the peak current (I_p), peak potential (E_p) and half-height width ($w_{1/2}$) values will define the sensitivity and selectivity in the voltammetric analysis, considering the pesticides analysis from different chemical class and in several sample types. Besides, the relations between each experimental and voltammetric parameters with peak current, peak potential and half-height width values allied to theoretical diagnoses criteria can provide analytical data about the kinetic and the electron transfer mechanism, indicating reversibility in the reactions, adsorption process, diffusion kinetic, the intermediaries species formed, final products and associated energies involved in the electrons transfer reactions, as indicated early.

Considering different chemical classes in many sample determinations, some steps involved are very well-defined, as shown in the flowchart presented in Fig. 1.7. These steps are operationally easy compared to traditional chromatographic techniques, which facilitates the analysis by less qualified professionals, contributing in the reduction of costs of chemical and time of the analysis. The following will be presented the basic instrumentation used in the pesticides electroanalysis, and a brief description of each step involved in the electroanalytical determination of any pesticide.

1.4.1 Basic Instrumentation

Electroanalysis of pesticides requires an electroactive pesticide or a redox reaction related with pesticide interactions, the use of a basic instrumentation, that is constituted by a voltammetric analyzer or potentiostat/galvanostat, an electrochemical cell, and a software that controls the system, receive and evaluate the analytical data.

1.4.1.1 Equipment

Comparing to other analytical techniques, the electroanalyses involves a simplified equipment, where it is not necessary the use of any signal converter obtained in electrical signal since the results obtained represent it, resulting in lower cost of the instrumentation. The equipment is usually sold as potentiostat/galvanostat, where the first controls the electric potential established between the working and reference electrodes, the second promote the control of the flux that circulate through an electrochemical cell. It is an equipment containing an electrical circuit able to perform several electrochemical techniques based only in the parameter changes, which are controlled by specific software.

The potentiostat/galvanostat can be represented by a simple electrical circuit, where occurs an accurately control of the potential between the auxiliary electrode, also known as counter electrode, and the working electrode when is applied a

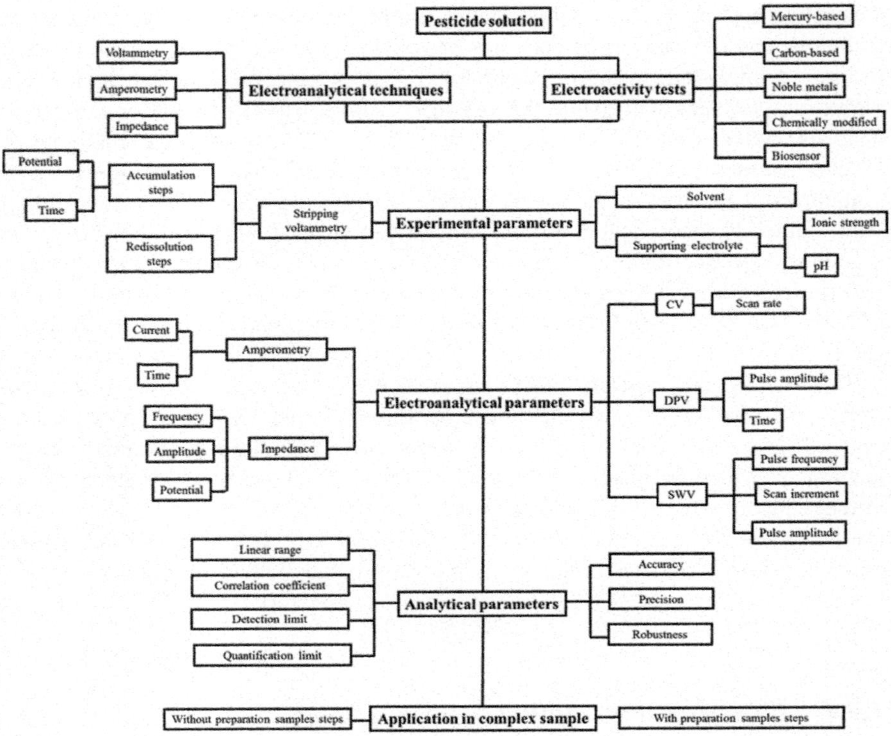

Fig. 1.7 Diagram showing all the steps involved in the analysis of pesticides using electroanalytical techniques. Electroactivity tests are presented using different types of electrodes, the choice of the adequate electroanalytical technique, the optimization of experimental parameters and the voltammetric parameters inherent to the chosen technique. Main analytical parameters that must be used to validate the electroanalytical methodology and finally the application in complex samples are shown in this diagram. Here, pH is hydrogen potential, CV is cyclic voltammetry, DPV is differential pulse voltammetry and SWV is square wave voltammetry

potential difference between the reference electrode and the working electrode, which is defined by analyst according to redox reaction of the interest. In this equipment, the potential difference between the reference and working electrodes and the current flows between the auxiliary and working electrodes are continuously monitored producing analytical signals or voltammograms.

Companies all over the world develop and sell these devices such as: Metrohm, CH Instruments, Ivium Technologies, Ametek Scientific Instruments, Palm Sense, Pine Research, Scribner Associates and more with prices ranging from an average of € 2.000,00 (two thousand euros) to € 20.000,00 (twenty thousand euros) and sizes ranging from a desktop computer to a smaller than a cell phone. They can be bought with the several electrochemical techniques such as amperometry, cyclic voltammetry, differential pulse voltammetry and square wave voltammetry, and with optional modulus to perform specific experiments as it is the case of low current module to specific use with microelectrodes, electrochemical quartz crystal microbalance and electrochemical impedance spectroscopy. The modulus prices range between two thousand euros (€ 2.000) to eight thousand euros (€ 8.000,00). The manipulation of these equipment is simple, and the software employed is very user friendly, allowing the use of a less qualified analyst, reducing sign, costs with instrumentation and labor.

1.4.1.2 Electrochemical Cell

The electrochemical cell is a device constituted by an ionic conductor, constituted by a known supporting electrolyte solution, and the electrical conductors, that are the working, reference and auxiliary electrodes, which generate an electrical current from the redox chemical reaction that take places in the interface of the working electrode/electrolyte, or potentiostatic techniques. The measurement can also occur through the establishment of an electrical potential in the indicator electrode interface, or potentiometric techniques. In practice, an electrochemical cell is any inert environment where the supporting electrolyte, the electrodes and the compound of interest can perform an electrochemical reaction or produce an electrical potential.

Electrochemical cell can present varied sizes and configurations, according to the experiment of interest and the nature of the samples. Usually, the cell is like a covered beaker with different volume capacity, containing the compound under study. Two compartments cells are designed when is necessary the products separation of the redox reaction, where in the one compartment contain the cathode where occurs the reduction reaction, and the other compartment contain the anode where occurs the oxidation reaction. One compartment is most common configuration to pesticides analysis, where sometimes can be used only two electrodes configuration (working and reference electrodes, when are employed microelectrodes or potentiometry) or four electrodes (two working electrodes or two auxiliary electrodes when are studied reactions that involves simultaneously reduction and oxidation reactions or corrosion studies, respectively) (Wang 2006).

Electrochemical cell can be constructed using a variety of materials that are easily processable, are completely inert to the electrochemical reactions, and present low cost, and for this, the glass, Teflon and Nylon, since the material used should. For this, in research lab, it is usually made of glass, that is easy to make and presents low cost. Sizes can change from some microliter, when the sample volume is limited, to 50 mL when are used bath injection analysis, as described early. For use in flow systems such as high performance liquid chromatography coupled to electrochemical detection device and flow injection analysis coupled to electrochemical detection device, the electrochemical cell design will depend specially of working electrode geometries, which is necessary a parallel, perpendicular and wall-jet flows (Zoski 2007; Escarpa et al. 2015).

There are commercially available electrochemical cells with temperature control capability, mainly used when the working electrode are biosensors and the temperature control maintain the efficiency of enzymatic activity. Additionally, sometimes

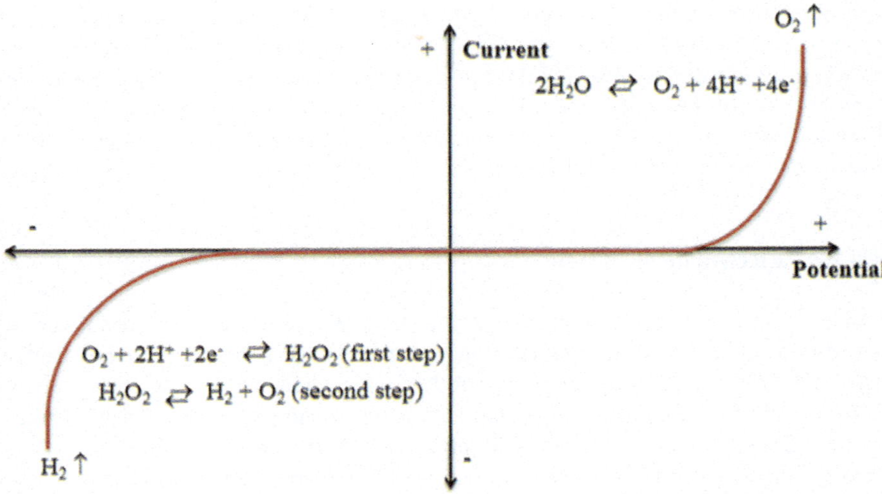

Fig. 1.8 Representation of the potential range of working electrodes, which are limited in positive direction by reaction of oxygen evolution and in cathodic direction by hydrogen evolution, both from water hydrolyze reaction, which is presented indicating all steps involved. (Modified after Wang 2006)

can be required the stirring of the solution between each experiment to promote a renovation in the working electrode surface, or during the experiment to increase the mass transference, and for this is necessary a cell containing mechanical stirrer. In specific experiments can be necessary the use of a cell that permits the removal of dissolved oxygen, mainly when the working electrodes are based on mercury use (hanging mercury dropping electrode and amalgam electrodes). In these surfaces, the oxygen is initially reduced producing hydrogen peroxide, as shown in Fig. 1.8, which produces a high background current, that can interfere in the analytical signals from reduction reaction of many reducible pesticides and for this, a deaeration step is necessary (Wang 2006). So, the most commons manner of oxygen removal is using a purified inert gas (argon or nitrogen), that is previously purged in the cell during a time, which depends of sample volume, before analysis.

1.4.1.3 Electrodes

In a conventional electrochemical cell, the operability will depend on the presence of working, reference and auxiliary electrodes. If low currents values are generated in the electrochemical system or if the potentiometry is used, a two-electrodes configuration, working (in potentiometry is called of indicator electrode) and reference electrodes, is adequate. Working electrode correspond to a surface where the reaction of interest occurs. Reference electrode is responsibly by manutention of the potential at the working electrode by means of a potentiostat, counter electrode is used as the current-carrying and for this, are known as auxiliary electrode (Wang 2006).

Working electrodes are a key factor in the pesticides electroanalysis. Therefore, the choice of the working electrode manufacture materials needs to be evaluated with great care to obtain a high signal-to-noise ration and reproducible responses. It should also be considered the potential range where the pesticide redox reaction occurs, and the properties related to working electrode surface such as: electrical conductivity, reproducibility in its surface, mechanical stability, low cost of manufacturing, easy availability of the material, and without toxicity (Wang 2006). However, it is very difficult for a working electrode to display all the above-mentioned characteristics, and for this, there are several commercially available or lab made, different working electrodes type, considering chemical composition, sizes or geometry, as will be described next.

Reference electrode indicates an electrical potential with steady and reproducible value that is independent of the sample composition and against which the working electrode potential is compared. For this, reference electrode must present the chemical composition and physical structure effectively constant, without variation during the experiments, to provide steady potential which no change, regardless of current, flows through it, i.e. it should be an electrode nonpolarizable allowing an adequate control in the working electrode potential. These electrodes are constituted by robust body with adequate top seal, junction, and an active component that defines the reference potential, with no alterations in its value during the experiment realization. Moreover, present the ability of the return to its value of potential after a current stress (Zoski 2007). These properties are obtained using redox couple constituted by a metal-metal salt, and the main used are silver-silver chloride (Ag/AgCl) and the saturated calomel reference (Hg/Hg_2Cl_2), both with different concentrations of chloride, which provide the potential values known.

These reference electrodes can be commercially obtained or lab made. A saturated calomel reference electrode with redox reaction presented in Eq. 1.28, is prepared using a small amount of mercury and Hg_2Cl_2 (calomel), that are mixed with KCl, and held in saturated KCl solution to moisten the mixture. The electrical contact between the paste and the small mercury pool formed is performed by platinum wire, that serv as an electrical contact, which is connected to one channel in the potential measuring device (Christian et al. 2014).

$$Hg_2Cl_2 + 2e^- \leftrightarrow 2Hg + Cl^- \; E^{\circ}_{Hg/Hg_2Cl_2} = 0.268 \text{ V vs.NHE} \left(25^{\circ}C\right) \tag{1.28}$$

where E° is the potential of a cell measured under standard conditions, which all chemical species are present in its standard states, 1.0 mol L^{-1}; NHE is normal hydrogen electrode (NHE) and correspond to the standard values of the reference in the standard reduction potentials, E°, assigned conventionally at all temperatures as 0.00 volts, V.

Due to environmental problems generated by toxic residues from reference electrode preparation steps, saturated calomel reference electrode is less favored today. Moreover, it presents alterations in the potential values with temperature changes. Therefore, in most pesticides electroanalysis, the Ag/AgCl is the most adequate reference electrode, that it is also the most simple, inexpensive and nontoxic. This electrode is prepared from chloritization of a silver wire in a solution containing around 1.0 mol L^{-1} of HCl or KCl, followed by application of +1.0 V for approximately 2 min, coating the wire, as shown in Eq. 1.29. The electrode is maintained in a filling solution of KCl.

$$AgCl + e^- \leftrightarrow Ag + Cl^- \; E^{\circ}_{Ag/AgCl} = 0.222 \text{ V vs.NHE } (25^{\circ}C) \tag{1.29}$$

In both reference electrodes, the internal solutions of KCl are usually separated from the compound under study using a salt bridge that minimize the liquid-junction potential and to make it reasonably constant. Besides, the Ag/AgCl electrode potential depends only on the chloride concentration also is calculated from the Nernst equation presented in Eq. 1.30:

$$E = E^0 + \frac{0.05916}{n} \; log \, C_{Cl^-} \tag{1.30}$$

where n is equal to 2 and 1, for saturated calomel reference electrode and Ag/AgCl, respectively, and $C_{Cl\text{-}}$ correspond to chloride concentration.

Finally, the counter electrode is prepared from an inert conductor material, such as platinum wire or graphite rod. In the cell configuration reference and counter electrode should be located on the downstream side of the working electrode, so that reaction products at the counter electrode or leakage from the reference electrode does not interfere with the working electrode.

1.4.1.4 Supporting Electrolyte

In an electrochemical cell the potential application promotes the working electrode polarization, that are positively or negatively charged. This charge in electrode surface can repulse or attract the compound of interest from the bulk solution to the working electrode surface, yielding a migration current that interfere in the resultant current signals, by decrease or increase in its values. So, the addition of an inert electrolyte that act as a supporting electrolyte, minimize the attraction or repulsion of the charged species to working electrode surface, reduce or eliminate migration flows. However, this occur only if the supporting electrolyte present high concentration, about a hundred times bigger than the compounds of interest. Furthermore, the supporting electrolyte allows a considerable increase in the solution conductivity and in the electron transfer rate, and consequently, can improve the quality of the electroanalytical signals (Zoski 2007).

1.4.2 Electroactivity Tests

As the redox reaction occurs in the interface between the working electrode surface and the thin layer, called as double electric layer, containing the pesticide under study and supporting electrolyte, the adequate choices of working electrode materials will influence the electroactivity of pesticides. It can occur by reduction and or oxidation reactions from functional groups present in the chemical structure or interaction with components of the medium or working electrode surface. Besides, the preliminary steps of electroactivity tests also involves a carefully potential scan in positive direction, evaluating the oxidation reactions, and the potential scan in negative direction, evaluating the reduction reactions.

In any selected working electrode, it is evaluated the potential range where occurs the electrolysis of solvent, that produce high background signals, limiting their use, as shown in Fig. 1.8. For this, preliminary studies should perform by running a blank voltammogram, with only the supporting electrolyte, at the voltammetric conditions that will be employed with the pesticide solutions, which will quickly establish the interval potential of the selected working electrode. The interval of the working electrode potential can be evaluated using different supporting electrolytes in different pH values, considering the acid, neutral and basic mediums.

All tests can be performed using cyclic voltammetry or other voltammetric techniques, according to the main objective of the analyst that can be identification, quantification and electrochemical redox studies, since each electroanalytical technique present different performance, as described in the previous section. The intensities, position and profiles of the analytical signals will be dependent mainly on the redox behavior of pesticide over working electrode surface, which can present different physical, chemical, and electronic properties.

These properties will influence in the ability of electrode to carrier electrons and adsorb reactants and or products, and for this, are responsible by peak current, peak potential and half height width values obtained from redox reaction of pesticides, also influencing in kinetics of the electron transfer and in the mechanisms of the redox system (Zoski 2007). The main properties necessary for the working electrode are high electrical conductivity, chemical inertness, hardness, durability, homogeneous microstructure in all over the surface, background currents with very low and stability values, surfaces with stability in the morphology and microstructure over a wide range of the potential, a fast kinetics in the electron transfer considering a great variety of redox systems, easily in the fabrication process, well defined shaped, low cost and nontoxic.

The great challenge in the electroanalysis area is to find a material that present all these properties. Due to this fact, there are varied materials available such as mercury-based, nobles metals, carbon forms, chemically modified and biosensors, and the appropriate choice depends mainly on the applicability of interest, such as the chemical group that reacts, aqueous or organic solvents medium, potential range, and pH of the medium. Sometimes the geometry and the size of the working electrode can also be considered in analytical practices (Wang 2006).

Each of these surfaces have their own characteristics, which include among the most important, potential range and pretreatment protocol to obtain efficiency in the activation, renovation and clean in its surface. The potential range is limited by the solvent used, that can also be reduced or oxidized, producing a high background current, as shown in Fig. 1.8. When the water is the solvent, the interval of the potential is limited in the anodic running by oxidation of the water and in cathodic running by the hydrogen ions reduction. The range of potential for use in pesticides analysis, without backgrounds currents from water hydrolyze or superficial process, such as oxides formation in the working electrode surface, can be changed in the cathodic or anodic direction by alterations in the pH of the medium. Besides, considering negative potential scan, the water reduction occurs with oxygen presence, which explains the necessity of oxygen removal in some voltammetric experiments, thus expanding the range of potential in the negative direction.

Contaminants in chemical structure, products and reactants adsorbed onto the surface and oxides formations in working electrode surface can block sites for electron transfer, promote alterations in the double-layer structure and increase the resistance of electronic transference. Therefore, each material used in the working electrode preparation present a specific procedure to surface cleanliness and activation steps to improve the sensitivity, selectivity and mainly the reproducibility in analytical signals. Therefore, before the electroanalysis, working electrode need to be firstly pretreated employing specific protocols, which depends only on the materials used in its preparation. These protocols can combine the surface clean by specific solvents or surfactants, alteration of the exposed microstructure by mechanical polishing, manipulation of the surface chemistry by electrochemical process that can introduce in working electrode surface carbon-oxygen or carbon-hydrogen terminations that improve the electron transfer kinetic.

1.4.2.1 Mercury-Based Electrodes

From the polarography discovery in 1922 by Jaroslav Heyrovsky until the implementation of the principles of Green Chemistry in the early 1990s mercury has the most used material in electroanalysis of pesticides that electrochemically react in the cathodic potential interval (Armenta et al. 2008). Its intensive use occurred due to inherent physical/chemical properties, such as its liquid state at room temperature, it presents high interfacial tension, hydrophobic surface, and it can produce amalgams with many metals and soluble salts such as halides, sulphide, sulphate, and some other anions. Besides, it is an ideal polarizable interface over a range of 2.0 V in the cathodic potential direction and responds precisely to all changes in the potential difference of an external source when coupled to an electrochemical cell.

Electrochemically, mercury-based electrodes present an excellent overvoltage, with the reduction of hydrogen ions from aqueous medium occurring around −1.1 V in strongly acid medium and around −2.8 V in strongly basic medium, both vs. saturated calomel reference electrode. This fact considerably expands the cathodic potential intervals when compared to noble metals and carbon-based electrodes. This characteristic is very important due to certain chemical groups in the pesticide structures, which present strongly cathodic redox potentials, and other compounds that are electro-inactive in the anodic potential range. In electroanalysis, the non-electroactive compounds can be converted into electroactive to its determination, but these conversions process is rarely used. Mercury-based electrodes can be classified as liquid electrodes, such as hanging mercury drop electrode and mercury film electrode, or solid electrodes, that are amalgam electrodes in the past or solid forms.

In hanging mercury drop electrode instrumentation, the mercury is pushed manually out of a tank sealed by the pressure of an inert gas or using a valve that can be drive by electrical discharge, promoting this way, a constant control in the mercury flow through the capillary with 0.07 mm diameter or less. This system is coupled to a potentiostat/galvanostat and specific software that controls the purge time to remove oxygen from the solution, drop size, drop dispensing, drop dislodgment, and the electroanalytical technique for potential scanning and current acquisition. The drop formed presents a constant time and size, promoting excellent renovation of the surface with suitable reproducibility in the mercury drop and, mainly, a decrease in residue generation due to the potential scan and measure of the current performed using a single drop. Despite all the advantages, the use of hanging mercury drop electrode is questioned due to the high toxicity of mercury.

Mercury film electrodes are prepared from the electro deposition of a thin layer of liquid mercury under a support, generally noble metals or carbon surfaces. The use of platinum, gold and silver as support results in difficult reproducibility of the electroanalytical responses due to the formation of a non-uniform surface in the formed amalgam. Iridium, a metal that present a low solubility in liquid mercury, is the most adequate support because it produces a film with suitable adherence promoting an excellent reproducibility in the analytical signal (Scholz 2010). Carbon surfaces such as pyrolytic graphite, carbon paste, carbon fiber and glassy carbon present chemical inertness, broad potential range, low cost, and easy surface renovation by mechanical polishing, making it a suitable material for mercury-coating (Economou and Fielden 2003).

Although mercury film electrode preparation is easy, it still presents difficulties in the reproducibility of film thickness, altering the electroactive area and, consequently, the reproducibility of the electroanalytical signals. Use of mercury film electrode minimize the mercury amount produced, but the toxicity of the electrode is not eliminated. Furthermore, like hanging mercury drop electrode, the mercury film electrode presents difficulties in the use in electrochemical detection devices due to the low stability of liquid mercury in flows systems. For this, research regarding other materials with similar analytical and electrochemical characteristics as traditional mercury electrodes is an area of great development in electroanalytical and materials science, resulting in the use of solid amalgam electrodes.

Several materials are used in preparation of solid amalgam electrodes, which are formed from the junction between mercury and another metal, such as silver, copper, gold, and bismuth. They present a solid structure and do not contain any liquid mercury but, they exhibit similar electrochemical properties as the hanging mercury drop electrode. If no chemical interactions occur between the pesticide and the electrode, the peak potential values are like those observed in the hanging mercury drop electrode, which are also strongly dependent of the medium considering the solvent, supporting electrolyte, and pH. Besides, the solid amalgam electrode present excellent chemical stability, electrical conductivity, reproducibility on the surface, mechanical properties, low cost of manufacture, minimum toxic waste generation and the possibility of its use in electrochemical detection devices. Researches demonstrated that silver is the most adequate metal for solid amalgam electrodes preparation due to its higher overpotential towards the hydrogen evolution reaction compared to other metals and the positive interval of oxidation that is close to mercury, promoting this way a working electrodes with a wide range of potential that can be variated since 0.5 V until −2.0 V vs. saturated calomel reference electrode, according to the supporting electrolyte employed (De Souza et al. 2011).

Despite all controversies involving mercury toxicity, in the last two decades, there still has been a considerable number of researches developed and published that employ the electroanalysis of pesticides using mercury-based electrodes. A search of scientific databases (Web of Sciences®, Scopus® and ScienceDirect®) using the keywords pesticides, amalgam electrodes, and mercury electrodes from 2000 to 2019 indicated the substitution of classical hanging mercury drop electrode by solid amalgam electrodes. All works has been shown that solid amalgam electrodes present good sensitivity in pesticide analysis, analytical stability, and minimize mercury waste and it does still reduce adsorptive problems related to the use of other solid surfaces. Thus, solid amalgam electrodes can be considered an eco-friendly tool and a very interesting alternative for analytical determination of pesticides from different chemical classes.

1.4.2.2 Noble Metals Electrodes

Gold, iridium, rhodium, platinum or the mixture of the metals in the form of alloys represent the main noble metals used as working electrode in the pesticides electroanalysis. These metals are adequate in the oxidation of pesticides at positive potentials and sometimes for easily reducible pesticides. Their use promotes electron transfer reactions in a large anodic potential interval, very suitable to organic compounds such as pesticides. The cathodic interval of the potential varies from −0.2 V until −0.5 V according to the pH of the supporting electrolyte, where in the most negative values occur the hydrogen overvoltage, which are lower than observed in mercury-based surfaces. In general, noble metals are used as working electrode when carbon forms electrodes are not applicable. Gold and platinum electrodes present suitable chemical inertness, compared to other metals, and for this are less inclinable to the stable oxide films formation or surface contamination.

The use of working electrode constituted by noble metals in aqueous solutions should be examined very closely because its potential interval are sometimes limited by intense background signals from the formation of surface oxides or adsorbed hydrogen layers (Wang 2006). Besides, the voltammetric response depends on the surface state, which is normally blocked by products of redox reactions, interfering with voltammetric reproducibility and, consequently, the analytical sensitivity. Silver, copper, and nickel can also be used in some specific applications, but the surface heterogeneity promotes non reproducibility in the analytical responses.

1.4.2.3 Carbon-Based Electrodes

Carbons forms such as carbon paste, carbon fiber, pyrolytic graphite, glassy-carbon, carbon nanotubes, screen-printed, carbon films, carbon composites, graphene, and diamond have also been used in the electroanalysis of pesticides. These electrodes present a broad anodic potential interval, with very low background current, present low cost, chemical inertness and mechanical stability. However, in some cases the electron-transfer reaction is slow and with intense dependence of the origin of the carbon surface, reducing the analytical sensitivity.

Carbon-based electrodes, from different forms and origins, have the electrochemical activity related to its microstructure surface, which present variability in function of carbon type. Therefore, cleanliness and pretreatment procedures of the surfaces and the presence of specific functional groups can promote suitable alterations in the electron transfer rates, promoting this way, a profound effect in the analytical performance (Rana et al. 2019). Additionally, these surfaces are suitable as supports for modification by organic polymers and biological compounds typical of the enzymes, antibodies, deoxyribonucleic acid (DNA), ribonucleic acid (RNA) structures compounds, due to the strong interaction between carbon surfaces and these specific modifiers (Wang 2006).

1.4.2.4 Chemically Modified Electrodes

The chemically modified electrode been used as a suitable alternative for the electroanalytical determination of pesticides. These electrodes are electrodic surfaces prepared from of a conducting or semiconducting material that is preliminary coated by a selected polymeric film that can be in the monomolecular, multimolecular, ionic forms, or modified by a chemical compounds and for this, exhibits chemical, electrochemical, and or optical properties of the modifiers due to some interfacial potential differences, or no net charge transfer, or the faradaic or charge-transfer reactions. The judicious choice of the modifier and the control of electrode potential can improve the electrochemical responses by enhance the selectivity and inhibit the surface fouling. Besides, the adequate modifications in working electrode surface, by use of immobilized electron-transfer mediators, can catalyze slow electron transfer reactions. The modifier can also yield some type of the biological activity on the electrode surface over enzymes use as modifiers and also permits that some anions and cations which no present electron transfer reactions should detected using ion-exchange coatings (Baldwin and Thomsen 1991).

The formation of a thin polymeric film through electropolymerization is one of the easiest ways of chemically modified electrode preparation. Based on the electric prospects from specific monomers and the knowledge of the most appropriate supporting electrolyte and solvent, the plunge of the electrode in a monomer solution allied with a suitable electrochemical technique is enough for polymer formation. Different from the chemical polymerization, the parameters of the techniques can be easily changed to shape the polymer over the electrode surface regarding its thickness, conductivity and doping properties. Additionally, the presence of a polymer over an electrode surface is useful for biosensor development to quantify pesticides. The formation of a conductive polymer can increase the electronic transference throughout the transducer and the presence of non-polymerized functional groups can be used to attach the biomolecules.

Modification by carbon forms also has been used to prepare chemically modified electrode to chemically react with pesticides, minimizing the activation energy in the redox reaction and, consequently, modifying the redox potential. For this, chemically modified electrode by carbon nanotubes in single or multi wall forms shows the properties of electrocatalytic activity and electro separation for the detection of pesticides in low concentrations or in complex samples. These chemically modified electrode by carbon nanotubes usually include modification that produce paste electrode, intercalated electrode, coating electrode and embedded polymer electrode (Wang et al. 2008).

Chemically modified electrode by graphene, graphene oxide, pristine and reduced graphene oxide, or combination of these types of graphene with inorganic or polymeric materials produce high electronic, electrical, and thermal conductivities and very high surface area sensors, that improve the analytical sensitivity to pesticides detection (Park et al. 2018; Lawal 2019). These electrodes have been used as the emerging platforms in development of new biosensors, with adequate sensitivity, selectivity, and rapid detection with low cost-effective to pesticides analysis, from different chemical class. The employ of metal nanoparticles at chemically modified electrode, mainly silver and gold nanoparticles, produce chemically modified electrode with suitable electrochemical properties, and for this, the modifications with nanoparticles has been used as an excellent procedure to obtain highly effective sensors, with some special functionalities and for this may enhance the operating range or selectivity compared with conventional materials. Electrode modifications by nanoparticles provide excellent transference of electrons between the pesticide and the surface of the chemically modified electrode, improving the analytical sensitivity (Schröfel et al. 2014; Malarkodi et al. 2017).

Chemically modified electrode by metallophthalocyanine, mainly containing cobalt, iron, manganese and nickel, are obtained by a simple adsorption on graphite and carbon surfaces or electropolymerized layers of the complexes from solutions containing a specific metallophthalocyanine. The produced sensors present roughness of the surface and the adhesion of the coating to the substrate increase the active catalysts for a large variety of electrochemical redox reactions from pesticides.

Molecularly imprinted polymer-based sensors, known as MIP, mainly containing metal organic frameworks, graphene, carbon nanotubes and quantum dots are the most used chemically modified electrode to replace antibodies in different electrochemical biosensors. Molecularly imprinted polymer-based sensors are prepared from use of some components by reactions that involves classical free radical polymerization at bulk, suspension, emulsion and precipitation polymerization (Włoch and Datta 2019). For this are used templates, functional monomers, where the most commons are carboxylic acids, sulfonic acids, and heteroaromatic bases, and cross-linkers such as dimethyl acrylate esters of diols, and initiators mainly azo-compounds and peroxides, and pyrogenic solvents.

Molecularly imprinted polymer-based sensors have been used in sensing platforms such as electrochemical, optical and mass-sensitive sensors, it consists of well-defined three-dimensional cavities, with affinity to a template molecule, in the polymer matrix (Florea et al. 2016). They present highly specific molecular recognition capacity and excellent stability in adverse chemical or physical conditions. So, are considerate as alternatives to bio-receptors for pesticides analysis. Molecularly imprinted polymer-based sensors are a class of working electrodes very efficient, sensitive, with low-cost and permits the miniaturization and the use in electrochemical detection devices.

1.4.2.5 Biosensor

The use of biosensors prepared from the immobilization of biological molecules such as antibodies, deoxyribonucleic acid, enzymes or ribonucleic acid, have also been used for monitoring pesticides with the advantage of being highly sensitive and selective, relatively easy to develop, accessible, and ready for use. In practice, are considerate as biosensor any apparatus that employ some specific biochemical reactions in the direct detection of chemical compounds employing its electrical, thermal or optical signals from the responses with the mediators that can be tissues, organelles, immunosystems, isolated enzymes, or some whole cells (Santoro and Ricciardi 2015). The electrochemical biosensor has a biological recognition layer originated from the immobilization of a specific biomolecule that can be enzyme, antibody, deoxyribonucleic acid fragments, over an electrochemical transducer. Some types of carbon and gold electrodes are the most used transducers. The electrochemical signal recorded from the biochemical reactions between the biological materials and the specific pesticides is used to quantify the different pesticide, mainly organophosphorus and carbamates (Abdulbari and Basheer 2017).

The major and most important step in the biosensor preparation is the immobilization of the biological recognition element on top one the transducer surface, that can employ the adsorption, or covalent bonding, or entrapment or cross-linking as immobilization techniques. The adsorption is the easiest to perform, usually relying on the drop-coating method, where the biomolecule is casting onto the transducer and allowed to dry, evaporating the solvent. The interaction consists in noncovalent bonds such as hydrophobic interactions, electrostatic or van der Waals forces. The main advantage is that it is not necessary the use of chemicals, although the electrode surface need some functionalization. The drawback is the difficulty to ensure reproducibility and robustness since the immobilization is susceptible to biomolecule lixiviation (Chung et al. 2011; Štěpánková and Vorčáková 2016; Jain and Cheng 2017).

A survey has demonstrated that around 78% of the works using biosensors developed enzymatic biosensors. The choice for this type of sensor is because the pesticides easily disturb the enzymatic reaction due to enzyme inhibition, especially the acetylcholinesterase. The first enzymatic electrochemical system developed for pesticide detection was performed in 1962, where the authors used two platinum electrodes to measure the inhibition of acetylcholinesterase by organophosphates pesticides systox, sarin, parathion and malathion. The substrate butyrylthiocholine iodide hydrolysis into the products thiocholine and butyric acid are hampered, and the potential signal is altered in comparison with a system without the pesticides. Since then, many authors have been working and improving electrochemical detection of pesticides based on enzyme inhibition (Guilbault et al. 1962).

One improvement in the enzymatic biosensor development occurs during the immobilization step using chitosan, a natural polyaminosaccharide analogue of cellulose, one of the few cationic polyelectrolytes found in nature. Chitosan is biodegradable and has biocompatibility with proteins. Because of its basicity, it forms a film over negatively charged surfaces and is often used to stabilize enzyme activity by adsorption or cross-linked with glutaraldehyde (Krajewska 2004).

Acetylcholinesterase is the main choice because it is a well-studied enzyme and their use results in obtention of the signals in low detection limits, with high sensitivity. Additionally, this enzyme presents a great relevance for human health as it is extremely important in the establishment human central nervous system normal running. In addition, thiocholine, the product formed from the enzymatic catalysis of acetylthiocholine is easily oxidized at lower potentials. The major disadvantage is that many organophosphates and carbamates pesticides present inhibitory ability towards acetylcholinesterase, so the biosensor lacks specificity (Lang et al. 2016; da Silva et al. 2018; Rajangam et al. 2018; Lu et al. 2019).

Another option is to use the pesticides inhibition ability on other enzymes even though they are less commonly used. Some enzymes can be inhibited by pesticides, mainly the tyrosinase, laccase, urease, aldehyde dehydrogenase and alkaline phosphatase and for this, different approaches were performed to develop novel strategies for its detection through biosensors. This is important because of the recognition of the inhibition ability of some pesticides for different enzymes than acetylcholinesterase, but the biosensors still present lack of selectivity.

Although many authors concentrate their efforts into the use of enzymatic inhibition, other approaches have been studied for pesticide detection. Direct enzymatic biosensor using the pesticide as substrate can be performed by the use of the enzyme organophosphate hydrolase, responsible to catalyze the breakdown of organophosphorus pesticides (Grimsley et al. 1997). The search for specificity droves some authors to search other approaches than enzymatic systems.

Immunosensors are a class of biosensors prepared from specific interaction that occurs among an antigen and an antibody. They have found a remarkable limit of detection in femtomole scale with specificity through an antigen binding. Immunosensors are showing promising results and they are attractive devices with excellent performance for the detection of pesticides.

The researches seeking for better devices for pesticide detection are amazingly increasing in the last few years. Alongside the enzymatic biosensors and immunosensor, there is also the genosensors, that have shown high specificity and sensitivity compared to antibodies. However, the use of aptamers has some advantages such as the ease of obtaining through synthesis with no need of a long and expensive immune response in hosts, it can be very easily chemically modified with several immobilization possibilities and present enhanced environmental stability (Hayat and Marty 2014).

Biosensors have been an excellent alternative for the development of novel strategies for pesticide detection in a sensitive and selective way. Much still needs to be done regarding selectivity and sensitivity, but recent advances have been showing that the ongoing researches are presenting promising results. However, they present some limitations, such as electrochemically active interferences in the sample, little long-term stability, excessive cost of enzymes, maintenance of enzymatic activity, and electron transfer problems.

1.4.3 Experimental Parameters

Solvent, supporting electrolyte, ionic strength and medium pH are initially evaluated because they influence directly in the position, intensity and voltammetric profile from electroanalysis of pesticides. Moreover, it is necessary the evaluation of the applicability of stripping steps, the potential, time for preconcentration and, scan potential directions.

1.4.3.1 Solvents

The solvent is need to promote the dissolution and dissociation of the salts in order to obtain a solution with high ionic conductivity, promote the solvation of the reactants and products in electrochemical reactions and cannot suffer reactions of the reduction or oxidation in any of the electrode used. As the solvent molecules are directly involved in the overall electrode reactions, its choice limits working electrode potential range, and consequently the redox potential of pesticides under study.

Water, methanol and ethanol are commonly protic solvents used because proton reduction is kinetically slow and promotes a high potential range in cathodic direction. However, in the anodic direction, the use of alcohols is limited to its oxidation reactions. Acetonitrile and dimethyl sulfoxide are aprotic solvents that present an excellent solvating ability in the dissolution of the reactants in bulk solution and the adsorbed products from the electrode reaction. However, in most pesticide analyzes using electroanalytical techniques the solvent is water due to its excellence as a solvent for salts and facility in the purification.

1.4.3.2 Supporting Electrolytes

The use of supporting electrolyte, in high concentration, is necessary to minimize or eliminate the electric field effects from electrode polarization on the movement of the electroactive molecules. Besides, it is important to provide ionic mobility in the electrochemical cell allowing electrode potential control and or measurements. The use of supporting electrolyte will ensure that the establishment of a uniform ionic strength in bulk solution, independent of the production or consumption of the ions under the electrodes.

Supporting electrolyte should present a low viscosity that permits a quick displacement of reactants and products to and from electrodes, needs to be electro inactive in all electrodes surfaces and no present reaction with reactive species which might be produced at electrodes. The most commons supporting electrolyte used in pesticides electroanalysis in aqueous medium are inorganic salts such as chloride, nitrate, sulphates and perchlorates salts of the potassium and sodium, mineral acid such as hydrochloric, perchloric, sulphuric and nitric acids, hydroxides of sodium and potassium, and some buffer systems such as phosphates, acetates, ammonium, borates, oxalates, among others, which cover the usual range of pH from 0 to 12. However, the electroanalysis in organic mediums employs as supporting electrolyte tetraalkylammonium salts, such as tetraethylammonium, tetrabutylammonium and tetraphenylphosphonium, due to their organic character and relatively low reactivity for reduction reactions (Wang 2006; Zoski 2007).

All salts used in the preparation of supporting electrolyte need to present high purity and should not be easily oxidized or reduced in the working electrode surface, and in the medium used in the pesticides electroanalyses. In all evaluation of solvents and supporting electrolyte it is performed a potential scan in a medium without addition of pesticide of interest to observe the presence of redox process from supporting electrolyte, the potential range and the electroactive contaminants presence in the medium.

1.4.3.3 Ionic Strength and pH

The pH values and the ionic strength of the supporting electrolyte directly influence the electron transfer mechanism because their values influence in the redox potential and current intensities. The ionic strength is related to the total electrolyte concentration and depends of the charge on each individual ion and its concentrations. Therefore, to supporting electrolyte containing ions with unitary charge, the ionic strength is equal to its concentration. The control in the ionic strength will ensure the establishment of a uniform distribution of charged species throughout the solution, and the stability in the kinetic electrode without significative alterations in analytical signals, even if some ions are consumed or formed at the electrodes surfaces.

The control of ionic force must be carefully carried out, due to the high ionic strength in the medium can cause alterations in the activity of pesticides, that deviates significatively from its concentration values. However, high ionic strength increases the solution conductivity, improving the electron transfers. The ionic strength can be adjusted employing a buffer, if the pH control is also desired, or using salts containing sodium (Na^+) or potassium (K^+) as counter ions, because their concentration corresponds to its activity coefficient. Additionally, in pesticides electroanalyses is necessary to add a high concentration of an electrolyte in the medium to maintain the ionic strength constant during electroanalytical experiments, due to generally the ionic strength of the sample containing pesticides is unknown. Considering the use of biosensors, the ionic strength strongly affects the enzymatic activity and for this, necessity of previously optimization.

The pH of the electrolyte will affect the position, the intensities and the half-height width of the obtained signals, because the H^+ ions can participate of the electrochemical reaction, either before or after the electrons transfer. Additionally, the pH of medium will define the potential range of the working electrode, in which occurs the reactions of hydrolyzes of water, producing an intense anodic current due to oxygen evolution or an intense cathodic current due to hydrogen overpotential. The pH variation is an adequate manner to promote dislodgment of potential redox values and, in this way, decrease the energy involved in the electron transfer.

In electrochemical mechanism involving protonation as determining steps, the pH is the most important parameters and it is initially evaluated. In some pesticides, the pH of the medium shifts the stability of the molecule and can promote hydrolyzes reactions. In some pesticides, only the molecule in the protonated form will react easily through electron transfer. In some cases, after the electron transfer, the radical produced are stabilized by reaction with protons of the medium, in a chemical step, producing a stable final product.

Therefore, in preliminary experiments, where the electrochemical behavior of pesticide is not known, it is necessary to observe the ionization constant of pesticide, pK_a. The electroanalytical experiments are initially performed using supporting electrolyte in pH with values like ionization constant and at values above and below the known ionization constant. In this way it is possible to observe the presence of protonation equilibriums or following the electron transfer and select the pH values where occurs the maximum of sensitivity and selectivity and mainly with lower redox potential values.

1.4.3.4 Stripping Steps

The viability of stripping voltammetry use in the pesticide analysis is evaluated by study the accumulation steps, considering potential and time of the preconcentration, and the redissolution step or measurement step, as describe in previous section. Therefore, after the solvent, electrolyte support and pH optimization, it is defined the accumulation potential, considering the values before, near the maximum, in the maximum, before the maximum and at final, where all redox reaction were completed, in a fixed time. After this, the potential selected is fixed and the accumulation time is evaluated from 0 to 600 s. In both steps, the current signals intensities and the profile of the voltammetric responses are considered, which will define the sensitivity and selectivity of the analysis.

1.4.4 Parameters Related to Electroanalytical Technique

Each electroanalytical technique described above has its own set of parameters, which need to be carefully and individually changed and the resultant responses evaluated. In the potentiometry is evaluated the potential measured by the device, and no parameters related to the techniques are considerate, since it involves only in the immersion of ion selective electrode or ion-sensitive field-effect transistor in the recipient containing the samples to analysis. In the amperometry experiment electric current in function of time are obtained, and for this, there are preliminary potential and time evaluation, normally given in volts and in minutes or seconds, respectively. In the electrochemical impedance spectroscopy experiment the current responses are measured in a sinusoidal shape but with a phase shift with time. For this, it is important perform a previously evaluation in the amplitude of the alternated current signal applied to the cell, the constant potential applied to the cell throughout the frequency sweep and the initial and final frequency sweep during data-acquisition.

Each voltammetric technique present different parameters, which should be individually assessed. Therefore, in the voltammetric technique the responses of peak current, peak potential and half height width obtained after the variation of each voltammetric parameters are considered, including interval and direction of potential scan. The relationships between each voltammetric parameters and peak current, peak potential and half height width are used to obtain suitable sensitivity and selectivity. Additionally, its use permits the obtention of information about the kinetic and mechanism of redox reactions, as described in previous section.

Cyclic voltammetry involves linearly change of an electrode potential considering the use of the two limits values which indicate the potential scan rate during the monitoring the current that develops in an electrochemical cell under conditions where voltage is higher than the predicted by the Nernst equation. Therefore, the main parameters that should be optimized in cyclic voltammetry experiments are the initial, switching and final potential values and potential scan rates. In the differential pulse voltammetry, the pulse amplitude or the height of the potential pulse given in millivolts, pulse width or the duration of the potential pulse and sample period or the time at the end of the pulse during which the current is measured, both given in milliseconds, should be individually evaluated. In square wave voltammetry the scan increment potential or the step height of the staircase wave form given in millivolts and main responsible by potential resolution, the pulse amplitude or the height of pulse potential given in millivolts and the pulse frequency or the length of each half-cycle, given in hertz, also should be individually and previously optimized.

1.4.5 Analytical Parameters

After the experimental and voltammetric optimization, the analytical parameters such as linear range, linearity, accuracy, precision, selectivity, sensitivity, detection limits, quantification limits and robustness are assessed as guidelines to validation process of the electroanalytical method, considering each pesticide analyzed (Christian et al. 2014). In pesticides electroanalysis, as the interaction pesticide with electrode surface is strongly dependent of the working electrode used, these analytical parameters can variate significatively from one pesticide to another, even if it has a similar chemical structure. The use of electroanalytical techniques, mainly square wave voltammetry or the use of stripping voltammetry steps, permit obtain these analytical parameters with values similar or lower than to those defined by government and international agencies, which employ high performance liquid chromatography or other conventional techniques.

1.4.6 Applications

From an analytical viewpoint, complex samples, such as natural water, foods, beverages, soil, biological fluid, among them, have several constituents that can interfere in the analytical responses, by decrease or increase in its signals. Therefore, prior to analyze the complex samples, it is required several steps of preparation for the elimination of interferences that can be performed through capillary electrophoresis and high performance liquid chromatographic techniques. However, the complexity of these samples and the interactions of pesticides with the other samples components requires differentiated sample preparation steps, which can involve liquid-liquid, solid-phase, enzymatic, cloud point and supercritical-fluid extractions procedures. These sample preparation steps present a high time of the analytical works and required large amount of chemicals, which are expensive, generate considerable waste and may contaminate the sample. Additionally, these pre-treatment steps require a highly qualified analyst to minimize errors in the end results.

Considering the use of electroanalytical techniques, analysis in complex samples present a considerable drawback regarding the previously or posteriorly adsorption of interferents under the working electrode surface, which may cause the overlapping of target signals and or an alteration in the ionic strength of the supporting electrolyte solution. These drawbacks have been controlled by application of procedure of the electrochemical activation and cleanness of working electrode between each experiment, and using buffer solutions, respectively.

In potentiometry, the ion selective electrode or ion-sensitive field-effect transistor is inserted in a solution holding the pesticide of interest in a complex samples solution. In others electroanalytical techniques, an aliquot of the sample solution is summed in electrochemical cell and the experiments are performed according to previously optimized experimental and voltammetric or electrochemical impedance spectroscopy or amperometry parameters. Sometimes, the presence of proteins and fats need to be extracted or precipitated by classical solvent extraction or salts. However, in general, the electroanalytical procedures are performed with the complex samples in very low concentration, and the interferences from samples constituents are practically negligible. For this, in electroanalysis, the sample preparations steps, if necessary, are simple, fast, solvent-free and inexpensive.

1.5 Electrochemical Behavior of Pesticides from Different Chemical Class

According to previously discussed in early sections, there are a substantial number of researches developed regarding the electroanalysis of pesticides with mercury-based, carbon-based, noble metals, chemically modified electrodes and biosensors using the electroanalytical techniques described in Sect. 1.3. The pesticides analyzed belong to different chemical classes, such as: organochlorine, organophosphorus, triazines, nitro pesticides and nitrophenol derivatives, carbamates, dithiocarbamates, neonicotinoids, sulphonylureas, bipyridiniums, pyrazoles, phenoxyacetic acid derivatives, among others. These pesticides contain, in its chemical structure, functional groups that are easily reducible or oxidable with a known electron-transfer mechanism or can react with a component in a chemically modified electrode or biosensor or can react with some components in the medium. In all situations, occurs the production of electroanalytical signals that can be used in the identification and quantification of pesticide of interest.

Voltammetric studies in pesticides solutions are based on the reduction reactions, such as single bond reduction from halogen in an aromatic ring, reduction of a carbon-carbon carbon-sulfur or carbon-nitrogen double bonds, and the reduction of different chemical substituents in chemical structure of pesticides, such as nitro, imine, imide, azomethine, thioether, dialkylthiophosphate, bipyridinium, pyrazinediium, carbonyl and dithiophosphate. However, voltammetric responses from oxidation reactions at carbon-based surfaces or noble metals surfaces are also used, where functional groups characteristic of pesticide chemical class or substituents groups such as dithiocarbamate, thioether, aromatic hydroxy, aromatic amino, heterocyclic ring, amino groups, among others, can be evaluated through electrochemical oxidation mechanism studies.

However, if the electrochemical mechanism is not known, a detailed study using the current and potential signals from variation in the experimental and voltammetric parameters can be performed. All data obtained can be compared to mathematical models, and the evaluation of the chemical structure of the pesticides, jointly previously knowledge about the reactivity and organic electrochemistry can provide information about electrochemical mechanisms. This is fundamentally important, because it provides information about toxicity and degradation steps, indicates the chemical species intermediaries and helps in obtaining experimental and voltammetric conditions to attain the best sensitivity and selectivity in electroanalysis of pesticides.

It is true that it is quite difficult to achieve suitable selectivity when several pesticides present similar functional groups, which is sometimes typical of the chemical class of the pesticides. However, by using adequate manipulation in the experimental and voltammetric parameters, previously described, it is possible to evaluate the redox potential of functional groups and to realize the suitable identification and quantification of pesticides, even in complex systems like natural waters, soil, and *in natura* foods.

1.6 Conclusions

The indiscriminate use of pesticides, regarding mainly the amount and chemical class of the pesticide, is a worldwide problem not only in the toxicity levels and remediation procedure, but also in the adequate choices of the analytical procedures to detection of these compounds in the contaminated food, water, soil and air. This chapter introduced a detailed presentation of the electroanalytical methods of pesticides analysis, one the easiest, efficient, robust, specific and low-cost way of pesticide determination in all its sources. A brief description of the most employed techniques used to achieve this goal was described combining their specific signals, such as current, potential, charge, resistance, with the pesticide's physicochemical properties such as redox functional groups, ionizable hydrogens and enzyme inhibition ability.

The equipment is equivalent to a desktop computer in size and can be miniaturized to a pocket device, especially for in situ use, it is cheap compared to the traditional chromatographic methods of analysis and does not depend on skilled-labor personal as most software present a friendly platform. Despite the easiness in the electroanalysis manipulation there are a series of conditions to be assessed to obtain the maximum efficiency depending on a deep electrochemistry knowledge to link the useful technique with the compound of interest.

We presented the main choices of these conditions as the type of unmodified working electrode such as mercury-based, carbon-based and noble metals, the supporting electrolyte in aqueous or organic medium, electrochemical cell type, the detection modes such as amperometric, voltammetric, impedimetric and potentiometric, as well as alternative methods through indirect detection using chemically modified electrodes through biosensor development such as genosensor, enzymatic sensor and immunosensor.

At finally, we indicated that choose the best approach by the selection of the most appropriate detection technique to be used and explain the results observed based on an extent survey regarding electrochemical mechanism will help to expand electroanalysis and make them the official method to pesticide determination in a near future. So, electroanalysis are certified as an efficient analytical method applied in several branches of science and it has been proving so far to be an excellent alternative for pesticide detection with many different approaches, low detection limits, high selectivity, robustness, easiness, low-cost and low chemical use and waste generation.

Acknowledgements The authors thank you FAPEMIG (Minas Gerais State Research Support Foundation), and PROPP-UFU (Pro rectory of Research and Graduate of the Federal University of Uberlândia) by financial supporting in the development of the all researches.

Dedication This chapter is dedicated to all young researchers that worked in the Laboratory of Electroanalytical Applied to Biotechnology and Food Engineering (LEABE), Campus of Patos de Minas, Federal Uberlândia University since its foundation in 2012 until 2019. De Souza dedicates this work to her children, Talisson for all the affection and Victoria who taught her the value of peace and equilibrium in her life.

References

Abdulbari HA, Basheer EAM (2017) Electrochemical biosensors: electrode development, materials, design, and fabrication. ChemBioEng Rev 4:92–105. https://doi.org/10.1002/cben.201600009

Abollino O, Giacomino A, Malandrino M (2018) Stripping voltammetry. Elsevier, New York

Abu Shawish HM, Ghalwa NA, Hamada M, Basheer AH (2012) Modified carbon paste electrode for potentiometric determination of diquat dibromide pesticide in water and urine samples. Mater Sci Eng C 32:140–145. https://doi.org/10.1016/j.msec.2011.10.008

Akyüz D, Koca A (2019) An electrochemical sensor for the detection of pesticides based on the hybrid of manganese phthalocyanine and polyaniline. Sensors Actuators B Chem 283:848–856. https://doi.org/10.1016/j.snb.2018.11.155

Akyüz D, Keleş T, Biyiklioglu Z, Koca A (2017) Electrochemical pesticide sensors based on electropolymerized metallophthalocyanines. J Electroanal Chem 804:53–63. https://doi.org/10.1016/j.jelechem.2017.09.044

Alves MF, Corrêa RAMS, da Cruz FS, Franco DL, Ferreira LF (2018) Electrochemical enzymatic fenitrothion sensor based on a tyrosinase/poly(2-hydroxybenzamide)-modified graphite electrode. Anal Biochem 553:15–23. https://doi.org/10.1016/j.ab.2018.05.014

Amine A, Mohammadi H (2018) Amperometry. In: Worsfold P, Townshend A, Poole C (eds) Encyclopedia of analytical sciences. Elsevier, New York, pp 70–79

Amorello D, Orecchio S (2013) Micro-determination of dithiocarbamates in pesticide formulations using voltammetry. Microchem J 110:334–339. https://doi.org/10.1016/j.microc.2013.05.002

Andrade FR, de Toledo RA, Pedro Vaz CM (2014) Electroanalytical methodology for the direct fetermination of 2,4-fichlorophenoxyacetic acid in soil samples using a graphite-polyurethane electrode. Int J Electrochem 2014:1–9. https://doi.org/10.1155/2014/308926

Anirudhan TS, Alexander S (2014) Multiwalled carbon nanotube based molecular imprinted polymer for trace determination of 2,4-dichlorophenoxyaceticacid in natural water samples using a potentiometric method. Appl Surf Sci 303:180–186. https://doi.org/10.1016/j.apsusc.2014.02.139

Arduini F, Cinti S, Caratelli V, Amendola L, Palleschi G, Moscone D (2019) Origami multiple paper-based electrochemical biosensors for pesticide detection. Biosens Bioelectron 126:346–354. https://doi.org/10.1016/j.bios.2018.10.014

Arip MNM, Heng LY, Ahmad M, Ujang S (2013) A cell-based potentiometric biosensor using the fungus Lentinus sajor-caju for permethrin determination in treated wood. Talanta 116:766–781. https://doi.org/10.1016/j.talanta.2013.07.065

Armenta S, Garrigues S, de la Guardia M (2008) Green analytical chemistry. Trends Anal Chem 27:497–511. https://doi.org/10.1016/j.trac.2008.05.003

Asadian E, Ghalkhani M, Shahrokhian S (2019) Electrochemical sensing based on carbon nanoparticles: a review. Sensors Actuators B Chem 293:183–209. https://doi.org/10.1016/j.snb.2019.04.075

Bahaghighat HD, Freye CE, Synovec RE (2019) Recent advances in modulator technology for comprehensive two dimensional gas chromatography. Trends Anal Chem 113:379–391. https://doi.org/10.1016/j.tra.2018.04.016

Bakirhan NK, Uslu B, Ozkan SA (2018) The detection of pesticide in foods using electrochemical sensors. In: Food safety and preservation. Elsevier, New York, pp 91–141

Baldwin RP, Thomsen KN (1991) Chemically modified electrodes in liquid chromatography detection: a review. Talanta 38:1–17. https://doi.org/10.1016/0039-9140(91)80004-J

Bandžuchová L, Šelešovská R, Navrátil T, Chýlková J (2013) Sensitive voltammetric method for determination of herbicide triasulfuron using silver solid amalgam electrode. Electrochim Acta 113:1–8. https://doi.org/10.1016/j.electacta.2013.09.077

Bao J, Huang T, Wang Z, Yang H, Geng X, Xu G, Samalo M, Sakinati M, Huo D, Hou C (2019) 3D graphene/copper oxide nano-flowers based acetylcholinesterase biosensor for sensitive detection of organophosphate pesticides. Sensors Actuators B Chem 219:95–101. https://doi.org/10.1016/j.snb.2018.09.118

Bard A, Faulkner L (1944) Electrochemical methods: fundamentals and applications. Wiley, New York

Batchelor-McAuley C, Yang M, Hall EM, Compton RG (2015) Correction factors for the analysis of voltammetric peak currents measured using staircase voltammetry. J Electroanal Chem 758:1–6. https://doi.org/10.1016/j.jelechem.2015.10.004

Bockris JOM, Reddy NAK, Gamboa-Aldeco M (2000) Modern electrochemistry, second. Spring Street, New York

Bolat G, Abaci S, Vural T, Bozdogan B, Denkbas EB (2018) Sensitive electrochemical detection of fenitrothion pesticide based on self-assembled peptide-nanotubes modified disposable pencil graphite electrode. J Electroanal Chem 809:88–97. https://doi.org/10.1016/j.jelechem.2017.12.060

Borah H, Gogoi S, Kalita S, Puzari P (2018) A broad spectrum amperometric pesticide biosensor based on glutathione S-transferase immobilized on graphene oxide-gelatin matrix. J Electroanal Chem 828:116–123. https://doi.org/10.1016/j.jelechem.2018.09.047

Brandão PF, Duarte AC, Duarte RMBO (2019) Comprehensive multidimensional liquid chromatography for advancing environmental and natural products research. Trends Anal Chem 116:186–197. https://doi.org/10.1016/j.trac.2019.05.016

Brycht M, Skrzypek S, Guzsvány V, Berenji J (2013) Conditioning of renewable silver amalgam film electrode for the characterization of clothianidin and its determination in selected samples by adsorptive square-wave voltammetry. Talanta 117:242–249. https://doi.org/10.1016/j.talanta.2013.08.048

Brycht M, Skrzypek S, Robak J, Guzsvány V, Vajdle O, Zbiljić J, Nosal-Wiercińska A, Guziejewski D, Andrijewski G (2015) Ultra trace level determination of fenoxanil by highly sensitive square wave adsorptive stripping voltammetry in real samples with a renewable silver amalgam film electrode. J Electroanal Chem 738:69–76. https://doi.org/10.1016/j.jelechem.2014.11.029

Brycht M, Burnat B, Nosal-Wiercińska A, Skrzypek S (2016) New sensitive square-wave adsorptive stripping voltammetric determination of pesticide chlornitrofen, and an evaluation of its corrosivity towards steel agricultural equipment. J Electroanal Chem 777:8–18. https://doi.org/10.1016/j.jelechem.2016.07.023

Caetano FR, Carneiro EA, Agustini D, Figueiredo-Filho LCS, Banks CE, Bergamini MF, Marcolino-Junior LH (2018) Combination of electrochemical biosensor and textile threads: a microfluidic device for phenol determination in tap water. Biosens Bioelectron 99:982–387. https://doi.org/10.1016/j.bios.2017.07.070

Castanho GM, de Toledo RA, Shim H, Lu Q, Vaz CMP (2016) Electroanalytical procedure and sorption studies for imazaquin in different soils. Int J Environ Anal Chem 96:530–541. https://doi.org/10.1080/03067319.2016.1172217

Chang BY, Park SM (2010) Electrochemical impedance spectroscopy. Annu Rev Anal Chem 3:207–229. https://doi.org/10.1146/annurev.anchem.012809.102211

Chen J, Chen C (2013) A new data analysis method to determine pesticide concentrations by cyclic voltammetry. Measurement 46:1828–1833. https://doi.org/10.1016/j.measurement.2013.02.001

Cheng X, Wang Q, Zhang S, Zhang W, He P, Fang Y (2007) Determination of four kinds of carbamate pesticides by capillary zone electrophoresis with amperometric detection at a polyamide-modified carbon paste electrode. Talanta 71:1083–1087. https://doi.org/10.1016/j.talanta.2006.06.001

Chicharro M, Bermejo E, Sánchez A, Zapardiel A, Fernandez-Gutierrez A, Arraez D (2005) Multiresidue analysis of phenylurea herbicides in environmental waters by capillary electrophoresis using electrochemical detection. Anal Bioanal Chem 382:519–526. https://doi.org/10.1007/s0216-004-2840-6

Chicharro M, Arribas AS, Moreno M, Bermejo E, Zapardiel A (2007) Comparative study of multi walled carbon nanotubes-based electrodes in micellar media and their application to micellar electrokinetic capillary chromatography. Talanta 74:376–386. https://doi.org/10.1016/j.talanta.2007.07.034

Cho YA, Lee HS, Cha GS, Lee YT (1999) Fabrication of butyrylcholinesterase sensor using polyurethane-based ion-selective membranes. Biosens Bioelectron 14:435–438. https://doi.org/10.1016/S0956-5663(99)00016-0

Christian GD, Dasgupta P, Schug K (2014) Analytical chemistry. Elsevier, New York

Christie JH, Lingane PJ (1965) Theory of staircase voltammetry. J Electroanal Chem 10:176–182. https://doi.org/10.1016/0022-0728(65)85021-5

Chung DJ, Kim KC, Choi SH (2011) Electrochemical DNA biosensor based on avidin-biotin conjugation for influenza virus (type A) detection. Appl Surf Sci 257:9390–9396. https://doi.org/10.1016/j.apsusc.2011.06.015

Cichosz S, Masek A, Zaborski M (2018) Polymer-based sensors: a review. Polym Test 67:342–348. https://doi.org/10.1016/j.polymertesting.2018.03.024

Compton RG, Banks CE (2007) Understanding voltammetry. Imperial College, Oxford

Costa DJE, Santos JCS, Sanches-Brandão FAC, Ribeiro WF, Salazar-Banda GR, Araujo MCU (2017) Boron-doped diamond electrode acting as a voltammetric sensor for the detection of methomyl pesticide. J Electroanal Chem 789:100–107. https://doi.org/10.1016/j.jelechem.2017.02.036

Crossley S (2004) Pesticide residues in food and drinking water human exposure and risks. Wiley, New York

da Silva MKL, Vanzela HC, Defavari LM, Cesarino I (2018) Determination of carbamate pesticide in food using a biosensor based on reduced graphene oxide and acetylcholinesterase enzyme. Sensors Actuators B Chem 277:555–561. https://doi.org/10.1016/j.snb.2018.09.051

Dahmen EAMF (1986) Electroanalysis: Theory and applications in aqueous and non-aqueous media and in automated chemical control. Elsevier, New York

de Carvalho LA, de Andrade AR, Bueno PR (2006) Espectroscopia de impedância eletroquímica aplicada ao estudo das reações heterogêneas em ânodos dimensionalmente estáveis. Quim Nova 29:797–804. https://doi.org/10.1590/s0100-40422006000400029

de Figueiredo-Filho LCS, Baccarin M, Janegitz BC, Fatibello-Filho O (2017) A disposable and inexpensive bismuth film minisensor for a voltammetric determination of diquat and paraquat pesticides in natural water samples. Sensors Actuators B Chem 240:749–756. https://doi.org/10.1016/j.snb.2016.08.157

de Jesus VO, Barbosa AMJ, Petroni JM, Lucca BG, Ferreira VS (2017) Voltammetric determination of herbicide molinate in river water and rice samples using solid silver amalgam electrode fabricated with nanoparticles. Int J Environ Anal Chem 97:468–483. https://doi.org/10.1080/03067319.2017.1325880

De Lima F, Gozzi F, Fiorucci AR, Cardoso CAL, Arruda GJ, Ferreira VS (2011) Determination of linuron in water and vegetable samples using stripping voltammetry with a carbon paste electrode. Talanta 83:1763–1768. https://doi.org/10.1016/j.talanta.2010.12.014

De Souza D, Machado SAS (2005) Electroanalytical method for determination of the pesticide dichlorvos using gold-disk microelectrodes. Anal Bioanal Chem 382:1720–1725. https://doi.org/10.1007/s00216-005-3324-z

De Souza D, Machado SAS (2006) Study of the electrochemical behavior and sensitive detection of pesticides using microelectrodes allied to square-wave voltammetry. Electroanalysis 18:862–872. https://doi.org/10.1002/elan.200603480

De Souza D, Machado SAS, Avaca LA (2003) Voltametria de onda quadrada. Primeira parte: aspectos teóricos. Quim Nova 26:81–89

De Souza D, Codognoto L, Malagutti AR, Toledo RA, Pedrosa VA, Oliveira RTS, Mazo LH, Avaca LA, Machado SAS (2004) Voltametria de onda quadrada. Segunda parte: Aplicações. Quim Nova 27:790–797

De Souza D, De Toledo RA, Suffredini HB, Mazo LH, Machado SAS (2006) Characterization and use of copper solid amalgam electrode for electroanalytical determination of triazines-based herbicides. Electroanalysis 18:605–612. https://doi.org/10.1002/elan.200503441

De Souza D, Mascaro LH, Fatibello-Filho O (2011) The effect of composition of solid silver amalgam electrodes on their electrochemical response. J Solid State Electrochem 15:20232029. https://doi.org/10.1007/s10008-010-1225-3

Deroco PB, Lourencao BC, Fatibello-Filho O (2017) The use of modified electrode with carbon black as sensor to the electrochemical studies and voltammetric determination of pesticide mesotrione. Microchem J 133:184–194. https://doi.org/10.1016/j.microc.2017.03.024

Díaz-González M, Gutiérrez-Capitán M, Niu P, Baldi A, Jiménez-Jorquera C, Fernández-Sánchez C (2016) Electrochemical devices for the detection of priority pollutants listed in the EU water framework directive. Trends Anal Chem 77:186–201. https://doi.org/10.1016/j.trac.2015.11.023

Ding J, Qin W (2009) Potentiometric sensing of butyrylcholinesterase based on in situ generation and detection of substrates. Chem Commun 8:971–973. https://doi.org/10.1039/b817064a

Dornellas RM, Franchini RAA, Aucelio RQ (2013) Determination of the fungicide picoxystrobin using anodic stripping voltammetry on a metal film modified glassy carbon electrode. Electrochim Acta 97:2012–2019. https://doi.org/10.1016/j.electacta.2013.02.125

Dornellas RM, Munoz RAA, Aucelio RQ (2015) Electrochemical determination of picoxystrobin on boron-doped diamond electrode: square-wave voltammetry versus BIA-multiple pulse amperometry. Microchem J 123:1–8. https://doi.org/10.1016/j.microc.2015.05.010

Dos Santos LBO, Abate G, Masini JC (2004) Determination of atrazine using square wave voltammetry with the hanging mercury drop electrode (HMDE). Talanta 62:667–674. https://doi.org/10.1016/j.talanta.2003.08.034

Duarte EH, Casarin J, Sartori ER, Tarley CRT (2018) Highly improved simultaneous herbicides determination in water samples by differential pulse voltammetry using boron-doped diamond electrode and solid phase extraction on cross-linked poly(vinylimidazole). Sensors Actuators B Chem 255:166–175. https://doi.org/10.1016/j.snb.2017.08.021

Dzyadevych SV, Soldatkin AP, El'skaya AV, Martelet C, Jaffrezic-Renault N (2006) Enzyme biosensors based on ion-selective field-effect transistors. Anal Chim Acta 568(1–2):248–258

Economou A, Fielden PR (2003) Mercury film electrodes: developments, trends and potentialities for electroanalysis. Analyst 128:205–212. https://doi.org/10.1039/b201130c

Eddleston M, Bateman DN (2012) Pesticides. Medicine 40:147–150. https://doi.org/10.1016/j.mpmed.2011.12.029

Edwards GA, Bergren AJ, Porter MD (2007) Chemically modified electrodes. In: Zoski CG (ed) Handbook of electrochemistry. Elsevier, New York, pp 295–327

El Harmoudi H, Achak M, Farahi A, Lahrich S, El Gaini L, Abdennouri M, Bouzidi A, Bakasse M, El Mhammedi MA (2013) Sensitive determination of paraquat by square wave anodic stripping voltammetry with chitin modified carbon paste electrode. Talanta 115:172–177. https://doi.org/10.1016/j.talanta.2013.04.002

Escarpa A, González MC, López MA (2015) Agricultural and food electroanalysis. Wiley, Chichester/West Sussex

Esteban M, Casassas E (1994) Stripping electroanalytical techniques in environmental analysis. Trends Anal Chem 13:110–117. https://doi.org/10.1016/0165-9936(94)87075-6

Estela JM, Tomás C, Cladera A, Cerdà V (1995) Potentiometric stripping analysis: a review. Crit Rev Anal Chem 35:91–141. https://doi.org/10.1080/10408349508050559

Facure MHM, Mercante LA, Mattoso LHC, Correa DS (2017) Detection of trace levels of organophosphate pesticides using an electronic tongue based on graphene hybrid nanocomposites. Talanta 167:59–66. https://doi.org/10.1016/j.talanta.2017.02.005

Farahi A, El Gaini L, Achak M, El Yamani S, El Mhammedi MA, Bakasse M (2014) Interaction study of paraquat and silver electrode using electrochemical impedance spectroscopy: Application in milk and tomato samples. Food Control 47:679–685. https://doi.org/10.1016/j.foodcont.2014.08.005

Fatibello-Filho O, Silva TA, Moraes FC, Janegitz BC (2019) Potenciometria: aspectos teóricos e práticos. Edufscar, São Carlos

Fischer J, Hájková A, Pereira M, Křeček M, Vyskočil V, Barek J (2016) Investigation of voltammetric behaviour of Insecticide chlorpyrifos on a mercury meniscus modified silver solid amalgam electrode. Electrochim Acta 216:510–516. https://doi.org/10.1016/j.electacta.2016.09.013

Florea A, Hosu O, Ciui B, Cristea C (2016) Molecularly imprinted polymer-based sensors for biomedical and environmental applications. In: Tiwari A, Uzun L (eds) Advanced moleculary imprinting materials. Scrivener Publishing LLC, Beverly, pp 283–325. https://doi.org/10.1002/9781119336181.ch7

Forster RJ, Cumba LR (2018) Cyclic voltammetry of organic compounds. Elsevier, New York

Francesquett JZ, Rizzetti TM, Cadaval TRS, Prestes OD, Adaime MB, Zanella R (2019) Simultaneous determination of the quaternary ammonium pesticides paraquat, diquat, chlormequat, and mepiquat in barley and wheat using a modified quick polar pesticides method, diluted standard addition calibration and hydrophilic interaction liquid chrom. J Chromatogr A 1592:101–111. https://doi.org/10.1016/j.chroma.2018.12.060

Galeano Díaz T, Guiberteau Cabanillas A, López Soto MD, Ortiz JM (2008) Determination of fenthion and fenthion-sulfoxide, in olive oil and in river water, by square-wave adsorptive-stripping voltammetry. Talanta 76:809–814. https://doi.org/10.1016/j.talanta.2008.04.037

Galli A, De Souza D, Machado SAS (2011) Pendimethalin determination in natural water, baby food and river sediment samples using electroanalytical methods. Microchem J 98:135–143. https://doi.org/10.1016/j.microc.2010.12.009

Galus Z (1994) Fundamentals of electrochemical analysis. Polish Scientific Publishers, New York

Gaweł M, Kiljanek T, Niewiadowska A, Semeniuk S, Goliszek M, Burek O, Posyniak A (2019) Determination of neonicotinoids and 199 other pesticide residues in honey by liquid and gas chromatography coupled with tandem mass spectrometry. Food Chem 282:36–47. https://doi.org/10.1016/j.foodchem.2019.01.003

Geto A, Noori JS, Mortensen J, Svendsen WE, Dimaki M (2019) Electrochemical determination of bentazone using simple screen-printed carbon electrodes. Environ Int 129:400–407. https://doi.org/10.1016/j.envint.2019.05.009

Ghindilis AL, Morzunova TG, Barmin AV, Kurochkin IN (1996) Potentiometric biosensors for cholinesterase inhibitor analysis based on mediatorless bioelectrocatalysis. Biosens Bioelectron 11:873–880. https://doi.org/10.1016/0956-5663(96)89436-X

Govindasamy M, Rajaji U, Chen SM, Kumaravel S, Chen TW, Al-Hemaid FMA, Ali MA, Elshikh MS (2018) Detection of pesticide residues (fenitrothion) in fruit samples based on niobium carbide/molybdenum nanocomposite: an electrocatalytic approach. Anal Chim Acta 1030:52–60. https://doi.org/10.1016/j.aca.2018.05.044

Grimsley JK, Scholtz JM, Pace CN, Wild JR (1997) Organophosphorus hydrolase is a remarkably stable enzyme that unfolds through a homodimeric intermediate. Biochemist 36:14366–14374. https://doi.org/10.1021/bi971596e

Gu S, Lu Y, Ding Y, Li L, Zhang F, Wu Q (2013) Droplet-based microfluidics for dose-response assay of enzyme inhibitors by electrochemical method. Anal Chim Acta 796:68–74. https://doi.org/10.1016/j.aca.2013.08.016

Gubanova O, Andrianova M, Saveliev M, Komarova N, Kuznetsov E, Kuznetsov A (2017) Fabrication and package of ISFET biosensor for micro volume analysis with the use of direct ink writing approach. Mater Sci Semicond Process 60:71–78. https://doi.org/10.1016/j.mssp.2016.12.007

Guiberteau A, Galeano T, Mora N, Parrilla P, Salinas F (2001) Study and determination of the pesticide imidacloprid by square wave adsorptive stripping voltammetry. Talanta 53:943–949. https://doi.org/10.1016/S0039-9140(00)00576-2

Guilbault GG, Kramer DN, Cannon PL (1962) Electrochemical determination of organophosphorous compounds. Anal Chem 34:1437–1439. https://doi.org/10.1021/ac60191a027

Halim NHA, Lee YH, Marugan RSPM, Hashim U (2017) Mediatorless impedance studies with titanium dioxide conjugated gold nanoparticles for hydrogen peroxide detection. Biosensors 7:38–49. https://doi.org/10.3390/bios7030038

Hance RJ (1970) Polarography of herbicides – a preliminary survey. Pestic Sci 1:112–113. https://doi.org/10.1002/ps.2780010308

Hashemi P, Karimian N, Khoshsafar H, Arduini F, Mesri M, Afkhami A, Bagheri H (2019) Reduced graphene oxide decorated on Cu/CuO-Ag nanocomposite as a high-performance material for the construction of a non-enzymatic sensor: application to the determination of carbaryl and fenamiphos pesticides. Mater Sci Eng C 102:764–772. https://doi.org/10.1016/j.msec.2019.05.010

Hayat A, Marty JL (2014) Aptamer based electrochemical sensors for emerging environmental pollutants. Front Chem 2:1–9. https://doi.org/10.3389/fchem.2014.00041

Hornsby AG, Wauchope RD, Herner AE (1996) Pesticide properties in the environment. Springer, Amsterdan

Itkes MPM, de Oliveira GG, Silva TA, Fatibello-Filho O, Janegitz BC (2019) Voltammetric sensing of fenitrothion in natural water and orange juice samples using a single-walled carbon nanohorns and zein modified sensor. J Electroanal Chem 840:21–26. https://doi.org/10.1016/j.jelechem.2019.03.055

Jafari M, Hasanzadeh M, Karimian R, Shadjou N (2019) Sensitive detection of Trifluralin in untreated human plasma samples using reduced graphene oxide modified by polyethylene imine and silver nanoparticles: a new platform on the analysis of pesticides and chemical injuries. Microchem J 147:741–748. https://doi.org/10.1016/j.microc.2019.03.092

Jain A, Cheng K (2017) The principles and applications of avidin-based nanoparticles in drug delivery and diagnosis. J Control Release 245:27–40. https://doi.org/10.1016/j.jconrel.2016.11.016

Jaya S, Rao TP (1982) Potentiometric stripping analysis. Rev Anal Chem 6:343–358

Jevtić S, Vukojević V, Djurdjić S, Pergal MV, Manojlović DD, Petković BB, Stanković DM (2018) First electrochemistry of herbicide pethoxamid and its quantification using electroanalytical approach from mixed commercial product. Electrochim Acta 277:136–142. https://doi.org/10.1016/j.electacta.2018.05.004

Jiang X, Li D, Xu X, Ying Y, Li Y, Ye Z, Wang J (2008) Immunosensors for detection of pesticide residues. Biosens Bioelectron 23:1577–1587. https://doi.org/10.1016/j.bios.2008.01.035

Jokerst JC, Emory JM, Henry CS (2012) Advances in microfluidics for environmental analysis. Analyst 137:24–34. https://doi.org/10.1039/clan15368d

Justino CIL, Freitas AC, Pereira R, Duarte AC, Rocha Santos TAP (2015) Recent developments in recognition elements for chemical sensors and biosensors. Trends Anal Chem 68:2–17. https://doi.org/10.1016/j.trac.2015.03.006

Kalliora C, Mamoulakis C, Vasilopoulos E, Stamatiades GA, Kalafati L, Barouni R, Karakousi T, Abdollahi M, Tsatsakis A (2018) Association of pesticide exposure with human congenital abnormalities. Toxicol Appl Pharmacol 346:58–75. https://doi.org/10.1016/j.taap.2018.03.025

Kesik M, Ekiz Kanik F, Turan J, Kolb M, Timur S, Bahadir M, Toppare L (2014) An acetylcholinesterase biosensor based on a conducting polymer using multiwalled carbon nanotubes for amperometric detection of organophosphorous pesticides. Sensors Actuators B Chem 205:3949. https://doi.org/10.1016/j.snb.2014.08.058

Khaled E, Hassan HNA, Mohamed GG, Ragab FA, Seleim AEA (2010) Disposable potentiometric sensors for monitoring cholinesterase activity. Talanta 83:357–363. https://doi.org/10.1016/j.talanta.2010.09.020

Khaled E, Kamel MS, Hassan HNA, Abdel-Gawad H, Aboul-Enein HY (2014) Performance of a portable biosensor for the analysis of ethion residues. Talanta 119:467–472. https://doi.org/10.1016/j.talanta.2013.11.001

Kharbouche L, Gil García MD, Lozano A, Hamaizi H, Galera MM (2019) Solid phase extraction of pesticides from environmental waters using an MSU-1 mesoporous material and determination by UPLC-MS/MS. Talanta 199:612–619. https://doi.org/10.1016/j.talanta.2019.02.092

Kim K-H, Kabir E, Jahan SA (2017) Exposure to pesticides and the associated human health effects. Sci Total Environ 575:525–535. https://doi.org/10.1016/j.scitotenv.2016.09.009

Krajewska B (2004) Application of chitin- and chitosan-based materials for enzyme immobilizations: a review. Enzym Microb Technol 35:126–139. https://doi.org/10.1016/j.enzmictec.2003.12.013

Kudr J, Zitka O, Klimanek M, Vrba R, Adam V (2017) Microfluidic electrochemical devices for pollution analysisa a review. Sensors Actuators B Chem 246:578–590. https://doi.org/10.1016/j.snb.2017.02.052

Kunert O, Goger B, Seger C, Wintersteiger R, Rinelli R (2002) Quantification of phenylurea pesticides by HPLC/ECD and photolysis. Electroanalysis 13:1335–1341. https://doi.org/10.1002/1521-4109(200111)13:16<1335::aid-elan1335>3.3.co;2-n

Lang Q, Han L, Hou C, Wang F, Liu A (2016) A sensitive acetylcholinesterase biosensor based on gold nanorods modified electrode for detection of organophosphate pesticide. Talanta 156–157:34–41. https://doi.org/10.1016/j.talanta.2016.05.002

Lawal AT (2019) Graphene-based nano composites and their applications. A reeview. Biosens Bioelectron 141:111384. https://doi.org/10.1016/j.bios.2019.111384

Li F, Yu Z, Han X, Lai RY (2019) Electrochemical aptamer-based sensors for food and water analysis: a review. Anal Chim Acta 1051:1–23. https://doi.org/10.1016/j.aca.2018.10.258

Liu G, Lin Y (2006) Biosensor based on self-assembling acetylcholinesterase on carbon nanotubes for flow injection/amperometric detection of organophosphate pesticides and nerve agents. Anal Chem 78:835–843. https://doi.org/10.1021/ac051559q

Liu M, Khan A, Wang Z, Liu Y, Yang G, Deng Y, He N (2019) Aptasensors for pesticide detection. Biosens Bioelectron 130:174–184. https://doi.org/10.1016/j.bios.2019.01.006

Llorent-Martínez EJ, Ortega-Barrales P, Fernández-de Córdova ML, Ruiz-Medina A (2011) Trends in flow-based analytical methods applied to pesticide detection: a review. Anal Chim Acta 17:21–30. https://doi.org/10.1016/j.aca.2010.10.036

López-Ruiz R, Romero-González R, Frenich AG (2019) Ultrahigh-pressure liquid chromatography-mass spectrometry: an overview of the last decade. Trends Anal Chem 118:170–181. https://doi.org/10.1016/j.trac.2019.05.044

Lu X, Tao L, Li Y, Huang H, Gao F (2019) A highly sensitive electrochemical platform based on the bimetallic Pd/Au nanowires network for organophosphorus pesticides detection. Sensors Actuators B Chem 284:103–109. https://doi.org/10.1016/j.snb.2018.12.125

Lucca BG, Petroni JM, Ferreira VS (2017) Voltammetric determination of insecticide thiodicarb through its electrochemical reduction using novel solid amalgam electrode fabricated with silver nanoparticles. Electrochim Acta 246:748–756. https://doi.org/10.1016/j.electacta.2017.06.056

Ma Y, He X, Qi K, Wang T, Qi Y, Cui L, Wang F, Song M (2019) Effects of environmental contaminants on fertility and reproductive health. J Environ Sci 77:210–217. https://doi.org/10.1016/j.jes.2018.07.015

Madej K, Kalenik TK, Piekoszewski W (2018) Sample preparation and determination of pesticides in fat-containing foods. Food Chem 269:527–541. https://doi.org/10.1016/j.foodchem.2018.07.007

Madianos L, Skotadis E, Tsekenis G, Patsiouras L, Tsigkourakos M, Tsoukalas D (2018) Impedimetric nanoparticle aptasensor for selective and label free pesticide detection. Microelectron Eng 189:39–45. https://doi.org/10.1016/j.mee.2017.12.016

Mahé E, Devilliers D, Dardoize F (2015) Boron doped diamond microelectrodes arrays for electrochemical detection in HPLC. Talanta 132:641–647. https://doi.org/10.1016/j.talanta.2014.10.028

Malarkodi C, Rajeshkumar S, Annadurai G (2017) Detection of environmentally hazardous pesticide in fruit and vegetable samples using gold nanoparticles. Food Control 132:641–647. https://doi.org/10.1016/j.foodcont.2017.04.023

Maleki N, Absalan G, Safavi A, Farjami E (2007) Ultra trace adsorptive stripping voltammetric determination of atrazine in soil and water using mercury film electrode. Anal Chim Acta 581:37–41. https://doi.org/10.1016/j.aca.2006.08.043

Malik AK, Faubel W (2001) A review of analysis of pesticides using capillary electrophoresis. Crit Rev Anal Chem 31:223–279. https://doi.org/10.1080/20014091076758

Malvano F, Albanese D, Pilloton R, Di Matteo M, Crescitelli A (2017) A new label-free impedimetric affinity sensor based on cholinesterases for detection of organophosphorous and carbamic pesticides in food samples: impedimetric versus amperometric detection. Food Bioprocess Technol 10:1834–1843. https://doi.org/10.1007/s11947-017-1955-7

Martinez RC, Gonzalo ER, García FG, Méndez JH (1993) Automated high-performance liquid chromatographic method for the determination of organophosphorus pesticides in waters with dual electrochemical (reductive-oxidative) detection. J Chromatogr A 644:49–58. https://doi.org/10.1016/0021-9673(93)80119-S

Medina-Sánchez M, Mayorga-Martinez CC, Watanabe T, Ivandini TA, Honda Y, Pino F, Nakata A, Fujishima A, Einaga Y, Merkoçi A (2016) Microfluidic platform for environmental contaminants sensing and degradation based on boron-doped diamond electrodes. Biosens Bioelectron 75:365–374. https://doi.org/10.1016/j.bios.2015.08.058

Mehta J, Vinayak P, Tuteja SK, Chhabra VA, Bhardwaj N, Paul AK, Kim KH, Deep A (2016) Graphene modified screen printed immunosensor for highly sensitive detection of parathion. Biosens Bioelectron 83:339–346. https://doi.org/10.1016/j.bios.2016.04.058

Melo LC, Salazar-Banda GR, Machado SAS, de Lima-Neto P, De Souza D, Correia AN (2009) A simple and sensitive detection of diquat herbicide using a dental amalgam electrode. A comparison using the chromatographic technique. Talanta 79:1216–1222. https://doi.org/10.1016/j.talanta.2009.04.024

Melo LC, Julião MSS, Milhome MAL, Ronaldo F, De Souza D, De Lima-neto P, Correia AN (2018) Square wave adsorptive stripping voltammetry determination of chlorpyriphos in irrigation agricultural water. J Anal Chem 73:707–715. https://doi.org/10.1134/S1061934818070109

Miller JN (2005) Chromatography: concepts and contrasts. Wiley, New York

Mirceski V, Komorsky-Lovric S, Lovric M (2007) Square wave voltammetry: theory and application. Springer, Amsterdan

Mirčeski V, Komorsky-Lovrić Š, Lovric M (2007) Square-wave voltammetry: theory and application. Springer, Berlin

Miró M, Frenzel W (2018) Flow injection analysis: detection techniques. Elsevier, New York

Moreno M, Bermejo E, Sánchez A, Chicharro M, Zapardiel A (2008) Application of matrix solid-phase dispersion to the determination of amitrole and urazole residues in apples by capillary electrophoresis with electrochemical detection. Anal Bioanal Chem 391:867–872. https://doi.org/10.1007/s00216-008-2058-0

Moreno MT, Rodríguez-Amaro R, Rodríguez Mellado JM, Mayén M, Jiménez Guardeño F (2018) Imidazolinone and triazine herbicides in soils in relation to the complexes formed with Cu (II) ions. Comptes Rendus Chim 21:884–889. https://doi.org/10.1016/j.crci.2018.05.007

Mostafalou S, Abdollahi M (2013) Pesticides and human chronic diseases: evidences, mechanisms, and perspectives. Toxicol Appl Pharmacol 15:157–177. https://doi.org/10.1016/j.taap.2013.01.025

Mross S, Pierrat S, Zimmermann T, Kraft M (2015) Microfluidic enzymatic biosensing systems: a review. Biosens Bioelectron 15:376–391. https://doi.org/10.1016/j.bios.2015.03.049

Muscalu AM, Górecki T (2018) Comprehensive two-dimensional gas chromatography in environmental analysis. Trends Anal Chem 106:225–245. https://doi.org/10.1016/j.trac.2018.07.001

Nagels LJ, Poels I (2000) Solid state potentiometric detection systems for LC, CE and μTAS methods. Trends Anal Chem 19:410–417. https://doi.org/10.1016/S0165-9936(00)00003-0

Narenderan ST, Meyyanathan SN, Karri VVSR (2019) Experimental design in pesticide extraction methods: a review. Food Chem 15:384–395. https://doi.org/10.1016/j.foodchem.2019.03.045

Néher-Neumann E (2009) Advanced potentiometry: potentiometric titrations and their systematic errors. Springer, Amsterdan

Nehra A, Pal Singh K (2015) Current trends in nanomaterial embedded field effect transistor-based biosensor. Biosens Bioelectron 15:731–743. https://doi.org/10.1016/j.bios.2015.07.030

Nikolelis DP, Simantiraki MG, Siontorou CG, Toth K (2005) Flow injection analysis of carbofuran in foods using air stable lipid film based acetylcholinesterase biosensor. Anal Chim Acta 537:169–177. https://doi.org/10.1016/j.aca.2004.12.086

Nowell LH, Norman JE, Moran PW, Martin JD, Stone WW (2014) Pesticide toxicity index-a tool for assessing potential toxicity of pesticide mixtures to freshwater aquatic organisms. Sci Total Environ 476-477:144–157. https://doi.org/10.1016/j.scitotenv.2013.12.088

Noyrod P, Chailapakul O, Wonsawat W, Chuanuwatanakul S (2014) The simultaneous determination of isoproturon and carbendazim pesticides by single drop analysis using a graphene-based electrochemical sensor. J Electroanal Chem 719:54–59. https://doi.org/10.1016/j.jelechem.2014.02.001

Nurdin M, Maulidiyah M, Salim LOA, Muzakkar MZ, Umar AA (2019a) High performance cypermethrin pesticide detection using anatase TiO2-carbon paste nanocomposites electrode. Microchem J 145:756–761. https://doi.org/10.1016/j.microc.2018.11.050

Nurdin M, Prabowo OA, Arham Z, Wibowo D, Maulidiyah M, Saad SKM, Umar AA (2019b) Highly sensitive fipronil pesticide detection on ilmenite (FeO.TiO2)-carbon paste composite electrode. Surf Interfaces 16:108–113. https://doi.org/10.1016/j.surfin.2019.05.008

Ochoa V, Maestroni B (2018) Pesticides in water, soil, and sediments. In: Integrated analytical approaches for pesticide management. Elsevier, New York, pp 133–147

Oliveira TMBF, Becker H, Longhinotti E, De Souza D, de Lima-Neto P, Correia AN (2013) Carbon-fibre microelectrodes coupled with square-wave voltammetry for the direct analysis of dimethomorph fungicide in natural waters. Microchem J 109:84–92. https://doi.org/10.1016/j.microc.2012.03.032

Pabbi M, Kaur A, Mittal SK, Jindal R (2018) A surface expressed alkaline phosphatase biosensor modified with flower shaped ZnO for the detection of chlorpyrifos. Sensors Actuators B Chem 258:215–227. https://doi.org/10.1016/j.snb.2017.11.079

Park J, Cho YS, Sung SJ, Byeon M, Yang SJ, Park CR (2018) Characteristics tuning of graphene-oxide-based-graphene to various end-uses. Energy Storage Mater 14:8–21. https://doi.org/10.1016/j.ensm.2018.02.013

Patel H, Rawtani D, Agrawal YK (2019) A newly emerging trend of chitosan-based sensing platform for the organophosphate pesticide detection using Acetylcholinesterase- a review. Trends Food Sci Technol 85:78–91. https://doi.org/10.1016/j.tifs.2019.01.007

Picó Y, Rodríguez R, Mañes J (2003) Capillary electrophoresis for the determination of pesticide residues. Trends Anal Chem 22:133–151. https://doi.org/10.1016/S0165-9936(03)00302-9

Pospíšil L, Sokolová R, Colombini MP, Giannarelli S, Fuoco R (2000) Electrochemical impedance study of reduction kinetics of the pesticide vinclozoline. Microchem J 67:305–312. https://doi.org/10.1016/S0026-265X(00)00077-1

Primel EG, Caldas SS, Marube LC, Escarrone ALV (2017) An overview of advances in dispersive liquid–liquid microextraction for the extraction of pesticides and emerging contaminants from environmental samples. Trends Environ Anal Chem 14:1–18. https://doi.org/10.1016/j.teac.2017.03.001

Pundir CS, Malik A, Preety CS (2019) Bio-sensing of organophosphorus pesticides: a review. Biosens Bioelectron 140:111348. https://doi.org/10.1016/j.bios.2019.111348

Qiu P, Ni YN (2008) Determination of ziram in vegetable samples by square wave voltammetry. Chin Chem Lett 19:1337–1340. https://doi.org/10.1016/j.cclet.2008.07.013

Qiu P, Ni Y-N, Kokot S (2013) Application of artificial neural networks to the determination of pesticides by linear sweep stripping voltammetry. Chin Chem Lett 24:246–248. https://doi.org/10.1016/j.cclet.2013.01.029

Qiu L, Lv P, Zhao C, Feng X, Fang G, Liu J, Wang S (2019) Electrochemical detection of organophosphorus pesticides based on amino acids conjugated nanoenzyme modified electrodes. Sensors Actuators B Chem 286:386–393. https://doi.org/10.1016/j.snb.2019.02.007

Rajangam B, Daniel DK, Krastanov AI (2018) Progress in enzyme inhibition based detection of pesticides. Eng Life Sci 18:4–19. https://doi.org/10.1002/elsc.201700028

Rana A, Baig N, Saleh TA (2019) Electrochemically pretreated carbon electrodes and their electroanalytical applications – a review. J Electroanal Chem 833:313–332. https://doi.org/10.1016/j.jelechem.2018.12.019

Rancan M, Rossi S, Sabatini AG (2006a) Determination of Thiamethoxam residues in honeybees by high performance liquid chromatography with an electrochemical detector and post-column photochemical reactor. J Chromatogr A 1123:30–65. https://doi.org/10.1016/j.chroma.2006.05.006

Rancan M, Sabatini AG, Achilli G, Galletti GC (2006b) Determination of imidacloprid and metabolites by liquid chromatography with an electrochemical detector and post column photochemical reactor. Anal Chim Acta 555:20–24. https://doi.org/10.1016/j.aca.2005.08.058

Randles JEB (1947) Kinetics of rapid electrode reactions. Faraday Discuss 1:11–19. https://doi.org/10.1039/DF9470100011

Randviir EP, Banks CE (2013) Electrochemical impedance spectroscopy: an overview of bioanalytical applications. Anal Methods 5:1098–1115. https://doi.org/10.1039/c3ay26476a

Requena M, Parrón T, Navarro A, García J, Ventura MI, Hernández AF, Alarcón R (2018) Association between environmental exposure to pesticides and epilepsy. Neurotoxicology 68:13–18. https://doi.org/10.1016/j.neuro.2018.07.002

Reybier K, Zairi S, Jaffrezic-Renault N, Fahys B (2002) The use of polyethyleneimine for fabrication of potentiometric cholinesterase biosensors. Talanta 56:1015–1020. https://doi.org/10.1016/S0039-9140(01)00588-4

Rhouati A, Majdinasab M, Hayat A (2018) A perspective on non-enzymatic electrochemical nanosensors for direct detection of pesticides. Curr Opin Electrochem 11:12–18. https://doi.org/10.1016/j.coelec.2018.06.013

Ribeiro FWP, Sousa CP, Morais S, de Lima-Neto P, Correia AN (2018) Sensing of formetanate pesticide in fruits with a boron-doped diamond electrode. Microchem J 142:24–29. https://doi.org/10.1016/j.microc.2018.06.012

Rivoira L, De Carlo RM, Cavalli S, Bruzzoniti MC (2015) Simple SPE-HPLC determination of some common drugs and herbicides of environmental concern by pulsed amperometry. Talanta 131:205–212. https://doi.org/10.1016/j.talanta.2014.07.070

Rocha DP, Cardoso RM, Tormin TF, de Araujo WR, Munoz RAA, Richter EM, Angnes L (2018) Batch-injection analysis better than ever: new materials for improved electrochemical detection and on-site applications. Electroanalysis 30:1386–1399. https://doi.org/10.1002/elan.201800042

Rodríguez E, Barrio RJ, Goicolea A, Gómez De Balugera Z (1999) Determination of diflubenzuron and its main metabolites in forestry matrices by liquid chromatography with on-line diode-array and electrochemical detection. Anal Chim Acta 384:63–70. https://doi.org/10.1016/S0003-2670(98)00805-8

Rohlman DS, Ismail A, Bonner MR, Rasoul GA, Hendy O, Ortega Dickey L, Wang K, Olson JR (2019) Occupational pesticide exposure and symptoms of attention deficit hyperactivity disorder in adolescent pesticide applicators in Egypt. Neurotoxicology 74:1–6. https://doi.org/10.1016/j.neuro.2019.05.002

Sabarwal A, Kumar K, Singh RP (2018) Hazardous effects of chemical pesticides on human health – cancer and other associated disorders. Environ Toxicol Pharmacol 63:103–114. https://doi.org/10.1016/j.etap.2018.08.018

Sajid M, Alhooshani K (2018) Dispersive liquid-liquid microextraction based binary extraction techniques prior to chromatographic analysis: a review. Trends Anal Chem 108:167–182. https://doi.org/10.1016/j.trac.2018.08.016

Sajini T, Gigimol MG, Mathew B (2019) A brief overview of molecularly imprinted polymers supported on titanium dioxide matrices. Mater Today Chem 11:283–295. https://doi.org/10.1016/j.mtchem.2018.11.010

Samsidar A, Siddiquee S, Shaarani SM (2018) A review of extraction, analytical and advanced methods for determination of pesticides in environment and foodstuffs. Trends Food Sci Technol 71:188–201. https://doi.org/10.1016/j.tifs.2017.11.011

Sanghavi BJ, Hirsch G, Karna SP, Srivastava AK (2012) Potentiometric stripping analysis of methyl and ethyl parathion employing carbon nanoparticles and halloysite nanoclay modified carbon paste electrode. Anal Chim Acta 735:37–45. https://doi.org/10.1016/j.aca.2012.05.029

Santoro K, Ricciardi C (2015) Biosensors. Encycl Food Health 143:430–436. https://doi.org/10.1016/B978-0-12-384947-2.00072-6

Sarigül T, Inam R (2009) Study and determination of the herbicide cyclosulfamuron by square wave stripping voltammetry. Electrochim Acta 54:5376–5380. https://doi.org/10.1016/j.electacta.2009.04.022

Scholz F (2010) Electroanalytical methods: guide to experiments and applications. Springer, Berlin

Schöning MJ, Arzdorf M, Mulchandani P, Chen W, Mulchandani A (2003) A capacitive field-effect sensor for the direct determination of organophosphorus pesticides. Sensors Actuators B Chem 91:92–97. https://doi.org/10.1016/S0925-4005(03)00071-6

Schröfel A, Kratošová G, Šafařík I, Šafaříková M, Raška I, Shor LM (2014) Applications of biosynthesized metallic nanoparticles – a review. Acta Biomater 10:4023–4042. https://doi.org/10.1016/j.actbio.2014.05.022

Selva TMG, de Araujo WR, Bacil RP, Paixão TRLC (2017) Study of electrochemical oxidation and quantification of the pesticide pirimicarb using a boron-doped diamond electrode. Electrochim Acta 246:588–596. https://doi.org/10.1016/j.electacta.2017.06.051

Seralathan M, Osteryoung RA, Osteryoung JG (1987) General equivalence of linear scan and staircase voltammetry. J Electroanal Chem 222:69–100. https://doi.org/10.1016/0022-0728(87)80278-4

Sgobbi LF, Machado SAS (2018) Functionalized polyacrylamide as an acetylcholinesterase-inspired biomimetic device for electrochemical sensing of organophosphorus pesticides. Biosens Bioelectron 100:290–297. https://doi.org/10.1016/j.bios.2017.09.019

Shan Lin M, Iuan Jan B, Leu HJ, Shing Lin J (1999) Trace measurement of dithiocarbamate based pesticide by adsorptive stripping voltammetry. Anal Chim Acta 388:111–117. https://doi.org/10.1016/S0003-2670(99)00026-4

Shetti NP, Malode SJ, Vernekarg PR, Nayak DS, Shetty NS, Reddy KR, Shukla SS, Aminabhavi TM (2019) Electro-sensing base for herbicide aclonifen at graphitic carbon nitride modified carbon electrode – water and soil sample analysis. Microchem J 149:103976. https://doi.org/10.1016/j.microc.2019.103976

Shi M, Xu J, Zhang S, Liu B, Kong J (2006) A mediator-free screen-printed amperometric biosensor for screening of organophosphorus pesticides with flow-injection analysis (FIA) system. Talanta 68:1089–1095. https://doi.org/10.1016/j.talanta.2005.07.007

Shin D, Sarada BV, Tryk DA, Fujishima A, Wang J (2003) Application of diamond microelectrodes for end-column electrochemical detection in capillary electrophoresis. Anal Chem 75:530–534. https://doi.org/10.1021/ac020513j

Shrivas K, Sahu S, Sahu B, Kurrey R, Patle TK, Kant T, Karbhal I, Satnami ML, Deb MK, Ghosh KK (2019) Silver nanoparticles for selective detection of phosphorus pesticide containing π-conjugated pyrimidine nitrogen and sulfur moieties through non-covalent interactions. J Mol Liq 275:297–303. https://doi.org/10.1016/j.molliq.2018.11.071

Sia SK, Whitesides GM (2003) Microfluidic devices fabricated in poly(dimethylsiloxane) for biological studies. Electrophoresis 24:3563–3576. https://doi.org/10.10002/elps.200305584

Silva LM, De Souza D (2017) Ziram herbicide determination using a polished silver solid amalgam electrode. Electrochim Acta 224:541–550. https://doi.org/10.1016/j.electacta.2016.11.133

Silva CCG, De Souza D (2018) Polished silver solid amalgam electrode and cationic surfactant as tool in electroanalytical determination of methomyl pesticide. Talanta 189:389–396. https://doi.org/10.1016/j.talanta.2018.07.029

Silva RO, da Silva EA, Fiorucci AR, Ferreira VS (2019) Electrochemically activated multi-walled carbon nanotubes modified screen-printed electrode for voltammetric determination of sulfentrazone. J Electroanal Chem 835:220–256. https://doi.org/10.1016/j.jelechem.2019.01.018

Simões FR, Vaz CMP, Brett CMA (2007) Electroanalytical detection of the pesticide paraquat by batch injection analysis. Anal Lett 40:1800–1810. https://doi.org/10.1080/00032710701380806

Sinha A, Huang Y, Zhao H (2019) Preparation of 3D assembly of mono layered molybdenum disulfide nanotubules for rapid screening of carbamate pesticide diethofencarb. Talanta 204:455–464. https://doi.org/10.1016/j.talanta.2019.06.040

Sipa K, Brycht M, Leniart A, Nosal-Wiercińska A, Skrzypek S (2018) Improved electroanalytical characteristics for the determination of pesticide metobromuron in the presence of nanomaterials. Anal Chim Acta 1030:61–69. https://doi.org/10.1016/j.aca.2018.05.068

Skládal P (1997) Advances in electrochemical immunosensors. Electroanalysis 9:737–745. https://doi.org/10.1002/elan.1140091002

Skrzypczyńska K, Kuśmierek K, Świątkowski A (2016) Carbon paste electrodes modified with various carbonaceous materials for the determination of 2,4-dichlorophenoxyacetic acid by differential pulse voltammetry. J Electroanal Chem 766:8–15. https://doi.org/10.1016/j.jelechem.2016.01.025

Soares GB, da Silva WTL, Vaz CMP (2012) Graphite-polyurethane composite electrode for the electroanalytical determination of herbicide diuron in soil solutions. Sens Lett 9:1786–1793. https://doi.org/10.1166/sl.2011.1712

Songa EA, Okonkwo JO (2016) Recent approaches to improving selectivity and sensitivity of enzyme-based biosensors for organophosphorus pesticides: a review. Talanta 155:289–304. https://doi.org/10.1016/j.talanta.2016.04.046

Stanković DM, Kalcher K (2016) Amperometric quantification of the pesticide ziram at boron doped diamond electrodes using flow injection analysis. Sensors Actuators B Chem 233:144–146. https://doi.org/10.1016/j.snb.2016.04.069

Starodub NF, Dzantiev BB, Starodub VM, Zherdev AV (2000) Immunosensor for the determination of the herbicide simazine based on an ion-selective field-effect transistor. Anal Chim Acta 424:37–43. https://doi.org/10.1016/S0003-2670(00)01143-0

Štěpánková Š, Vorčáková K (2016) Cholinesterase-based biosensors. J Enzyme Inhib Med Chem 31:180–193. https://doi.org/10.1080/14756366.2016.1204609

Tertiş M, Ciui B, Suciu M, Săndulescu R, Cristea C (2017) Label-free electrochemical aptasensor based on gold and polypyrrole nanoparticles for interleukin 6 detection. Electrochim Acta 258:1208–1218. https://doi.org/10.1016/j.electacta.2017.11.176

Thanh CT, Binh NH, Van Tu N, Thu VT, Bayle M, Paillet M, Sauvajol JL, Thang PB, Lam TD, Minh PN, Van Chuc N (2018) An interdigitated ISFET-type sensor based on LPCVD grown graphene for ultrasensitive detection of carbaryl. Sensors Actuators B Chem 260:78–85. https://doi.org/10.1016/j.snb.2017.12.191

Timur S, Telefoncu A (2004) Acetylcholinesterase (AChE) electrodes based on gelatin and chitosan matrices for the pesticide detection. Artif Cells Blood Substit Immobil Biotechnol 32:427–442. https://doi.org/10.1081/BIO-200027497

Tormin TF, Cunha RR, Da Silva RAB, Munoz RAA, Richter EM (2014) Combination of screen-printed electrodes and batch injection analysis: a simple, robust, high-throughput, and portable electrochemical system. Sensors Actuators B Chem 202:93–98. https://doi.org/10.1016/j.snb.2014.04.096

Tran HV, Yougnia R, Reisberg S, Piro B, Serradji N, Nguyen TD, Tran LD, Dong CZ, Pham MC (2012) A label-free electrochemical immunosensor for direct, signal-on and sensitive pesticide detection. Biosens Bioelectron 31:62–68. https://doi.org/10.1016/j.bios.2011.09.035

Trojanowicz M (2011) Recent developments in electrochemical flow detections-A review Part II. Liquid chromatography. Anal Chim Acta 688(1):8–35. https://doi.org/10.1016/j.aca.2010.12.024

Trojanowicz M (2016) Impact of nanotechnology on design of advanced screen-printed electrodes for different analytical applications. Trends Anal Chem 84:22–47. https://doi.org/10.1016/j.trac.2016.03.027

Tsiafoulis CG, Nanos CG (2010) Determination of azinphos-methyl and parathion-methyl in honey by stripping voltammetry. Electrochim Acta 56:566–574. https://doi.org/10.1016/j.electacta.2010.09.009

Uwizeyimana H, Wang M, Chen W, Khan K (2017) The eco-toxic effects of pesticide and heavy metal mixtures towards earthworms in soil. Environ Toxicol Pharmacol 55:20–29. https://doi.org/10.1016/j.etap.2017.08.001

Vaghela C, Kulkarni M, Haram S, Aiyer R, Karve M (2018) A novel inhibition based biosensor using urease nanoconjugate entrapped biocomposite membrane for potentiometric glyphosate detection. Int J Biol Macromol 108:32–40. https://doi.org/10.1016/j.ijbiomac.2017.11.136

Valdés-Ramírez G, Gutiérrez M, del Valle M, Ramírez-Silva MT, Fournier D, Marty JL (2009) Automated resolution of dichlorvos and methylparaoxon pesticide mixtures employing a Flow Injection system with an inhibition electronic tongue. Biosens Bioelectron 24:1103–1108. https://doi.org/10.1016/j.bios.2008.06.022

Vaz CMP, Silva PRV, Prado I, Castanho G, Simões F, Machado SAS (2008) Electrochemical behavior of parent and photodegradation products of some selected pesticides. Quim Nova 31:1310–1314. https://doi.org/10.1590/S0100-40422008000600007

Wang J (1985) Stripping analysis: principles, instrumentation, and applications. VCH Publishers Inc, New York

Wang J (2002) Electrochemical detection for microscale analytical systems: a review. Talanta 56:223–231. https://doi.org/10.1016/S0039-9140(01)00592-6

Wang J (2006) Analytical electrochemistry. Wiley, New York

Wang J, Chatrathi MP, Mulchandani A, Chen W (2001) Capillary electrophoresis microchips for separation and detection of organophosphate nerve agents. Anal Chem 73:1804–1808. https://doi.org/10.1021/ac001424e

Wang YR, Hu P, Liang QL, Luo GA, Wang YM (2008) Application of carbon nanotube modified electrode in bioelectroanalysis. Chin J Anal Chem 36:1011–1016. https://doi.org/10.1016/S1872-2040(08)60052-3

Wei Y, Li Y, Qu Y, Xiao F, Shi G, Jin L (2009) A novel biosensor based on photoelectro-synergistic catalysis for flow-injection analysis system/amperometric detection of organophosphorous pesticides. Anal Chim Acta 643:13–18. https://doi.org/10.1016/j.aca.2009.03.045

Włoch M, Datta J (2019) Synthesis and polymerisation techniques of molecularly imprinted polymers. Compr Anal Chem. https://doi.org/10.1016/bs.coac.2019.05.011

Wu S, Lan X, Cui L, Zhang L, Tao S, Wang H, Han M, Liu Z, Meng C (2011) Application of graphene for preconcentration and highly sensitive stripping voltammetric analysis of organophosphate pesticide. Anal Chim Acta 699:170–176. https://doi.org/10.1016/j.aca.2011.05.032

Xu M, Obodo D, Yadavalli VK (2019) The design, fabrication, and applications of flexible biosensing devices. Biosens Bioelectron 124–125:96–114. https://doi.org/10.1016/j.bios.2018.10.019

Yildiz E, Camurlu P, Tanyeli C, Akhmedov I, Toppare L (2008) A soluble conducting polymer of 4-(2,5-di(thiophen-2-yl)-1H-pyrrol-1-yl)benzenamine and its multichromic copolymer with EDOT. J Electroanal Chem 612:247–256. https://doi.org/10.1016/j.jelechem.2007.10.004

Yola ML (2019) Electrochemical activity enhancement of monodisperse boron nitride quantum dots on graphene oxide: Its application for simultaneous detection of organophosphate pesticides in real samples. J Mol Liq 277:50–57. https://doi.org/10.1016/j.molliq.2018.12.084

Yoon Y, Jo J, Kim S, Lee I, Cho B, Shin M, Hwang W (2017) Impedance spectroscopy analysis and equivalent circuit modeling of graphene oxide solutions. Nano 7:447–452. https://doi.org/10.3390/nano7120446

Yuqing M, Jianguo G, Jianrong C (2003) Ion sensitive field effect transducer-based biosensors. Biotechnol Adv 21:527–534. https://doi.org/10.1016/S0734-9750(03)00103-4

Zhao F, Wu J, Ying Y, She Y, Wang J, Ping J (2018) Carbon nanomaterial-enabled pesticide biosensors: Design strategy, biosensing mechanism, and practical application. Trends Anal Chem 106:62–83. https://doi.org/10.1016/j.trac.2018.06.017

Zheng Q, Chen Y, Fan K, Wu J, Ying Y (2017) Exploring pralidoxime chloride as a universal electrochemical probe for organophosphorus pesticides detection. Anal Chim Acta 982:78–83. https://doi.org/10.1016/j.aca.2017.06.006

Zhu Z, Zhang Y, Wang J, Li X, Wang W, Huang Z (2019) Sugaring-out assisted liquid-liquid extraction coupled with high performance liquid chromatography-electrochemical detection for the determination of 17 phenolic compounds in honey. J Chromatogr A 1601:104–114. https://doi.org/10.1016/j.chroma.2019.06.023

Zoski CG (2007) Handbook of electrochemistry. Elsevier, New Mexico

Chapter 2
Protocols for Extraction of Pesticide Residues

María Luisa Fernández de Córdova and Eulogio José Llorent Martínez

Abstract Over the past few years, the presence of pesticides in foods has received growing attention as one of the main food safety problems. Many countries and health organizations have dictated increasingly strict regulations on the maximum pesticide residue limits on foodstuffs. Pesticides can also reach surface water through runoff from treated plants and soil, contaminate groundwater, soil, ground and grass, and be toxic to non-target organisms. Therefore, reliable methods for the analysis of pesticides need to be developed that can ensure accurate and precise results with sufficiently low detection limits for the determination of a wide range of these compounds. In most cases, these methods will need to include a sample pretreatment to extract, isolate, and concentrate the target compounds, as it is generally not possible to directly manipulate the sample.

We have reviewed the main classical and advanced methodologies available for the treatment of food and environmental samples for analysis of pesticide residues. Conventional sample treatment techniques have shown to be efficient, but they entail drawbacks such as the employ of toxic organic solvents, complicated procedures that require a lot of time, and difficulty in automation. The numerous liquid-phase and solid-phase microextraction techniques developed in recent years together with the use of alternative green extractants as ionic liquids and new solid sorbents as molecularly imprinted polymers, carbon nanotubes or magnetic materials have allowed overcoming these problems, developing highly selective extraction methods, and simplifying analytical procedures. On the other hand, QuEChERs method has become in one of the sample treatment mostly applied as it allows performing pesticide extraction and cleanup in a single step, which is evidenced by the fact that organizations such as AOAC International and the European Committee for Standardization have adopted it in two official methods for pesticide analysis. All these methodologies, their advantages, drawbacks and future perspectives have been described and discussed in this chapter.

M. L. Fernández de Córdova (✉) · E. J. Llorent Martínez
Department of Physical and Analytical Chemistry, Faculty of Experimental Sciences, University of Jaén, Jaén, Spain
e-mail: mferna@ujaen.es

Inamuddin et al. (eds.), *Sustainable Agriculture Reviews 48*, Sustainable Agriculture Reviews 48, https://doi.org/10.1007/978-3-030-54719-6_2

Keywords Pesticides · Extraction · Cleanup · Sample preparation · Microextraction techniques · Liquid-phase extraction techniques · Solid-phase extraction techniques · Food samples · Environmental samples · Green analytical chemistry

2.1 Introduction

In recent years, the application of pesticides has rapidly increased mainly in agriculture thus their use allows having abundant and high quality vegetables available at a relatively low price. However, the presence of pesticide residues in foods, soils, plants and water has become a major issue and could have considerable impact on health and the environment (Samsidar et al. 2018). To ensure that pesticide residues found in foods do not represent an unacceptable risk for human consumption, most governments have set maximum residue limits in vegetables and fruits. Maximum residue limits of pesticides in food and feed are internationally restricted by the Regulations of the European Union and Codex Alimentarius Commissions (European Union 2005; Codex Alimentarius: http://www.fao.org/fao-who-codexalimentarius/codex-texts/dbs/pestres/en/), which reflect the highest amount of residues expected in food when plant protection products are applied correctly in accordance with authorized conditions of use. They are not safety limits, and are always set below, often far below, levels that would present a risk to consumers. The European Union maximum residue limits apply to most food commodities of animal and plant origin, concretely to the product as placed on the market. In the last European Union report on pesticide residues in foods, which provides an insight into the official control activities carried out by European Union Member States, Iceland and Norway in 2016, pesticide concentrations in 96.2% of the 84,657 samples analyzed met the established legal limits. Similar to the results obtained in 2015, processed grape leaves, rice, wild fungi, tomatoes, sweet peppers, sweet corn, table grapes, and table olives were the commodities that most frequently exceeded the maximum residue limits. Teas, strawberries, pineapples, milk (goat) and muscle (swine) were also found to exceed frequently the maximum residue limits (more than 2% of the samples) (European Food Safety Authority 2018). On the other hand, in the European Union pesticide residues in water are subject to the Water Framework Directive, whose purpose is to establish a framework for the protection of inland surface waters, transitional waters, coastal waters and groundwater (European Union 2000). In accordance with the above, it is very important to provide accurate, precise and reliable methods for pesticide residue analysis for ensuring food safety and preserve environment.

Currently, the determination of pesticides has received more attention because governments have reduced the maximum residue limits, and hence the detection of trace levels of pesticides needs more efficient techniques for pre-treatment of samples. The analytical procedure usually consists of numerous stages, being the sample collection and its preparation for analysis the most important ones. Sample preparation is often a neglected area even though it plays an essential role in the whole analytical procedure, particularly in the analysis of solid samples. Most analytical instruments are unable to handle matrices directly and, therefore, some form of sample pretreatment is required to extract and enrich analytes. Environmental and food samples are complex and very diverse and analytes are usually found at trace and ultra-trace amounts (Stocka et al. 2011) so that sample preparation has a special relevance. From an analytical point of view, a sample pretreatment for pesticide residue analysis should to have the following features: (a) to allow the analysis of a number of pesticides as large as possible in a single run (multiresidue method); (b) to provide recovery percentages as close as possible to 100%; (c) efficient removal of potential interfering species from the matrix; (d) high sensitivity; (e) good precision; (f) good ruggedness; (g) low cost; and (h) high throughput. Moreover, sample treatment should be environmentally friendly so that it implies the use of as low solvent volumes as possible (Hercegová et al. 2007).

Typical sample preparation for pesticide analysis includes the following steps: sampling/homogenization, extraction of target compounds from the bulk of the matrix, and cleanup of analytes from matrix co-extractives. Among them, the extraction of the target compounds and cleanup of the extracts are the most time-consuming and challenging steps and play a key role in the success of analysis. Special attention must be paid to high-fat content matrices, since it is necessary to avoid high amounts of fat residues in the final extracts, which would affect to the proper functioning of the chromatographic system. Although conventional methods of extraction, especially liquid-liquid extraction, have been widely used for analysis of pesticide residues, they are subject to the loss of pesticides due to the long and laborious experimental procedure, require large amounts of organic solvents, and are time-consuming. Therefore, over the past years analytical developments have attempted to minimize the solvent volumes, the use of toxic solvents, the physical and chemical manipulations, and have aimed to automate the extraction and cleanup procedures as far as possible (LeDoux 2011). Figure 2.1 shows the desired analytical characteristics and current trends in the treatment of samples for pesticide residue analysis.

To date, a lot of techniques have been exploited for determination of pesticide residues. Following the extraction/cleanup steps, the detection of pesticides is usually performed by mass spectrometric techniques (i.e. tandem mass spectrometry using ion-trap or triple quadrupole instruments) in combination with gas chromatography or liquid chromatography, owing to their high sensitivity, selectivity and throughput. Most of the traditional methods for analysis of pesticide residues have been applied to the determination of a single component or a class of pesticides. On the contrary, multiresidue methods of analysis are used for analysis not only of several components of the same class of pesticides, but also of several components of different classes of pesticides. Taking into account the large number of pesticides on the market, the development of multiresidue methods capable of analyzing large number of pesticides in one single run is the most efficient and common approach. These multiresidue methodologies, in addition to allow the determination of a large number of compounds in one run, should provide large recoveries of target compounds, high sensitivity and good precision, ensure maximum removal of interfering

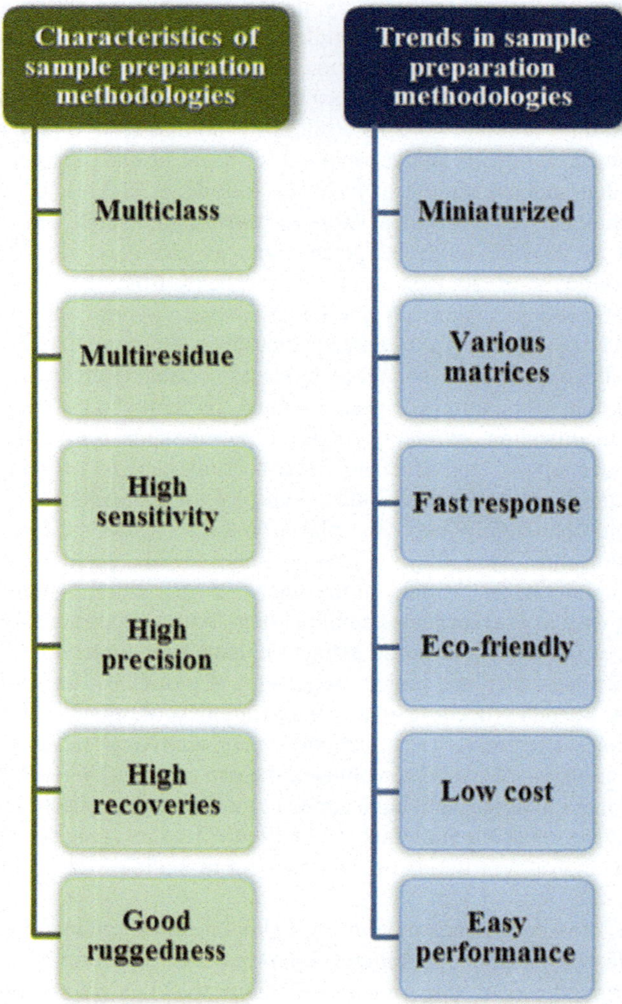

Fig. 2.1 Analytical characteristics and trends for sample preparation methods in pesticide residue analysis

species from extracts, and be environmentally-friendly, quick, cheap and easy to carry out.

Many works have been reported over the years to achieve this objective. The present chapter describes a review of the main techniques used for extraction and cleanup in field of pesticide residue analysis. The basic principles and main features of both classical and modern techniques are described, compared their advantages and disadvantages, and their ability and applicability for pesticide residue analysis. A distinction has not been made between techniques for extraction and purification since in practice most of them are used for both purposes. After introducing these techniques and their applications, particular attention will be paid on the recent developments and future trends. The main focus has been paid on microextraction techniques, since currently they constitute one of the main trends in the development and enhancement of sample treatment methods. Finally, the applications addressed in the present chapter cover food samples and aqueous and solid environmental samples.

2.2 Liquid-Phase Extraction Techniques

2.2.1 Liquid-Liquid Extraction

Liquid-liquid extraction, also called solvent partitioning, is a conventional technique used for sample preparation and preconcentration. The extraction of the organic compounds is based on their relative solubilities in two immiscible liquids, water and an organic solvent or two organic solvents. The factors influencing efficiency of the extraction are: solubility of the compounds in the extraction solvent, extraction volume and number of extractions. In the conventional extraction of organic compounds the solid or liquid sample is homogenized and repeatedly extracted with an organic solvent; then, the extracts are centrifuged, concentrated and/or purified to remove matrix co-extractives before the final analysis. The techniques used for cleanup of the extracts usually are gel permeation chromatography and/or solid-phase extraction (Guardia-Rubio et al. 2006; Zheng et al. 2014). This is an efficient extraction technique in terms of recovery, except for a few pesticides such as hexachlorobenzene for instance which is very volatile (LeDoux 2011). In this technique solvents or solvent mixtures are used for extraction, according to the polarity of target compounds. In the development of multiresidue methods it is necessary to take into account that the solvents used have to be able to extract pesticides with a wide range of polarities from the same matrix. Moreover, pesticides can be accompanied by their metabolites or degradation products, whose extraction requires different solvent systems according to their polarity.

The usual solvents for liquid-liquid extraction of pesticides in food and environmental matrices have been ethyl acetate (Fernandez-Alba et al. 1994; Mol et al. 2007; Banerjee et al. 2007), acetone (Guan et al. 2010), methanol (Hanot et al. 2015), hexane (Cabras et al. 2001), dichlorometane and acetonitrile (Fillion et al. 2000; Lehotay et al. 2001; Guardia Rubio et al. 2007). Mixtures of these solvents with different composition have also been widely used, such as hexane/dichloromethane (Tian et al. 2019), hexane/acetone (Rezić et al. 2005), hexane/acetonitrile (Zayats et al. 2013), or dichlorometane/acetone (Pose-Juan et al. 2006). The use of medium polarity solvents such as ethyl acetate or acetone can increase the polarity of a non-polar solvent or decrease the polarity of a polar solvent in the extraction procedure. Acetone is completely miscible with water and allows an adequate penetration in the aqueous part of the sample. However, to induce its separation from the water phase a non-polar solvent has to be added, which leads to dilution and can originate low recoveries of the most polar analytes. Ethyl acetate is sufficiently miscible with water to allow good penetration into matrices and its polarity is sufficient to extract the most polar pesticides. Moreover, after extraction no extra partition step is required, since water can be simply removed by adding sodium sulfate. Acetonitrile is miscible with water, but it can be separated from the aqueous phase by salting-out effect. In addition, due to its hydrophobicity, it is efficient in extracting both polar and non-polar pesticide residues with low amounts of matrix co-extractives. Therefore, acetonitrile has been one of the most widely used solvents for liquid-liquid extraction of pesticide residues. Acetonitrile also presents an additional advantage, the possibility to analyze in the same extract both gas chromatography- and liquid chromatography-amenable analytes (Anastassiades et al. 2003; Lehotay et al. 2005a), since it is compatible with reversed-phase liquid chromatography and can also be served as a medium for gas chromatography injection. As a rule, less polar organic solvents such as dichlorometane or hexane favor extraction of more non-polar pesticides and can be occasionally used for the extraction of non-polar pesticides or to modify the polarity of other non-polar solvents (Zhang et al. 2014).

Classical liquid-liquid extraction is aided by shaking, but ultrasonic extraction has also been widely applied for extraction of pesticides from solid food and environmental samples (Gonçalves and Alpendurada 2005; Lambropoulou et al. 2006). Acoustic energy is used for enhancing mechanical washing of the sample, providing a high efficient contact between sample matrix and solvent. Several studies show that ultrasonic extraction is an effective technique during sample preparation for pesticide analysis in soil, sediments, oil, coffee, honey and fruits and leafy vegetables (Beyer and Biziuk 2008; Pan et al. 2008; Da Silva Souza and Navickiene 2019; Tian et al. 2019). Moreover, it was reported that ultrasonic extraction can allow excellent results without cleanup if solvent type and extraction conditions are optimized (Rezić et al. 2005).

Conventional liquid-liquid extraction is an accepted and popular technique for sample preparation due to its advantages like simplicity, efficiency, and ruggedness. This technique mostly provides recovery percentages for pesticide residue analysis in the range 70–120%, as advised by the European Union. Multiresidue methods also show recoveries in this range for over 80% of the analytes (LeDoux 2011). However, liquid-liquid extraction presents many inherent drawbacks, e.g., it requires a relatively large amount of matrix, and the evaporation of a large quantity of toxic and flammable solvents to get to required pre-concentration coefficient. Moreover, it is time-consuming, laborious, expensive regarding the consumption of solvents and other materials, and difficult to automate. These drawbacks have been overcome by the introduction of simpler and faster sample preparation methods for pesticide residue analysis in foods and environmental samples. The advent of novel modified liquid-liquid extraction methods in the sample treatment field has caused the decrease of organic solvent consumption, more effective extraction, higher extraction yields and on-line connecting to analytical instrumentation. These techniques are described in the next Sections.

2.2.2 Soxhlet Extraction

Conventional Soxhlet has been used for decades as one of the most relevant techniques in the environmental pesticide extraction field from solid samples. In this technique, a glass extractor consisting of three parts is used: a distillation flask, a main chamber and a coolant. The sample is placed inside a thimble-holder and deposited in the main chamber, covering it with the extraction solvent. The solvent is then heated to reflux so that its vapors condense into the coolant and drop down on the sample cartridge. When the liquid in the central chamber reaches an overflow level, it passes through a siphon to the distillation flask by dragging with it the compounds extracted from the sample. This operation is repeated until the extraction is completed. Therefore, this performance makes Soxhlet a hybrid discontinuous-continuous technique. Soxhlet extraction has been mostly applied to the extraction of organochlorine pesticides from different matrices, such as soils (Wang et al. 2007b), river sediments (Tan and Vijayaletchumy 1994), fatty foods (Doong and Lee 1999), mollusks and crabs (Yang et al. 2006), and eggs, chicken and meat (Ahmad et al. 2010). The solvents used usually are either *n*-hexane or *n*-hexane/acetone, although other solvents such as acetone/dichlorometane or ethyl acetate have also been reported (Luque de Castro and Priego-Capote 2010). In the application of Soxhlet for the extraction of pesticides from soils, it is necessary to take into account that the strong organic solvents used are not able to solubilize humate matter, which may contain pesticides with a high affinity to organic humic acids, such as organochlorine pesticides (Andreu and Picó 2004).

Soxhlet extraction integrates the advantages of the reflux extraction and percolation, which uses the principle of reflux and siphoning to continuously extract samples with fresh solvent. It provides high extraction efficiency and requires less time and solvent consumption than maceration or percolation. The most significant drawbacks of Soxhlet are: large amounts of solvent usage (300–500 mL per sample), the need for evaporation solvent to concentrate analytes before analysis, it is expensive in terms of energy, cleanup of extracts is usually needed, and long extraction times are required (e.g. up to 24–48 h). In addition, this technique cannot be used for thermolabile compounds such as *N*-methylcarbamates, sulfonyl urea, and chlorophenoxy acid herbicides, since prolonged heating may lead to their degradation.

2.2.3 Microwave-Assisted Extraction

Microwave-assisted extraction has been demonstrated to be a versatile extraction technique for pesticide residue analysis in solid samples, with the advantages of rapidity, simplicity and low cost of operation (Hercegová et al. 2007; Merdassa et al. 2013). This technique has also been successfully applied to the extraction of liquid samples (Fuentes et al. 2009). Microwave energy is a non-ionizing radiation (frequency 300–300000 MHz), which can get in certain materials to interact with the polar components by ionic conduction and rotation of dipoles, and generate heat (Sparr Eskilsson and Björklund 2000). No drying of sample is required since water absorbs microwave energy quickly and can help to heat up sample. The most common procedure in microwave-assisted extraction involves extraction in a closed vessel under controlled temperature and pressure, although a focused microwave oven at atmospheric pressure is also used (Falqui-Cao et al. 2001; Fuentes et al. 2009).

When compared to conventional heating, microwave-assisted extraction presents advantages such a significant decreasing of time of analysis because the heat is transferred directly to the solvent, and an easy control of extraction parameters (temperature, power, time). Moreover, thermolabile compounds can be extracted by appropriately controlling working conditions and trace analytes can be extracted from very low amounts of sample. However, microwave-assisted extraction has several drawbacks such as: partial decomposition of some compounds in high pressure and temperature conditions can occur; volatile compounds can be lost at the end of the process, so that the vessels must be cooled at room temperature before being opened; extracts must be filtered after extraction; polar solvents are necessary; and cleanup of extracts is usually needed since microwave-assisted extraction is very efficient. Some of these drawbacks can be overcome when this technique is carried out in on-line mode by coupling the instrument to a flow system consisting of a high-pressure quaternary gradient pump for solvent delivery (Carvalho et al. 2008). In this modality of microwave-assisted extraction the extract can be on-line filtered and subjected to other coupled subsequent pre-treatments. The degradation of thermolabile compounds can be avoided by using dynamic mode. In dynamic microwave-assisted extraction fresh solvent is continuously pumped through the extraction vessel and analytes are transferred out of the vessel as soon as they are extracted (Wang et al. 2012a; Wu et al. 2016).

The choice of an appropriate solvent in microwave-assisted extraction is essential for getting an optimal extraction yield. The pesticide extraction process may occur through two different heating mechanisms (Sparr Eskilsson and Björklund 2000): immersion of sample in a high dielectric constant solvent or mixture of solvents that absorb microwave energy efficiently (Font et al. 1998); and immersion of sample in a combination of low and high dielectric constant solvents (Chee et al. 1996).

In recent years, there has been a general interest for the development of microwave-assisted extraction methods based on the use of extraction media alternative to toxic organic solvents, such as micellar (surfactant-rich) systems (Moreno et al. 2007) or ionic liquids (Wang et al. 2013). Ionic liquids are liquid salts with melting points close to or below room temperature, which act as surfactants and absorb microwave radiation extremely well and transfer energy quickly by ionic conduction to the matrix, thus improving extraction efficiency and speed of analysis. Aqueous solutions or steam have also been used as green solvents to extract pesticides from solid samples. Song et al. (2014) developed an on-line and green microwave-assisted extraction approach to extract carbamate pesticides in rice with water steam. The extract was subjected to an additional cleanup and concentration step by solid-phase extraction on an octadecylsilane cartridge. Recoveries ranging from 66% to 117% were obtained with relative standard deviation values below 9.1% for all the pesticides. Limits of detection of 1.1–4.2 ng g^{-1} were achieved by liquid chromatography-tandem mass spectrometry analysis.

2.2.4 Pressurized Liquid Extraction

Pressurized liquid extraction, also known as accelerated solvent extraction, is a well-known technique in the field of pesticide residue analysis that involves the extraction of residues from solid or semi-solid samples (Vazquez-Roig and Picó 2015; Khan et al. 2018; Andreu and Picó 2019). In this technique the extraction is carried out with small amounts of water and organic solvents (15–40 mL) at high temperatures (up to 200 °C) and pressures (up to 3000 psi) for short time periods, which results in better extraction efficiency and reduction of solvent volume compared with traditional extraction techniques such as liquid-liquid and Soxhlet extraction. Sample (5–100 g) is usually dried to enhance the solvent penetration into the matrix, mixed with a solid sorbent (e.g., Hydromatrix, diatomaceous earth, or sea sand) that reduces particle clumping and solvent channeling in the extraction process, and loaded into the stainless-steel closed-cell (Juan-García et al. 2010; Choi et al. 2016). Pressurized liquid extraction is usually carried out in static mode, although it can also be operated in dynamic mode by continuously pumping small volumes of solvent through the extraction cell at a specified flow rate (0.5–2.5 mL min^{-1}), which accelerates the mass-transfer rate (Vazquez-Roig and Picó 2015). The extraction time of both modes is usually similar and the extraction efficiency of dynamic mode is equal or even higher than when using static mode. The drawback of dynamic mode is that it requires a larger volume of fluid than the static mode, so that analytes will be diluted in the extract, and a pre-concentration step will be required before their chromatographic analysis.

The majority of instruments allow loading up to 24 cells, no filtration is required, and the extraction can be done in unattended operation. Elevated temperature breaks matrix-analyte interactions and achieves a higher diffusion rate, improving the solubility of the analytes in the extraction solvent. On the other hand, elevated pressure raises the solvent to the near-supercritical region, where it shows better extraction properties. Moreover, at high temperature and pressure, viscosity and surface tension of solvent decrease, so that it penetrates into solid sample pores more efficiently (Beyer and Biziuk 2008; Vazquez-Roig and Picó 2015). Extraction time hardly depends on the sample amount, whereas extraction efficiency principally depends on temperature. During the development of a pressurized liquid extraction method it is necessary to optimize several parameters, including extraction time and temperature, type of solvent and sorbent, and number of extraction cycles. Although high temperatures increase extraction efficiency, they may cause the degradation of thermolabile compounds, and also increase the amount of matrix co-extractives. Thus, in the case of fatty matrices, the presence of large amounts of co-extracted lipids constitutes a major problem, and a further cleanup of the extract is usually required (Choi et al. 2016). Therefore, it will be necessary to make a compromise between achieving maximum extraction efficiency and minimizing the amount of co-extractive species. Solvents used for extraction of pesticide residues include hexane (Pinto et al. 2014), acetone (Moreno-González and León 2017), ethyl acetate (Khan et al. 2018), acetonitrile (Masiá et al. 2015), and water (Juan-García et al. 2010). Mixtures of solvents such as hexane-dichlorometane (Choi et al. 2016), acetonitrile-water (Homazava et al. 2014), acetone-dichlorometane (You et al. 2010; Köck-Schulmeyer et al. 2013; Clark et al. 2015), and hexane-acetone (Celeiro et al. 2014) have also been used.

The small volumes of extracts facilitate further concentration and cleanup, usually performed by solid-phase extraction (You et al. 2010; Wang et al. 2010; Köck-Schulmeyer et al. 2013). Fairly recently, a new approach, called selective pressurized liquid extraction, that combines extraction and '*in-cell*' cleanup step, has been reported. Several sorbents placed at the bottom of the pressurized liquid extraction cell are used for cleanup of the extracts, so that the need for additional manipulation of sample is avoided (Homazava et al. 2014; Pinto et al. 2014; Masiá et al. 2015). Choi et al. (2016) studied which sorbent was better for '*in-cell*' cleanup in the extraction of organochlorine pesticides in fish. The efficiency in the removal of lipids in decreasing order was: Alumina > acid-treated silica gel > alumina and acid-treated silica gel > Florisil > alumina and silica gel.

Pressurized liquid extraction is mostly combined with liquid chromatography or gas chromatography for analysis of pesticide residues, coupled with both low and high resolution mass spectrometry (Andreu and Picó 2019), although capillary electrophoresis has also been used (Juan-García et al. 2010). Currently, pressurized liquid extraction is already a routine technique for trace analysis of pesticides and other organic contaminants, used in many laboratories because of its short extraction time, easy manipulation, low sample-amount requirements and good performance. However, it requires sophisticated and expensive equipment and higher consumption of energy than other traditional extraction techniques.

2.2.5 Supercritical Fluid Extraction

Supercritical fluid extraction is usually an efficient technique, mainly applicable to solid samples. Fluids are defined as supercritical when they are maintained at pressures and temperatures above the critical point. Supercritical fluids densities are greater than those of gases and close to those of liquids and, consequently, their solvation properties are similar to those of liquid solvents. Moreover, they have lower viscosity than traditional liquid solvents, which increases the diffusion of solutes in the solvent. CO_2 is most widely used as a supercritical fluid because of it is a nonflammable, nontoxic, and environmentally-friendly solvent. Supercritical fluid extraction does not use hazardous solvents and is easily automated, so reducing the time, cost and manual labor required for extraction. Furthermore, selectivity of extraction can be easily controlled by varying the density of the supercritical fluid, which depends on pressure and temperature. Taking into account that the supercritical CO_2 fluid has low polarity, at a constant temperature, it will be capable to extract low polar analytes at low pressure, and very polar analytes at high pressure (Zougagh et al. 2004). Supercritical fluid extraction is usually carried out at pressures that are not high enough to efficiently extract polar analytes. Therefore, in supercritical fluid extraction, the range of analytes that can be extracted with CO_2 is limited due to its low polarity, being considered a nonpolar solvent with a liquid solubility equal to that of hexane. To overcome this problem, polar solvents (modifiers or co-solvents) are often added to modify the CO_2 fluid or sample matrices for optimal extraction of polar pesticides. The modifier can influence the extraction process in three ways, which depends on the type of sample matrix and the analyte's retaining nature on the matrix: (a) increasing the solubility of the analyte in the supercritical fluid, because of the interaction modifier-analyte in the fluid phase; (b) competing with the analytes for the active sites in the matrix, so facilitating the desorption of the analyte; (c) distorting the matrix, which favors the penetration of the supercritical fluid inside the matrix. Nemoto et al. (1997) showed that water and alcohols were effective modifiers for improving the recovery of pesticides. Methanol is the most common solvent added to CO_2 although other solvents such as acetonitrile (Rissato et al. 2004) have also been used.

The application of supercritical fluid extraction to pesticide residue analysis has been demonstrated for some pesticides in samples such as sediments (Janda et al. 1989), soil (Rissato et al. 2005), oils (Hopper 1999), and agricultural products (Howard et al. 1993; Valverde-García et al. 1995). Nevertheless, supercritical fluid extraction technique has not been widely used for pesticide residue analysis in foods of animal origin since, in the first attempts carried out in these matrices, the recoveries obtained were unacceptable for many pesticides. However, other authors as Fiddler et al. (1999) reported very good recoveries in matrices such as eggs, ranged from 81.8% to 108.3%. Several applications of supercritical fluid extraction to multiresidue analysis of pesticides have also been reported (Lehotay and Eller 1995; Snyder et al. 1993). The most recent studies have shown than supercritical fluid extraction methods combined with an additional cleanup step by solid-phase extraction (Norman and Panton 2001; Rissato et al. 2005; Ono et al. 2006) or QuEChERs (Sartori et al. 2017) meet the strict criteria established for the analysis of pesticide residues.

Large scale supercritical fluid extraction can compete economically with traditional extraction and separation processes, but at the laboratory scale it has serious disadvantages such as high equipment and upkeep cost, difficult optimization, and high blank and noise values (LeDoux 2011).

2.2.6 *Liquid-Phase Microextraction Techniques*

To overcome the drawbacks of conventional liquid-liquid extraction and provide fast, simple and green procedures, several liquid-phase microextraction techniques by reducing organic solvent consumption and simplifying sample preparation procedures have been developed. These techniques are the result of the miniaturization of conventional liquid-liquid extraction by greatly reducing the solvent volume to just a few microliters, and they combine in one step extraction, concentration, and sample introduction. As consequence of the small volume of solvent used for extraction, in liquid-phase microextraction techniques almost all of the obtained organic extract can be injected into the separation instruments, unlike in liquid-liquid extraction where only part of the extract is injected (Lambropoulou and Albanis 2007). In liquid-phase microextraction the extraction is performed between an aqueous phase containing the analytes (donor phase) and a small volume of water-immiscible solvent (acceptor phase) (Sarafraz-Yazdi and Amiri 2010). Two modes of working are possible: to immerse directly the acceptor phase in the sample solution or hang it above the sample for headspace extraction. Most methods reported for the liquid-phase microextraction of pesticides focus on liquid samples, although several applications of liquid-phase microextraction to solid matrices have also been developed. These latter are usually based on the combined use of liquid-phase microextraction with a previous treatment that implies a conventional extraction with organic solvent or another extraction technique such as microwave-assisted extraction (Basheer et al. 2005), ultrasonic extraction or pressurized liquid extraction (Lüthje et al. 2005). However, liquid-phase microextraction has also been applied to the direct extraction of pesticide residues in solid samples such as vegetables (Romero-González et al. 2006) or soils (Hou and Lee 2004) without a previous treatment of sample.

Three major categories of liquid-phase microextraction techniques have been developed to date, which differ in the way in which sample and extraction solvent are contacted: (a) single-drop microextraction; (b) hollow-fiber microextraction; and (c) dispersive liquid–liquid microextraction. All them will be review from here on out.

2.2.6.1 Single-Drop Microextraction

In single-drop microextraction the extraction phase is a micro-drop of organic solvent, so that analyte extraction and preconcentration take place in a single run. The extraction is assisted by stirring, so that the surface of the drop is continuously exposed to fresh aqueous sample and the thickness of the static layer is reduced, which enhances extraction efficiency. A relevant feature of single-drop microextraction is that the drops generated can be renewed for each extraction, so that there is no sample carryover. After extraction, the micro-drop is withdrawn back into the syringe and usually transferred to a gas chromatography or liquid

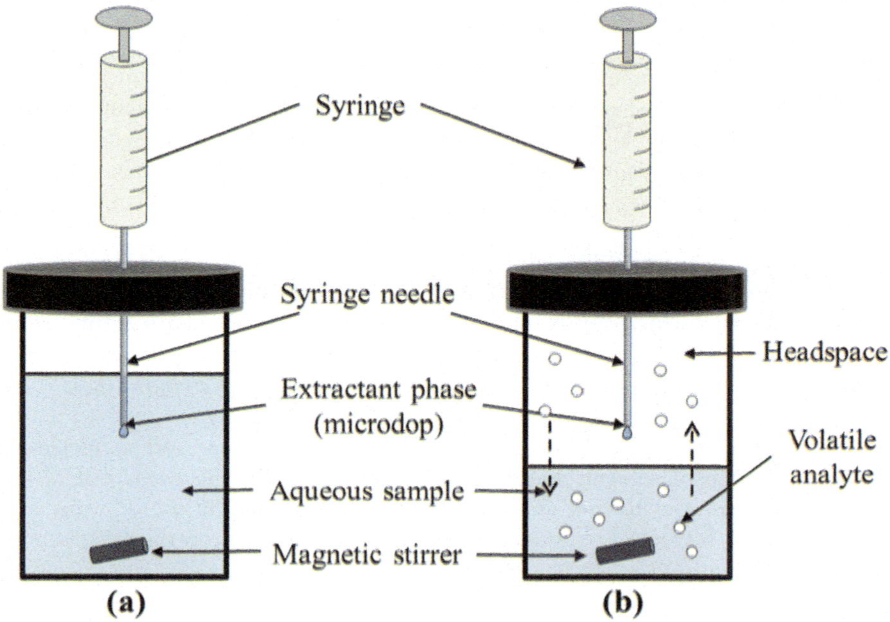

Fig. 2.2 Principle of single-drop microextraction: (a) Direct immersion; (b) Headspace. (**a**) A microdrop of organic solvent is suspended at the tip of a syringe needle in a stirred aqueous sample; (**b**) the organic microdrop is held above the aqueous sample solution

chromatography instrument for further analysis. Single-drop microextraction, of all the solvent microextraction techniques, is also the most easily fully automated method developed to date (Kokosa 2015), and it has been widely applied to analysis of pesticide residues in foods and environmental samples containing target compounds in trace amounts (Zhao et al. 2006; Xiao et al. 2006; Tian et al. 2014).

In practice, single-drop microextraction can be performed by two main approaches (Fig. 2.2):

1. direct immersion
2. headspace

In direct immersion single-drop microextraction, a droplet of a water-immiscible solvent hangs from the tip of a syringe needle immersed in an aqueous sample. Applications of direct immersion single-drop microextraction are usually limited to nonpolar and medium polarity analytes, since the organic solvents used have to be immiscible with water (Zhao et al. 2006; Sarafraz-Yazdi and Amiri 2010; Garbi et al. 2010; Pano-Farias et al. 2017). Two important drawbacks of direct immersion single-drop microextraction are the limited size (typically 1–3 μL) and instability of the droplet when fast stirring is used. A number of recent advances in this technique have focused to overcome these problems, such as the use of support materials to stabilize the extraction solvent and adaptors for the normal syringe needle (Kokosa 2015). One of the strategies developed is the use of bubble-in-drop technique, which consists in the incorporation of a constant volume of air into the solvent drop (Williams et al. 2011). The solvent is drawn into a microsyringe, followed by air, and when the syringe is depressed to expel the contents into the sample solution, a bubble of solvent containing an air bubble is produced at the tip. As long as the solvent is immiscible in the sample solution, it will remain at the end of the tip until it is drawn back into the syringe. For solvents denser than water, the bubble incorporated into the drop stabilizes it due to its buoyancy, making it less likely to fall from the needle. Moreover, the surface area of the drop is increased, which enhances the extraction efficiency. This methodology was used for the analysis of metolochlor and atrazine herbicides using gas chromatography-mass spectrometry. A 0.5 μL air bubble in a 1 μL volume of chloroform was used for the extraction procedure. The results gave limits of detection from 0.024 ng L^{-1} for atrazine to 0.013 ng L^{-1} for metolochlor (Williams et al. 2014).

In headspace single-drop microextraction, volatile compounds are extracted by placing the drop of solvent generated in a microsyringe in the headspace of the sample solution or a flowing air sample stream. Therefore, the use of water and water-miscible solvents is possible in this mode of single-drop microextraction, which significantly increases the range of both extractable analytes and analytical methods that can be coupled to headspace single-drop microextraction. Moreover, in headspace single-drop microextraction technique a very efficient cleanup of the sample is carried, so that it is a suitable extraction technique for complex matrices. Salemi et al. (2013) successfully applied this technique for extraction of organophosphorus pesticides from soil and compared the results obtained with those provided by using direct immersion single-drop microextraction. According to the authors, direct contact of the extraction phase with the sample would lead to at least three main drawbacks: increasing risk of solvent droplet contamination with interfering non-volatile sample components and also solid particles; decreasing of stability of the solvent drop; and limitation in the range of some experimental parameters, such as agitation power or speed, water/soil ratio and modifying solvent due to the presence of the solvent drop in the sample phase. The limits of detection achieved ranged from 0.1 to 2.0 ng g^{-1} and the method was also reproducible with relative standard deviations of 2.1–6.9%.

Originally, relatively low-viscosity/low-boiling organic solvents, such as toluene, chloroform, cyclohexane, and 1-octanol, were used in single-drop microextraction because of being compatible with gas chromatography. Solvent volatility resulted in the evaporation of the drop, especially when elevated extraction temperatures were used. Later, the replacement of organic solvents for microextraction of organic compounds by ionic liquids, ionic media containing organic and inorganic anions coupled with organic cations, was proposed. These solvents have high viscosity and high boiling point, which reduces evaporation, allows the use of larger and more reproducible extraction volumes, and allows long-time headspace extraction even at high temperature (Liu et al. 2003). Moreover, ionic liquids can be considered as '*green*' solvents and are easily synthesized or commercially available. Headspace single-drop microextraction employing ionic liquids as extraction solvents has been successfully used for extracting pesticides. 1-butyl-3-methylimidazolium hexafluorophosphate ([C4MIM][PF6]) was employed as an extraction solvent for the analysis of dichlorodiphenyltrichloroethane (*p,p'*-DDT and *o,p'*-DDT) and its metabolites in water samples by high-performance liquid chromatography with ultraviolet detection (Ye et al. 2006). Zhang and Lee (2010) also developed a headspace single-drop microextraction method using the same ionic liquid for extraction, followed by gas chromatography with electron capture detection, which was applied to the analysis of organochlorine pesticides in soils.

Recently, single-drop microextraction has emerged as one of the simplest and most easily implemented forms of micro-scale sample pre-concentration and cleanup for pesticide analysis because it provides high selectivity, good quantitation, low limits of detection, no carryover, and minimal sample preparation.

2.2.6.2 Hollow-Fiber Microextraction

Pedersen-Bjergaard and Rasmussen (1999) introduced an innovative alternative to single-drop microextraction with the aim of improving the reliability of this technique and avoid one of its main shortcomings, instability of the extracting phase. They proposed to use a porous polypropylene hollow-fiber membrane as a support for the extraction solvent and as an interface between the donor and acceptor phases. In this technique, called hollow-fiber membrane liquid-phase microextraction, hollow-fiber membrane may be either a u-shape with both ends connected to guiding tubes or a rod with a closed bottom. Hollow-fiber membrane liquid-phase microextraction can be performed in either the dynamic or static mode. In the static mode, the hollow-fiber membrane is immersed into the aqueous sample with the aid of a syringe, whereas in the dynamic mode the syringe with the hollow-fiber membrane is connected to a syringe pump to perform the extraction (Sarafraz-Yazdi and Amiri 2010).

Prior the extraction, the hollow-fiber membrane is immersed in an organic solvent for a few seconds to fill the wall pores with the solvent, so that it provides a supported liquid membrane. An aqueous acceptor solution is then held within its lumen. To ensure that the organic solvent remains in the wall pores of the hollow-fiber membrane during extraction without leakage to the aqueous sample, it has to be water-immiscible. Analytes are extracted first into the intermediary organic phase, represented by the membrane, and then subsequently into the aqueous phase. This system can be considered three-phase hollow-fiber membrane liquid-phase microextraction (liquid-liquid-liquid) and is preferable for extraction of ionizable compounds, which usually are then analyzed by liquid chromatography or capillary electrophoresis (Fig. 2.3b). Several applications of three-phase hollow-fiber membrane liquid-phase microextraction employing the dynamic mode have been reported for the determination of phenoxyacetic acid herbicides (Wu et al. 2005; Chen et al. 2006).

Another mode of hollow-fiber membrane liquid-phase microextraction is based on a two-phase system in which the organic solvent fills both the hollow-fiber

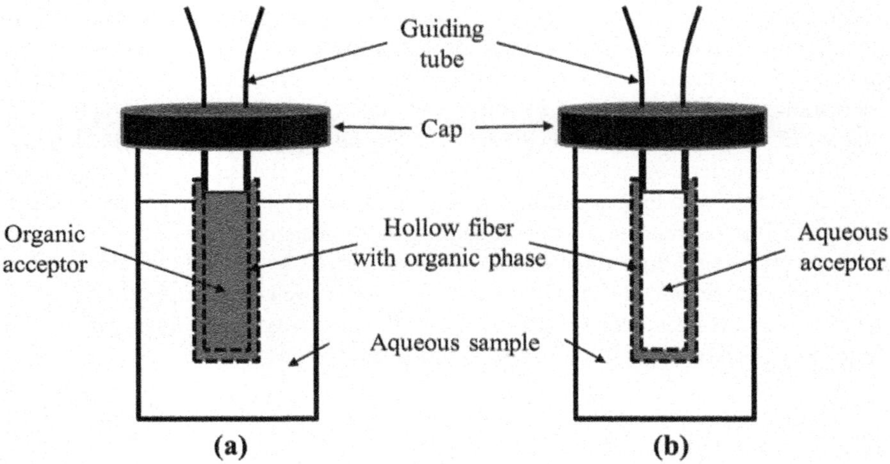

Fig. 2.3 Principle of hollow-fiber liquid-phase microextraction: (a) two-phase; (b) three phase (modified after Han and Row 2012**).** (**a**) The acceptor phase solution is the same as organic solvent of the hollow-fiber and both are immiscible in water; (**b**) an aqueous–organic–aqueous system in which the immobilized organic solvent is exposed to two aqueous phases of sample solution and aqueous acceptor phase located inside the hollow-fiber

membrane wall pores and lumen (liquid-liquid). The two-phase mode is generally employed for hydrophobic compounds, following by gas chromatography analysis (Lee et al. 2008; Sun et al. 2011) (Fig. 2.3a). An interesting two-phase hollow-fiber membrane liquid-phase microextraction methodology is the use of the organic solvent immobilized only in the hollow-fiber membrane pores as acceptor phase, which is desorbed after the extraction for further gas chromatography analysis. This procedure could be an alternative to the three-phase system when the fiber can be desorbed in a compatible solvent for liquid chromatography analysis, minimizing fiber handling because of the acceptor is not injected into the lumen, and only a desorption step is necessary prior chromatographic analysis (Bolaños et al. 2008). Two-phase hollow-fiber membrane liquid-phase microextraction has been applied to the analysis of pyrethroid (San Román et al. 2012), carbamate (Zhang and Lee 2006), and organochlorine (Basheer et al. 2002) pesticides in water samples. Dynamic hollow-fiber membrane liquid-phase microextraction has also been employed for the extraction of pesticides. Chen and Huang (2006) proposed a dynamic hollow-fiber membrane liquid-phase microextraction method for the extraction of organophosphorus pesticides from lake water. The extraction only required 3.5 μL of extraction solvent and 20 mL of water sample, and recoveries ranging from 83.1% to 107.0% were achieved. The proposed procedure showed great capability and stability for the analysis of field samples and the extraction process was affected little by the sample matrix.

Solvents such as toluene, cyclohexane, 1-octanol and *n*-hexane are usually used as organic solvents in hollow-fiber membrane liquid-phase microextraction for pesticide extraction. Toluene is most often used as an acceptor phase in this technique due to it is immiscible with water, remains stable in the hollow-fiber membrane over the extraction time, and provides high solubility for a wide range of analytes (Lambropoulou and Albanis 2007). This solvent has been widely applied for extraction of all types pesticides, such as organophosphorus (Lambropoulou and Albanis 2005), organochlorine (Basheer et al. 2004), carbamates (Lambropoulou and Albanis 2005), and triazine herbicides (Shen and Lee 2002). Octanol has been used for extraction of more polar pesticides such as carbamates (Asensio-Ramos et al. 2012) or phenoxy herbicides (Peng et al. 2011). More recently, ionic liquids have also been proposed as extracting solvents for hollow-fiber membrane liquid-phase microextraction (Ebrahimi et al. 2011).

All hollow-fiber membrane liquid-phase microextraction methodologies have proven to be attractive as alternatives to single-drop microextraction since, besides being nearly solvent-free, simple, fast, disposable, and low cost, the hollow-fiber membrane can accommodate larger volumes of acceptor phase than droplets, which usually provides higher sensitivity and reproducibility. In addition, the small pores of the membrane function as a filter that prevents larger molecules and interfering compounds in the sample solution from being extracted into the organic solvent. Therefore, hollow-fiber membrane liquid-phase microextraction is not only a good sample enrichment technique, but also an interesting sample cleanup procedure (Sun et al. 2011; Menezes et al. 2016).

2.2.6.3 Dispersive Liquid-Liquid Microextraction

Dispersive liquid-liquid microextraction was originally introduced by Rezaee et al. (2006) for water samples, although afterwards it has been widely demonstrated its applicability to different matrices, such as foodstuffs and soils. This technique has also been successfully applied to the specific case of extraction and concentration of a wide variety of pesticides. Dispersive liquid-liquid microextraction is based on the formation of a cloudy solution after quickly injecting and stirring a blend of extraction and disperser solvents into an aqueous sample solution (Fig. 2.4). The large contact surface between the extractant droplets and the sample solution achieve an effective and rapid mass transfer (Ahmad et al. 2015). The extraction operation can be accomplished using a cone bottom centrifugal pipe. After centrifuging the mixture, the extraction solvent containing the analytes settles at the bottom of the pipe or collects on the top of the extraction vessel based on its density. Dispersive liquid-liquid microextraction is usually coupled with gas chromatography or liquid chromatography.

In dispersive liquid-liquid microextraction, the extraction and disperser solvents used usually are nonpolar water-immiscible and polar water-miscible, respectively. Water immiscible and high density extracting solvents have been the commonly used extraction solvents for pesticides, such as chlorobenzene (Berijani et al. 2006; Zhao et al. 2007), chloroform (Liu et al. 2009), and carbon tetrachloride (Zhao et al. 2008). Because these solvents are highly toxic and are limited in extracting various

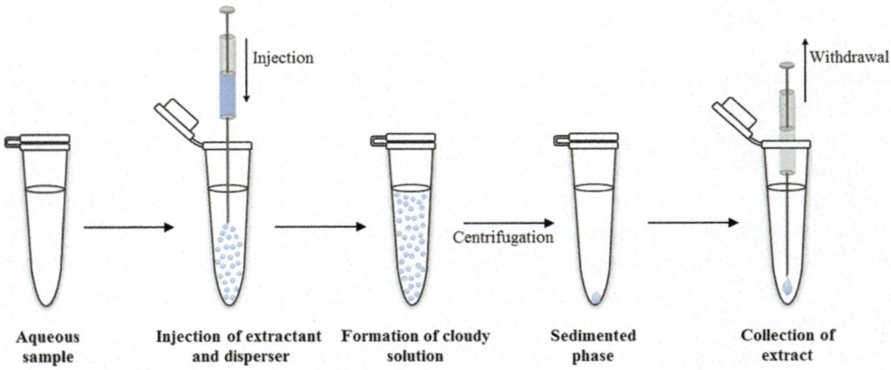

Fig. 2.4 Principle of dispersive liquid-liquid microextraction. A mixture of the extraction solvent and the disperser solvent with high miscibility in both organic and aqueous phases is rapidly injected into the aqueous sample and a cloudy solution is then formed. This allows an instantaneous partitioning of analytes from the aqueous sample into a small volume of the extraction solvent due to the large surface area created by the numerous fine droplets. The cloudy solution is centrifuged and the extraction solvent is easily recovered for analysis

analytes with different polarities, with the development of dispersive liquid-liquid microextraction polar and low-density organic solvents and some environmentally friendly ionic liquids have also been used as extractants, such as *n*-hexane (Moinfar and Hosseini 2009), cyclohexane (Farajzadeh et al. 2009), undecanol (Sanagi et al. 2012), 1-octyl-3-methylimidazolium hexafluorophosphate (Liang et al. 2013), and 1-benzyl-3-methylimidazolium bis[(trifluoromethyl)sulfonyl]imide (Wang et al. 2016). Several homemade special extraction devices were used for collecting organic solvents with a density lower than water after extraction. The task of dispersant in dispersive liquid-liquid microextraction is to disperse the extraction solvent into small organic droplets, so that the volume of the extractant droplet decreases, and the surface area of the extractant increases. Acetone (Berijani et al. 2006) and acetonitrile (Zhao et al. 2007) have been the main options as disperser solvents.

A new mode of dispersive liquid-liquid microextraction based on solidification of floating organic droplet appeared to solve the extraction device problem for low-density organic solvents (Wang et al. 2019). The extractant used in this mode is a low density and low melting point organic solvent (10–30 °C), and no specific holder is necessary to collect it. After the extraction process, the extraction device is placed in an ice bath to solidify the extractant phase, and this latter is collected and melted for further analysis (Leong and Huang 2009; Wu et al. 2010). However, an additional cooling step is required in this methodology.

The main drawback of dispersive liquid-liquid microextraction is the consumption of large volumes (i.e. mL) of disperser solvent and that its presence increases solubility of analytes in the aqueous phase, which results in relatively low extraction efficiency. Several progresses have been made to overcome this, such as the introduction of ultrasound-assisted dispersive liquid-liquid microextraction. In this variant of dispersive liquid-liquid microextraction, under the action of ultrasonic waves, an emulsion system of extractant/water is formed only by ultrasonic treatment without dispersant (Regueiro et al. 2008; Liang et al. 2013). A new dispersive liquid-liquid microextraction mode called vortex-assisted liquid–liquid microextraction was introduced by Yiantzi et al. (2010), which made use of vortex mixing for the dispersion of low density extraction solvents in water. The floating extractant phase could restore its first single-drop shape after extraction and be collected. However, among the solvents tested, only 1-octanol was able to restore its single-drop shape. This approach was later successfully applied for the extraction of pesticides in water samples by using hexane as an extraction solvent. Acceptable recoveries over the range of 72–106.3% were obtained (Jia et al. 2010). Another interesting modification of dispersive liquid-liquid microextraction was developed, which consisted in performing the dispersion of the extraction solvent in the sample solution during suction/injection cycles with a glass syringe. This disperser solvent-free methodology was called air-assisted liquid-liquid microextraction (Farajzadeh and Mogaddam 2012). Air-assisted liquid-liquid microextraction was applied for the assessment of triazole pesticides residues in surface water, cucumber, tomato and grape juices samples (Farajzadeh and Khoshmaram 2013). The proposed methodology eliminates both drawbacks of dispersive liquid-liquid microextraction by using a low-density extraction solvent along with the formation of a cloudy state in the absence of a disperser solvent. Other recent applications of air-assisted liquid-liquid microextraction are the extraction of organophosphorus pesticide residues in fruit juice samples (You et al. 2013), fungicides from environmental water and juice samples (Wu et al. 2015), triazole and triazine pesticides from honey (Farajzadeh et al. 2014) or triazole pesticide residues in edible oils (Farajzadeh et al. 2015).

The large number of publications is evidence that dispersive liquid-liquid microextraction is principally applied to pesticide residue analysis (Ahmad et al. 2015) and that they are mostly analyzed in aqueous samples. Gas chromatography is the most appropriate technique for further separation and determination of pesticides after dispersive liquid-liquid microextraction, although liquid chromatography has also become a popular technique for this purpose. However, in liquid chromatography a reconstitution of the extract in a solvent more compatible with liquid chromatography columns is usually required. Other techniques like micellar electrokinetic chromatography have also been successfully used in combination with dispersive liquid-liquid microextraction (Zhang et al. 2010, 2011).

Dispersive liquid-liquid microextraction has also been successfully applied for extraction of pesticides in solid samples. Examples of these applications are the analysis of carbamate pesticides in apple (Zhang et al. 2010), organophosphorus pesticides in fruits and vegetables (Ho et al. 2013), sulfonylurea herbicides in soil (Zhang et al. 2011), neonicotinoid insecticides in cucumber (Zhang et al. 2012a), carbamate pesticides in watermelon and tomato (Liu et al. 2012) or pyrethroids in soil (Wang et al. 2012b). In most cases when samples are solid such as vegetables and fruits, their juices are extracted and the extraction procedure is performed on these juices. However, a portion of analytes may remain in the refuse of samples and be discarded. Very recently, to overcome this difficulty some authors proposed to combine QuEChERs and dispersive liquid-liquid microextraction for the extraction and preconcentration of analytes (Farajzadeh et al. 2019), so that the adsorbed analytes onto the refuse of the samples are desorbed by the same solvent that is used as dispersant (acetonitrile) in the following preconcentration procedure. Thus, the combination of dispersive liquid-liquid microextraction with QuEChERs for sample preparation widens its application to those samples with complex matrices.

Dispersive liquid-liquid microextraction is a very versatile technique, since it can be applied to nearly all classes of compounds and combined directly or after solvent replacement with most pre-concentration methodologies. On the other hand, dispersive liquid-liquid microextraction for analysis of pesticides shows several advantages, such as high recoveries and high enrichment factors, low consumption of organic solvents, low cost, ease to perform, and short time of analysis.

2.3 Solid-Phase Extraction Techniques

2.3.1 Solid-Phase Extraction

Solid-phase extraction was introduced in the mid-1970s and came to prominence in the 1980s. In the last years, solid-phase extraction has been one of the most commonly used sorbent techniques for pesticide residue analysis. It has been proven that solid-phase extraction shows several important advantages over traditional liquid-liquid extraction, such as reduction of solvent consumption, shorter time of analysis, no phase emulsions, better enhancement factors, and more effective removal of interfering species. Both solid-phase extraction and liquid-liquid extraction are frequently combined to achieve better enrichment and cleanup. Extraction is performed by passing the sample through a particulate or monolithic sorbent packed into columns of short length, usually called 'cartridges', or immobilized in the form of a thin disc. Solid-phase extraction can be used either manually or with greater throughput using automated workstations.

There is a wide choice of sorbents for solid-phase extraction, including polar, nonpolar and mixed mode sorbents that provide the selectivity necessary to obtain clean extracts for pesticide analysis. The conventional sorbents used in solid-phase extraction are chemically modified silica gel (C_2, C_8, C_{18}) graphitized or porous carbon, and polymer sorbents (Rastrelli et al. 2002; Poole 2003; Zheng et al. 2014), which show low retention for polar compounds. More recently, hydrophilic polymeric sorbents, such as Oasis HLB or Porapak RDX, have been used which, as a general rule, have a high specific surface area and improve the extraction efficiency of polar compounds (Fontanals et al. 2005; Bratkowska et al. 2010).

Most developments in solid-phase extraction technique have been directly related to the improvement of sorbent materials. Efforts have been made to make magnetic the sorbent material, which permits an easy removal of the extraction medium from the sample solution by using a magnet (Wan Ibrahim et al. 2015; Bagheri et al. 2016). Further developments incorporated nanoparticles that were shown effective due to the high surface area to volume ratio, which allows rapid adsorption of analytes (Yu and Yang 2017). Rashidi Nodeh et al. (2017) synthesized a graphene-based tetraethoxysilane-methyltrimethoxysilane sol-gel hybrid magnetic (Fe_3O_4@G-TEOS-MTMOS) nanocomposite that was applied as adsorbent in solid-phase extraction for simultaneous determination of polar and nonpolar organophosphorus pesticides from different water samples. Good recoveries between 83% and 105% were obtained. New materials as carbon nanotubes, both multi-walled and single-walled carbon nanotubes, have also been successfully applied for cleanup purposes (López-Feria et al. 2009). These materials have an extremely large surface area and unique structure, so that they have excellent adsorption ability. Carbon nanotubes have been proposed as a kind of sorbents in a packed column, and applied for extraction of pesticides from water samples (Zhou et al. 2006; Wang et al. 2007a; El-Sheikh et al. 2008). Du et al. (2008) developed a new analytical method to determinate organophosphate pesticides in garlic by extraction with multi-walled carbon nanotubes. Ravelo-Pérez et al. (2008b) also demonstrated that multi-walled carbon nanotubes can be used as effective solid-phase extraction materials for the extraction of pesticides from apple, grape, orange and pineapple fruit juices.

On the other hand, molecularly imprinted polymers have also been used for selective solid-phase extraction, since they yield specific binding sites within a polymeric matrix and are also helpful in cleaning up complex matrices. The first attempt on the development of a dual methodology for the trace analysis of pesticide residues based on molecularly imprinted polymer technology was recently reported. The method was developed for the simultaneous analysis of dimethoate and terbuthylazine residues in olive oil samples, and was based on the use of a dual layer of 'tailor-made' molecularly imprinted polymers as solid-phase extraction sorbent. High recovery rates for dimethoate (95%) and terbuthylazine (94%) were achieved with good accuracy and precision (Garcia et al. 2016).

The introduction of new sorbents with more selective modes of attraction, novel product formats such as the solid-phase extraction disc, and the proliferation of automated techniques for performing the extractions, ensure that solid-phase extraction will continue to be a preferred technique for sample preparation in many different analytical disciplines.

2.3.2 *Dispersive Solid-Phase Extraction: QuEChERS*

Solid-phase extraction technique includes several steps, is time-consuming and requires much organic solvent volume than other modern techniques recently reported. Moreover, sometimes several packed columns with different sorbents are required to obtain good results. To overcome these drawbacks, Anastassiades et al. (2003) developed a simple and rapid cleanup technique for fruits and vegetables named dispersive solid-phase extraction. In this new approach, the sample extract was not passed through a solid-phase extraction column but the matrix interfering species were removed by adding an amount of sorbent material to the extract. Then, the sorbent was separated from the extract by centrifugation. The main advantages of this new methodology compared with solid-phase extraction were: the possibility of using a much smaller quantity of sorbent and solvent, a shorter time of analysis, and a simpler procedure (Fontana et al. 2011). In the original method proposed by Anastassiades et al. dispersive solid-phase extraction was combined with a previous extraction, giving rise to the QuEChERS method, which is defined as a simple, fast, safe, rugged, effective and inexpensive method for the multiclass or multiresidue analysis of pesticides. Over the years, dispersive solid-phase extraction has been used as an independent cleanup methodology (Nguyen et al. 2010; Dias et al. 2016), although in most of the cases, has been applied as a part of QuEChERS method.

QuEChERS combines the extraction/isolation and extract cleanup of both non-polar and polar pesticides simultaneously from food matrices. The optimized procedure consisted of several steps: (1) an initial extraction step in which the minced and homogenized sample (10 g) was extracted with acetonitrile (10 mL); (2) an extraction/partitioning step by means of the addition of salt mixture (4 g of anhydrous $MgSO_4$ and 1 g of NaCl) and; (3) a dispersive solid-phase extraction cleanup step of an aliquot of the extract (1 mL) with 25 mg of N-propylethylenediamine (primary secondary amine) and 125 mg of $MgSO_4$. Finally, the extract was directly transferred to a vial for gas chromatography-mass spectrometry analysis. In these working conditions, recovery values between 85% and 101% for more than 95% of pesticides were achieved. Although the usual solvents for analysis of pesticides in food matrices have been ethyl acetate (Fernandez-Alba et al. 1994; Mol et al. 2007; Banerjee et al. 2007), acetone (Guan et al. 2010) and acetonitrile (Fillion et al. 2000; Lehotay et al. 2001), this latter was selected in the QuEChERS methodology as the most appropriate solvent due to its clear advantages over the others. Although acetonitrile is miscible with water, a phase separation is possible by addition of salts. Besides, the remaining water can be easily eliminated by adding a drying agent like $MgSO_4$. However, acetone cannot be separated from water without using nonpolar solvents. As for ethyl acetate, it can be easily separated from water since both solvents are partially miscible, but it is not a good solvent for most very polar pesticides. Additionally, many lipophilic compounds such as fats, pigments and waxes are not extracted in acetonitrile compared to acetone and ethyl acetate (Wilkowska and Biziuk 2011). Different salts were evaluated to induce a phase separation in the extraction/partitioning step. The most effective salt with respect the recovery of analytes resulted to be $MgSO_4$. Moreover, the addition of this salt facilitated the partition of very polar pesticides like acephate, methamidophos or omethoate into the organic phase, since the volume of the aqueous phase was effectively reduced. Selectivity of the extraction procedure, that is, the polarity range of the method, can be controlled through the amount of NaCl added in the partitioning step with $MgSO_4$. The cleanup of the extract using dispersive solid-phase extraction with primary secondary amine allowed efficiently removing many polar interfering species coming from the matrix like organic acids, fatty acids, some sugars and polar pigments.

Liquid chromatography was not used in the initial QuEChERS method, thus the applicability of this approach to very polar and thermally labile pesticides was not evaluated. Later, Lehotay et al. (2005a) continued the researches of Anastassiades et al. and introduced some modifications to validate the application of the QuEChERS methodology to the analysis of 229 pesticides in lettuce and orange matrices using concurrent large volume injection (LVI)/gas chromatography-mass spectrometry and liquid chromatography-mass spectrometry in tandem detection. In the original method the first extraction step was carried out in unbuffered solutions. Under such conditions, the use of primary secondary amine in the cleanup step can originate losses of pH-sensitive pesticides (base-labile compounds) due to their hydrolization under basic conditions. To overcome this drawback, Lehotay et al. (2005a) added 0.1% glacial acetic or formic acid solutions after the extraction process. Later, different modifications of the QuEChERS method were proposed for improving stability of pH-dependent pesticides in the final extracts. As a consequence of these efforts, two official methods for pesticide residue analysis were developed, AOAC Official Method 2007.01 (AOAC 2007) and CEN Standard Method EN 15662 (European Committee for Standardization 2008), in which acetate and citrate buffers (pH = 5) are included in the extraction step, respectively. Both methods allowed obtaining recoveries higher than 70% for acid-sensitive pesticides without affecting the base-sensitive pesticides.

Since its introduction, the QuEChERS method has undergone many other modifications with the purpose to improve sensitivity and recovery of specific types of pesticides, and reduce the matrix effect. There are two factors mutually related to each other that play a relevant impact on the recovery of the target compounds in QuEChERS, as follows: (a) the solvent used for extraction and (b) the type and amount of salts and sorbents for dispersive solid-phase extraction. Other solvents as substitutes for acetonitrile have been tested for the extraction step, such as hexane/acetone or acetonitrile/hexane (Cheng et al. 2016). Regarding the salts for partitioning and salting-out effects, $MgSO_4$ (0.5–10 g) and NaCl (0.2–6 g) are the most frequently used (Santana-Mayor et al. 2019) although other salts such as $CaCl_2$ and Na_2SO_4 have also been tested. For dispersive solid-phase extraction step, the sorbent is chosen to retain the matrix components and allow the analytes of interest to remain in the liquid phase.

Although the QuEChERS method was initially developed for samples with high content in water (80–95%) such as vegetables and fruits, afterward it has been applied to other matrices. In the case of samples with water content <80% it is necessary the addition of water at the beginning of the process. A method was developed for the analysis of pesticide residues in foods containing up to 20% fat (milk, avocado and eggs). In this kind of foods, both hydrophilic and lipophilic pesticides could be found and, therefore, the development of a methodology able to simultaneously determine analytes with a wide range of polarity was essential (Lehotay et al. 2005b). The authors demonstrated that the combined use of primary secondary amine with octadecylsilane in these matrices was especially effective, since this latter removes non-polar interfering species like lipids. An additional freeze-out step to reduce interfering lipids and, in addition, other matrix component co-extractives with poor solubility was also proposed. Nevertheless octadecylsilane demonstrated to be faster, easier, and equally effective in removing lipids

(Walorczyk 2008). Another interesting alternative is the use of chitin, a new cleanup sorbent introduced by Cerqueira et al. (2014), which exhibits various bioactivities and functionalities. The authors proposed it for the dispersive solid-phase extraction step of the QuEChERS process for the simultaneous extraction of simazine, atrazine, clomazone and tebuconazole, and other organic contaminants in drinking water treatment sludge samples by employing liquid chromatography-tandem mass spectrometry. This sorbent has shown to be more efficient than the combination primary secondary amine/octadecylsilane for multiclass pesticide residue analysis in matrices such as fish liver and muscle, since it reduced background effectively without significantly reducing recovery of lipophilic pesticides such as organochlorine and some pyrethroids (Kaczyński et al. 2017). Moreover, it is cheaper because it is obtained from natural sources.

Another important aspect to take into account was the co-extraction of pigments in some samples, such as spinach and other highly pigmented vegetables. The addition of a small amount of graphitized carbon black (as in the CEN Standard Method EN 15662 (European Committee for Standardization 2008) combined with primary secondary amine improved the results, although the strong affinity of graphitized carbon black for planar molecules (chlorophyll, carotenoids or sterols) also reduced in approximately a 25% the recovery of structurally planar pesticides (i.e. thiabendazol, terbufos, hexachlorobenzene, etc.). An alternative to graphitized carbon black, a cross-linked polymeric sorbent called ChloroFiltr®, was designed that effectively eliminated chlorophyll without affecting the planar pesticides (Łozowicka et al. 2017; Rutkowska et al. 2018). Multi-walled carbon nanotubes have also been proposed as effective sorbents in the dispersive solid-phase extraction step of the QuEChERS method to remove pigments from vegetables (Zhao et al. 2012) owing to its unique structure and the huge surface area (Iijima 1991; Ren et al. 1998). The method was validated for determination of 30 representative pesticides in cabbage, spinach, grape and orange spiked at concentration levels of 0.02 and 0.2 mg kg^{-1}. The recoveries were between 71% and 110%, with relative standard deviations <15%.

The QuEChERS method has also been applied to the pesticide residue analysis in high fat content foods such as almond, avocado, edible vegetable oils or olives (Cunha et al. 2007; Lozano et al. 2014; He et al. 2017; Manav et al. 2019). In these matrices, analytes are mainly nonpolar, so there is no need to develop analytical methods for determining polar pesticides. The main components of vegetable oils are lipids, pigments and fatty acids. Therefore, the sample treatment in these highly complex matrices for pesticide residue analysis is still a challenge, since the high molecular-mass fat from the sample has to be completely removed in order to avoid damages to the chromatographic system. A modified QuEChERS method using amine modified graphene (CH_3NH-G) material for the dispersive solid-phase extraction step was proposed by Guan et al. (2013) and applied to the determination of pesticide residues in four oil crops obtaining recovery percentages between 70.5% and 100% and low limits of detection (0.1–8.3 μg kg^{-1}). These amine modified graphene materials have a high surface area and amine groups forming strong hydrogen bonds with carboxylic acid groups. Therefore, they are effective cleanup sorbent in pesticide multiresidue analysis in high-fat (>20%) commodities and show better performance than primary secondary amine, graphitized carbon black and multi-walled carbon nanotubes (Hakme et al. 2018).

Recently, new sorbents based on zirconium oxide such as zirconium oxide-based sorbent (Z-Sep), zirconium oxide-based sorbent/octadecylsilane (Z-Sep$^+$) and yttria-stabilized zirconium dioxide nanoparticles (ZrO_2/Y_2O_3) have demonstrated to have a higher capacity than primary secondary amine, octadecylsilane and graphitized carbon black sorbents to remove fatty acids, esters of fatty acids, carboxylic acids, sterols and pigments from high oil matrices (Lozano et al. 2014). The new lipid-removal sorbent zirconium oxide-based sorbent/octadecylsilane (Z-Sep$^+$) was evaluated for the analysis of carbamate pesticides in edible vegetable oils and high-fat cheeses by tandem mass spectrometry combined with ultra-high performance liquid chromatography, providing a significant removal of co-extractive interferences and excellent recoveries (Moreno-González et al. 2014; Hamed et al. 2017). In this commercially available product zirconium binds to the polar groups of the lipids through Lewis acid-base interactions and octadecylsilane binds hydrophobic chains of the lipids (Hakme et al. 2018). It is necessary to take into account that with this type of sorbents nonpolar pesticides could also be removed in the cleanup step (Moreno-González et al. 2014; Tuzimski and Rejczak 2016).

One of the most novel sorbents for dispersive solid-phase extraction, EMR-Lipid, was introduced by a vendor for removing interfering lipids from samples while leaving analytes behind. The structure of EMR-Lipid is a proprietary secret, and its action mechanism in dispersive solid-phase extraction process involves a combination of hydrophobic interactions and size exclusion between the long chain of lipids and the sorbent. It dissolves to saturation in the extracts and then, the lipid-EMR-Lipid complex is precipitated out or remains in the aqueous phase during the salting-out step (Han et al. 2016). Parrilla-Vázquez et al. (2016) and Dias et al. (2016) used EMR-Lipid as a highly selective sorbent to remove lipids in the multiresidue analysis of pesticides in edible vegetable oils. The proposed procedures showed good recoveries (70–20%), low limits of quantitation (10–50 μg kg^{-1}) and low relative standard deviations (<20%). The EMR-Lipid showed as the only disadvantage the use of a higher amount of the extract when compared to the other approaches (primary secondary amine and zirconium oxide-based sorbent (Z-Sep)). Han et al. (2016) also evaluated the novel commercial EMR-Lipid product for cleanup of kale, salmon, avocado, and pork extracts for multiclass, multiresidue analysis of 65 pesticides and other environmental contaminants. The sorbent efficiently removed 79–98% of co-extracted matrix compounds, providing clean extracts and low background in gas chromatography-tandem mass spectrometry analysis. Moreover, up to a 76% of the co-extracted chlorophyll was removed without loss of co-planar pesticides. The use of 'Analyte Protectants' after dispersive solid-phase extraction with EMR-Lipid sorbent has been evaluated in order to minimize the matrix effect during injection into the gas chromatography system. Analyses of pesticides by gas chromatography are frequently affected by degradation problems and peak tailing associated with the active sites on the surface of the stationary phases. 'Analyte Protectants' are mixtures of compounds containing multiple hydroxyl groups, whose function is to mask these active sites by making hydrogen bonds through their hydroxyl groups and thus protecting the analytes (Sanchez Costa et al. 2018). Other novel dispersive solid-phase sorbents have been very recently developed by different companies, such as Cleanert® NANO or LipiFiltr®, whose efficiency in the improvement of the dispersive solid-phase step in the QuEChERS procedure will have to be assessed in the next years.

Nowadays, the expansion of the QuEChERS method and its establishment practically as a universal method for routine analysis of pesticide residues in food products has led to the commercialization by several companies of the different products involved in the procedure (salts, buffers, sorbents) in the form of kits, so that the time of analysis is significantly reduced.

2.3.3 Matrix Solid-Phase Extraction

Matrix solid-phase dispersion was developed by Barker et al. (1989) to carry out extraction of target analytes from solid matrices, and overcome the serious restrictions on solid-phase extraction methods for this kind of samples. The key characteristics for the success of matrix solid-phase dispersion are its feasibility, flexibility, versatility, low cost and rapidity and, over the past years, it has been widely used for extraction of pesticides from both vegetable and animal matrices (Wang et al. 2015). In this technique, the solid or semi-solid sample is mixed with an appropriate adsorbent with the aid of a mortar and pestle, which causes the disruption and dispersion of the analytes on the solid support. Then, the blend sample-adsorbent is added to a column/cartridge and the analytes are eluted with an appropriate solvent. An additional cleanup of the sample can be carried out by packing in the bottom of the same column/cartridge another adsorbent. Thus, this technique allows performing, on a small sample aliquot, extraction and cleanup in a single step. Even if matrix solid-phase dispersion has many characteristics in common with classical solid-phase extraction, the mechanism governing the two processes is quite different. The main difference between both techniques is the physical state of samples applied to the column, since whereas in solid-phase extraction samples are liquid (solution) in matrix solid-phase dispersion samples are solid. Generally, after matrix solid-phase dispersion, a liquid chromatography or gas chromatography separation is followed by mass spectrometry determination. Although matrix solid-phase dispersion was initially introduced for disrupting and extracting solid samples, it has also been applied to semi-solid and viscous samples. Otherwise from other classical extraction methodologies, matrix solid-phase dispersion does not require a large amount of sample, sorbents and organic solvents, is not laborious, is rapid, and more environmentally-friendly. An additional cleanup step can be required or not after the extraction depending on the analytes and the instrumental analytical technique used for the analysis.

Carbon based sorbent materials such as octylsilane and octadecylsilane are the most frequently used sorbents, since they provide a good disruption and dispersion of the sample because their lipophilic characteristics (Rodrigues et al. 2010). Other adsorbents such as Florisil (Łozowicka et al. 2012), graphitized carbon black (Barriada-Pereira et al. 2010) and silica have also been frequently used. Aminopropyl, followed by cleanup with Florisil, has also been successfully applied as sorbent material for multiresidue pesticide analysis in olives by gas chromatography–mass spectrometry and liquid chromatography–tandem mass spectrometry determination. The most recent innovation introduced in matrix solid-phase dispersion over the past few years has been the use of novel supporting materials, such as multi-walled carbon nanotubes and nanomaterials (Capriotti et al. 2013). Fang et al. (2009) decided to evaluate the efficiency of multi-walled carbon nanotubes in matrix solid-phase dispersion applications, and they use them as dispersant to extract 31 pesticides from agriculture samples before gas chromatography-mass spectrometry determination. The authors found that the extracts obtained by using multi-walled carbon nanotubes were cleaner than those obtained by using diatomite and octadecylsilane as dispersant materials, with recoveries above 71% for both apple and potato samples. Another interesting alternative to the classical matrix solid-phase dispersion is the use of hydrophobic magnetic octadecylsilane-based nanoparticles, which was applied for the multiresidue determination of 26 pesticides in carrots (Binellas and Stalikas 2015). The magnetic nanoparticles could be removed from the solution by *n*-octanol and application of a magnetic field. The extracts were then directly analyzed by gas chromatography-mass spectrometry without needing a previous cleanup step. The average extraction recoveries were 77–107% and the limits of quantitation for most pesticides were in the low μg kg^{-1} level. Very recently, other authors proposed, for the first time, the integration of magnetic ionic liquids into a matrix solid-phase dispersion procedure for the analysis of multi-class pesticide residues in potatoes and other vegetables (Chatzimitakos et al. 2018). The viscous nature of the selected magnetic ionic liquids assists in blending with the matrix, while their hydrophobic properties facilitate easier separation and retrieval. Moreover, they can be removed magnetically. The raw vegetables were analyzed without any other pretreatment prior to or after extraction, without the need for solid dispersing materials or co-sorbent for cleanup.

Of all the features of matrix solid-phase dispersion above described, it can be considered as a valid alternative to Soxhlet and microwave-assisted extraction, as well as pressurized liquid extraction and supercritical fluid extraction. In fact, compared to these techniques, matrix solid-phase dispersion requires mild extraction conditions (room temperature and atmospheric pressure) as well as it provides suitable yield and selectivity. Moreover, matrix solid-phase dispersion is a good alternative to solid-phase extraction for the analysis of solid samples. Development of new sorbent materials for improving the capacity or selectivity is still the exciting research area in this technique.

2.3.4 Solid-Phase Microextraction

Solid-phase microextraction is a sorption-based microextraction methodology in which a concentration equilibrium between the target analyte(s) in a sample and a sorptive phase occurs. It was introduced in 1990 (Arthur and Pawliszyn 1990), and since then it has undergone a rapid development. In most traditional approaches to solid-phase microextraction, the extraction of organic compounds can be performed in two different modes, by immersing an externally polymer-coated fiber in sample solution, the so-called direct immersion solid-phase microextraction, or exposing this fiber to the headspace above the samples, the so-called headspace solid-phase microextraction (Ai 1997). The solid-phase microextraction device is constructed of a thin fiber (<50 μm), which is used as substrate/support for a thin layer (5–100 μm) of a suitable sorptive (extraction) phase. Solid-phase microextraction uses a fiber coated with a liquid (polymer), a solid (sorbent), or a combination of both. Analytes are extracted from sample by absorption or adsorption, depending on the nature of the fiber coating. Coating thickness determines the volume and surface area of sorptive phase and consequently, the amount and rate of absorption. For further chromatographic analysis, the fiber can be directly put into the injection port of a gas chromatography instrument (thermal desorption) or immersed into an appropriate solvent (chemical desorption) for gas chromatography or liquid chromatography analysis. In-tube solid-phase microextraction is an automated version of solid-phase microextraction that allows its easy coupling with liquid chromatography, and it has proven to be a very efficient extraction method for the analysis of polar and thermally labile pesticides (Eisert and Pawliszyn 1997). This mode of solid-phase microextraction typically uses a gas chromatography capillary column with an appropriate extraction phase coated on its internal surface to extract the analytes (Gou et al. 2000).

Popularity of solid-phase microextraction within the analytical community has been continuously increasing since its first introduction, generating applications encompassing a multitude of compounds and matrices in a large variety of fields (Lashgari and Yamini 2019). This technique integrates sampling, extraction, concentration and sample introduction into a single step. Moreover, solid-phase microextraction requires minimal or zero amounts of solvent, which distinguishes it a 'greener', more environmentally-friendly technique in comparison to other sample preparation techniques. Therefore, solid-phase microextraction gathers most of the required features for pesticide analysis in food and environmental matrices, that is, to cover a wide scope of pesticide–matrix combinations, provide a certain grade of selectivity, be rapid, and be easy to carry out (Souza-Silva et al. 2015a, b). In fact, it has become a routine technique in many laboratories due to it can be fully automated, is easy to miniaturize, and is convenient in coupling with chromatographic instruments.

One of the key steps in the development of a solid-phase microextraction method for analysis of pesticides is the selection of an appropriate fiber coating based on their nature and volatility. Thin fibers are used for pesticides with high boiling points and thicker fibers are suitable for volatile pesticides. Polydimethylsiloxane (Neuwirthová et al. 2019) and poly(acrylate) (Fernandez-Alvarez et al. 2008) coated fibers have been used for years as general-purpose sorbents for pesticide residue analysis. Nowadays, a great variety of more specific fiber coatings based on solid sorbents is available in the market, that include: Carbowax/template resin (Blasco et al. 2011), polydimethylsiloxane/divinylbenzene (Ravelo-Pérez et al. 2008a; Zhang et al. 2018), and polydimethylsiloxane/poly(acrylate) (Fernandez-Alvarez et al. 2009) coated fibers. For over a decade since the introduction of direct immersion solid-phase microextraction, it only could be applied to extraction of analytes from very simple matrices or complex matrices subjected to multiple cleanup steps, since the available coating surfaces suffered an irreversible attachment of macromolecules from the matrix samples, which decreased the extraction efficiency. Solid coatings frequently suffer from an irreversible fouling effect or sorption of unwanted matrix components, which is less frequent in liquid coatings. Thus, for years the liquid polydimethylsiloxane coating was the most used sorbent for the direct immersion solid-phase microextraction of pesticide residues in food matrices (Beltran et al. 2000). Therefore, many solid-phase microextraction applications to food and environmental samples had to be carried out by headspace solid-phase microextraction mode since matrix compatible coatings were not available.

Liquid coatings are better than solid coatings in regards to the fouling effect, but these latter show higher extraction efficiency because the rough morphology of their surfaces. Moreover, liquid coatings are not good sorbents for semi-polar and polar compounds. To overcome these limitations and take advantage of the properties of both types of coatings, in the last years a great variety of porous sorbents have been prepared by combining liquid and solid coatings, which can withstand direct exposure to complex matrices and are described as 'matrix compatible' (Godage and Gionfriddo 2019). Souza Silva and Pawliszyn (2012) developed a new type of solid-phase microextraction fiber coating, polydimethylsiloxane/divinylbenzene/polydimethylsiloxane, which was obtained modifying commercial solid-phase microextraction fiber coatings by incorporation of an additional protective thin layer of polydimethylsiloxane. This coating was more robust, prevented wicking of matrix components and analytes into the fiber core, and allowed up to 100 extractions of triazole pesticides in water samples and grape pulp by direct immersion solid-phase microextraction with no sample treatment. The polydimethylsiloxane/divinylbenzene/polydimethylsiloxane coating also showed excellent robustness and matrix compatibility with oily matrices for the extraction of multiclass pesticides and other contaminants by direct immersion solid-phase microextraction in avocado pulp (De Grazia et al. 2017). Its extraction efficiency remained constant for more than 100 extraction runs, whereas the extraction efficiency of divinylbenzene/polydimethylsiloxane coating showed a severe decrease after only 20 successive extractions. Polydimethylsiloxane/divinylbenzene/polydimethylsiloxane coating has also been employed in the direct coupling of solid-phase microextraction to mass spectrometry for pesticide residue analysis in food and environmental samples (Mirabelli et al. 2018). This coating showed matrix compatibility and excellent antifouling properties, taking into account that this coupling solid-phase microextraction-mass spectrometry does not involve chromatography and, therefore, it is mandatory that the coatings prevent sorption of matrix components which could drastically affect the ionization process.

The use of solid-phase microextraction fibers presents some other limitations: coatings have usually poor thermal stability, so that the gas chromatography injector temperature has to be maintained below the maximum operating temperature of the coating (usually 240–280 °C); in organic solvents coatings are unstable and suffer swelling, which restricts their use with liquid chromatography; and fibers have to be conditioned before using, are fragile and have a short lifetime due to the irreversible sorption of high molecular weight compounds. In the past decade, there has been a growing interest in the use of polymeric ionic liquids, a novel class of solid-phase microextraction coatings (Yu et al. 2013). These sorbents present interesting features as high viscosity, thermal stability, and compatibility with gas chromatography. Moreover, ionic liquids are often referred as 'designer solvents', since millions of them can be formed through different combinations of anions, cations, and other functional groups. This unique property allows developing ionic liquids with an enhanced selectivity towards the target pesticides. In recent years, several applications of direct immersion solid-phase microextraction to analysis of pesticides in food and environmental matrices have been reported. Recently, Zhang et al. (2012b) reported a new ionic liquid, namely, 1-vinyl-3-hexadecylimidazolium hexafluorophosphate, for the determination of seven pyrethroids in vegetables by direct immersion into hexane extracts, followed by gas chromatography-electron capture detection. Multi-walled carbon nanotubes have also found application as solid-phase microextraction coating fibers. In a recent study, pyrethroid pesticide residues in tea brew were extracted without prior sample treatment with multi-walled carbon nanotubes solid-phase microextraction fibers, which were formed by a stainless steel core coated with polyacrylonitrile solution as a binder, providing recoveries between 94.2% and 107.3% for spiked samples (Ren et al. 2018). Nevertheless, the matrix compatibility of this new solid-phase microextraction coating still needs to be assessed.

The application of solid-phase microextraction for pretreatment samples with a complex composition, as a previous step before the chromatographic analysis of pesticides and other organic contaminants, is continuously increasing. Moreover, solid-phase microextraction is an environmentally-friendly sample preparation technique that meets the sustainability requirements, and concretely those derived from the application of green analytical chemistry. Owing to all these features, to its flexibility and possibility of being combined with a great variety of analytical techniques, it is expected that the evolution of this technique in the coming years will not have limits.

2.3.5 Stir Bar Sorptive Extraction

Stir bar sorptive extraction was introduced by Baltussen et al. (1999) to deliver more sorptive-phase mass and surface area. The extraction mechanism is similar to that of solid-phase microextraction, that is, a sorptive interaction of usually a coating of polydimethylsiloxane deposited on a magnetic glass with the target compounds (Fig. 2.5). Stir bar sorptive extraction was developed by Gerstel GmbH & Co. KG (Mülheim an der Ruhr, Germany) and commercialized under the trade name Twister®. The stir bar is immersed in the liquid sample to be analyzed, which is shaken for a given time. While stirring, the bar adsorbs the target compounds, and then it is removed from the sample, rinsed with deionized water and dried. Then, the retained compounds can be desorbed with a suitable solvent for gas chromatography or liquid chromatography analysis. On-line thermal desorption can be also used for thermally stable semi-volatile and volatile compounds and gas chromatography analysis.

Stir bar sorptive extraction overcomes the main drawback of solid-phase microextraction, which is the low amount of available polydimethylsiloxane phase (~0.5 μL; 100 μm film thickness). In stir bar sorptive extraction the amount of phase typically coated is 24–126 μL and, consequently, much higher recoveries can be reached and sensitivity is increased by a factor of 50 and 250, reducing detection limits to sub-ng L^{-1} levels (Chaves et al. 2007). Moreover, stir bar sorptive extraction shows good blank levels and no degradation after 100 extractions (Zuin et al. 2006). In recent years, stir bar sorptive extraction has become one of the most studied sample extraction techniques for analysis of organic compounds. Originally, this extraction technique was intended to screen priority organic micro-pollutants in water samples (León et al. 2003; Sandra et al. 2003), but over time, applications for almost every field of analytical chemistry have been developed (Campillo et al. 2010; Li et al. 2012). Organophosphorus pesticides in vegetables were determined

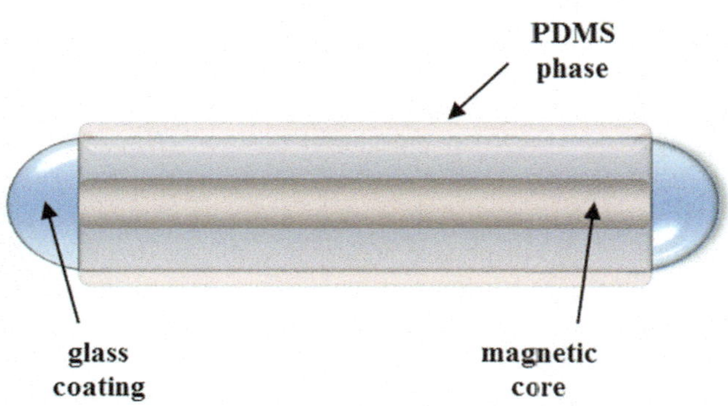

Fig. 2.5 Stir bar sorptive extraction device. Sampling is done by directly introducing the device into the aqueous sample and the solutes are extracted into a polymer coated on the stir bar. Analyte desorption is performed by thermal desorption or alternatively by solvent desorption using a small amount of an adequate organic solvent. *PDMS* polydimethylsiloxane

by stir bar sorptive extraction and gas chromatography with thermionic specific detection by using as extraction phase a hydroxy-terminated polydimethylsiloxane prepared by sol–gel methodology. The linear ranges were 0.05–50 ng g^{-1} with detection limits ≤0.15 ng g^{-1} (Liu et al. 2005). Sandra et al. (2001) used stir bar sorptive extraction followed by thermal desoption or liquid desorption for the analysis of dicarboximide fungicides in wines. A multiresidue method based on stir bar sorptive extraction followed by gas chromatography-mass spectrometry, by applying a new thermal desorption unit, fully automated, allowed the determination of more than 300 pesticides in baby food, vegetables and fruits (Sandra et al. 2003). Micellar electrokinetic chromatography–diode array analysis has also been applied in combination with stir bar sorptive extraction for the simultaneous extraction of pesticides (acrinathrin, bitertanol, cyproconazole, fludioxonil, flutriafol, myclobutanil, pyriproxyfen, and tebuconazole) in vegetables samples (Juan-García et al. 2005).

Stir bar sorptive extraction presents some drawbacks such as the limited range of analyte polarities for the available stationary phases, the presence of strong matrix effects, the requirement of a strict control of extraction parameters (Camino-Sánchez et al. 2014) and the complexity in the automation of the desorption step, especially to liquid chromatography (Chaves et al. 2007). At present, polydimethylsiloxane and ethylene glycol-silicone are the only commercially available phases in stir bar sorptive extraction. A formerly introduced poly(acrylate) coated Twister is no longer commercially available, since its applicability and robustness were limited (David et al. 2019). In order to overcome the limitations of polydimethylsiloxane phase, that reduces its use to nonpolar analytes, new coating materials are being developed, so widening the applicability of stir bar sorptive extraction. This interest is catalyzed in part by the need to develop methods for more polar emerging pollutants, such as polar pesticides. These novel materials include carbon adsorbents, molecularly imprinted polymers, sol–gel prepared coatings, restricted access materials, ionic liquids, microporous monoliths and dual phase materials (Abdulra'uf and Tan 2014). Unfortunately, most of the new coating materials can only be used in combination with solvent desorption, which limits the sensitivity of the method since only a fraction of the extract can be analyzed by gas chromatography or liquid chromatography. Large volume injection can partly compensate for this loss.

Recently, Benedé et al. (2014) introduced the use of magnetic nanoparticles coated dynamically with a magnetic stir bar, this technique being termed stir bar sorptive-dispersive microextraction. It combines the principles and benefits of stir bar sorptive extraction and dispersive solid-phase microextraction based on magnetic nanoparticles. At high stirring rate magnetic nanoparticles are dispersed into the sample solution (or into the desorbing solvent) and return to the magnetic bar when the stirring is stopped. Stir bar sorptive-dispersive microextraction with magnetically modified graphene (G-Fe_3O_4) as the sorbent was applied in isolation of seven model pesticides from water samples (boscalid, chlorpyrifos, deltamethrin, dimethenamid-P, dimoxystrobin, metazachlor and tebuconazol) (Madej et al. 2019). The developed method resulted more convenient than magnetic solid-phase extraction, performed in analogous conditions, in terms of sorbent collection and effective mixing of larger sample volumes. The multiclass pesticides analyzed could be detected in concentration of several nanograms per milliliter. Molecularly imprinted stir bars based upon the entrapment of modified magnetic nanoparticles within an imprinted polymer monolith have also been proposed for the stir bar sorptive extraction of triazines in soil sample extracts (Díaz-Álvarez et al. 2016). Although low recoveries were obtained (2.4–8.7%), the high selectivity provided by the new molecularly imprinted stir bars allowed reaching detection limits below 7.5 ng g^{-1} by liquid chromatography coupled to UV detection.

In the last decade, multiresidue methods using stir bar sorptive extraction have been described for pesticides, together with endocrine disruptors and other priority pollutants, either with thermal desorption combined with gas chromatography-mass spectrometry (Camino-Sánchez et al. 2012) or liquid desorption combined with liquid chromatography-mass spectrometry (Margoum et al. 2013) and gas chromatography-mass spectrometry (Camino-Sánchez et al. 2012). Recently, Lerch et al. (2018) developed a fully validated method for the analysis of pesticides and other priority pollutants in water samples at ng L^{-1} to sub-ng L^{-1} levels, according to the European Union Water Framework Directives (European Union 2008, 2013, 2015).

By comparing solid-phase microextraction and stir bar sorptive extraction, apart from other advantages, we can conclude that these solventless techniques combine simultaneously extraction and concentration of the analytes in a single step, present easy and fast manipulation, require low sample volume, allow reutilization, can operate overnight, do not present the breakthrough phenomena, are cost-effective, can be easily combined with modern analytical instrumentation and have demonstrated their applicability to a wide range of samples (Nogueira 2015). Although solid-phase microextraction has lower capacity due the limited amount of sorbent phase, it is definitely more appropriate for analysis of semi-volatile to volatile analytes than stir bar sorptive extraction when the headspace mode followed by gas chromatography is used.

2.4 Gel Permeation Chromatography

Gel permeation chromatography is an efficient technique that was first used in 1970s for extraction and cleanup of pesticides (Holding 1984). The gel permeation chromatography separation mechanism is based on the principle of size exclusion where larger molecules are eluted from the stationary phase (gel), followed by smaller molecules. Gel permeation chromatography columns are packed with a cross-linked polymer material that is very porous, so that small molecules can get into the pores and are thus retained for longer periods of time on the column, whereas large molecules cannot and therefore pass through the column very quickly. Cross-linked dextrans such as Sephadex or agarose are commonly used materials. Because of the robustness, rapidity, convenience, and cleanup efficiency, gel permeation chromatography has been used as a universal technique for the pretreatment of complex samples before multiresidue pesticide analysis (Huang et al. 2007; Cajka et al. 2008), and especially to purify fat-rich extracts, such as those coming from olive oil (Guardia-Rubio et al. 2006), milk (Zheng et al. 2014), avocado (Fernández Moreno et al. 2006), or pork and lamb meat (Garrido Frenich et al. 2006). Lipids, which are often co-extracted along with pesticides, may jam in the injector and at the top of the column of gas chromatography. However, lipids are too big and cannot go through the pores of the gel, so that they are eluted first from the column. Different mobile phases can be used for the gel permeation chromatography cleanup: ethyl acetate-cyclohexane (Cajka et al. 2008) or acetone-cyclohexane and hexane-ethyl acetate (Huang et al. 2007).

The analysis of pesticides in oil samples by gas chromatography requires a complete removal of the high molecular weight fat components to lengthen the life of the columns and keep the chromatographic system working properly. Gel permeation chromatography has demonstrated to be a very useful technique to separate the main components of oils (triglycerides) from the low molecular weight of pesticides, being one of the more widely spread techniques (Sánchez et al. 2006; Garrido Frenich et al. 2007). It is generally used after a preliminary solvent partitioning with acetonitrile for the subsequent gas chromatography analysis (Guardia-Rubio et al. 2006). However, there are methods based on direct extraction and cleanup of pesticides from olive oil without a liquid-liquid partition step, using only gel permeation chromatography (Patel et al. 2005). In most cases, the recoveries in gel permeation chromatography are satisfactory and it is applicable to both polar and nonpolar pesticides within the same injection on a fully automated system. A drawback of this technique is the partial overlapping between the pesticide fraction and the components of the matrix (mainly triglycerides), which results in loss of some nonpolar pesticides such as bromopropylate, phenotrin and acrinatrin (Hakme et al. 2018). Therefore, an additional cleanup step is usually included after gel permeation chromatography before chromatographic analysis, for example by solid-phase extraction (Huang et al. 2007). Other drawbacks of gel permeation chromatography are the use of large volumes of organic toxic solvents, the requirement of an additional instrumentation, and that it is time-consuming.

2.5 Conclusions and Perspectives

In most analytical methods, the complexity of sample and/or the concentration and nature of analyte(s) make the sample treatment an unavoidable step, so that sample preparation remains the weakest and time-determining stage of the analysis. In many cases, extraction and cleanup techniques are labor intensive, involve the use of organic solvents, energy, and most of them are hazardous to both, laboratory workers and the environment. The particular case of solid samples usually requires, especially for trace analysis, the use of an extraction process to dissolve the target analyte(s) in an aqueous or organic solvent solution previous to analyte pre-concentration.

In the last years, improvements in sample preparation techniques are the result of modifying existing techniques and developing new techniques, always with the aim of saving time and reducing consumption of chemicals and waste generation, and thus to improve overall process performance. High efficacy extraction approaches for solid samples have been developed, such as matrix solid-phase dispersion, microwave-assisted extraction, pressurized liquid extraction, and supercritical fluid extraction. Nevertheless, some of these techniques usually involve a treatment at high temperature and pressure to improve the extraction yield with reduced time and solvent consumption. A special attention deserves the use of headspace extraction for analysis of volatile compounds from solid samples, which is solvent-free, carried out at moderate temperature, and can be combined with solid-phase microextraction for significantly improving extraction efficacy. All these techniques have grown in their maturity with increased applications to pesticide residue analysis in food and environmental samples.

Regarding liquid or dissolved samples, the miniaturization of liquid-liquid extraction has resulted in more efficient extraction techniques with solvent and time savings. Single-drop microextraction, hollow-fiber membrane liquid-phase microextraction, and dispersive liquid-liquid microextraction offer green alternatives to solve problems of sensitivity and/or selectivity of methods to be applied directly to analyze complex samples. Moreover, the use of alternative extractants like ionic liquids, with high solvation properties and versatility, has been explored as potential candidates to replace conventional solvents in a green perspective. As a consequence, and due to the advantageous features they present compared to conventional techniques, currently liquid-phase microextraction techniques play a very important role in sample preparation for analysis, and specifically in pesticide analysis. This is evident from the relatively large number of publications in this regard in recent years.

Other strategies based on the miniaturization of solid-phase extraction, such as solid-phase microextraction and stir bar sorptive extraction have been developed in the last years to reduce substantially sample and solvent consumption. The use of new solid sorbents, such as molecularly imprinted polymers or carbon nanotubes has allowed the development of high selective extraction methods and contributed to reduce the number of steps of analytical procedures. On the other hand, the introduction of magnetic materials has contributed to the development of solid-phase extraction procedures simpler, faster than the conventional methods, achieving sorbent separation by the use of a magnet, which reduces time and energy consumption. On the other hand, QuEChERs method has become in one of the sample treatment mostly applied, not only for extraction of pesticides from food samples, but also from environmental matrices, since it allows performing extraction and cleanup in a single step. This is evident from the fact that important international organizations, the AOAC International and the European Committee for Standardization, have approved two different official methods based on method QuEChERs for pesticide analysis.

In the future, it is expected that the development of new techniques as well as all improvements in current sample preparation techniques will be aimed at improving the efficiency of extraction and removal of matrix co-extractives, increasing throughput, and reducing the use of organic solvents. Therefore, as a general rule, these improvements should be directed towards the miniaturization and automation of the analytical process.

References

Abdulra'uf LB, Tan GH (2014) Review of SBSE technique for the analysis of pesticide residues in fruits and vegetables. Chromatographia 77:15–24. https://doi.org/10.1007/s10337-013-2566-8

Ahmad R, Salem NM, Estaitieh H (2010) Occurrence of organochlorine pesticide residues in eggs, chicken and meat in Jordan. Chemosphere 78:667–671. https://doi.org/10.1016/j.chemosphere.2009.12.012

Ahmad W, Al-Sibaai AA, Bashammakh AS, Alwael H, El-Shahawi MS (2015) Recent advances in dispersive liquid-liquid microextraction for pesticide analysis. TrAC Trends Anal Chem 72:181–192. https://doi.org/10.1016/J.TRAC.2015.04.022

Ai J (1997) Headspace solid phase microextraction. Dynamics and quantitative analysis before reaching a partition equilibrium. Anal Chem 69:3260–3266. https://doi.org/10.1021/ac970024x

Anastassiades M, Lehotay SJ, Stajnbaher D, Schenck FJ (2003) Fast and easy multiresidue method employing acetonitrile extraction/partitioning and dispersive solid-phase extraction for the determination of pesticide residues in produce. J AOAC Int 86:412–431

Andreu V, Picó Y (2004) Determination of pesticides and their degradation products in soil: Critical review and comparison of methods. TrAC Trends Anal Chem 23:772–789. https://doi.org/10.1016/j.trac.2004.07.008

Andreu V, Picó Y (2019) Pressurized liquid extraction of organic contaminants in environmental and food samples. TrAC Trends Anal Chem 118:709–721. https://doi.org/10.1016/j.trac.2019.06.038

AOAC Official Method 2007.01 (2007) Pesticide residues in foods by acetonitrile extraction and partitioning with magnesium sulfate gas chromatography/mass spectrometry and liquid chromatography/tandem mass spectrometry first action 2007

Arthur CL, Pawliszyn J (1990) Solid phase microextraction with thermal desorption using fused silica optical fibers. Anal Chem 62:2145–2148. https://doi.org/10.1021/ac00218a019

Asensio-Ramos M, Hernández-Borges J, González-Hernández G, Rodríguez-Delgado MÁ (2012) Hollow-fiber liquid-phase microextraction for the determination of pesticides and metabolites in soils and water samples using HPLC and fluorescence detection. Electrophoresis 33:2184–2191. https://doi.org/10.1002/elps.201200138

Bagheri H, Yamini Y, Safari M, Asiabi H, Karimi M, Heydari A (2016) Simultaneous determination of pyrethroids residues in fruit and vegetable samples via supercritical fluid extraction coupled with magnetic solid phase extraction followed by HPLC-UV. J Supercrit Fluids 107:571–580. https://doi.org/10.1016/j.supflu.2015.07.017

Baltussen E, Sandra P, David F, Cramers C (1999) Stir bar sorptive extraction (SBSE), a novel extraction technique for aqueous samples: Theory and principles. J Microcolumn Sep 11:737–747. https://doi.org/10.1002/(SICI)1520-667X(1999)11:10<737::AID-MCS7>3.0.CO;2-4

Banerjee K, Oulkar DP, Dasgupta S, Patil SB, Patil SH, Savant R, Adsule PG (2007) Validation and uncertainty analysis of a multi-residue method for pesticides in grapes using ethyl acetate extraction and liquid chromatography–tandem mass spectrometry. J Chromatogr A 1173:98–109. https://doi.org/10.1016/j.chroma.2007.10.013

Barker SA, Long AR, Short CR (1989) Isolation of drug residues from tissues by solid phase dispersion. J Chromatogr A 475:353–361. https://doi.org/10.1016/S0021-9673(01)89689-8

Barriada-Pereira M, González-Castro MJ, Muniategui-Lorenzo S, López-Mahía P, Prada-Rodríguez D (2010) Sample preparation based on matrix solid-phase dispersion and solid-phase extraction cleanup for the determination of organochlorine pesticides in fish. J AOAC Int 93:992–998

Basheer C, Lee HK, Obbard JP (2002) Determination of organochlorine pesticides in seawater using liquid-phase hollow fibre membrane microextraction and gas chromatography-mass spectrometry. J Chromatogr A 968:191–199. https://doi.org/10.1016/s0021-9673(02)00793-8

Basheer C, Suresh V, Renu R, Lee HK (2004) Development and application of polymer-coated hollow fiber membrane microextraction to the determination of organochlorine pesticides in water. J Chromatogr A 1033:213–220. https://doi.org/10.1016/j.chroma.2004.01.050

Basheer C, Obbard JP, Lee HK (2005) Analysis of persistent organic pollutants in marine sediments using a novel microwave assisted solvent extraction and liquid-phase microextraction technique. J Chromatogr A 1068:221–228. https://doi.org/10.1016/j.chroma.2005.01.099

Beltran J, López FJ, Hernández F (2000) Solid-phase microextraction in pesticide residue analysis. J. Chromatogr A 885:389–404. https://doi.org/10.1016/S0021-9673(00)00142-4

Benedé JL, Chisvert A, Giokas DL, Salvador A (2014) Development of stir bar sorptive-dispersive microextraction mediated by magnetic nanoparticles and its analytical application to the determination of hydrophobic organic compounds in aqueous media. J Chromatogr A 1362:25–33. https://doi.org/10.1016/j.chroma.2014.08.024

Berijani S, Assadi Y, Anbia M, Milani Hosseini M-R, Aghaee E (2006) Dispersive liquid–liquid microextraction combined with gas chromatography-flame photometric detection: Very simple, rapid and sensitive method for the determination of organophosphorus pesticides in water. J Chromatogr A 1123:1–9. https://doi.org/10.1016/j.chroma.2006.05.010

Beyer A, Biziuk M (2008) Applications of sample preparation techniques in the analysis of pesticides and PCBs in food. Food Chem 108:669–680. https://doi.org/10.1016/j.foodchem.2007.11.024

Binellas CS, Stalikas CD (2015) Magnetic octadecyl-based matrix solid-phase dispersion coupled with gas chromatography with mass spectrometry in a proof-of-concept determination of multiclass pesticide residues in carrots. J Sep Sci 38:3575–3581. https://doi.org/10.1002/jssc.201500519

Blasco C, Vazquez-Roig P, Onghena M, Masia A, Picó Y (2011) Analysis of insecticides in honey by liquid chromatography-ion trap-mass spectrometry: Comparison of different extraction procedures. J Chromatogr A 1218:4892–4901. https://doi.org/10.1016/j.chroma.2011.02.045

Bolaños PP, Romero-González R, Frenich AG, Vidal JLM (2008) Application of hollow fibre liquid phase microextraction for the multiresidue determination of pesticides in alcoholic beverages by ultra-high pressure liquid chromatography coupled to tandem mass spectrometry. J Chromatogr A 1208:16–24. https://doi.org/10.1016/j.chroma.2008.08.059

Bratkowska D, Fontanals N, Borrull F, Cormack PAG, Sherrington DC, Marcé RM (2010) Hydrophilic hypercrosslinked polymeric sorbents for the solid-phase extraction of polar contaminants from water. J Chromatogr A 1217:3238–3243. https://doi.org/10.1016/j.chroma.2009.08.091

Cabras P, Angioni A, Garau VL, Pirisi FM, Cabitza F, Pala M, Farris GA (2001) Fenhexamid residues in grapes and wine. Food Addit Contam 18:625–629. https://doi.org/10.1080/02652030120571

Cajka T, Hajslova J, Lacina O, Mastovska K, Lehotay SJ (2008) Rapid analysis of multiple pesticide residues in fruit-based baby food using programmed temperature vaporiser injection-low-pressure gas chromatography-high-resolution time-of-flight mass spectrometry. J Chromatogr A 1186:281–294. https://doi.org/10.1016/j.chroma.2007.12.009

Camino-Sánchez FJ, Zafra-Gómez A, Cantarero-Malagón S, Vílchez JL (2012) Validation of a method for the analysis of 77 priority persistent organic pollutants in river water by stir bar sorptive extraction in compliance with the European Water Framework Directive. Talanta 89:322–334. https://doi.org/10.1016/j.talanta.2011.12.037

Camino-Sánchez FJ, Rodríguez-Gómez R, Zafra-Gómez A, Santos-Fandila A, Vílchez JL (2014) Stir bar sorptive extraction: Recent applications, limitations and future trends. Talanta 130:388–399. https://doi.org/10.1016/j.talanta.2014.07.022

Campillo N, Viñas P, Aguinaga N, Férez G, Hernández-Córdoba M (2010) Stir bar sorptive extraction coupled to liquid chromatography for the analysis of strobilurin fungicides in fruit samples. J Chromatogr A 1217:4529–4534. https://doi.org/10.1016/j.chroma.2010.05.006

Capriotti AL, Cavaliere C, Laganà A, Piovesana S, Samperi R (2013) Recent trends in matrix solid-phase dispersion. TrAC Trends Anal Chem 43:53–66. https://doi.org/10.1016/j.trac.2012.09.021

Carvalho PN, Rodrigues PNR, Alves F, Evangelista R, Basto MCP, Vasconcelos MTSD (2008) An expeditious method for the determination of organochlorine pesticides residues in estuarine sediments using microwave assisted pre-extraction and automated headspace solid-phase microextraction coupled to gas chromatography–mass spectrometry. Talanta 76:1124–1129. https://doi.org/10.1016/j.talanta.2008.05.035

Celeiro M, Llompart M, Lamas JP, Lores M, Garcia-Jares C, Dagnac T (2014) Determination of fungicides in white grape bagasse by pressurized liquid extraction and gas chromatography tandem mass spectrometry. J Chromatogr A 1343:18–25. https://doi.org/10.1016/j.chroma.2014.03.057

Cerqueira MBR, Caldas SS, Primel EG (2014) New sorbent in the dispersive solid phase extraction step of quick, easy, cheap, effective, rugged, and safe for the extraction of organic contaminants in drinking water treatment sludge. J Chromatogr A 1336:10–22. https://doi.org/10.1016/j.chroma.2014.02.002

Chatzimitakos TG, Anderson JL, Stalikas CD (2018) Matrix solid-phase dispersion based on magnetic ionic liquids: An alternative sample preparation approach for the extraction of pesticides from vegetables. J Chromatogr A 1581–1582:168–172. https://doi.org/10.1016/j.chroma.2018.11.008

Chaves AR, Silva SM, Queiroz RHC, Lanças FM, Queiroz MEC (2007) Stir bar sorptive extraction and liquid chromatography with UV detection for determination of antidepressants in plasma samples. J Chromatogr B 850:295–302. https://doi.org/10.1016/j.jchromb.2006.11.042

Chee KK, Wong MK, Lee HK (1996) Determination of organochlorine pesticides in water by membranous solid-phase extraction, and in sediment by microwave-assisted solvent extraction with gas chromatography and electron-capture and mass spectrometric detection. J Chromatogr A 736:211–218. https://doi.org/10.1016/0021-9673(95)01354-7

Chen P-S, Huang S-D (2006) Determination of ethoprop, diazinon, disulfoton and fenthion using dynamic hollow fiber-protected liquid-phase microextraction coupled with gas chromatography–mass spectrometry. Talanta 69:669–675. https://doi.org/10.1016/j.talanta.2005.10.042

Chen C-C, Melwanki MB, Huang S-D (2006) Liquid–liquid–liquid microextraction with automated movement of the acceptor and the donor phase for the extraction of phenoxyacetic acids prior to liquid chromatography detection. J Chromatogr A 1104:33–39. https://doi.org/10.1016/j.chroma.2005.11.122

Cheng Z, Dong F, Xu J, Liu X, Wu X, Chen Z, Pan X, Zheng Y (2016) Atmospheric pressure gas chromatography quadrupole-time-of-flight mass spectrometry for simultaneous determination of fifteen organochlorine pesticides in soil and water. J Chromatogr A 1435:115–124. https://doi.org/10.1016/j.chroma.2016.01.025

Choi M, Lee I-S, Jung R-H (2016) Rapid determination of organochlorine pesticides in fish using selective pressurized liquid extraction and gas chromatography–mass spectrometry. Food Chem 205:1–8. https://doi.org/10.1016/j.foodchem.2016.02.156

Clark AE, Yoon S, Sheesley RJ, Usenko S (2015) Pressurized liquid extraction technique for the analysis of pesticides, PCBs, PBDEs, OPEs, PAHs, alkanes, hopanes, and steranes in atmospheric particulate matter. Chemosphere 137:33–44. https://doi.org/10.1016/j.chemosphere.2015.04.051

Codex Alimentarius. Pesticides residues in food online database. Available on: http://www.fao.org/fao-who-codexalimentarius/codex-texts/dbs/pestres/en/. Accessed on Sept 2019

Cunha SC, Lehotay SJ, Mastovska K, Fernandes JO, Beatriz M, Oliveira PP (2007) Evaluation of the QuEChERS sample preparation approach for the analysis of pesticide residues in olives. J Sep Sci 30:620–632. https://doi.org/10.1002/jssc.200600410

Da Silva Souza NR, Navickiene S (2019) Multiresidue determination of carbamate, organophosphate, neonicotinoid and triazole pesticides in roasted coffee using ultrasonic solvent extraction and liquid chromatography-tandem mass spectrometry. J AOAC Int 102:33–37. https://doi.org/10.5740/jaoacint.18-0294

David F, Ochiai N, Sandra P (2019) Two decades of stir bar sorptive extraction: A retrospective and future outlook. TrAC Trends Anal Chem 112:102–111. https://doi.org/10.1016/j.trac.2018.12.006

De Grazia S, Gionfriddo E, Pawliszyn J (2017) A new and efficient solid phase microextraction approach for analysis of high fat content food samples using a matrix-compatible coating. Talanta 167:754–760. https://doi.org/10.1016/j.talanta.2017.01.064

Dias JV, Cutillas V, Lozano A, Pizzutti IR, Fernández-Alba AR (2016) Determination of pesticides in edible oils by liquid chromatography-tandem mass spectrometry employing new generation materials for dispersive solid phase extraction clean-up. J Chromatogr A 1462:8–18. https://doi.org/10.1016/j.chroma.2016.07.072

Díaz-Álvarez M, Turiel E, Martín-Esteban A (2016) Molecularly imprinted polymer monolith containing magnetic nanoparticles for the stir-bar sorptive extraction of triazines from environmental soil samples. J Chromatogr A 1469:1–7. https://doi.org/10.1016/j.chroma.2016.09.051

Doong R, Lee C (1999) Determination of organochlorine pesticide residues in foods using solid-phase extraction clean-up cartridges. Analyst 124:1287–1289. https://doi.org/10.1039/a902722j

Du D, Wang M, Zhang J, Cai J, Tu H, Zhang A (2008) Application of multiwalled carbon nanotubes for solid-phase extraction of organophosphate pesticide. Electrochem Commun 10:85–89. https://doi.org/10.1016/j.elecom.2007.11.005

Ebrahimi M, Es'haghi Z, Samadi F, Hosseini M-S (2011) Ionic liquid mediated sol–gel sorbents for hollow fiber solid-phase microextraction of pesticide residues in water and hair samples. J Chromatogr A 1218:8313–8321. https://doi.org/10.1016/j.chroma.2011.09.058

Eisert R, Pawliszyn J (1997) Automated in-tube solid-phase microextraction coupled to high-performance liquid chromatography. Anal Chem 69:3140–3147. https://doi.org/10.1021/ac970319a

El-Sheikh AH, Sweileh JA, Al-Degs YS, Insisi AA, Al-Rabady N (2008) Critical evaluation and comparison of enrichment efficiency of multi-walled carbon nanotubes, C_{18} silica and activated carbon towards some pesticides from environmental waters. Talanta 74:1675–1680. https://doi.org/10.1016/j.talanta.2007.09.005

European Committee for Standardization (CEN) Standard Method EN 15662:2008 (2008) Multimethod for the determination of pesticide residues using GC- and LC-based analysis following acetonitrile extraction/partitioning and clean-up by dispersive SPE - Modular QuEChERS-method

European Food Safety Authority (2018) The 2016 European Union report on pesticide residues in food. EFSA J 16:5348. https://doi.org/10.2903/j.efsa.2018.5348

European Union (2000) Directive 2000/60/EC of the European Parliament and of the Council of 23 October 2000 establishing a framework for Community action in the field of water policy. Off J Eur Union Legis 327(1)

European Union (2005) Regulation (EC) No 396/2005 of the European Parliament and of the Council of 23 February 2005 on maximum residue levels of pesticides in or on food and feed of plant and animal origin and amending Council Directive 91/414/EEC. Off J Eur Union Legis 70 (1)

European Union (2008) Directive 2008/105/EC of the European Parliament and of the Council of 16 December 2008 on environmental quality standards in the field of water policy, amending and subsequently repealing Council Directives 82/176/EEC, 83/513/EEC, 84/156/EEC, 84/491/EEC. Off J Eur Union Legis 348(84)

European Union (2013) Directive 2013/39/EU of the European Parliament and of the Council of 12 August 2013 amending Directives 2000/60/EC and 2008/105/EC as regards priority substances in the field of water policy. Off J Eur Union Legis 226(1)

European Union (2015) Decision 2015/495. Commission Implementing Decision (EU) 2015/495 of 20 March 2015 establishing a watch list of substances for Union-wide monitoring in the field of water policy pursuant to Directive 2008/105/EC of the European Parliament and of the Council. Off J Eur Union Legis 78:40

Falqui-Cao C, Wang Z, Urruty L, Pommier JJ, Montury M (2001) Focused microwave assistance for extracting some pesticide residues from strawberries into water before their determination by SPME/HPLC/DAD. J Agric Food Chem 49:5092–5097. https://doi.org/10.1021/jf010519u

Fang G, Min G, He J, Zhang C, Qian K, Wang S (2009) Multiwalled carbon nanotubes as matrix solid-phase dispersion extraction absorbents to determine 31 pesticides in agriculture samples by gas chromatography-mass spectrometry. J Agric Food Chem 57:3040–3045. https://doi.org/10.1021/jf803913q

Farajzadeh MA, Khoshmaram L (2013) Air-assisted liquid-liquid microextraction-gas chromatography-flame ionisation detection: A fast and simple method for the assessment of triazole pesticides residues in surface water, cucumber, tomato and grape juices samples. Food Chem 141:1881–1887. https://doi.org/10.1016/j.foodchem.2013.05.088

Farajzadeh MA, Mogaddam MRA (2012) Air-assisted liquid-liquid microextraction method as a novel microextraction technique; Application in extraction and preconcentration of phthalate esters in aqueous sample followed by gas chromatography-flame ionization detection. Anal Chim Acta 728:31–38. https://doi.org/10.1016/j.aca.2012.03.031

Farajzadeh MA, Seyedi SE, Shalamzari MS, Bamorowat M (2009) Dispersive liquid-liquid microextraction using extraction solvent lighter than water. J Sep Sci 32:3191–3200. https://doi.org/10.1002/jssc.200900109

Farajzadeh MA, Feriduni B, Mogaddam MRA (2014) Extraction and enrichment of triazole and triazine pesticides from honey using air-assisted liquid-liquid microextraction. J Food Sci 79: H2140–H2148. https://doi.org/10.1111/1750-3841.12597

Farajzadeh MA, Feriduni B, Mogaddam MRA (2015) Determination of triazole pesticide residues in edible oils using air-assisted liquid-liquid microextraction followed by gas chromatography with flame ionization detection. J Sep Sci 38:1002–1009. https://doi.org/10.1002/jssc.201400818

Farajzadeh MA, Sohrabi H, Mohebbi A (2019) Combination of modified QuEChERS extraction method and dispersive liquid–liquid microextraction as an efficient sample preparation approach for extraction and preconcentration of pesticides from fruit and vegetable samples. Food Anal Methods 12:534–543. https://doi.org/10.1007/s12161-018-1384-x

Fernández Moreno JL, Arrebola Liébanas FJ, Garrido Frenich A, Martínez Vidal JL (2006) Evaluation of different sample treatments for determining pesticide residues in fat vegetable matrices like avocado by low-pressure gas chromatography-tandem mass spectrometry. J Chromatogr A 1111:97–105. https://doi.org/10.1016/j.chroma.2006.01.108

Fernandez-Alba AR, Valverde A, Agüera A, Contreras M (1994) Gas chromatographic determination of organochlorine and pyrethroid pesticides of horticultural concern. J Chromatogr A 686:263–274. https://doi.org/10.1016/0021-9673(94)00735-7

Fernandez-Alvarez M, Llompart M, Lamas JP, Lores M, Garcia-Jares C, Cela R, Dagnac T (2008) Simultaneous determination of traces of pyrethroids, organochlorines and other main plant protection agents in agricultural soils by headspace solid-phase microextraction-gas chromatography. J Chromatogr A 1188:154–163. https://doi.org/10.1016/j.chroma.2008.02.080

Fernandez-Alvarez M, Llompart M, Garcia-Jares C, Dagnac T, Lores M (2009) Investigation of the photochemical behaviour of pyrethroids lacking the cyclopropane ring by photo-solid-phase microextraction and gas chromatography/mass spectrometry. Rapid Commun Mass Spectrom 23:3673–3687. https://doi.org/10.1002/rcm.4301

Fiddler W, Pensabene JW, Gates RA, Donoghue DJ (1999) Supercritical fluid extraction of organochlorine pesticides in eggs. J Agric Food Chem 47:206–211. https://doi.org/10.1021/jf980436m

Fillion J, Sauvé F, Selwyn J (2000) Multiresidue method for the determination of residues of 251 pesticides in fruits and vegetables by gas chromatography/mass spectrometry and liquid chromatography with fluorescence detection. J AOAC Int 83:698–713. PMID: 10868594

Font N, Hernández F, Hogendoorn E, Baumann R, van Zoonen P (1998) Microwave-assisted solvent extraction and reversed-phase liquid chromatography–UV detection for screening soils for sulfonylurea herbicides. J Chromatogr A 798:179–186. https://doi.org/10.1016/S0021-9673(97)01194-1

Fontana AR, Camargo A, Martinez LD, Altamirano JC (2011) Dispersive solid-phase extraction as a simplified clean-up technique for biological sample extracts. Determination of polybrominated diphenyl ethers by gas chromatography–tandem mass spectrometry. J Chromatogr A 1218:2490–2496. https://doi.org/10.1016/j.chroma.2011.02.058

Fontanals N, Marcé RM, Borrull F (2005) New hydrophilic materials for solid-phase extraction. TrAC Trends Anal Chem 24:394–406. https://doi.org/10.1016/j.trac.2005.01.012

Fuentes E, Báez ME, Díaz J (2009) Microwave-assisted extraction at atmospheric pressure coupled to different clean-up methods for the determination of organophosphorus pesticides in olive and avocado oil. J Chromatogr A 1216:8859–8866. https://doi.org/10.1016/j.chroma.2009.10.082

Garbi A, Sakkas V, Fiamegos YC, Stalikas CD, Albanis T (2010) Sensitive determination of pesticides residues in wine samples with the aid of single-drop microextraction and response surface methodology. Talanta 82:1286–1291. https://doi.org/10.1016/j.talanta.2010.06.046

Garcia R, Carreiro EP, Nunes J, Da Silva MG, Freitas AMC, Burke AJ, Cabrita MJ (2016) Dual-layer solid-phase extraction based on molecular imprinting technology: Seeking a route to enhance selectivity for trace analysis of pesticide residues in olive oil. Electrophoresis 37:1916–1922. https://doi.org/10.1002/elps.201600073

Garrido Frenich A, Martínez Vidal JL, Cruz Sicilia AD, González Rodríguez MJ, Plaza Bolaños P (2006) Multiresidue analysis of organochlorine and organophosphorus pesticides in muscle of chicken, pork and lamb by gas chromatography-triple quadrupole mass spectrometry. Anal Chim Acta 558:42–52. https://doi.org/10.1016/j.aca.2005.11.012

Garrido Frenich A, Fernández Moreno JL, Martínez Vidal JL, Arrebola Liébanas FJ (2007) Application of gas chromatography coupled to triple quadrupole mass spectrometry for the multiresidue analysis of pesticides in olive oil. J Agric Food Chem 55:8346–8352. https://doi.org/10.1021/jf071615j

Godage NH, Gionfriddo E (2019) A critical outlook on recent developments and applications of matrix compatible coatings for solid phase microextraction. TrAC Trends Anal. Chem. 111:220–228. https://doi.org/10.1016/j.trac.2018.12.019

Gonçalves C, Alpendurada MF (2005) Assessment of pesticide contamination in soil samples from an intensive horticulture area, using ultrasonic extraction and gas chromatography–mass spectrometry. Talanta 65:1179–1189. https://doi.org/10.1016/j.talanta.2004.08.057

Gou Y, Eisert R, Pawliszyn J (2000) Automated in-tube solid-phase microextraction-high-performance liquid chromatography for carbamate pesticide analysis. J Chromatogr A 873:137–147. https://doi.org/10.1016/S0021-9673(99)01125-5

Guan H, Brewer WE, Garris ST, Craft C, Morgan SL (2010) Multiresidue analysis of pesticides in fruits and vegetables using disposable pipette extraction (DPX) and micro-Luke method. J Agric Food Chem 58:5973–5981. https://doi.org/10.1021/jf903448w

Guan W, Li Z, Zhang H, Hong H, Rebeyev N, Ye Y, Ma Y (2013) Amine modified graphene as reversed-dispersive solid phase extraction materials combined with liquid chromatography–tandem mass spectrometry for pesticide multi-residue analysis in oil crops. J Chromatogr A 1286:1–8. https://doi.org/10.1016/j.chroma.2013.02.043

Guardia Rubio M, Ruiz Medina A, Pascual Reguera MI, Fernández de Córdova ML (2007) Multiresidue analysis of three groups of pesticides in washing waters from olive processing by solid-phase extraction-gas chromatography with electron capture and thermionic specific detection. Microchem J 85:257–264. https://doi.org/10.1016/j.microc.2006.06.005

Guardia-Rubio M, Fernández-De Córdova ML, Ayora-Cañada MJ, Ruiz-Medina A (2006) Simplified pesticide multiresidue analysis in virgin olive oil by gas chromatography with thermoionic specific, electron-capture and mass spectrometric detection. J Chromatogr A 1108:231–239. https://doi.org/10.1016/j.chroma.2006.01.006

Hakme E, Lozano A, Ferrer C, Díaz-Galiano FJ, Fernández-Alba AR (2018) Analysis of pesticide residues in olive oil and other vegetable oils. TrAC Trends Anal Chem 100:167–179. https://doi.org/10.1016/j.trac.2017.12.016

Hamed AM, Moreno-González D, Gámiz-Gracia L, García-Campaña AM (2017) Evaluation of a new modified QuEChERS method for the monitoring of carbamate residues in high-fat cheeses by using UHPLC-MS/MS. J Sep Sci 40:488–496. https://doi.org/10.1002/jssc.201600845

Han D, Row KH (2012) Trends in liquid-phase microextraction, and its application to environmental and biological samples. Microchim Acta 76:1–22. https://doi.org/10.1007/s00604-011-0678-0

Han L, Matarrita J, Sapozhnikova Y, Lehotay SJ (2016) Evaluation of a recent product to remove lipids and other matrix co-extractives in the analysis of pesticide residues and environmental contaminants in foods. J Chromatogr A 1449:17–29. https://doi.org/10.1016/j.chroma.2016.04.052

Hanot V, Goscinny S, Deridder M (2015) A simple multi-residue method for the determination of pesticides in fruits and vegetables using a methanolic extraction and ultra-high-performance liquid chromatography-tandem mass spectrometry: Optimization and extension of scope. J Chromatogr A 1384:53–66. https://doi.org/10.1016/j.chroma.2015.01.040

He Z, Wang Y, Wang L, Peng Y, Wang W, Liu X (2017) Determination of 255 pesticides in edible vegetable oils using QuEChERS method and gas chromatography tandem mass spectrometry. Anal Bioanal Chem 409:1017–1030. https://doi.org/10.1007/s00216-016-0016-9

Hercegová A, Dömötörová M, Matisová E (2007) Sample preparation methods in the analysis of pesticide residues in baby food with subsequent chromatographic determination. J Chromatogr A 1153:54–73. https://doi.org/10.1016/j.chroma.2007.01.008

Ho YM, Tsoi YK, Leung KSY (2013) Highly sensitive and selective organophosphate screening in twelve commodities of fruits, vegetables and herbal medicines by dispersive liquid-liquid microextraction. Anal Chim Acta 775:58–66. https://doi.org/10.1016/j.aca.2013.02.043

Holding S (1984) Gel permeation chromatography. Endeavour 8:17–20. https://doi.org/10.1016/0160-9327(84)90124-8

Homazava N, Gachet Aquillon C, Vermeirssen E, Werner I (2014) Simultaneous multi-residue pesticide analysis in soil samples with ultra-high-performance liquid chromatography–tandem mass spectrometry using QuEChERS and pressurised liquid extraction methods. Int J Environ Anal Chem 94:1085–1099. https://doi.org/10.1080/03067319.2014.954558

Hopper ML (1999) Automated one-step supercritical fluid extraction and clean-up system for the analysis of pesticide residues in fatty matrices. J Chromatogr A 840:93–105. https://doi.org/10.1016/s0021-9673(99)00228-9

Hou L, Lee HK (2004) Determination of pesticides in soil by liquid-phase microextraction and gas chromatography–mass spectrometry. J Chromatogr A 1038:37–42. https://doi.org/10.1016/j.chroma.2004.03.012

Howard AL, Braue C, Taylor LT (1993) Feasibility of thiocarbamate pesticide analysis in apples by supercritical fluid extraction and high-performance liquid chromatography. J Chromatogr Sci 31:323–329. https://doi.org/10.1093/chromsci/31.8.323

Huang Z, Li Y, Chen B, Yao S (2007) Simultaneous determination of 102 pesticide residues in Chinese teas by gas chromatography–mass spectrometry. J Chromatogr B 853:154–162. https://doi.org/10.1016/j.jchromb.2007.03.013

Iijima S (1991) Helical microtubules of graphitic carbon. Nature 354:56–58. https://doi.org/10.1038/354056a0

Janda V, Steenbeke G, Sandra P (1989) Supercritical fluid extraction of s-triazine herbicides from sediment. J Chromatogr A 479:200–205. https://doi.org/10.1016/S0021-9673(01)83334-3

Jia C, Zhu X, Wang J, Zhao E, He M, Chen L, Yu P (2010) Extraction of pesticides in water samples using vortex-assisted liquid–liquid microextraction. J Chromatogr A 1217:5868–5871. https://doi.org/10.1016/j.chroma.2010.07.055

Juan-García A, Picó Y, Font G (2005) Capillary electrophoresis for analyzing pesticides in fruits and vegetables using solid-phase extraction and stir-bar sorptive extraction. J Chromatogr A 1073:229–236. https://doi.org/10.1016/j.chroma.2004.09.028

Juan-García A, Font G, Juan C, Picó Y (2010) Pressurised liquid extraction and capillary electrophoresis–mass spectrometry for the analysis of pesticide residues in fruits from Valencian markets, Spain. Food Chem 120:1242–1249. https://doi.org/10.1016/j.foodchem.2009.11.071

Kaczyński P, Łozowicka B, Perkowski M, Szabuńko J (2017) Multiclass pesticide residue analysis in fish muscle and liver on one-step extraction-cleanup strategy coupled with liquid chromatography tandem mass spectrometry. Ecotoxicol Environ Saf 138:179–189. https://doi.org/10.1016/j.ecoenv.2016.12.040

Khan Z, Kamble N, Bhongale A, Girme M, Bahadur Chauhan V, Banerjee K (2018) Analysis of pesticide residues in tuber crops using pressurised liquid extraction and gas chromatography-tandem mass spectrometry. Food Chem 241:250–257. https://doi.org/10.1016/j.foodchem.2017.08.091

Köck-Schulmeyer M, Olmos M, López de Alda M, Barceló D (2013) Development of a multiresidue method for analysis of pesticides in sediments based on isotope dilution and liquid chromatography-electrospray–tandem mass spectrometry. J Chromatogr A 1305:176–187. https://doi.org/10.1016/j.chroma.2013.07.036

Kokosa JM (2015) Recent trends in using single-drop microextraction and related techniques in green analytical methods. TrAC Trends Anal Chem 71:194–204. https://doi.org/10.1016/j.trac.2015.04.019

Lambropoulou DA, Albanis TA (2005) Application of hollow fiber liquid phase microextraction for the determination of insecticides in water. J Chromatogr A 1072:55–61. https://doi.org/10.1016/j.chroma.2004.11.076

Lambropoulou DA, Albanis TA (2007) Liquid-phase micro-extraction techniques in pesticide residue analysis. J Biochem Biophys Methods 70:195–228. https://doi.org/10.1016/j.jbbm.2006.10.004

Lambropoulou DA, Konstantinou IK, Albanis TA (2006) Coupling of headspace solid phase microextraction with ultrasonic extraction for the determination of chlorinated pesticides in bird livers using gas chromatography. Anal Chim Acta 573–574:223–230. https://doi.org/10.1016/j.aca.2006.05.048

Lashgari M, Yamini Y (2019) An overview of the most common lab-made coating materials in solid phase microextraction. Talanta 191:283–306. https://doi.org/10.1016/j.talanta.2018.08.077

LeDoux M (2011) Analytical methods applied to the determination of pesticide residues in foods of animal origin. A review of the past two decades. J Chromatogr A 1218:1021–1036. https://doi.org/10.1016/j.chroma.2010.12.097

Lee J, Lee HK, Rasmussen KE, Pedersen-Bjergaard S (2008) Environmental and bioanalytical applications of hollow fiber membrane liquid-phase microextraction: A review. Anal Chim Acta 624:253–268. https://doi.org/10.1016/j.aca.2008.06.050

Lehotay SJ, Eller KI (1995) Development of a method of analysis for 46 pesticides in fruits and vegetables by supercritical fluid extraction and gas chromatography/ion trap mass spectrometry. J AOAC Int 78:821–830. PMID: 7756898

Lehotay SJ, Lightfield AR, Harman-Fetcho JA, Donoghue DJ (2001) Analysis of pesticide residues in eggs by direct sample introduction/gas chromatography/tandem mass spectrometry. J Agric Food Chem 49:4589–4596. https://doi.org/10.1021/jf0104836

Lehotay SJ, de Kok A, Hiemstra M, Van Bodegraven P (2005a) Validation of a fast and easy method for the determination of residues from 229 pesticides in fruits and vegetables using gas and liquid chromatography and mass spectrometric detection. J AOAC Int 88:595–614. PMID: 15859089

Lehotay SJ, Mastovská K, Yun SJ (2005b) Evaluation of two fast and easy methods for pesticide residue analysis in fatty food matrixes. J AOAC Int 88:630–638. PMID: 15859091

León V, Álvarez B, Cobollo M, Muñoz S, Valor I (2003) Analysis of 35 priority semivolatile compounds in water by stir bar sorptive extraction–thermal desorption–gas chromatography–mass spectrometry. I. Method optimisation. J Chromatogr A 999:91–101. https://doi.org/10.1016/S0021-9673(03)00600-9

Leong MI, Huang SD (2009) Dispersive liquid-liquid microextraction method based on solidification of floating organic drop for extraction of organochlorine pesticides in water samples. J Chromatogr A 1216:7645–7650. https://doi.org/10.1016/j.chroma.2009.09.004

Lerch O, Zboron J, Brady K Aybar J-L (2018) Highly sensitive determination of contaminants in surface water in the context of the EU Water Framework Directive using stir bar sorptive extraction (SBSE) and GC-MS/MS. Gerstel Appl Note No 196. Available on-line at: http://www.gerstel.es/pdf/AppNote-196.pdf. Accessed on Sept 2019

Li B, Zeng F, Dong Q, Cao Y, Fan H, Deng C (2012) Rapid determination method for 12 pyrethroid pesticide residues in tea by stir bar sorptive extraction-thermal desorption-gas chromatography. Phys Procedia 25:1776–1780. https://doi.org/10.1016/j.phpro.2012.03.310

Liang P, Wang F, Wan Q (2013) Ionic liquid-based ultrasound-assisted emulsification microextraction coupled with high performance liquid chromatography for the determination of four fungicides in environmental water samples. Talanta 105:57–62. https://doi.org/10.1016/j.talanta.2012.11.065

Liu J, Jiang G, Chi Y, Cai Y, Zhou Q, Hu J-T (2003) Use of ionic liquids for liquid-phase microextraction of polycyclic aromatic hydrocarbons. Anal Chem 75:5870–5876. https://doi.org/10.1021/ac034506m

Liu W, Hu Y, Zhao J, Xu Y, Guan Y (2005) Determination of organophosphorus pesticides in cucumber and potato by stir bar sorptive extraction. J Chromatogr A 1095:1–7. https://doi.org/10.1016/j.chroma.2005.07.107

Liu ZM, Zang XH, Liu WH, Wang C, Wang Z (2009) Novel method for the determination of five carbamate pesticides in water samples by dispersive liquid-liquid microextraction combined with high performance liquid chromatography. Chinese Chem Lett 20:213–216. https://doi.org/10.1016/j.cclet.2008.10.047

Liu Z, Liu W, Rao H, Feng T, Li C, Wang C, Wang Z (2012) Determination of some carbamate pesticides in watermelon and tomato samples by dispersive liquid-liquid microextraction combined with high performance liquid chromatography. Int J Environ Anal Chem 92:571–581. https://doi.org/10.1080/03067311003628638

López-Feria S, Cárdenas S, Valcárcel M (2009) One step carbon nanotubes-based solid-phase extraction for the gas chromatographic–mass spectrometric multiclass pesticide control in virgin olive oils. J Chromatogr A 1216:7346–7350. https://doi.org/10.1016/j.chroma.2009.02.060

Lozano A, Rajski Ł, Uclés S, Belmonte-Valles N, Mezcua M, Fernández-Alba AR (2014) Evaluation of zirconium dioxide-based sorbents to decrease the matrix effect in avocado and almond multiresidue pesticide analysis followed by gas chromatography tandem mass spectrometry. Talanta 118:68–83. https://doi.org/10.1016/j.talanta.2013.09.053

Łozowicka B, Jankowska M, Kaczyński P (2012) Pesticide residues in *Brassica* vegetables and exposure assessment of consumers. Food Control 25:561–575. https://doi.org/10.1016/j.foodcont.2011.11.017

Łozowicka B, Rutkowska E, Jankowska M (2017) Influence of QuEChERS modifications on recovery and matrix effect during the multi-residue pesticide analysis in soil by GC/MS/MS and GC/ECD/NPD. Environ Sci Pollut Res 24:7124–7138. https://doi.org/10.1007/s11356-016-8334-1

Luque de Castro MD, Priego-Capote F (2010) Soxhlet extraction: Past and present panacea. J Chromatogr A 1217:2383–2389. https://doi.org/10.1016/j.chroma.2009.11.027

Lüthje K, Hyötyläinen T, Rautiainen-Rämä M, Riekkola M-L (2005) Pressurised hot water extraction-microporous membrane liquid-liquid extraction coupled on-line with gas chromatography-mass spectrometry in the analysis of pesticides in grapes. Analyst 130:52–58. https://doi.org/10.1039/b409276g

Madej K, Jonda A, Borcuch A, Piekoszewski W, Chmielarz L, Gil B (2019) A novel stir bar sorptive-dispersive microextraction in combination with magnetically modified graphene for isolation of seven pesticides from water samples. Microchem J 147:962–971. https://doi.org/10.1016/j.microc.2019.04.002

Manav ÖG, Dinç-Zor Ş, Alpdoğan G (2019) Optimization of a modified QuEChERS method by means of experimental design for multiresidue determination of pesticides in milk and dairy products by GC–MS. Microchem J 144:124–129. https://doi.org/10.1016/j.microc.2018.08.056

Margoum C, Guillemain C, Yang X, Coquery M (2013) Stir bar sorptive extraction coupled to liquid chromatography-tandem mass spectrometry for the determination of pesticides in water samples: Method validation and measurement uncertainty. Talanta 116:1–7. https://doi.org/10.1016/j.talanta.2013.04.066

Masiá A, Vásquez K, Campo J, Picó Y (2015) Assessment of two extraction methods to determine pesticides in soils, sediments and sludges. Application to the Túria River Basin. J Chromatogr A 1378:19–31. https://doi.org/10.1016/j.chroma.2014.11.079

Menezes HC, Paulo BP, Paiva MJN, Cardeal ZL (2016) A simple and quick method for the determination of pesticides in environmental water by HF-LPME-GC/MS. J Anal Methods Chem 2016:7058709. https://doi.org/10.1155/2016/7058709

Merdassa Y, Liu J, Megersa N (2013) Development of a one-step microwave-assisted extraction method for simultaneous determination of organophosphorus pesticides and fungicides in soils by gas chromatography–mass spectrometry. Talanta 114:227–234. https://doi.org/10.1016/j.talanta.2013.04.035

Mirabelli MF, Gionfriddo E, Pawliszyn J, Zenobi R (2018) A quantitative approach for pesticide analysis in grape juice by direct interfacing of a matrix compatible SPME phase to dielectric barrier discharge ionization-mass spectrometry. Analyst 143:891–899. https://doi.org/10.1039/c7an01663h

Moinfar S, Hosseini M-RM (2009) Development of dispersive liquid–liquid microextraction method for the analysis of organophosphorus pesticides in tea. J Hazard Mater 169:907–911. https://doi.org/10.1016/j.jhazmat.2009.04.030

Mol HGJ, Rooseboom A, van Dam R, Roding M, Arondeus K, Sunarto S (2007) Modification and re-validation of the ethyl acetate-based multi-residue method for pesticides in produce. Anal Bioanal Chem 389:1715–1754. https://doi.org/10.1007/s00216-007-1357-1

Moreno DV, Ferrera ZS, Rodríguez JJS (2007) SPME and SPE comparative study for coupling with microwave-assisted micellar extraction in the analysis of organochlorine pesticides residues in seaweed samples. Microchem J 87:139–146. https://doi.org/10.1016/j.microc.2007.07.002

Moreno-González R, León VM (2017) Presence and distribution of current-use pesticides in surface marine sediments from a Mediterranean coastal lagoon (SE Spain). Environ Sci Pollut Res 24:8033–8048. https://doi.org/10.1007/s11356-017-8456-0

Moreno-González D, Huertas-Pérez JF, García-Campaña AM, Gámiz-Gracia L (2014) Determination of carbamates in edible vegetable oils by ultra-high performance liquid chromatography-tandem mass spectrometry using a new clean-up based on zirconia for QuEChERS methodology. Talanta 128:299–304. https://doi.org/10.1016/j.talanta.2014.04.045

Nemoto S, Sasaki K, Toyoda M, Saito Y (1997) Effect of extraction conditions and modifiers on the supercritical fluid extraction of 88 pesticides. J Chromatogr Sci 35:467–477. https://doi.org/10.1093/chromsci/35.10.467

Neuwirthová N, Trojan M, Svobodová M, Vašíčková J, Šimek Z, Hofman J, Bielská L (2019) Pesticide residues remaining in soils from previous growing season(s) – can they accumulate in non-target organisms and contaminate the food web? Sci Total Environ 646:1056–1062. https://doi.org/10.1016/j.scitotenv.2018.07.357

Nguyen TD, Lee MH, Lee GH (2010) Rapid determination of 95 pesticides in soybean oil using liquid–liquid extraction followed by centrifugation, freezing and dispersive solid phase extraction as cleanup steps and gas chromatography with mass spectrometric detection. Microchem J 95:113–119. https://doi.org/10.1016/j.microc.2009.11.009

Nogueira JMF (2015) Stir-bar sorptive extraction: 15 years making sample preparation more environment-friendly. TrAC Trends Anal Chem 71:214–223. https://doi.org/10.1016/j.trac.2015.05.002

Norman KN, Panton SH (2001) Supercritical fluid extraction and quantitative determination of organophosphorus pesticide residues in wheat and maize using gas chromatography with flame photometric and mass spectrometric detection. J Chromatogr A 907:247–255. https://doi.org/10.1016/S0021-9673(00)01081-5

Ono Y, Yamagami T, Nishina T, Tobino T (2006) Pesticide multiresidue analysis of 303 compounds using supercritical fluid extraction. Anal Sci 22:1473–1476. https://doi.org/10.2116/analsci.22.1473

Pan J, Xia X-X, Liang J (2008) Analysis of pesticide multi-residues in leafy vegetables by ultrasonic solvent extraction and liquid chromatography-tandem mass spectrometry. Ultrason Sonochem 15:25–32. https://doi.org/10.1016/j.ultsonch.2007.06.005

Pano-Farias NS, Ceballos-Magaña SG, Muñiz-Valencia R, Jurado JM, Alcázar Á, Aguayo-Villarreal IA (2017) Direct immersion single drop micro-extraction method for multi-class pesticides analysis in mango using GC–MS. Food Chem 237:30–38. https://doi.org/10.1016/j.foodchem.2017.05.030

Parrilla-Vázquez P, Hakme E, Uclés S, Cutillas V, Martínez Galera M, Mughari AR, Fernández-Alba AR (2016) Large multiresidue analysis of pesticides in edible vegetable oils by using efficient solid-phase extraction sorbents based on quick, easy, cheap, effective, rugged and safe methodology followed by gas chromatography–tandem mass spectrometry. J Chromatogr A 1463:20–31. https://doi.org/10.1016/j.chroma.2016.08.008

Patel K, Fussell RJ, Hetmanski M, Goodall DM, Keely BJ (2005) Evaluation of gas chromatography–tandem quadrupole mass spectrometry for the determination of organochlorine pesticides in fats and oils. J Chromatogr A 1068:289–296. https://doi.org/10.1016/j.chroma.2005.01.040

Pedersen-Bjergaard S, Rasmussen KE (1999) Liquid−liquid−liquid microextraction for sample preparation of biological fluids prior to capillary electrophoresis. Anal Chem 71:2650–2656. https://doi.org/10.1021/ac990055n

Peng X, Pang J, Deng A (2011) Determination of seven phenoxyacid herbicides in environmental water by high performance liquid chromatography coupled with three phase hollow fiber liquid phase microextraction. Se Pu 29:1199–1204. PMID: 22500447

Pinto MI, Micaelo C, Vale C, Sontag G, Noronha JP (2014) Screening of priority pesticides in *Ulva* sp. seaweeds by selective pressurized solvent extraction before gas chromatography with electron capture detector analysis. Arch Environ Contam Toxicol 67:547–556. https://doi.org/10.1007/s00244-014-0038-2

Poole CF (2003) New trends in solid-phase extraction. TrAC Trends Anal Chem 22:362–373. https://doi.org/10.1016/S0165-9936(03)00605-8

Pose-Juan E, Cancho-Grande B, Rial-Otero R, Simal-Gándara J (2006) The dissipation rates of cyprodinil, fludioxonil, procymidone and vinclozoline during storage of grape juice. Food Control 17:1012–1017. https://doi.org/10.1016/j.foodcont.2005.07.009

Rashidi Nodeh H, Wan Ibrahim WA, Kamboh MA, Sanagi MM (2017) New magnetic graphene-based inorganic–organic sol-gel hybrid nanocomposite for simultaneous analysis of polar and non-polar organophosphorus pesticides from water samples using solid-phase extraction. Chemosphere 166:21–30. https://doi.org/10.1016/j.chemosphere.2016.09.054

Rastrelli L, Totaro K, De Simone F (2002) Determination of organophosphorus pesticide residues in Cilento (Campania, Italy) virgin olive oil by capillary gas chromatography. Food Chem 79:303–305. https://doi.org/10.1016/S0308-8146(02)00143-7

Ravelo-Pérez LM, Hernández-Borges J, Borges-Miquel TM, Rodríguez-Delgado MA (2008a) Pesticide analysis in tomatoes by solid-phase microextraction and micellar electrokinetic chromatography. J Chromatogr A 1185:151–154. https://doi.org/10.1016/j.chroma.2008.01.069

Ravelo-Pérez LM, Hernández-Borges J, Rodríguez-Delgado MÁ (2008b) Multi-walled carbon nanotubes as efficient solid-phase extraction materials of organophosphorus pesticides from apple, grape, orange and pineapple fruit juices. J Chromatogr A 1211:33–42. https://doi.org/10.1016/j.chroma.2008.09.084

Regueiro J, Llompart M, Garcia-Jares C, Garcia-Monteagudo JC, Cela R (2008) Ultrasound-assisted emulsification–microextraction of emergent contaminants and pesticides in environmental waters. J Chromatogr A 1190:27–38. https://doi.org/10.1016/j.chroma.2008.02.091

Ren ZF, Huang ZP, Xu JW, Wang JH, Bush P, Siegal MP, Provencio PN (1998) Synthesis of large arrays of well-aligned carbon nanotubes on glass. Science 282:1105–1107. https://doi.org/10.1126/science.282.5391.1105

Ren D, Sun C, Ma G, Yang D, Zhou C, Xie J, Li Y (2018) Determination of pyrethroids in tea brew by GC-MS combined with SPME with multiwalled carbon nanotube coated fiber. Int J Anal Chem 2018:8426598. https://doi.org/10.1155/2018/8426598

Rezaee M, Assadi Y, Milani Hosseini M-R, Aghaee E, Ahmadi F, Berijani S (2006) Determination of organic compounds in water using dispersive liquid–liquid microextraction. J Chromatogr A 1116:1–9. https://doi.org/10.1016/j.chroma.2006.03.007

Rezić I, Horvat AJM, Babić S, Kaštelan-Macan M (2005) Determination of pesticides in honey by ultrasonic solvent extraction and thin-layer chromatography. Ultrason Sonochem 12:477–481. https://doi.org/10.1016/j.ultsonch.2004.07.004

Rissato SR, Galhiane MS, Knoll FRN, Apon BM (2004) Supercritical fluid extraction for pesticide multiresidue analysis in honey: determination by gas chromatography with electron-capture and mass spectrometry detection. J Chromatogr A 1048:153–159. https://doi.org/10.1016/j.chroma.2004.07.053

Rissato SR, Galhiane MS, Apon BM, Arruda MSP (2005) Multiresidue analysis of pesticides in soil by supercritical fluid extraction/gas chromatography with electron-capture detection and confirmation by gas chromatography-mass spectrometry. J Agric Food Chem 53:62–69. https://doi.org/10.1021/jf048772s

Rodrigues SA, Caldas SS, Primel EG (2010) A simple; efficient and environmentally friendly method for the extraction of pesticides from onion by matrix solid-phase dispersion with liquid chromatography–tandem mass spectrometric detection. Anal Chim Acta 678:82–89. https://doi.org/10.1016/j.aca.2010.08.026

Romero-González R, Pastor-Montoro E, Martínez-Vidal JL, Garrido-Frenich A (2006) Application of hollow fiber supported liquid membrane extraction to the simultaneous determination of pesticide residues in vegetables by liquid chromatography/mass spectrometry. Rapid Commun Mass Spectrom 20:2701–2708. https://doi.org/10.1002/rcm.2653

Rutkowska E, Łozowicka B, Kaczyński P (2018) Modification of multiresidue QuEChERS protocol to minimize matrix effect and improve recoveries for determination of pesticide residues in dried herbs followed by GC-MS/MS. Food Anal Methods 11:709–724. https://doi.org/10.1007/s12161-017-1047-3

Salemi A, Rasoolzadeh R, Nejad MM, Vosough M (2013) Ultrasonic assisted headspace single drop micro-extraction and gas chromatography with nitrogen-phosphorus detector for determination of organophosphorus pesticides in soil. Anal Chim Acta 769:121–126. https://doi.org/10.1016/j.aca.2013.01.054

Samsidar A, Siddiquee S, Shaarani SM (2018) A review of extraction, analytical and advanced methods for determination of pesticides in environment and foodstuffs. Trends Food Sci Technol 71:188–201. https://doi.org/10.1016/j.tifs.2017.11.011

San Román I, Alonso ML, Bartolomé L, Alonso RM (2012) Hollow fibre-based liquid-phase microextraction technique combined with gas chromatography–mass spectrometry for the determination of pyrethroid insecticides in water samples. Talanta 100:246–253. https://doi.org/10.1016/j.talanta.2012.04.039

Sanagi MM, Abbas HH, Ibrahim WAW, Aboul-Enien HY (2012) Dispersive liquid-liquid microextraction method based on solidification of floating organic droplet for the determination of triazine herbicides in water and sugarcane samples. Food Chem 133:557–562. https://doi.org/10.1016/j.foodchem.2012.01.036

Sanchez Costa L, Rodríguez Martínez P, Medina Sala M (2018) Determination of 23 organochlorine pesticides in animal feeds by GC-MS/MS after QuEChERS with EMR-lipid clean-up. Anal Methods 10:5171–5180. https://doi.org/10.1039/c8ay01436a

Sánchez R, Cortes JM, Vazquez A, Villén-Altamirano J, Villén J (2006) Analysis of pesticide residues by online reversed-phase liquid chromatography-gas chromatography in the oil from olives grown in an experimental plot. Part II. J Sci Food Agric 86:1926–1931. https://doi.org/10.1002/jsfa.2565

Sandra P, Tienpont B, Vercammen J, Tredoux A, Sandra T, David F (2001) Stir bar sorptive extraction applied to the determination of dicarboximide fungicides in wine. J Chromatogr A 928:117–126. https://doi.org/10.1016/S0021-9673(01)01113-X

Sandra P, Tienpont B, David F (2003) Multi-residue screening of pesticides in vegetables, fruits and baby food by stir bar sorptive extraction–thermal desorption–capillary gas chromatography–mass spectrometry. J Chromatogr A 1000:299–309. https://doi.org/10.1016/S0021-9673(03)00508-9

Santana-Mayor Á, Socas-Rodríguez B, Herrera-Herrera AV, Rodríguez-Delgado MÁ (2019) Current trends in QuEChERS method. A versatile procedure for food, environmental and biological analysis. TrAC Trends Anal Chem 116:214–235. https://doi.org/10.1016/j.trac.2019.04.018

Sarafraz-Yazdi A, Amiri A (2010) Liquid-phase microextraction. TrAC Trends Anal Chem 29:1–14. https://doi.org/10.1016/j.trac.2009.10.003

Sartori RB, Higino ML, Bastos LHP, Mendes MF (2017) Supercritical extraction of pesticides from banana: Experimental and modeling. J Supercrit Fluids 128:149–158. https://doi.org/10.1016/j.supflu.2017.05.027

Shen G, Lee HK (2002) Hollow fiber-protected liquid-phase microextraction of triazine herbicides. Anal Chem 74:648–654. https://doi.org/10.1021/ac010561o

Snyder JL, Grob RL, McNally ME, Oostdyk TS (1993) The effect of instrumental parameters and soil matrix on the recovery of organochlorine and organophosphate pesticides from soils using supercritical fluid extraction. J Chromatogr Sci 31:183–191. https://doi.org/10.1093/chromsci/31.5.183

Song W, Zhang Y, Li G, Chen H, Wang H, Zhao Q, He D, Zhao C, Ding L (2014) A fast, simple and green method for the extraction of carbamate pesticides from rice by microwave assisted steam extraction coupled with solid phase extraction. Food Chem 143:192–198. https://doi.org/10.1016/j.foodchem.2013.07.101

Souza Silva ÉA, Pawliszyn J (2012) Optimization of fiber coating structure enables direct immersion solid phase microextraction and high-throughput determination of complex samples. Anal Chem 84:6933–6938. https://doi.org/10.1021/ac301305u

Souza-Silva ÉA, Gionfriddo E, Pawliszyn J (2015a) A critical review of the state of the art of solid-phase microextraction of complex matrices II. Food analysis. TrAC Trends Anal Chem 71:236–248. https://doi.org/10.1016/j.trac.2015.04.018

Souza-Silva ÉA, Ruifen J, Rodríguez-Lafuente A, Gionfriddo E, Pawliszyn J (2015b) A critical review of the state of the art of solid-phase microextraction of complex matrices I. Environmental analysis. Trends Anal Chem 71:224–235. https://doi.org/10.1016/j.trac.2015.04.016

Sparr Eskilsson C, Björklund E (2000) Analytical-scale microwave-assisted extraction. J Chromatogr A 902:227–250. https://doi.org/10.1016/S0021-9673(00)00921-3

Stocka J, Tankiewicz M, Biziuk M, Namieśnik J (2011) Green aspects of techniques for the determination of currently used pesticides in environmental samples. Int J Mol Sci 12:7785–7805. https://doi.org/10.3390/ijms12117785

Sun X, Zhu F, Xi J, Lu T, Liu H, Tong Y, Ouyang G (2011) Hollow fiber liquid-phase microextraction as clean-up step for the determination of organophosphorus pesticides residues in fish tissue by gas chromatography coupled with mass spectrometry. Mar Pollut Bull 63:102–107. https://doi.org/10.1016/j.marpolbul.2011.03.038

Tan GH, Vijayaletchumy K (1994) Determination of organochlorine pesticide residues in river sediments by Soxhlet extraction with hexane-acetone. Pestic Sci 40:121–126. https://doi.org/10.1002/ps.2780400206

Tian F, Liu W, Fang H, An M, Duan S (2014) Determination of six organophosphorus pesticides in water by single-drop microextraction coupled with GC-NPD. Chromatographia 77:487–492. https://doi.org/10.1007/s10337-013-2609-1

Tian L, Huang D, Shi Y, Han F, Wang Y, Ye H, Tang Y, Yu H (2019) Method for the analysis of 7 indictor polychlorinated biphenyls (PCBs) and 13 organochlorine pesticide residues in sediment by gas chromatography (GC). IOP Conf Ser Earth Environ Sci 237:022053. https://doi.org/10.1088/1755-1315/237/2/022053

Tuzimski T, Rejczak T (2016) Application of HPLC-DAD after SPE/QuEChERS with ZrO_2-based sorbent in d-SPE clean-up step for pesticide analysis in edible oils. Food Chem 190:71–79. https://doi.org/10.1016/j.foodchem.2015.05.072

Valverde-García A, Fernández-Alba AR, Agüera A, Contreras M (1995) Extraction of methamidophos residues from vegetables with supercritical fluid carbon dioxide. J AOAC Int 78:867–873. PMID: 7756903

Vazquez-Roig P, Picó Y (2015) Pressurized liquid extraction of organic contaminants in environmental and food samples. TrAC Trends Anal Chem 71:55–64. https://doi.org/10.1016/j.trac.2015.04.014

Walorczyk S (2008) Development of a multi-residue method for the determination of pesticides in cereals and dry animal feed using gas chromatography–tandem quadrupole mass spectrometry. II. Improvement and extension to new analytes. J Chromatogr A 1208:202–214. https://doi.org/10.1016/j.chroma.2008.08.068

Wan Ibrahim WA, Nodeh HR, Aboul-Enein HY, Sanagi MM (2015) Magnetic solid-phase extraction based on modified ferum oxides for enrichment, preconcentration, and isolation of pesticides and selected pollutants. Crit Rev Anal Chem 45:270–287. https://doi.org/10.1080/10408347.2014.938148

Wang S, Zhao P, Min G, Fang G (2007a) Multi-residue determination of pesticides in water using multi-walled carbon nanotubes solid-phase extraction and gas chromatography–mass spectrometry. J Chromatogr A 1165:166–171. https://doi.org/10.1016/j.chroma.2007.07.061

Wang W, Meng B, Lu X, Liu Y, Tao S (2007b) Extraction of polycyclic aromatic hydrocarbons and organochlorine pesticides from soils: A comparison between Soxhlet extraction, microwave-assisted extraction and accelerated solvent extraction techniques. Anal Chim Acta 602:211–222. https://doi.org/10.1016/j.aca.2007.09.023

Wang D, You J, Lydy MJ (2010) Sediment matrix effects in analysis of pyrethroid insecticides using gas chromatography–mass spectrometry. Arch Environ Contam Toxicol 59:382–392. https://doi.org/10.1007/s00244-010-9506-5

Wang H, Li G, Zhang Y, Chen H, Zhao Q, Song W, Xu Y, Jin H, Ding L (2012a) Determination of triazine herbicides in cereals using dynamic microwave-assisted extraction with solidification of floating organic drop followed by high-performance liquid chromatography. J Chromatogr A 1233:36–43. https://doi.org/10.1016/j.chroma.2012.02.034

Wang H, Yan H, Qiao J (2012b) Miniaturized matrix solid-phase dispersion combined with ultrasound-assisted dispersive liquid-liquid microextraction for the determination of three pyrethroids in soil. J Sep Sci 35:292–298. https://doi.org/10.1002/jssc.201100753

Wang R, Su P, Zhong Q, Zhang Y, Yang Y (2013) Ionic liquid-based microwave-assisted extraction of organochlorine pesticides from soil. J Liq Chromatogr Relat Technol 36:687–699. https://doi.org/10.1080/10826076.2012.673207

Wang Y, Sun Y, Xu B, Li X, Wang X, Zhang H, Song D (2015) Matrix solid-phase dispersion coupled with magnetic ionic liquid dispersive liquid–liquid microextraction for the determination of triazine herbicides in oilseeds. Anal Chim Acta 888:67–74. https://doi.org/10.1016/j.aca.2015.07.028

Wang Y-L, You L-Q, Mei Y-W, Liu J-P, He L-J (2016) Benzyl functionalized ionic liquid as new extraction solvent of dispersive liquid-liquid microextraction for enrichment of organophosphorus pesticides and aromatic compounds. Chinese J Anal Chem 44:942–949. https://doi.org/10.1016/S1872-2040(16)60937-4

Wang Q, Chen R, Shatner W, Cao Y, Bai Y (2019) State-of-the-art on the technique of dispersive liquid-liquid microextraction. Ultrason Sonochem 51:369–377. https://doi.org/10.1016/j.ultsonch.2018.08.010

Wilkowska A, Biziuk M (2011) Determination of pesticide residues in food matrices using the QuEChERS methodology. Food Chem 125:803–812. https://doi.org/10.1016/j.foodchem.2010.09.094

Williams DBG, George MJ, Meyer R, Marjanovic L (2011) Bubbles in solvent microextraction: the influence of intentionally introduced bubbles on extraction efficiency. Anal Chem 83:6713–6716. https://doi.org/10.1021/ac201323z

Williams DBG, George MJ, Marjanovic L (2014) Rapid detection of atrazine and metolachlor in farm soils: gas chromatography–mass spectrometry-based analysis using the bubble-in-drop single drop microextraction enrichment method. J Agric Food Chem 62:7676–7681. https://doi.org/10.1021/jf502411t

Wu J, Ee KH, Lee HK (2005) Automated dynamic liquid–liquid–liquid microextraction followed by high-performance liquid chromatography-ultraviolet detection for the determination of phenoxy acid herbicides in environmental waters. J Chromatogr A 1082:121–127. https://doi.org/10.1016/j.chroma.2005.05.077

Wu C, Liu H, Liu W, Wu Q, Wang C, Wang Z (2010) Determination of organophosphorus pesticides in environmental water samples by dispersive liquid–liquid microextraction with solidification of floating organic droplet followed by high-performance liquid chromatography. Anal Bioanal Chem 397:2543–2549. https://doi.org/10.1007/s00216-010-3790-9

Wu S, Jin T, Cheng J, Zhou H, Cheng M (2015) Air-assisted liquid liquid-microextraction for the analysis of fungicides from environmental water and juice samples. J Chromatogr Sci 53:1007–1012. https://doi.org/10.1093/chromsci/bmu136

Wu L, Hu M, Li Z, Song Y, Yu C, Zhang H, Yu A, Ma Q, Wang Z (2016) Dynamic microwave-assisted extraction combined with continuous-flow microextraction for determination of pesticides in vegetables. Food Chem 192:596–602. https://doi.org/10.1016/j.foodchem.2015.07.055

Xiao Q, Hu B, Yu C, Xia L, Jiang Z (2006) Optimization of a single-drop microextraction procedure for the determination of organophosphorus pesticides in water and fruit juice with gas chromatography-flame photometric detection. Talanta 69:848–855. https://doi.org/10.1016/j.talanta.2005.11.024

Yang Y, Liu M, Xu S, Hou L, Ou D, Liu H, Cheng S, Hofmann T (2006) HCHs and DDTs in sediment-dwelling animals from the Yangtze Estuary, China. Chemosphere 62:381–389. https://doi.org/10.1016/j.chemosphere.2005.04.102

Ye C-L, Zhou Q-X, Wang X-M (2006) Headspace liquid-phase microextraction using ionic liquid as extractant for the preconcentration of dichlorodiphenyltrichloroethane and its metabolites at trace levels in water samples. Anal Chim Acta 572:165–171. https://doi.org/10.1016/j.aca.2006.05.052

Yiantzi E, Psillakis E, Tyrovola K, Kalogerakis N (2010) Vortex-assisted liquid-liquid microextraction of octylphenol, nonylphenol and bisphenol-A. Talanta 80:2057–2062. https://doi.org/10.1016/j.talanta.2009.11.005

You J, Wang D, Lydy MJ (2010) Determination of pyrethroid insecticides in sediment by gas chromatography—Ion trap tandem mass spectrometry. Talanta 81:136–141. https://doi.org/10.1016/j.talanta.2009.11.050

You X, Xing Z, Liu F, Jiang N (2013) Air-assisted liquid-liquid microextraction used for the rapid determination of organophosphorus pesticides in juice samples. J Chromatogr A 1311:41–47. https://doi.org/10.1016/j.chroma.2013.08.080

Yu X, Yang H (2017) Pyrethroid residue determination in organic and conventional vegetables using liquid-solid extraction coupled with magnetic solid phase extraction based on polystyrene-coated magnetic nanoparticles. Food Chem 217:303–310. https://doi.org/10.1016/j.foodchem.2016.08.115

Yu H, Ho TD, Anderson JL (2013) Ionic liquid and polymeric ionic liquid coatings in solid-phase microextraction. TrAC Trends Anal Chem 45:219–232. https://doi.org/10.1016/j.trac.2012.10.016

Zayats MF, Leschev SM, Petrashkevich NV, Zayats MA, Kadenczki L, Szitás R, Dobrik HS, Keresztény N (2013) Distribution of pesticides in *n*-hexane/water and *n*-hexane/acetonitrile systems and estimation of possibilities of their extraction isolation and preconcentration from various matrices. Anal Chim Acta 774:33–43. https://doi.org/10.1016/j.aca.2013.03.003

Zhang J, Lee HK (2006) Application of liquid-phase microextraction and on-column derivatization combined with gas chromatography–mass spectrometry to the determination of carbamate pesticides. J Chromatogr A 1117:31–37. https://doi.org/10.1016/j.chroma.2006.03.102

Zhang J, Lee HK (2010) Headspace ionic liquid-based microdrop liquid-phase microextraction followed by microdrop thermal desorption-gas chromatographic analysis. Talanta 81:537–542. https://doi.org/10.1016/j.talanta.2009.12.039

Zhang S, Li C, Song S, Feng T, Wang C, Wang Z (2010) Application of dispersive liquid-liquid microextraction combined with sweeping micellar electrokinetic chromatography for trace analysis of six carbamate pesticides in apples. Anal Methods 2:54–62. https://doi.org/10.1039/b9ay00115h

Zhang S, Yin X, Yang Q, Wang C, Wang Z (2011) Determination of some sulfonylurea herbicides in soil by a novel liquid-phase microextraction combined with sweeping micellar electrokinetic chromatography. Anal Bioanal Chem 401:1071–1081. https://doi.org/10.1007/s00216-011-5138-5

Zhang S, Yang X, Yin X, Wang C, Wang Z (2012a) Dispersive liquid-liquid microextraction combined with sweeping micellar electrokinetic chromatography for the determination of some neonicotinoid insecticides in cucumber samples. Food Chem 133:544–550. https://doi.org/10.1016/j.foodchem.2012.01.028

Zhang Y, Wang X, Lin C, Fang G, Wang S (2012b) A novel SPME fiber chemically linked with 1-vinyl-3-hexadecylimidazolium hexafluorophosphate ionic liquid coupled with GC for the simultaneous determination of pyrethroids in vegetables. Chromatographia 75:789–797. https://doi.org/10.1007/s10337-012-2244-2

Zhang Y, Zhang X, Jiao B (2014) Determination of ten pyrethroids in various fruit juices: Comparison of dispersive liquid-liquid microextraction sample preparation and QuEChERS method combined with dispersive liquid-liquid microextraction. Food Chem 159:367–373. https://doi.org/10.1016/j.foodchem.2014.03.028

Zhang L, Gionfriddo E, Acquaro V, Pawliszyn J (2018) Direct immersion solid-phase microextraction analysis of multi-class contaminants in edible seaweeds by gas chromatography-mass spectrometry. Anal Chim Acta 1031 83–97. https://doi.org/10.1016/j.aca.2018.05.066

Zhao E, Han L, Jiang S, Wang Q, Zhou Z (2006) Application of a single-drop microextraction for the analysis of organophosphorus pesticides in juice. J Chromatogr A 1114:269–273. https://doi.org/10.1016/j.chroma.2006.03.011

Zhao E, Zhao W, Han L, Jiang S, Zhou Z (2007) Application of dispersive liquid–liquid microextraction for the analysis of organophosphorus pesticides in watermelon and cucumber. J Chromatogr A 1175:137–140. https://doi.org/10.1016/j.chroma.2007.10.069

Zhao R-S, Diao C-P, Wang X, Jiang T, Yuan J-P (2008) Rapid determination of amide herbicides in environmental water samples with dispersive liquid–liquid microextraction prior to gas chromatography–mass spectrometry. Anal Bioanal Chem 391:2915–2921. https://doi.org/10.1007/s00216-008-2208-4

Zhao P, Wang L, Zhou L, Zhang F, Kang S, Pan C (2012) Multi-walled carbon nanotubes as alternative reversed-dispersive solid phase extraction materials in pesticide multi-residue analysis with QuEChERS method. J Chromatogr A 1225:17–25. https://doi.org/10.1016/j.chroma.2011.12.070

Zheng G, Han C, Liu Y, Wang J, Zhu M, Wang C, Shen Y (2014) Multiresidue analysis of 30 organochlorine pesticides in milk and milk powder by gel permeation chromatography-solid phase extraction-gas chromatography-tandem mass spectrometry. J Dairy Sci 97:6016–6026. https://doi.org/10.3168/jds.2014-8192

Zhou Q, Wang W, Xiao J (2006) Preconcentration and determination of nicosulfuron, thifensulfuron-methyl and metsulfuron-methyl in water samples using carbon nanotubes packed cartridge in combination with high performance liquid chromatography. Anal Chim Acta 559:200–206. https://doi.org/10.1016/j.aca.2005.11.079

Zougagh M, Valcárcel M, Ríos A (2004) Supercritical fluid extraction: A critical review of its analytical usefulness. TrAC Trends Anal Chem 23:399–405. https://doi.org/10.1016/S0165-9936(04)00524-2

Zuin VG, Schellin M, Montero L, Yariwake JH, Augusto F, Popp P (2006) Comparison of stir bar sorptive extraction and membrane-assisted solvent extraction as enrichment techniques for the determination of pesticide and benzo[a]pyrene residues in Brazilian sugarcane juice. J Chromatogr A 1114:180–187. https://doi.org/10.1016/j.chroma.2006.03.035

Chapter 3
Analysis of Pesticide Residues by on Line Coupled Liquid Chromatography-Gas Chromatography Using the Through Oven Transfer Adsorption Desorption Interface

Rosa M. Toledano Torres, Jesús Villén Altamirano, and Ana M. Vázquez Moliní

Abstract Pesticides, compounds of many different chemical natures, are used to protect crops from pests, and more than 800 are currently in use all over the word. However, while pesticides help improve crop production, the presence of pesticide residues in the environment and food cause harm to both the environment itself and human health. For this reason, pesticide residues need to be monitored and controlled in food and in the environment by sensitive and, if possible, rapid analytical methods. Gas Chromatography has been the most widely used technique to date, while large volume injection and on line coupled liquid chromatography-gas chromatography are considered suitable to reduce the time needed for analysis and increase sensitivity. For its part, the Through Oven Transfer Adsorption Desorption interface can be used to inject large volume of polar and non-polar solvents and also on line coupled liquid chromatography-gas chromatography when liquid chromatography is carried out in normal or reversed phase.

Using this interface, several methods have been developed to analyse pesticide residues in food and environmental samples by Large Volume Injection or by on line coupled liquid chromatography-gas chromatography. This chapter focuses on these methods, after presenting a description of the Through Oven Transfer Adsorption Desorption interface and its operation mode. All the presented methods practically eliminate the time-consuming sample preparation step and provide high sensitivity and good repeatability.

R. M. Toledano Torres · A. M. Vázquez Moliní (✉)
Facultad de Educación de Albacete, Departamento de Química-Física, Universidad de Castilla-La Mancha, Albacete, Spain
e-mail: rosam.toledano@uclm.es; ana.vazquez@uclm.es

J. V. Altamirano
Escuela Técnica Superior de Ingenieros Agrónomos y de Montes, Departamento de Ciencia y Tecnología Agroforestal y Genética, Universidad de Castilla-La Mancha, Albacete, Spain
e-mail: jesus.villen@uclm.es

Inamuddin et al. (eds.), *Sustainable Agriculture Reviews 48*, Sustainable Agriculture Reviews 48, https://doi.org/10.1007/978-3-030-54719-6_3

Keywords Food · Environmental safety · Pesticide residue analysis · Olive oil · Nuts · Environmental water · Vegetables · Large volume injection · On line coupled liquid chromatography-gas chromatography · Through oven transfer adsorption desorption Interface · Mass-spectrometry

3.1 Introduction

With global population projected to increase to above 9 billion by 2050, producing food and taking steps to ensure food safety to feed this growing population is a challenge (Carvalho 2006).

Pesticides are compounds with varying chemical natures that are used to protect crops and reduce losses caused by pests, pathogens and weeds, increasing crop productivity and so helping feed the world's population. Nowadays, around 800 diferrent molecules are used as pesticides all over the word (Picó et al. 2006). However, the use of pesticides in crop protection may lead to the appearance of residues, that is, chemical substances resulting from the use of the agrochemicals, including their metabolites and the compounds obtained in their degradation or reaction.

Pesticides residues cause harm to both the environment itself and human health. The main source of people exposure to pesticides are the food and the water we consume that may be contaminated due to wrong pesticide application in agriculture, the use of overdoses and harvest collection before the necessary time for their elimination (Chen et al. 2011; Bakirci et al. 2014; Handford et al. 2015).

Pesticide residues, especially those from organochlorine and organophosphorous compounds, are also found in soils, the atmosphere and in the aquatic environment in relatively high concentrations (Carvalho et al. 1997). Consequently, controlling their presence in food (Ahmed 2001; Hogendoorn and van Zoonen 2000; Tadeo et al. 2000, 2004) or the environment plays an important role in human health and biodiversity protection. Worldwide laboratories carry out more than 20,000 analyses every day to monitor pesticide residues in food, which are regulated by many legislative authorities throughout the world, all of which are concerned with the quality, efficacy, and safety in the use of pesticides (Picó et al. 2006). For instance, maximum residue levels for pesticides in food and water have been set by the European Union for more than 150 products (Harris 2002). In general, maximum residue levels are in the range of 0.01–10 mg/kg; for example, the maximum permitted concentration of pesticides in drinking water set by the European Union is 0.1 mg/l (EC 98/83/EC, 1998 Directive 80/778/EEC). However, it goes without saying that reliable analytical methods are necessary to enforce legal maximum residue levels.

One separation technique that is commonly used for the analysis of pesticide residues is gas chromatography because of its high power of separation. The availability of a variety of sensitive and selective detectors, such as electron capture detector and nitrogen-phosphorus detector adds to its attractions, while mass spectrometry provides structural information to recognise non-target contaminants, increases the specificity of target-pesticide identification and permits highly sensitive trace-level determination (García-Reyes et al. 2007; Picó et al. 2000, 2004). However, the sample cannot be introduced directly into the chromatography system, and must first be submitted to a preparation procedure, including extraction, clean-up and concentration. Sample treatment depends on the composition of the matrix and the polarity of the pesticide being analysed. There is no suitable method for extracting all pesticides from all matrices, and sample preparation normally requires the use of large amounts of organic solvents. Moreover, the sample preparation step is the principal source of error and consumes most of the analytical process time (Xu et al. 2017).

In analytical chemistry manual sample pre-treatment for analyte detection is increasingly cut to a minimum by the use of fully automated methods, such as large volume injection and on-line coupled liquid chromatography-gas chromatography. Many hundred microlitres can be introduced into gas chromatography when using large volume injection (López et al. 1998), which increases sensitivity and/or reduces the need for extract or sample concentration steps (Hoh and Mastovska 2008). In on-line coupled liquid chromatography-gas chromatography, the specific components of a complex matrix are pre-fractionated using liquid chromatography before being transferred on-line to the very efficient and sensitive gas chromatography system for analytical separation. Liquid chromatography deals with sample preparation, including extraction, clean up and preconcentration, and the final separation is carried out by the more efficient gas chromatography (Toledano et al. 2010).

The main drawback of large volume injection and on-line coupled liquid chromatography-gas chromatography is basically the same: how to inject into the gas chromatograph a volume much higher than that usually introduced in this chromatographic system. The problem is the same without regard the origin of the volume: from the sample, the extract or the eluent from the liquid chromatography. This critical problem is the huge solvent vapour volume resulting from the expansion of the large liquid volume of the injected solvent. To overcome this drawback a robust and, if possible, automated interface is required. Any such interface should fulfil some requirements: do no cause losses or contaminations of the analytes when transferring then from the liquid chromatography to the gas chromatography, be able to remove practically the totally of the eluent of liquid chromatography and refocuse the analytes at the head of the gas chromatography column to obtain sharp peaks in the gas chromatography chromatogram (Cortés et al. 2010). The Through Oven Transfer Adsorption Desorption interface, first described by Villen et al. in 1999 (Pérez et al. 1999), allows the large volume injection of polar and non-polar solvents. On-line coupled liquid chromatography-gas chromatography can then be used, with liquid chromatography in normal or reversed phase. The Through Oven Transfer Adsorption Desorption interface allows the injection of much larger volumes than other injection systems, while maintaining good chromatographic conditions. Several analytical methods for analysing pesticide residues in food and environmental samples have been developed using this interface with large volume injection or on line coupled liquid chromatography-gas chromatography. This chapter focuses on these methods, but first a description of the Through Oven Transfer Adsorption Desorption interface and its operation mode is presented.

3.2 Through Oven Transfer Adsorption Desorption Interface

Several methods to analyse pesticides at trace level concentrations in different complex matrices have been developed using the Through Oven Transfer Adsorption Desorption interface (De la Peña et al. 2010).

The Through Oven Transfer Adsorption Desorption interface, is based on a widely modified programmed temperature vaporizer injector (Pérez et al. 1999, 2000). The changes introduced affect the pneumatic systems, sample introduction, solvent elimination and the operation mode. See Fig. 3.1 for a scheme of the Through Oven Transfer Adsorption Desorption interface. As can be seen, the modified programmed temperature vaporizer injector constitutes the body of the interface. A standard glass insert (liner) packed with an adsorbent or absorbent material held at both ends with silanized glass wool is placed inside the body of the interface. Two electro valves and a six-port valve also form part of the interface.

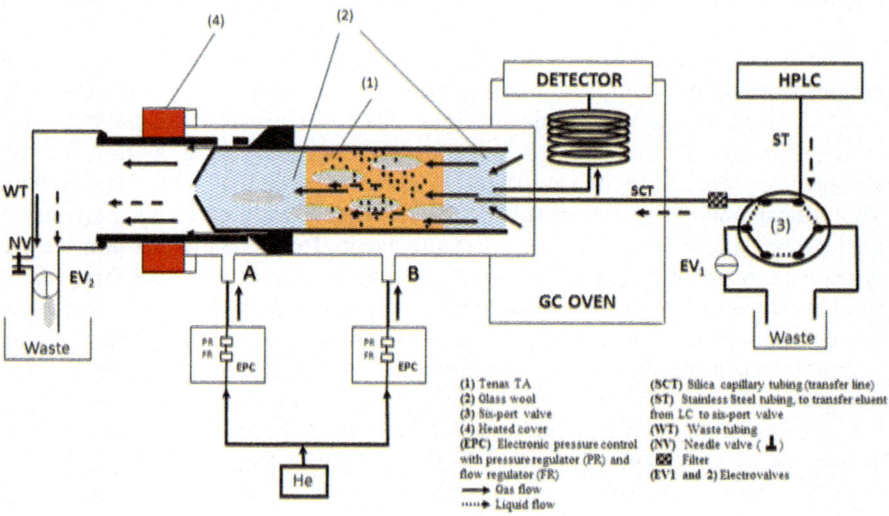

Fig. 3.1 Scheme of the Through Oven Transfer Adsorption Desorption interface during the injection step. Symbols: (1) sorbent (Tenax TA); (2) glass wool; (3) six-port valve; (4) heated cover; (EV1 and EV2) electrovalves 1 and 2; (EPC) electronic pressure control; (PR) pressure regulator; (FR) flow regulator; (solid arrows) gas flow; (dotted arrows) liquid flow; (ST) stainless steel tubing, 0.25 mm I.D., to transfer eluent from the liquid chromatograph to the six-port valve; (WT) waste tubing, to allow the exit of liquids and gases; (SCT) silica capillary tubing, 0.32 mm I. D.; NV (⊥) needle valve

The liquid chromatograph exit is connected to a six-port valve, which, is connected to waste through two of its ports, one of them with an electro valve called electrovalve 1. Another port is connected by a silica capillary tube to the body of the interface through the gas chromatography oven. The gas chromatography column is introduced through this side into the body of the interface. The gas chromatography column is placed less deeply in the body of the interface than the capillary tube which is connected to the six-port valve. On the other side of the body of the interface a waste tube, with an electrovalve called electrovalve 2, eliminates liquid and vapour to the waste when it is open. The interface has two gas inlets, one that is commonly found in a programmed temperature vaporizer injector (A) and another one that is normally used as split exit (B) in a programmed temperature vaporizer injector. At the beginning of the liquid chromatography process, the eluent is sent to waste through the six-port valve, while the carrier gas enters in the body of the interface through the two gas inlets mentioned above. Most of the carrier gas enters through gas inlet B, passes through the liner and exits through the waste tube. This gas flows in the opposite direction to the carrier gas into the gas chromatography column. A portion of this gas passes to the gas chromatography column. When the liquid chromatography fraction containing the target analyte reaches the six-port valve, it is switched and the liquid chromatography eluent is sent to the body of the interface. The adsorbent or absorbent material inside the liner retains the analytes, while the solvent is eliminated to the waste by the carrier gas that passes through the liner. The carrier gas entering through the usual inlet (A) prevents the solvent from condensing in the upper part of the body of the interface. The solvent is prevented from entering the gas chromatography column because the column is placed less deeply inside the liner than the silica capillary tube.

With the analytes retained in the liner, the six-port valve is switched and the electrovalve 1 is opened, so that the eluent from the liquid chromatography is sent to waste and the solvent remaining in the capillary tube is impelled by the gas pressure into the body of the interface. At the same time, the gas entering by the split exit (B) evaporates and drags the solvent through the liner to the waste. Once all the remaining solvent has been eliminated, electrovalves 1 and 2 are closed and the carrier gas entering the split exit (B) is closed. The carrier gas can now only enter through the usual inlet (A) and while it exits through the gas chromatography column. In these conditions, the body of the interface is heated to carry out the thermal desorption of the retained analytes. The carrier gas impels the analytes to the column, where they are separated and analysed (Cortés et al. 2010).

3.3 Pesticide Residue Analysis

Our research group has developed several analytical methods to determine pesticide residues in vegetable, edible oils, river water and nuts, using the above described Through Oven Transfer Adsorption Desorption interface by large volume injection into the gas chromatography and/or by on line coupled liquid chromatography-gas chromatography, depending on the nature of the matrix. Both techniques have a high sensitivity because they substantially increase the volume introduced into the gas chromatograph. However, while large volume injection allows any volume of sample or extract to be introduced, the volume injected using on line coupled liquid chromatography-gas chromatography depends on the load capacity of the liquid chromatography column. On line coupled liquid chromatography-gas chromatography eliminates the matrix components and therefore possible interfering substances in the liquid chromatography step, but in large volume injection possible interfering substances are concentrated, together with the analytes, into the body of the interface and consequently non-volatile compounds can be introduced into the system. Bearing this in mind, large volume injection can be regarded as a suitable technique for analysing pesticide residues present at trace level in a simple matrix, whereas on line coupled liquid chromatography-gas chromatography is more suitable for more complex matrixes.

Since vegetables are solid matrixes their analysis requires an extraction procedure to be carried out prior to chromatographic analysis, and as the obtained extract is quite clean, it can be analysed by large volume injection. In the case of river water samples, however, there are two possible courses of action: the samples used in this study came from the upper reaches of the river and so were relatively clean, making it possible to analyse them by direct injection using large volume injection. However, for samples taken from very contaminated river zones (for example, from stretches near cities or cultivated fields), large volume injection is not suitable and on line coupled liquid chromatography-gas chromatography should be used.

Edible oils are liquid samples with a high content of fat (98–99%) and they must be analysed by on line liquid chromatography-gas chromatography so that lipids from the matrix can be separated from the pesticides in the liquid chromatography step.

Nuts are solid samples with a high fat content (50–70%), so it is necessary to carry out a prior extraction procedure and then analyse the obtained extract by on line liquid chromatography-gas chromatography because the extract will usually be contaminated with small quantities of fat that must be separated from the pesticide before the gas chromatography analysis.

3.4 Pesticide Residue Analysis by Large Volume Injection

Large volume injection techniques allow the introduction of much higher volumes than are usually injected in gas chromatography, that is 1–2 μl (Teske and Engewald 2002). These techniques are very useful for analysing compounds at very low concentration levels because sensitivity is increased due to the higher amount of analyte that reaches the detector (Hoh and Mastovska 2008). Another advantage of large volume injection is that it reduces or even eliminates the extraction/concentration steps. Several authors have described the different techniques and methods possible with large volume injection in capillary gas chromatography, along with their advantages and disadvantages (Santos et al. 2019; Ferrari and Caldas 2018; Kurth et al. 2017). A sample or an extract can be injected. The most critical problem in large volume injection is the large amount of solvent vapour resulting from the large liquid volume injected. The vapour volume obtained from polar solvents is much greater than that obtained from non-polar solvent, meaning that large volume injection is more difficult with the former. For its part the Through Oven Transfer Adsorption Desorption interface allows the injection of up to several millilitres of sample or extract and it does not matter whether the solvent is non-polar or polar (Alario et al. 2001; Aragón et al. 2015).

3.4.1 *Pesticide Residue Analysis in Vegetables*

In this method, a small amount of sample (2.5 g) is mixed with 5 ml of ethyl acetate and 2 g of anhydrous sodium sulphate. After extracting for 1 min using a high-speed blender, the extract is filtered and then injected into the gas chromatograph by large volume injection using the Through Oven Transfer Adsorption Desorption interface. Using such a small sample size (in this case 2.5 g) means that a very small quantity of solvent (5 ml) is necessary for the extraction (Cortés et al. 2006a), whereas the majority of multiresidue methods use a 50–100 g sample, so that the volume of solvent used must be greater (about 100–250 ml). This provide an extract volume that then must be concentrated (up to 1 ml) to obtain the sensitivity necessary if only about 1 μl is to be sampled in the gas chromatograph (Podhorniak et al. 2001; Lehotay 2008; Gamón et al. 2001). The amount of solvent cannot be lower because of the difficulty in handling volumes of concentrated extract below 1 ml. The problem is that 99% of the analyte is wasted in such methods because it is not injected into the gas chromatography system, while in the analytical method developed by our research group, only 5 ml of extracting solvent is used because there is no concentration step after the extraction and the whole extract can be injected into the gas chromatograph.

A tomato and a cucumber sample fortified at 10 mg/kg together with a blanck trace provided the chromatograms shown in Fig. 3.2 when the flame ionisation detector was used. One peak, corresponding to malathion, eluted at the same retention time as another peak that corresponded to a matrix compound, suggesting that this pesticide might give a false positive if a flame ionisation detector is used. However, a nitrogen-phosphorus detector can be used to avoid this problem because of its greater selectivity. Figure 3.3 shows the chromatogram obtained when a tomato sample fortified at 0.05 mg/kg was analysed using an nitrogen phosphorus detector. The improvement in selectivity is clear from Figs. 3.2 to 3.3. Another important parameter to consider in the determination of pesticide residue is sensitivity, which increases with the volume injected. When a 50 μl extract was sampled the limit of detection was sufficient to monitor the pesticides at maximum residue levels. Table 3.1 shows the validation parameters when using an nitrogen phosphorus detector. As can be seen, good linearity was found for all the pesticides, the

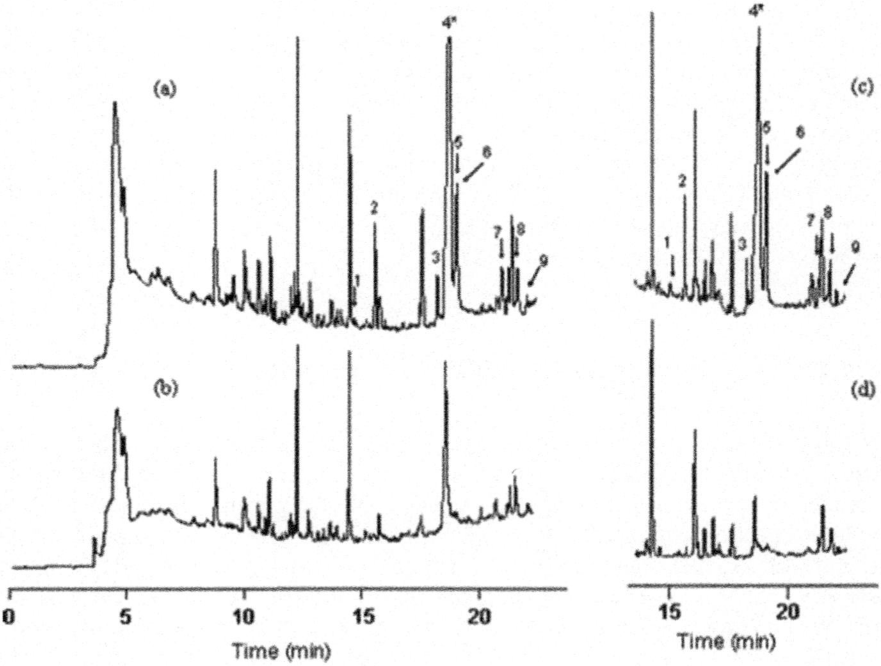

Fig. 3.2 Gas Chromatography chromatograms obtained in the Large Volume Injection-Gas Chromatography-Flame Ionisation Detector analysis of vegetal samples. On the left, the gas chromatography chromatograms correspond to (**a**) a tomato sample fortified at 10 mg/kg and (**b**) a blank trace. On the right the gas chromatography chromatograms correspond to (**c**) a cucumber sample fortified at 10 mg/kg and (**d**) a blank trace. The peaks identified correspond to the following: 1, dimethoate; 2, diazinon; 3, fenitrothion; 4, malathion plus matrix compounds; 5, fenthion; 6, chlorpyrifos; 7, chlorfenvinphos; 8, methidathion; 9, tetraclorvinphos. In both cases the volume of extract injected was 20 μL. Reprinted from Cortés et al. (2006a) with permission of ACS Publications and Copyright Clearance Center

determination coefficients ranging from 0.969 to 0.996. The coefficients of variation described in Table 3.1 correspond to the whole analytical procedure (extraction and gas chromatography analysis) and can be considered extremely good. Other vegetables (eggplant, lettuce, pepper, and cucumber) were also analysed using this procedure. The fact that the gas chromatography chromatograms were very similar to that of the tomato sample means that the matrix effect was low suggesting that the proposed method can be applied to different vegetables without having to be modified.

When the method was applied to the analysis of real tomato samples harvested from an experimental plot that had been treated with pesticides (Cortés et al. 2006a), the following pesticides were identified (μg/kg): dimethoate (13), diazinon (62), fenitrothion (24), chlorpyrifos (73), and methidathion (14). However, all were

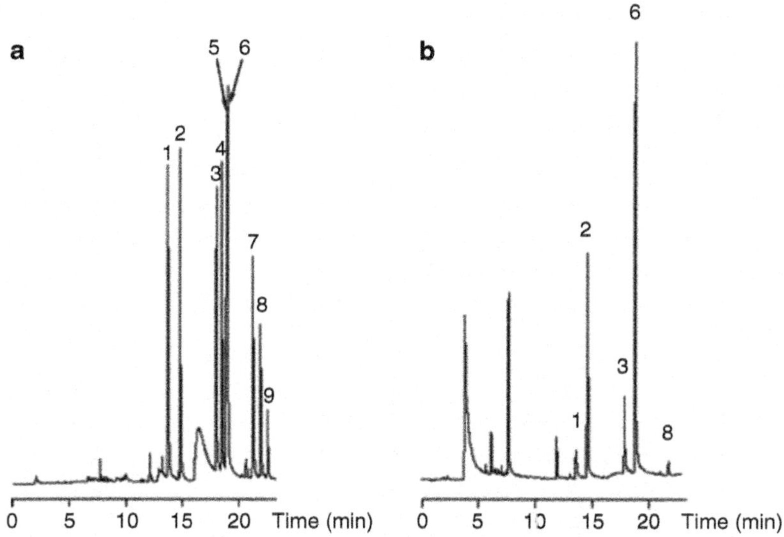

Fig. 3.3 (**a**) Gas Chromatography chromatogram obtained in the Large Volume Injection-Gas Chromatography-Nitrogen Phosphorus Detector analysis of a tomato sample fortified at 0.05 mg/kg. (**b**) Gas Chromatography chromatogram obtained in the analysis of a real tomato sample containing some pesticide residues. The identification of the peaks is the same as in Fig. 3.2. The volume of extract injected is 50 μL. Reprinted from Cortés et al. (2006a) with permission of ACS Publications and Copyright Clearance Center

Table 3.1 The coefficients of variation (CV) from the Absolute Peak Areas and from the Retention Times, repetition 3 times, of a Tomato Sample Fortified at 0.05 mg/kg with each pesticide, for the whole analytical procedure (Extraction and Gas Chromatography Analysis) when an Nitrogen Phosphorus Detector was used[a]

Pesticide	CV (Area)	CV (tr)	LOD (μg/kg)	R^2
Dimethoate	2.4	0.16	0.07	0.996
Diazinon	0.3	0.17	0.07	0.977
Fenitrothion	4.4	0.20	0.08	0.969
Malathion	3.6	0.21	0.07	0.991
Fenthion	8.4	0.21	0.06	0.969
Chlorpyrifos	2.0	0.20	0.06	0.977
Chlorfenvinphos	4.7	0.19	0.10	0.989
Methidathion	4.3	0.20	0.15	0.987
Tetrachlorvinphos	2.4	0.17	0.34	0.988

[a]Detection limits (LODs) calculated as the amount of product giving a signal equal to 5 times the background noise and correlation coefficients for the linear calibration (R^2)
Reprinted from Cortés et al. (2006a) with permission of ACS Publications and Copyright Clearance Center

present at lower concentrations than the maximum residues limits established by Spanish legislation Royal Decree 280/1994 (BOE 1994). When samples purchased from a local market were analysed, no residues of the target pesticides were found.

3.4.2 Pesticide Residue Analysis in River Water

Large volume injection can be used to analyse pesticide residues in river water if it is not very contaminated. To reach the detection limits required (normally expressed in parts-per-billion), at least 0.1 to 1 ml of water needs to be introduced into the gas chromatography system, but the polar nature of water makes the injection of such a high volume much more difficult than is the case with non-polar solvents. Fortunately, with the Through Oven Transfer Adsorption Desorption interface, 1 ml of water (and even higher volumes) can be injected, since the solvent is eliminated in evaporative and non-evaporative mode. The only condition is that the water has to be filtered prior to injection into the gas chromatography system (Alario et al. 2001). Figure 3.4 shows the chromatogram obtained when sampling 1 mL of a standard solution containing 10 μg/L of each pesticide. The small solvent peak in the chromatogram demonstrates that solvent elimination is almost complete. This means that a water-sensitive detector, such as an nitrogen phosphorus detector or even a mass spectrometry detector, can be used. The limits of detection, which ranged from 0.5 to 8.1 ng/l are much lower than the maximum concentration allowed for drinking water by European Union legislation (0.1 μg/l) (EC 98/83/EC, 1998). The coefficients of variations for the absolute peak areas varied from 5.7% to 11.7%, although this can be improved by 0.7–10.7% if normalized areas are considered.

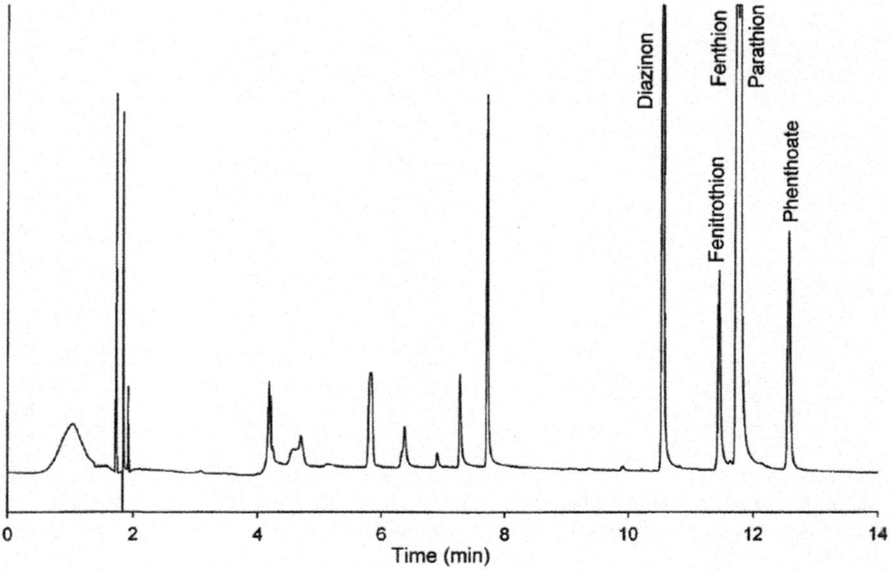

Fig. 3.4 Chromatogram obtained by sampling 1 ml of a standard solution containing 10 μg/l of each pesticide in methanol-water (10:90). Reprinted from Alario et al. (2001) with permission of Oxford University Press and Copyright Clearance Center

Because it provides unambiguous identification of pesticides, an mass spectrometry detector should be used whenever possible. However, it is important to totally eliminate the water because of its high degree of reactivity even a very small amounts, which can damage the mass spectrometer. A multiresidue analysis of pesticides in water by large volume injection in gas chromatography-mass spectrometry using the Through Oven Transfer Adsorption Desorption interface has been developed (Toledano et al. 2010), in which organophosphorus, organochlorine and triazine can determined in one run. As an mass spectrometry detector is less sensitive than an nitrogen phosphorus detector, a larger volume of water had to be injected. There was no problem with the volume injected even when a great volume of water (up to 5 mL) was injected, no solvent peak was observed. Figure 3.5 shows the chromatogram obtained in full scan mode and in selected-ion monitoring mode for some of the pesticides.

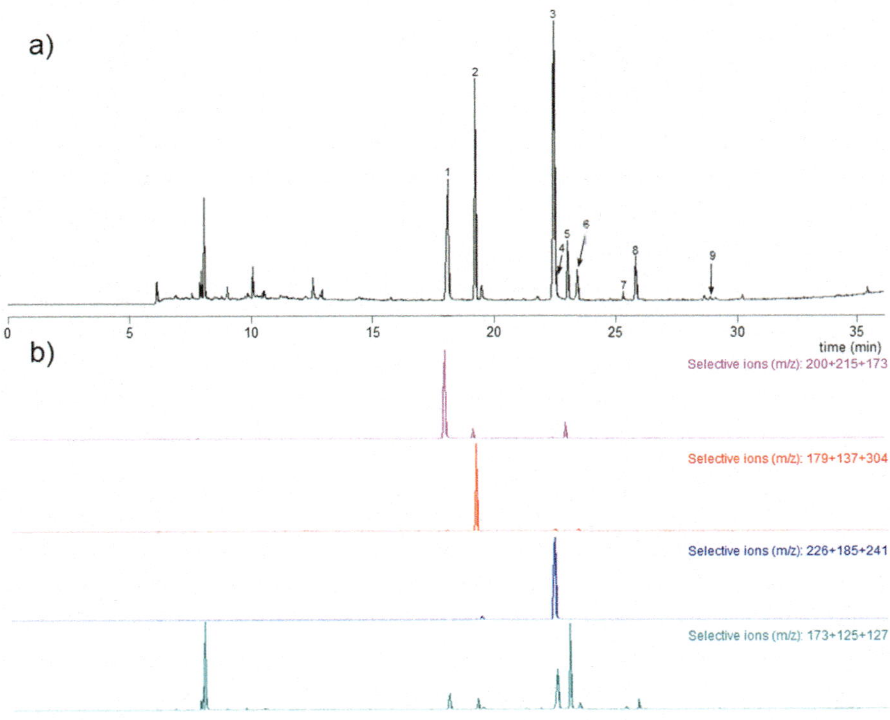

Fig. 3.5 Chromatograms obtained by Large Volume Injection-Gas Chromatography-Mass Spectrometry from a water sample spiked at 50 g/l of each pesticide: (1) atrazine, (2) diazinon, (3) terbutryne, (4) fenitrothion, (5) malathion, (6) parathion, (7) phenthoate, (8) methidathion and (9) Dichloro diphenil trichloroethane (DDT). Volume injected 5 ml. (a) Full scan mode and (b) Selected ion monitoring mode used for pesticide quantification for atrazine, diazinon, terbutryne and malathion. Reprinted from Toledano et al. (2010) with permission of Elservier and Copyright Clearance Center

3.5 Pesticide Residue Analysis by on Line Coupled Liquid Chromatography-Gas Chromatography

On line coupled liquid chromagraphy-gas chromatography is a multidimensional chromatographic technique that combines the effectiveness of pre-separation by liquid chromatography with the high efficiency of gas chromatography (Grob 1991). In the hyphenation of liquid chromatography and gas chromatography, in a heartcut system, the liquid chromatography process acts as a sample preparation step and target analytes are separated from the bulk of the complex matrix for selective clean-up, concentration or fractionation of the sample (Purcaro et al. 2013). The liquid chromatography fractions containing the compounds of interest are selected and then transferred to the gas chromatography system, avoiding the presence of non-volatile components in the gas chromatography system. To couple liquid chromatography to gas chromatography it is necessary to connect both chromatographic systems through an interface (Sciarrone et al. 2012). Such an interface should allow transfer of the analytes of interest included in the fraction from liquid chromatography, eliminate the eluent or solvent used in liquid chromatography with a volume varying from a few microlitres up to several millilitres, retain the analytes and transfer them to the gas chromatography system. However, if reversed phase is used in the liquid chromatography process, the polar nature of the solvent used makes it difficult to remove (Hyötyläinen and Riekkola 1998). For these reasons, most applications of on-line liquid chromatography-gas chromatography use normal phase liquid chromatography (Qi et al. 2014; Purcaro et al. 2012; Grob 2000). However, the majority of the liquid chromatography separation (70–80%) is performed in reversed-phase mode and, anyway, for the analysis of aqueous samples it is essential to use the reversed phase mode in the liquid chromatography system. The advantage of the Through Oven Transfer Adsorption Desorption interface is that it allows on-line coupled liquid chromatography-gas chromatography when liquid chromatography is carried out in normal (Aragón et al. 2011), as well in reversed phase (Flores et al. 2008; Cortés et al. 2006b).

On line coupled liquid chromatography-gas chromatography is a very appropieate techniques for the analysis trace level compounds in complex matrices. Liquid chromatography step allows to carry out the clean-up and the fractionation of the sample without the manipulation of the sample decreasing the risck of contamination and analyte losses (Hyötyläinen and Riekkola 2003; Purcaro et al. 2013).

3.5.1 Pesticide Residue Analysis in Water

To illustrate the applicability of the Through Oven Transfer Adsorption Desorption interface for analysing pesticide residues in water a method to analyse parathion was developed. The analysis involves off-line solid phase extraction and on-line reversed phase liquid chromatography-gas chromatography (Pérez et al. 1999). Figure 3.6 shows the gas chromatography chromatogram obtained when analysing a real water sample from the Ebro (a polluted Spanish river), in which parathion was quantified at 4 ng/l. Later, more pesticides were analysed (Pérez et al. 2000), fortifying real water samples collected from the River Ebro with the pesticides in question before carrying out solid phase extraction. When the solid phase extraction extract was directly injected into the gas chromatograph by large volume injection, the chromatogram obtained presented many peaks other than pesticides, which interfered with the analysis. In the analysis of dirty water sample, an on-line liquid chromatography-gas chromatography system must be used, because other volatile contaminants interfere in the analysis if large volume injection is used. On line coupled liquid chromatography-gas chromatography provides better separation of the pesticides from interference on the part of the matrix. The liquid chromatography step is not used as an analytical step, but as sample preparation. The first step is to establish the start and end times of the fraction to be transferred from liquid chromatography to gas chromatography in the liquid chromatography chromatogram. Figure 3.7 shows the liquid chromatography chromatogram obtained when 50 μl of a standard solution containing 100 mg/l of the target pesticides was analysed. This high concentration was necessary for the pesticides to be detected by the ultraviolet detector. The eluent volume transferred was calculated based on the flow rate in the liquid chromatography system (1 ml/min). In this case, volumes ranging from 500 to 1400 μl could be used, with methanol/water (70/30) as solvent. Since lowering the sample introduction rate increases sensitivity (Villén et al. 1999), the flow rate was reduced to 0.1 ml/min for the transfer step in order achieve the necessary sensitivity. In this way, the transfer time ranged from 5 to 14 min. The liquid chromatography chromatogram (Fig. 3.7) shows that atrazine eluted alone, and consequently the peak corresponding to this pesticide was the only important peak in the gas chromatography chromatogram when the liquid chromatography fraction containing atrazine was transferred to gas chromatography (Fig. 3.8a). However, in the liquid chromatography chromatogram, the peaks corresponding to phenthoate and parathion partially overlapped (Fig. 3.7). If only phenthoate is transferred to the gas chromatograph, the peak of this pesticide is the highest, although the parathion peak is still evident (Fig. 3.8b). If the selected cut corresponds to parathion, its peak is the highest, but phenthoate is also eluted (Fig. 3.8c). The retention times of diazinon and fenthion totally overlap in the liquid chromatography chromatogram while, two peaks of similar height can be seen in the gas chromatography chromatogram (Fig. 3.8d). The use of a multidimensional chromatographic system allows the peaks that overlap or elute together in the first dimension (liquid chromatography) to be separated in the second dimension of the system (gas chromatography).

The detection limits for the pesticides analysed ranged from 0.04 to 0.33 ng/l, except in the case of fenitrothion (1.54 ng/l). The injection of 50 μl of the solid phase extraction extract into the liquid chromatography system and then using on line liquid chromatography-gas chromatography for the analysis resulted in a high degree of high sensitivity, although the solid phase extraction step can be omitted

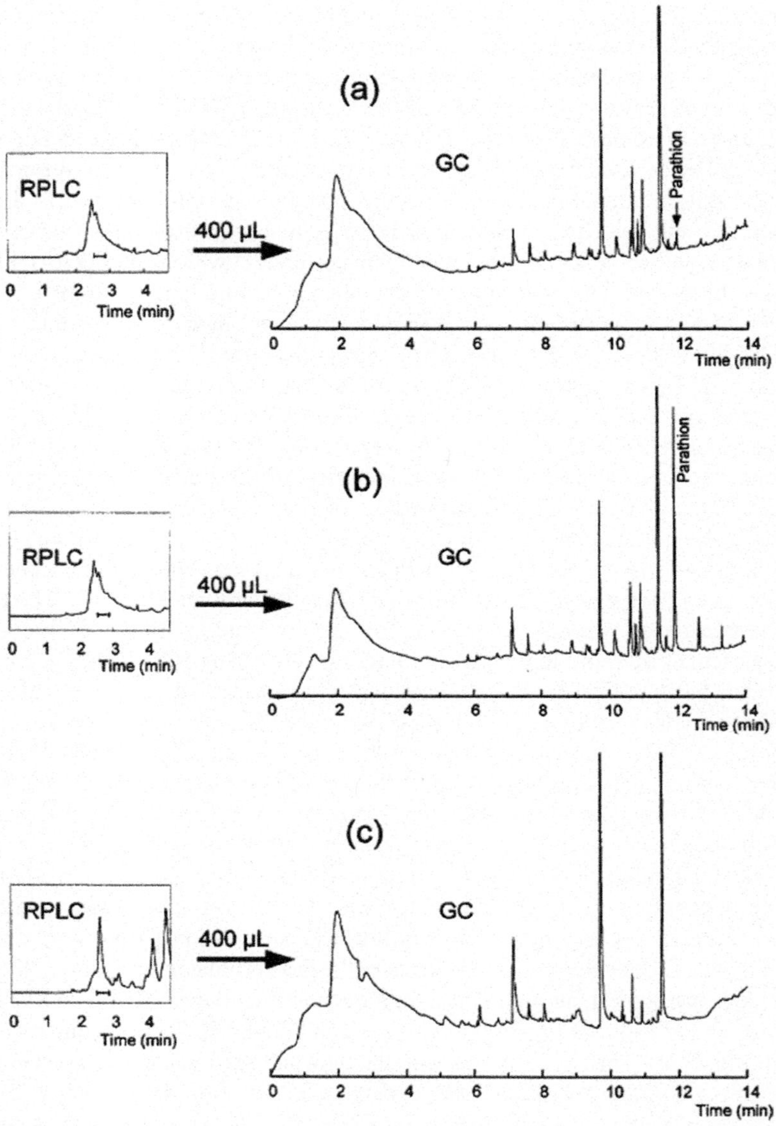

Fig. 3.6 Liquid and gas chromatograms of real water samples analysed by off-line solid phase extraction and on-line reversed phase liquid chromatography-gas chromatography. The parathion peak cannot be seen in the liquid chromatography chromatograms. The line below the liquid chromatography chromatogram indicates the fraction containing parathion, which was transferred to the gas chromatography. (**a**) Sample collected near the mouth of the River Ebro (Parathion gas chromatography peak corresponds to 4 ng/l). (**b**) The same SPE extract spiked with 50 µg/l of parathion. (**c**) sample collected upstream of the same river. Reprinted from Pérez et al. (1999) with permission of John Wiley and Sons and Copyright Clearance Center

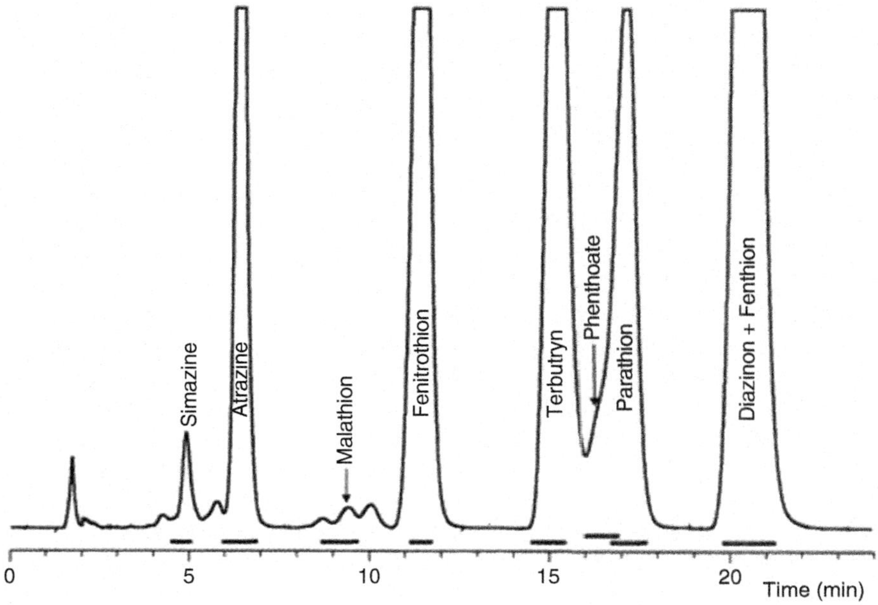

Fig. 3.7 Liquid chromatogram of a standard solution containing 100 mg/l of each of the nine pesticides selected for the experimentation. The phenthoate peak appears as a shoulder of the parathion peak. The thick lines between the time axis and the chromatogram indicate the liquid chromatography fractions which have been transferred from the liquid chromatograph to the gas chromatograph. Reprinted from Pérez et al. (2000) with permission of ACS Publications and Copyright Clearance Center

if lower sensitivity is required. The gas chromatography retention time did not vary from one liquid chromatography-gas chromatography analysis to another when the Through Oven Transfer Adsorption Desorption interface was used, but did vary when other interfaces were used (Grob 1991).

3.5.2 Pesticides Residue analysis in Edible Oils

Many of the pesticides used in the control of pests in olive trees are fat-soluble and non-polar and tend to concentrate in the oil. Analysing pesticide residues in edible oils is a difficult task because of the complexity of the matrix, which is mainly composed of triglycerides. Although many multiresidue methods have been developed (García-Reyes et al. 2007), most of them use an extraction procedure (solid phase extraction; gel permeation chromatography; liquid extraction; matrix solid phase dispersion). However, this produces a dirty extract containing small amount of fat and an additional clean-up step, usually based on liquid-liquid partitioning or gel-permeation chromatography, is required prior to the analytical process (Guardia-

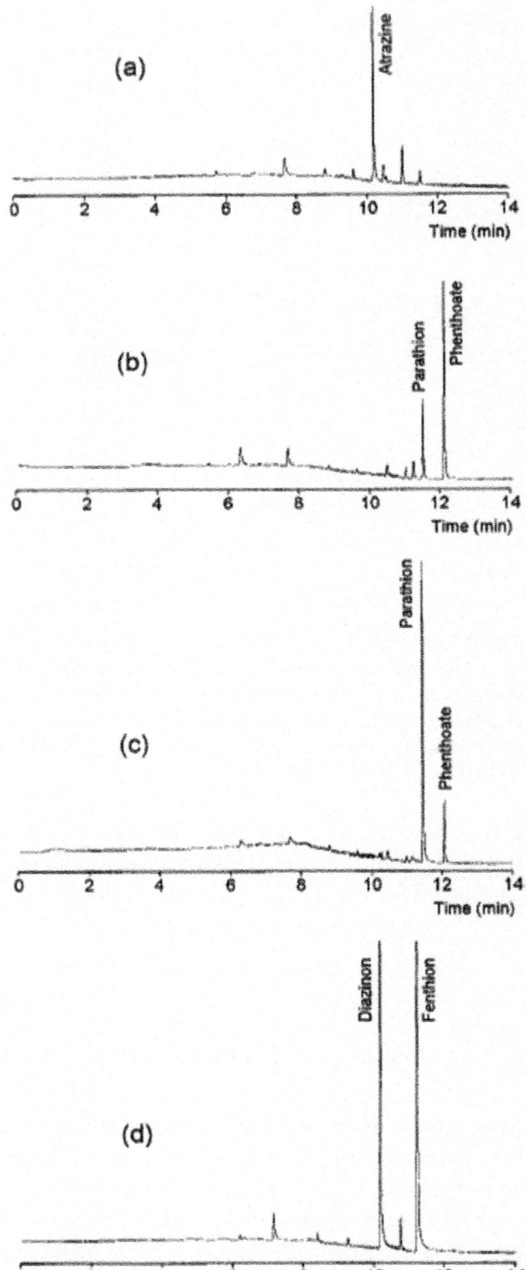

Fig. 3.8 Gas chromatograms of real-water samples from the River Ebro analysed by off-line solid phase extraction and on-line reversed phase liquid chromatography-gas chromatography: (**a**) atrazine analysis; (**b**) phenthoate analysis; (**c**) parathion analysis; (**d**) diazinon and fenthion analysis. Reprinted from Pérez et al. (2000) with permission of ACS Publications and Copyright Clearance Center

Rubio et al. 2006; Barrek et al. 2003). It is crucial that the analytical proceeding should be able to separate the pesticide residues from the lipidic fraction which constitutes the greater part of the matrix due to even very small quantities of lipids will spoil the gas chromatography system.

On line liquid chromatography-gas chromatography methods for analysing pesticide residues in edible oils have been developed but, due to solvent compatibility, most of them use normal phase in the liquid chromatography separation step (Grob and Kaelin 1991; Jongenotter et al. 1999; Qi et al. 2014). One of the problems when normal phase liquid chromatography is used is that the triglyceride fraction elutes before the pesticide fraction and, owing to the high tailing peak of the triglycerides, some fat is transferred to the gas chromatograph, harming the gas chromatography systems.

Reverse phase liquid chromatography is an interesting alternative but transferring a polar solvent to gas chromatography is difficult due to the large volume of vapour generated during the transfer of a water-based eluent from the liquid chromatography to the gas chromatography. The Through Oven Transfer Adsorption Desorption interface eliminates this high volume of vapour in both evaporative and non-evaporative mode. When reverse phase liquid chromatography and short columns (C4, 50 x 4.6 mm) are used, the pesticides elute in the first 2 minutes, and reverse phase liquid chromatography pre-separation guarantees the complete separation of the pesticides from the triglyceride fraction.

Figure 3.9 shows the liquid chromatography and gas chromatography chromatograms obtained for the direct analysis of an olive oil sample by reverse phase liquid chromatography-gas chromatography using the Through Oven Transfer Adsorption Desorption interface. A flame ionization detector was used (Sánchez et al. 2003, 2004), although this has two drawbacks: flame ionization detector is neither sensitive nor selective. In the gas chromatography chromatogram can be observed that peaks other than the pesticide peaks appear but do not match those of the target pesticides, enabling the pesticides to be quantified. The gas chromatography retention time of parathion and chlorpyrifos overlap. Both pesticides may be quantified together, or they can be quantified separately by transferring the liquid chromatography fraction of each pesticide to the gas chromatography individually because the on line couple liquid chromatography-gas chromatography is a multidimensional system and the peaks do not overlap in the liquid chromatography chromatogram. Figure 3.10 shows the chromatograms obtained when either the parathion fraction or the chlorpyrifos fraction is transferred. As can be seen, solvent elimination is almost complete, meaning that a water-sensitive detector can be used. The use of a more selective detector such as a nitrogen phosphorus detector provides a cleaner chromatogram (Fig. 3.11) and lower detection limits. When a volume of 20 μl was injected into the liquid chromatograph and the flow rate during the transfer step was 0.1 ml/min, the limits of detection obtained were lower than 7.0 μg/l for the triazines and 0.4 μg/l in the case of organophosphorus pesticides (Sánchez et al. 2005). When an electron capture detector was used to determine endosulfan (Sánchez et al. 2006b), the limits of detection obtained were 7 μg/l for α and β endosulfan and 134 μg/l for endosulfan sulphate. These low limits of detection (much lower than the

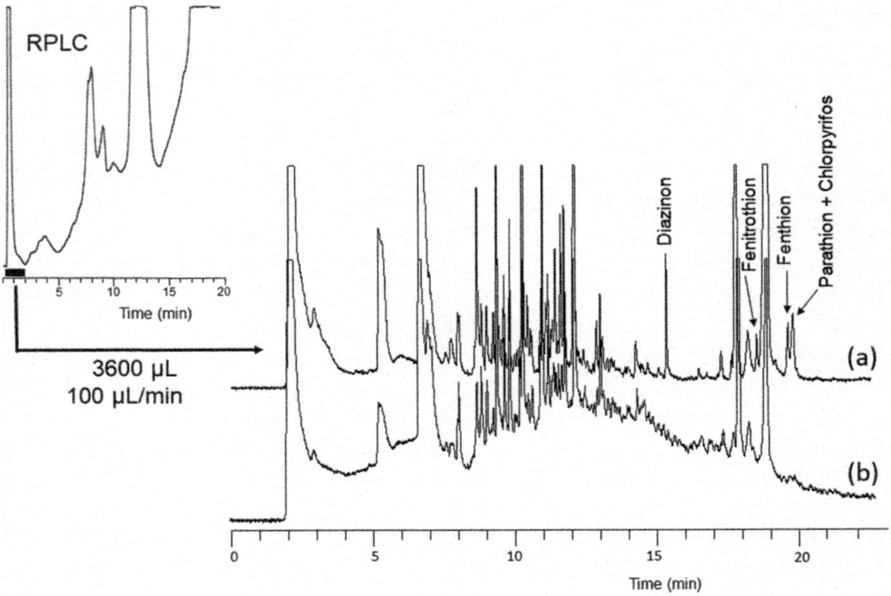

Fig. 3.9 Liquid and gas chromatogram obtained from the direct on line coupled liquid chromatography-gas chromatography-Flame Ionisation Detector analysis of (**a**) an olive oil sample spiked with diazinon, fenitrothion, fention, parathion and chlorpyrifos at 1 mg/l; (**b**) and an unfortfied olive oil sample. 20 µl olive oil was injected into the liquid chromatograph. The thick line between the time axis and the chromatogram indicates the liquid chromatography fraction that has been transferred from the liquid chromatograph to gas chromatograph. Reprinted from Sánchez et al. (2005) with permission of Jennifer Diatz (Director of Publications, AOAC International)

maximum residue levels established in the legislation) means that the flux at the end of the gas chromatography column can be divided by means of a "Y" piece in order to use two different detectors simultaneously. Organophosphorus, organochlorine and triazines were determined in a single run using an nitrogen phosphorus detector and an electron capture detector acting simultaneously (Díaz-Plaza et al. 2007). The electron capture detector detected organochlorines pesticides, while the nitrogen phosphorus detector detected organophosphorus and triazines.

These methods were used to analyse the olive oil obtained from olives from a grove treated with some of the most frequently used pesticides in olive trees (Sánchez et al. 2006a, b). The aim of the study was to determine the effect of the pesticide dose and the date of treatment on the residues present in the oil. The olive trees were treated in summer (June–July) and/or in autumn (October–November) at the recommended doses and twice the recommended doses. No residues of the water-soluble pesticides (dimethoate and trichlorphon) were detected because they passed into the aqueous phase during the extraction of the oil (Lentza-Rizos and Avramides 1995). Clorfenvinphos was not detected probably due to its rapid degradation. The fat-soluble pesticides used tended to concentrate in the oil. Treatments at the recommended dose carried out in summer did not leave residues higher than

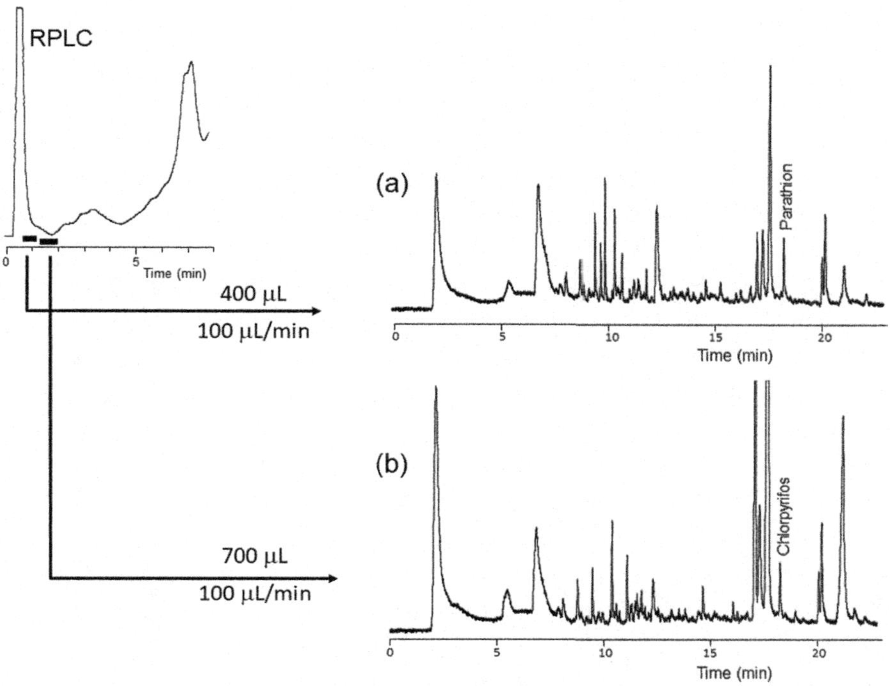

Fig. 3.10 Liquid and gas chromatograms resulting from the direct on line coupled liquid chromatography-gas chromatography-Flame Ionisation Detector analysis of an olive oil spiked with parathion and chlorpyrifos at 1 mg/l. The thick lines in the liquid chromatogram indicate the two different fractions transferred to the gas chromatograph: (**a**) Parathion LC fraction transferred to the GC (**b**) Chlorpyrifos LC fraction transferred to the GC. 20 μl olive oil was injected into the liquid chromatograph. Reprinted from Sánchez et al. (2005) with permission of Jennifer Diatz (Director of Publications, AOAC International)

the maximum residue levels. Malathion showed the lowest residue values in the oil, and olives with two treatments presented more pesticides residues in the oil than those with one treatment.

3.5.3 Pesticide Residue Analysis in Nuts

Nuts, like oils, have a lipidic matrix (fat content 50–70%) and tend to concentrate the hydrophobic pesticides used in their cultivation. Unlike the oil, nuts are solid so, likewise vegetable samples, an extraction procedure must be carried out before gas chromatography analysis. As mentioned above (see Sect. 4.1), vegetable samples can be analysed by large volume injection after an extraction step but, in this case, the fatty nature of nuts means that the extract contains a large amount of fat, and so large volume injection cannot be used. Instead, samples are subjected to an

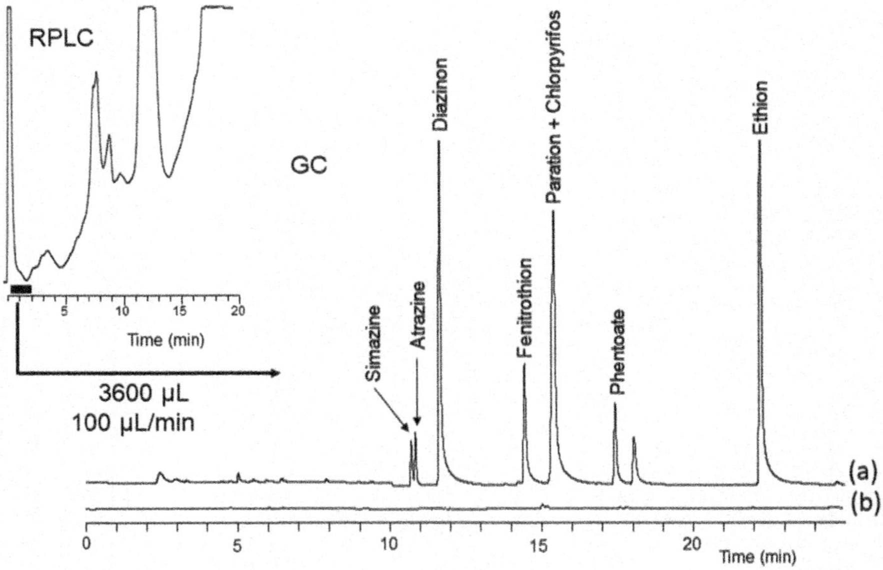

Fig. 3.11 Liquid and gas chromatogram obtained from the direct on line coupled liquid chromatography-gas chromatography-Nitrogen Phosphorus Detector analysis of (**a**) an olive oil sample spiked with each organophosphorus pesticide at 0.1 mg/l and each triazine at 1 mg/l. (**b**) reagent blank. The thick line situated between the time axis and the chromatogram indicated the liquid chromatography fraction that has been transferred from the liquid chromatograph to gas chromatograph. Reprinted from Sánchez et al. (2005) with permission of Jennifer Diatz (Director of Publications, AOAC International)

extraction procedure similar to that described for vegetable samples and the extract is analysed by on line reverse phase liquid chromatography-gas chromatography (Cortés et al. 2008). The liquid chromatography step replaces the clean-up and concentrations steps and the pesticide residues are separated from the co-extracted fatty material. The pesticide residues elute in the first minute and the fatty material elutes after 2 min because it is more strongly retained since reversed phase is used. The pesticides liquid chromatography fraction is automatically transferred to the gas chromatograph for analysis. Organophosphorus pesticides can be detected using an nitrogen phosphorus detector and organochlorine pesticides using an electron capture detector. When 50 µl of extract was injected into the liquid chromatograph and the flow during the transfer step was 0.1 ml/min, the limits of detection obtained for organophosphorus pesticides were lower than for organochlorine pesticides, but, in all cases, the maximum residue levels established by European legislation are much higher. Figure 3.12 shows the gas chromatography chromatograms obtained for the analysis of different nuts: walnuts, hazelnuts, peanuts and sunflower seeds.

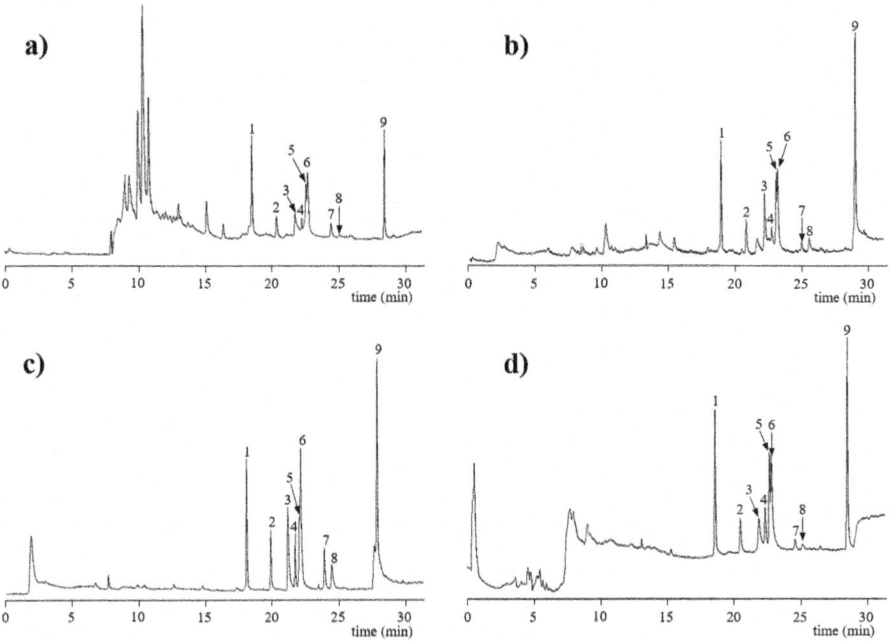

Fig. 3.12 Gas chromatograms obtained from the direct on line coupled liquid chromatography-gas chromatography-Nitrogen Phosphorus Detector analysis of extracts from different nuts: (**a**) peanut; (**b**) sunflower seed; (**c**) walnut; (**d**) hazelnut. The samples were fortified at 0.05 mg/kg of each pesticide. Identification peaks: (1) diazinon; (2) methylchlorpyrifos; (3) fenitrothion; (4) malathion; (5) chlorpyrifos; (6) parathion; (7) phenthoate and chlorfenvinphos; (8) methidathion; (9) ethion. Reprinted from Cortés et al. (2008) with permission of ACS Publications and Copyright Clearance Center

References

Ahmed FE (2001) Analyses of pesticides and their metabolites in foods and drinks. Trends Anal Chem 20:649–661. https://doi.org/10.1016/S0165-9936(01)00121-2

Alario J, Pérez M, Vázquez A, Villén J (2001) Very-large-volume sampling of water in gas chromatography using the through oven transfer adsorption desorption (TOTAD) Interface for pesticide-residue analysis. J Chromatogr Sci 39:65–69. https://doi.org/10.1093/chromsci/39.2.65

Aragón A, Cortés JM, Toledano RM, Villén J, Vázquez A (2011) Analysis of wax esters in edible oils by automated on-line coupling liquid chromatography–gas chromatography using the through oven transfer adsorption desorption (TOTAD) interface. J Chromatogr A 1218:4960–4965. https://doi.org/10.1016/j.chroma.2010.12.042

Aragón A, Toledano RM, Vázquez A, Villén J, Cortés JM (2015) Analysis of polycyclic aromatic hydrocarbons in aqueous samples by large volume injection gas chromatography-mass spectrometry using the through oven transfer adsorption desorption interface. Talanta 139:1–5. https://doi.org/10.1016/j.talanta.2015.02.028

Bakirci GT, Acay DBY, Bakirci F, Otles S (2014) Pesticides residues in fruits and vegetables from the Aegean region, Turkey. Food Chem 160:379–392. https://doi.org/10.1016/j.foodchem.2014.02.051

Barrek S, Paisse O, Grenier-Loustalot MF (2003) Determination of residual pesticides in olive oil by GC-MS and HPLC-MS after extraction by size-exclusion chromatography. Anal Bioanal Chem 376:355–359. https://doi.org/10.1007/s00216-003-1917-y

BOE (1994) Real Decreto 280/1994 de 18 de Febrero (BOE 9/3/94) y sucesivas modificaciones: Establece los límites máximos de residuos (LMR) de plaguicidas y su control en determinados productos de origen vegetal

Carvalho FP (2006) Agriculture, pesticides, food security and food safety. Environ Sci Pol 9:685–692. https://doi.org/10.1016/j.envsci.2006.08.002

Carvalho FP, Fowler SW, Villeneuve JP, Horvat M (1997) Pesticides residues in the marine environment and analytical quality assurance of the results. In: Proceedings of an international FAO/IAEA symposium on the environmental behaviour of crop protection chemicals. IAEA, Vienna, pp 35–57

Chen C, Qian T, Tao C, Li C, Li Y (2011) Evaluation of pesticide residues in fruits and vegetables from Xiamen, China. Food Control 22:1114–1120. https://doi.org/10.1016/j.foodcont.2011.01.007

Cortés JM, Sánchez R, Díaz-Plaza EM, Villén J, Vázquez A (2006a) Large volume GC injection for the analysis of Organophosphorus pesticides in vegetables using the through oven transfer adsorption desorption (TOTAD) Interface. J Agric Food Chem 54:1997–2002. https://doi.org/10.1021/jf0524675

Cortés JM, Sánchez R, Villén J, Vázquez A (2006b) Analysis of unsaponifiable compounds of edible oils by automated on-line coupling reversed-phase liquid chromatography–gas chromatography using the through oven transfer adsorption desorption interface. J Agric Food Chem 54:6963–6968. https://doi.org/10.1021/jf060956e

Cortés JM, Toledano RM, Villén J, Vázquez A (2008) Analysis of pesticides in nuts by online reversed-phase liquid chromatography-gas chromatography using the through-oven transfer adsorption/desorption interface. J Agric Food Chem 56:5544–5549. https://doi.org/10.1021/jf800773k

Cortés JM, Toledano RM, Andini JC, Villén J, Vázquez AM (2010) Automated on-line coupling liquid chromatography-gas chromatography and large volume injection by using the TOTAD interface. In: Quintin TJ (ed) Chromatoghapy types, techniques and methods. Nova, New York, pp 347–368

De la Peña MF, Patricia Blanch G, Flores G, Ruiz del Castillo ML (2010) Development of a method based on on-line reversed phase liquid chromatography and gas chromatography coupled by means of an adsorption–desorption interface for the analysis of selected chiral volatile compounds in methyl jasmonate treated strawberries. J Chromatogr A 1217:1083–1088. https://doi.org/10.1016/j.chroma.2009.10.037

Díaz-Plaza EM, Cortés JM, Vázquez A, Villén J (2007) Automated determination of pesticides residues in olive oil by on-line reversed-phased liquid chromatography-gas chromatography using the through oven transfer adsorption desorption interface with electron-capture and nitrogen-phosphorus detector operating simultaneously. J Chromatogr A 1174:145–150. https://doi.org/10.1016/j.chroma.2007.06.073

European Commission (1998) Council directive 98/83/EC of 3 November 1998 on the quality of water intended for human consumption. Off J Eur Communities 41:32–54

Ferrari E, Caldas ED (2018) Simultaneous determination of drugs and pesticides in postmortem blood using dispersive solid-phase extraction and large volume injection-programmed temperature vaporization-gas chromatography-mass spectrometry. Forensic Sci Int 290:318–326. https://doi.org/10.1016/j.forsciint.2018.07.031

Flores G, Díaz-Plaza EM, Cortés JM, Villén J, Herraiz M (2008) Use of absorbent materials in on-line coupled reversed-phase liquid chromatography-gas chromatography via the through oven transfer adsorption desorption interface. J Chromatogr A 1211:99–103. https://doi.org/10.1016/j.chroma.2008.09.075

Gamón M, Lleó C, Ten A, Mocholí F (2001) Multiresidue determination of pesticides in fruit and vegetables by gas chromatography-tandem mass spectrometry. J AOAC Int 84(4):1209–1216

García-Reyes JF, Ferrer C, Gómez-Ramos MJ, Molina-Díaz A, Fernández-Alba A (2007) Determination of pesticides residues in olive oil and olive. Trends Anal Chem 26(3):239–251. https://doi.org/10.1016/j.trac.2007.01.004

Grob K (1991) On-line coupled LC-GC. Hüthig, Heidelberg

Grob K (2000) Efficiency through combining high-performance liquid chromatography and high resolution gas chromatography: progress 1995-1999. J Chromatogr A 892:407–420. https://doi.org/10.1016/S0021-9673(00)00048-0

Grob K, Kaelin I (1991) Attempt for an on-line size exclusion chromatography-gas chromatography method for analysing pesticide residues in foods. J Agric Food Chem 39:1950–1953. https://doi.org/10.1021/jf00011a012

Guardia-Rubio M, Fernández-De Córdova ML, Ayora-Cañada MJ, Ruiz-Medina A (2006) Simplified pesticide multiresidue analysis in virgin olive oil by gas chromatography with thermoionic specific, electron-capture and mass spectrometric detection. J Chromatogr A 1108:231–239. https://doi.org/10.1016/j.chroma.2006.01.006

Handford CE, Elliott CT, Campbell K (2015) A review of the global pesticide legislation and the scale of challenge in reaching the global harmonization of food safety standards. Integr Environ Assess Manag 11:525–536. https://doi.org/10.1002/ieam.1635

Harris CA (2002) The regulation of pesticides in Europe – directive 91/414. J Environ Monit 4:28–31. https://doi.org/10.1039/B202153F

Hogendoorn E, van Zoonen P (2000) Recent and future development of liquid chromatography in pesticide trace analysis. J Chromatogr A 892:435–453. https://doi.org/10.1016/S0021-9673(00)00151-5

Hoh E, Mastovska K (2008) Large volume injection techniques in capillary gas chromatography. J Chromatogr A 1186:2–15. https://doi.org/10.1016/j.chroma.2007.12.001

Hyötyläinen T, Riekkola ML (1998) Direct coupling of reversed-phase liquid chromatography to gas chromatography. J Chromatogr A 819:13–24. https://doi.org/10.1016/S0021-9673(98)00539-1

Hyötyläinen T, Riekkola ML (2003) On-line coupled liquid chromatography–gas chromatography. J Chromatogr A 1000:357–384. https://doi.org/10.1016/S0021-9673(03)00181-X

Jongenotter GA, Kerkhoff MAT, van der Knapp HCM, Vandeginste BGM (1999) Automated on-lin GPC-GC-FPD involving co-solvent trapping and the on-column interface for the -determination of organophosphorus pesticides in olive oils. HRC J High Resolut Chromatogr 22:17–23. https://doi.org/10.1002/(SICI)1521-4168(19990101)22:1<17::AID-JHRC17>3.0.CO;2-7

Kurth D, Krauss M, Schulze T, Brack W (2017) Measuring the internal concentration of volatile organic compounds in small organisms using micro-QuEChERS coupled to LVI-GC-MS/MS. Anal Bioanal Chem 409:6041–6052. https://doi.org/10.1007/s00216-017-0532-2

Lehotay SJ (2008) Analysis of pesticide residues in mixed fruit and vegetable extracts by direct sample introduction-gas chromatography-tandem mass spectrometry. J AOAC Int 83 (3):680–697

Lentza-Rizos C, Avramides EJ (1995) Pesticide residues in olive oil. In: Ware GW, Gunther FA (eds) Reviews of environmental contamination and toxicology, vol 141. Springer, New York, pp 111–134. https://doi.org/10.1007/978-1-4612-2530-0_4

López FJ, Beltran J, Forcada M, Hernández F (1998) Comparison of simplified methods for pesticide resiude analysis – use of large-volume injection in capillary gas chromatography. J Chromatogr A 823:25–33. https://doi.org/10.1016/S0021-9673(98)00393-8

Pérez M, Alario J, Vázquez A, Villén J (1999) On-line reversed phase LC-GC by using the new TOTAD (through oven transfer adsorption desorption) interface: application to parathion residue analysis. J Microcolumn Sep 11:582–589. https://doi.org/10.1002/(SICI)1520-667X(1999)11:8<582::AID-MCS3>3.0.CO;2-E

Pérez M, Alario J, Vázquez A, Villén J (2000) Pesticide residue analysis by off-line SPE and on-line reversed-phase LC-GC using the through-oven-transfer adsorption/desorption Interface. Anal Chem 72:846–852. https://doi.org/10.1021/ac990438l

Picó Y, Font G, Moltó JC, Mañes J (2000) Pesticide residue determination in fruit and vegetables by liquid chromatography-mass spectrometry. J Chromatogr A 882:153–173. https://doi.org/10.1016/S0021-9673(00)00294-6

Picó Y, Blasco C, Font G (2004) Environmental and food applications of LC-tandem mass spectrometry in pesticide-residue analysis: an overview. Mass Spectrom Rev 23:45–85. https://doi.org/10.1002/mas.10071

Picó Y, Font G, Ruiz MJ, Fernández M (2006) Control of pesticide residues by liquid chromatography-mass spectrometry to ensure food safety. Mass Spectrom Rev 25:917–960. https://doi.org/10.1002/mas.20096

Podhorniak LV, Negron JF, Griffith FD Jr (2001) Gas chromatography with pulsed flame photometric detection multiresidue method for organophosphate pesticide and metabolite residues at the parts-per-billion level in representative commodities of fruit and vegetable crop groups. J AOAC Int 84(3):873–890

Purcaro G, Moret S, Conte L (2012) Hyphenated liquid chromatography-gas chromatography technique: recent evolution and applications. J Chromatogr A 1255:100–111. https://doi.org/10.1016/j.chroma.2012.02.018

Purcaro G, Moret S, Conte L (2013) Sample pre-fractionation of environmental and food samples using LC-GC multidimensional techniques. Trends Anal Chem 43:149–160. https://doi.org/10.1016/j.trac.2012.10.007

Qi D, Sha Y, Wang L, Li G, Wu D, Lui B (2014) A novel fully automated on-line coupled liquid chromatography-gas chromatography technique used for the determination of organochlorine pesticide residues in tobacco and tobacco products. J Chromatogr A 1374:273–277. https://doi.org/10.1016/j.chroma.2014.11.040

Sánchez R, Vázquez A, Riquelme D, Villén J (2003) Direct analysis of pesticide residues in olive oil by on-line reversed phase liquid chromatography-gas chromatography using an automated through oven transfer adsorption desorption (TOTAD) interface. J Agric Food Chem 51:6098–6102. https://doi.org/10.1021/jf030182r

Sánchez R, Vázquez A, Andini JC, Villén J (2004) Automated multiresidue analysis of pesticides in olive oil by on-line reversed-phase liquid chromatography-gas chromatography using the through oven transfer adsorption-desorption interface. J Chromatogr A 1029:167–172. https://doi.org/10.1016/j.chroma.2003.12.018

Sánchez R, Cortés JM, Villén J, Vázquez A (2005) Determination of organophsphorus and triazine pesticides in oilve oil by on-line coupling reversed-phase liquid chromatography/gas chromatography with nitrogen-phosphorus detection and an automated through oven transfer adsorption desorption interface. J AOAC Int 88(4):1255–1260

Sánchez R, Vázquez A, Villén-Altamirano J, Villén J (2006a) Analysis of pesticide residues by on-line reversed-phase liquid chromatography-gas chromatography in the oil from olives grown in an experimental plot. J Sci Food Agric 86:129–134. https://doi.org/10.1002/jsfa.2267

Sánchez R, Cortés JM, Vázquez A, Villén-Altamirano J, Villén J (2006b) Analysis of pesticide residues by online reversed-phase liquid chromatography – gas chromatography in the oil from olives grown in an experimental plot. Part II. J Sci Food Agric 86:1926–1931. https://doi.org/10.1002/jsfa.2565

Santos IC, Smuts J, Crawford ML, Grant RP, Schug KA (2019) Large-volume injection gas chromatography-vacuum ultraviolet spectroscopy for the qualitative and quantitative analysis of fatty acids in blood plasma. Anal Chim Acta 1053:169–177. https://doi.org/10.1016/j.aca.2018.12.007

Sciarrone D, Tranchida PQ, Dugo P, Mondello L (2012) Advances in multidimensional LC-GC for food analysis. LCGC Europe:31–35

Tadeo JL, Sánchez-Brunete C, Pérez RA, Fernández M (2000) Analysis of herbicide residues in cereals, fruits and vegetables. J Chromatogr A 882:175–191. https://doi.org/10.1016/S0021-9673(00)00103-5

Tadeo JL, Sánchez-Brunete C, Albero B, González L (2004) Analysis of pesticide residues in juice and beverages. Crit Rev Anal Chem 34:121–131. https://doi.org/10.1080/10408340490491887

Teske J, Engewald W (2002) Methods for and applications of large-volume injection in capillary gas chromatography. Trends Anal Chem 21(9–10):584–593. https://doi.org/10.1016/S0165-9936(02)00801-4

Toledano RM, Cortés JM, Andini JC, Villén J, Vázquez A (2010) Large volume injection of water in gas chromatography-mass spectrometry using the through oven transfer adsorption desorption interface: application to multiresidue analysis of pesticides. J Chromatogr A 1217:4738–4742. https://doi.org/10.1016/j.chroma.2010.05.011

Villén J, Señoráns FJ, Herraiz M (1999) Very large volume sample introduction in capillary gas chromatography using a programmed temperature injector for pesticide analysis. J Microcolumn Sep 19:89–95. https://doi.org/10.1002/(SICI)1520-667X(1999)11:2<89::AID-MCS1>3.0.CO;2-X

Xu Y, Song N, Zhan Q, Lui J, Chen G, Shi L, Wang J (2017) A strategy for the evaluation of an analytical approach for selected pesticide residues in complex agricultural product matrices – a case study of leek. Food Chem 221:205–213. https://doi.org/10.1016/j.foodchem.2016.10.009

Chapter 4
Detectors for the Analysis of Pesticides Residues

Hitika Patel

Abstract The modern crop-farming has raised demands for the increasing use of synthetic pesticides. These newly introduced pesticides have exhibited an enhanced specific action, polarity, and portray a reduced persistence. These have paved the way for their augmented use. Pesticides penetrate the atmosphere when they are poorly adsorbed onto the applied surface of the soil. This partial adsorption leads to the presence of these pesticides in the surrounding water bodies and plants. These find their way up the food chain and reach the food products. Pesticides exist in either ppt or ppb levels. Subsequently, a highly specific analytical procedure needs to be applied for the target pesticides to be detected in a single run.

We reviewed that, currently, gas chromatography (GC), gas chromatography-mass spectrometry (GC-MS), gas chromatography-ion trap mass spectrometry (GC-ITMS) and gas chromatography-tandem mass spectrometry (GC–MS-MS) are the most preferred methods due to their noteworthy selectivity, specificity and separation ability. Supplementary to the gas chromatography-mass spectrometry (GC-MS) methods, there exist quantification methods like liquid chromatography-mass spectrometry (LC-MS), liquid chromatography-tandem mass spectrometry (LC–MS-MS), high-performance liquid chromatography (HPLC) and low-pressure gas chromatography-mass spectrometry (LP–GC/MS). Conclusively, the pesticide residues can be resolved only up to an extent by gas chromatography or liquid chromatography-based technique. There is a need to develop more efficient techniques while keeping these techniques still in use, for the satisfactory separation of the pesticide residues.

Keywords Pesticides detection · Air contamination · Water contamination · Soil penetration · Food products · Gas chromatography · High-performance liquid chromatography · Gas chromatography-mass spectrometry · Liquid chromatography-mass spectrometry · Tandem mass spectrometry

---

H. Patel (✉)
Food Chemist and Microbiologist, Analytical and Environmental Services, Vadodara, Gujarat, India

Inamuddin et al. (eds.), *Sustainable Agriculture Reviews 48*, Sustainable Agriculture Reviews 48, https://doi.org/10.1007/978-3-030-54719-6_4

Abbreviations

CE	Capillary electrophoresis
CEC	Capillary electrochromatography
GC	Gas chromatography
GC-MS	Gas chromatography-mass spectrometry
HILIC	Hydrophilic interaction liquid chromatography
HPLC	High-performance liquid chromatography
HPLC-MS	High-performance liquid chromatography-mass spectrometry
IA	Immunoassays
LC	Liquid chromatography
LC-GC	Liquid chromatography-gas chromatography
MRM	Multiple-reaction monitoring
QqQ	Triple quadrupole
TLC	Thin-layer chromatography
UPLC	Ultra-performance liquid chromatography

4.1 Introduction

After World War II, the pesticide usage in agriculture accelerated due to an increase in world food production. Ordinarily, insecticides and herbicides have been used before harvesting of a crop, rodenticides during post-harvest phase, while the application of fungicides is random and controlled by the nature of the crop. Herbicides are the most commonly used pesticides, followed by insecticides, fungicides and the rest follow. The contemporary intensive crop-based agriculture has led to the need for an augmented use of synthetic pesticides. (Jones and De Voogt 1999).

However, pesticides were used not only in food production but also for a variety of other uses. The most common organochlorine pesticide, 1, 1, 1-Trichloro-2, 2-bis (4-chlorophenyl) ethane (DDT), was used in preventing the outbreak of malaria along with some vector-borne diseases such as leishmaniasis and dengue, by preventing the mosquito growth. Lindane was another extensively organochlorine pesticide that was used in the treatment of children with head lice (Jones and De Voogt 1999).

Pesticides enter the atmosphere due to the poor adsorption onto the applied surface, because of the spray drift, through post-application processes like re-entrainment of soil particles onto the plant surfaces or by volatilization. Due to the partial adsorption of pesticides on the surface of the soil, the residues are commonly found scattered in food products, water bodies and soil (Picó et al. 2007; Morais et al. 2009; Andreu and Picó 2004). Additionally, when the industrial emissions from the production of pesticides reach the atmosphere, the traces of these residues can be found in the fruits, vegetables, and crops as well. In recent years, new

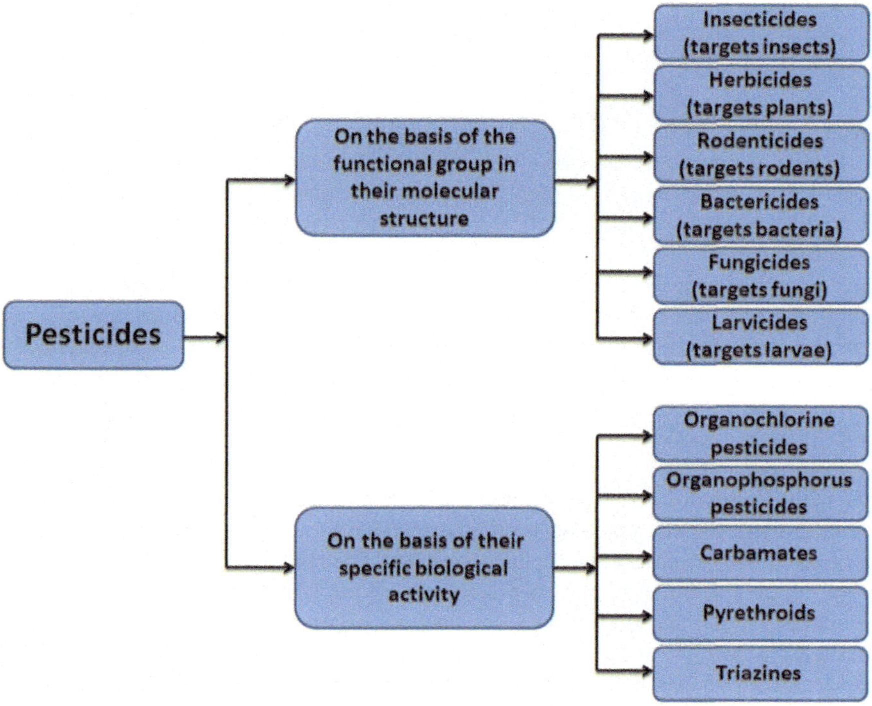

Fig. 4.1 Pesticides can be classified based on the functional groups in their molecular structure and on the basis of their specific biological activity on the target compounds

pesticides that exhibit a greater specific action and have an enhanced polarity and reduced persistence than the old ones have come into existence (Bai et al. 2006).

Over 800 pesticides of numerous chemical groups and functions have been registered to restrict the growth of the unwanted pests and weeds on the crops. Pesticides can be classified based on the functional groups in their molecular structure and of their specific biological activity on the target compounds. This is shown in Fig. 4.1. Thus, pesticides have diverse physical and chemical properties. For a wide variety of the pesticide and food commodity combinations, the European Union has set Maximum Residue Limits (MRLs) to address the health concerns on exposure to these chemicals (Kosikowska and Biziuk 2010). The presence of agrochemical compounds in the farmer's blood has also been used as an indicator to estimate the occupational exposure to these compounds (Dagnac et al. 2002).

Pesticides can be altered into the environment into a vast number of degradation products that may also be known as transformation products. When compared to the original compound, the transformation products show greater insecticidal activity, are more polar, less volatile and they are also thermolabile. These can also be more toxic and persistent than the parent compound, for instance, carbamates or certain other organophosphorus pesticides like Malathion and fenitrothion (Uygun et al.

2007). As a matter of fact, these transformation products can have additional properties that enable them to reach the environmental areas not reached by the parent compound itself. For instance, some of the degradation products can access the groundwater effectively than the parent compound due to their enhanced soil-water mobility (Belfroid et al. 1998).

From the agricultural and the environmental point of view, the organophosphorus insecticides like chlorpyrifos-methyl, chlorpyrifos or chlorfenvinphos and triazine herbicides such as simazine, atrazine, and metribuzin are the most routinely used pesticides (Carro and Lorenzo 2001). Although the use of chlorinated pesticides has been restricted long ago, their use is still prevalent in the remote regions where their residues have been found (Manirakiza et al. 2001). These residues are of great concern because these pesticides are the most toxic and remain persistent in the environment. Organophosphorus pesticides too are routinely used. Studies indicate that these are found to be mutagenic, genotoxic, cytotoxic, carcinogenic, teratogenic and immunotoxic (Hercegová et al. 2007; Sarabia et al. 2009).

Pesticides exist within the bounds of the environment in trace amounts of parts per trillion (ppt) or parts per billion (ppb) levels. Hence, to detect them in any sample, a specific analytical procedure corresponding to the target compound needs to be applied. For most of the products, the pesticide residues need to be first isolated from the complex-matrix and then cleaned and pre-concentrated for further determination. For enhanced efficiency, the analysis of target compounds in agricultural products is brought about by the multiresidue methods (MRMs) than the single-residue methods (SRMs). This ensures that numerous compounds are detected in a single run. Multiresidue methods (MRMs) can be run in two modes. First is the multiclass Multiresidue Method mode wherein the residues of numerous groups of pesticides are covered. Secondly, the numerous residues of chemically related pesticides such as carboxylic acids, N-methyl carbamate pesticides (NMCs), and phenols, are taken into consideration (Kosikowska and Biziuk 2010).

Of late, gas chromatography (GC), gas chromatography-ion trap mass spectrometry (GC-ITMS), gas chromatography-mass spectrometry (GC-MS), and gas chromatography-tandem mass spectrometry (GC–MS-MS) have paved their way in the field of scanning pesticides because of their effective selectivity, separation capabilities and identification powers of mass spectrometry (Vidal et al. 2002; Arrebola et al. 2003). Supplementary to the gas chromatography-mass spectrometry (GC-MS) methods, there exist quantification methods like high-performance liquid chromatography (HPLC), liquid chromatography-tandem mass spectrometry (LC–MS-MS), low-pressure gas chromatography-mass spectrometry (LP–GC-MS), and liquid chromatography-mass spectrometry (LC-MS) (López-Blanco et al. 2002; Hernández et al. 2006; Mol et al. 2003; Walorczyk and Gnusowski 2006). The commonly used detectors in Gas chromatography, Liquid chromatography, and Mass Spectrometry are summarized in Table 4.1.

Table 4.1 The commonly used detectors in gas chromatography, liquid chromatography, and mass spectrometry are tabulated here along with their specific targets

	Types of detectors employed	Uses
Gas chromatography (GC)	Electron capture detector (ECD) and electrolytic conductivity detection (ELCD)	For the detection of halogenated compounds
	Nitrogen-phosphorus detector (NPD)	For the detection of nitrogen- and phosphorus-containing pesticides
	Flame photometric detection (FPD)	For the detection of sulfur or phosphorus-containing pesticides
	Atomic emission detector (AED)	For the detection of metallic and non-metallic organic pesticides
	Microwave induced plasma atomic emission detection	For the detection of halogenated hydrocarbons in pesticide containing water samples
Liquid chromatography (LC) and high performance liquid chromatography (HPLC)	UV/Vis detector	For the detection of compounds with UV chromophore
	Fluorescence detector (FI)	For the detection of groups with native fluorescence or a fluorescence tag in pesticides
	Electrochemical detector	For the detection of electro-active groups in pesticides
Mass spectrometry (MS)	Single quadrupole detector	For pesticide analysis in food and water
	Time of flight (TOF) and quadrupole time of flight (QqTOF)	For pesticide analysis in food and water
	Triple quadrupole (QqQ) detector	For pesticide analysis in food and water
	Quadrupole ion trap (IT) and triple quadrupole linear ion trap (QqLIT)	For pesticide analysis in food and water
	Orbitrap mass spectrometer	For pesticide analysis in food and water

This review has been compiled to summarize the current analytical techniques available to detect the pesticides that are being used heavily in recent times. The routinely used quantification procedures have been described below in general. The subsequent sections give an insight into the specific and additional detection methods employed for water, air, soil, and food products.

4.2 General Methods of Detection for Pesticides

4.2.1 *Gas Chromatography (GC)*

The pesticides that possess low-polarity, are volatile or have thermal stability are readily detected by gas chromatography. The column here is selected depending on the nature of the pesticide to be identified (Ahmed 2001). Largely, organochlorine pesticides occur in a non-ionic form along with being highly volatile. Thus, gas chromatography can be the best suited in the determination of these. The commonly used stationary phases for the organochlorine pesticides detection include 100% dimethylpolysiloxane, methyl polysiloxane containing 5, 35 or 50% phenyl polysiloxane or 14 or 50% cyanopropyl phenyl polysiloxane. The non-polar stationary phases are favored due to their stability (Chung and Chen 2011).

Furthermore, a variety of conventional and sensitive detectors coupled to gas chromatography improved the quantification procedures of pesticide residues from different matrices and enhanced the detection. For instance, electron capture detection (ECD) is the approach for detecting compounds that are organochlorine, since they have a noteworthy response for this pesticides' group. Nitrogen-phosphorus detector (NPD) or a flame photometric detector (FPD) is preferably used over electron capture detection because of the increasingly sensitive and specific response obtained. Electron capture detection possesses a reduced selectivity and thus necessitates the confirmation of the identity of the residue to discard the false-positive residue, particularly the ones at low levels in juices and beverages. (Salvador et al. 2006; Gebara et al. 2005; Parveen et al. 2005).

The detectors are occasionally connected in parallel to produce results with a single injection. In capillary gas chromatography, the injector used for analyzing the pesticides can be either operated in a split mode, in which the gas stream is divided, or a splitless mode wherein the gas stream is not divided. Although the split injection mode is widely used, there is a considerable loss of sample and hence the trace analysis of pesticides in the air cannot be accomplished. Here, the splitless injection is more suitable where the entire sample is introduced for analysis. Nonetheless, there is a much notable thermal degradation as compared to that of a split injection, since the sample is in the heated inlet for a longer time. An on-column injector provides the substitute as the samples of known volumes can be directly injected onto the column. Subsequently, the chances of discrimination and thermal degradation are highly reduced. The splitless injection technique is picked over the on-column injection of the sample into the gas chromatography due to its cleanliness. The recently introduced autosamplers allow for the introduction of large volumes of extracts to be introduced at lower speeds and produce enhanced resolutions (Kosikowska and Biziuk 2010).

In addition to these conventionally accessible element-specific detection methods, gas chromatography hyphened with mass spectrometry detections include ion trap, single quadrupole, as well as triple quadrupole mass spectrometers. Their adoption particularly involves the pesticide residue analysis in foods of animal origin

(Ahmed 2001). Nowadays, mass spectrometry coupled to gas chromatography has been used to analyze pesticides of different classes at trace levels. Mass spectrometry detection helps provide a full scan as a universal detector including specific detector operating with the selected ion monitoring (SIM) mode. When specific major ions of a mass spectrum of pesticide are supplemented to a certain retention time in gas chromatography selective to every complex, an excellent confirmation of the identity of residue is obtained. These can also provide the identity of its metabolism products as well as the products of deterioration while in the same acquisition run.

Two-dimensional gas chromatography segregation- Two-dimensional gas chromatography (GC×GC), also known as comprehensive gas chromatography, achieves dissociation of targeted compounds after placing the two capillary columns orthogonally, wherein the succeeding column has a shorter length and is smaller in diameter. Henceforth, the co-eluting peaks resulting from the preceding column are efficiently split up by the succeeding column. Subsequently, sharper peaks with higher intensity and smaller peak width are obtained (Tadeo et al. 2004).

There are limited numbers of applications of two-dimensional gas chromatography –mass spectrometry (GCxGC-MS) methods for pesticide detection in food and water since the two-dimensional gas chromatography (GC×GC) instrumentation is complex, requires high maintenance, and is expensive. Along with the operation being complicated, the two-dimensional gas chromatography (GC×GC) tool gives chromatographic peaks that are very narrow and thus high acquisition rates provided by high-speed time of flight (TOF) can be integrated with it. Enhanced resolution by the two-dimensional gas chromatography (GC×GC) would reduce if mass spectrometry detection is employed (Górecki et al. 2006; Mondello et al. 2008). For satisfactory separation of all organochlorine pesticides analytes, two-dimensional gas chromatography (GC×GC) conditions need development by repositioning modulation time as well as hot and cold-pulse duration. One column has been generally used for the same class of pesticides. However, a pair of columns with varying polarities along with distinct geometric parameters with similar or different chromatographic conditions can also be used. Recurrently, the preceding column has been used in analytical determination, while the succeeding column helps to confirm the identification of the peak (Dasgupta et al. 2010).

4.2.2 Liquid Chromatography (LC) and High-Performance Liquid Chromatography (HPLC)

The high-performance liquid chromatography (HPLC) can be operated in either normal-phase or reversed-phase. The normal-phase uses a polar stationary phase like that of organic moieties with cyano and amino functional groups or silica. The mobile phase is non-polar consisting of a very non- polar solvent of hexane or heptane mixed with a slightly more polar solvent of isopropanol, ethyl acetate or chloroform. To determine the analytes that can easily be soluble in non-ionic medium, operating the high-performance liquid chromatography (HPLC) in this mode is adequate.

Pesticides which are thermally labile or non-volatile like triazines and carbamates, along with their metabolism products and products of degradation, are determined by high-performance liquid chromatography (HPLC). Since their detection is hindered by gas chromatography, they are detected by the application of a reverse-phase high-performance liquid chromatography comprising of a non-polar C18 stationary phase bonded to silica and composite containing acetonitrile or tetrahydrofuran or methanol with water being the mobile phase. C8-bonded materials on silica stationary phases are used only to an extent. The compounds that elute at high organic solvent concentration have an enhanced sensitivity due to the enhancement of the ionization efficiency by the organic solvent composition (Kostiainen and Kauppila 2009).

Ultraviolet-visible (UV), diode-array detection (DAD) or fluorescence detection (FLD) is a part of the liquid chromatography based detectors for the determination of parent as well as its transformation products (Alder et al. 2006). Additionally, the compounds that fluoresce can be detected with a fluorescence detector (FD) encompassing higher sensitivity and selectivity. Recently, liquid chromatography has been coupled to numerous mass spectrometry detectors, for determining pesticide residue levels and to resolve their structures in solid and liquid samples in the environment along with baby food and foods of plant origin (Ahmed 2001). Tandem MS-MS integrated with high-performance liquid chromatography (HPLC) improves the selectivity of the detection while allowing the verification of trace level residues. This is ensured by eradicating the interferences through parent ion segregation and the underlying fragmentation before ion measurement (Tadeo et al. 2004). This analytical technique has replaced various specific, burdensome methods for the analysis of N-methyl carbamates, by encompassing the target compounds in a multiresidue method (Pihlström et al. 2007).

High-performance liquid chromatography (HPLC) is presently being used for multiple reaction monitoring (MRMs) of compounds that have low volatility like bipyridylium herbicides as well as heat-sensitive compounds like NMCs or benzoylurea. These no more require extensive clean-up procedures or derivatization. The limits of detection (LOD) are in the range of μg/kg or even lower when various detectors are employed. These particularly depend on the combination of pesticide and matrix and the employed detection technique. The limits of detection are lowered and the selectivity is highly enhanced when mass spectrometry or tandem MS-MS detection is used (Tadeo et al. 2004).

The development of new liquid chromatography packing material for the column has led to the development of immunoaffinity sorbents, molecularly imprinted polymer, and restricted access medium columns. Owing to their feasibility, stability, and ease of preparation; molecularly imprinted polymer materials have gained much popularity. The affinities and selectivity obtained by the binding of the prepared polymeric receptors to that of the small molecules in these columns are similar to those of the immuno-based interactions. Thus, the target analyte can be selected and retained in the course of the imprinting process. The principle holding the restricted access medium columns is the analyte retention. There is the plausibility of an enhanced screening using the Ultra-Violet detection in reversed-phase Liquid chromatography of the polar pesticides when using these recently developed columns (Ahmed 2001).

Apart from the use of conventional Liquid chromatography or ultra-performance liquid chromatography (UHPLC), there exist other methods for the distinct separation of transformation products. As these transformation products are small ionic compounds, their separation becomes difficult using the typical reversed-phase liquid chromatography. The use of ionic or ion pair chromatography is the best suited for such ionic compounds like dialkyl phosphates, that is only partially separated by Liquid chromatography. This results in a very efficient separation (Botitsi et al. 2011).

4.2.3 Gas Chromatography-Mass Spectrometry (GC-MS) and High-Performance Liquid Chromatography-Mass Spectrometry (HPLC-MS)

The determination of both the pesticide residues and its transformation products requires that chromatographic techniques hyphenated to mass spectrometry as the detection system is used. Mass spectrometry can work in a hyphenated system with gas chromatography or liquid chromatography to give more sensitive analytical techniques of gas chromatography-mass spectrometry (GC-MS) and liquid chromatography-mass spectrometry (LC-MS) (Botitsi et al. 2011). These can determine a wider range of concentrations. Further sensitivity can be increased by the use of tandem mass spectrometry like LC–MS-MS and GC–MS-MS (Kosikowska and Biziuk 2010). Mass spectrometry has been a robust detection tool which may use several analyzers, as shown in Table 4.1, for the identification, quantification, detection, and confirmation of the pesticide residues and its transformation products. The development of a robust technique of ion trap MS-MS provides a substitute in determination as well as verification of pesticides present in minute quantities. Tandem mass spectrometry reduces intrusion by matrix, enhances selectivity, attains higher signal-to-noise ratio and consequently enhances the limit of detection (Chung and Chen 2011).

Two categories of bench-top mass spectrometry detectors are used together with capillary gas chromatography, a quadrupole mass spectrometry and, an ion trap. These offer the capabilities of commonly used GC–MS-MS without the added cost and offer the unambiguous identification and quantification of numerous compounds in complex fatty matrixes. Thus, the prevailing methods use mass spectrometry in the analysis of trace residues due to its inherent capabilities like selectivity, sensitivity, avoidable need for derivatization and identification-confirmation capability in a single step in different types of complex matrices. The amalgamation of HPLC with mass spectrometry (HPLC-MS) has overcome the disadvantages of high performance liquid chromatography (HPLC) detectors having reduced sensitivity and selectivity (Lehotay 2006).

Liquid chromatography –mass spectrometry (LC-MS) has been employed in the detection of non-volatile, ionic compounds like triazines, chlorinated phenoxy acids, and carbamates that are readily available for thermal degradation or cannot be vaporized in gas chromatography separations (Chaves et al. 2008). Some of the modern pesticides are not amenable to gas chromatography and if they do not contain any strong chromophore for ultraviolet or visible absorption or if they do not fluoresce, then LC–MS-MS becomes the only choice to detect the pesticide in its unmodified form. Liquid chromatography –mass spectrometry (LC-MS) can provide constitutional make-up too, for the verification of the identity of a sample analyte (Rosen 2007). A selected ion from the target compound can be fragmented by collision and hence all of the potential interferences from the matrix of sample, stationary and mobile phases can be eliminated (Rodrigues et al. 2007).

Diode array detection, when used in a multi-wavelength scanning program, may register spectra from 190 to 400 nm concurrently. This helps ensuring the detection of all of the UV-Vis absorbing constituents. The introduction of Atmospheric pressure photoionization (APPI) technology is enabling liquid chromatography – mass spectrometry (LC-MS) in turning out to be amenable to non-polar compounds. LC–MS-MS has been a potent mechanism in the detection of pesticide residues in numerous complex matrices due to its distinguished advantages. Some of them include the enhanced sensitivity and selectivity, reduction in sample pre-treatment steps, satisfactory quantification and the reliable confirmation that can be achieved even for trace amounts of pesticides. Until now, all organochlorine pesticides and their degradation products were analyzed by coupling with a selective detector or mass spectrometry detector. Anyhow, chlordecone is insensitive even with GC–MS-MS. Thus, LC–MS-MS might be taken into consideration while determining chlordecone (Chung and Chen 2011).

The newly developed ambient-ionization mass spectrometry techniques like desorption electrospray ionization (DESI) and direct analysis in real time (DART), hyphenated with high-resolution mass spectrometry platforms without liquid chromatographic separation, require minimum pretreatment. Conclusively, these have shown potential for the screening of pesticide residues (Botitsi et al. 2011). Although numerous pesticides can be analyzed by both gas chromatography –mass spectrometry (GC-MS) and liquid chromatography –mass spectrometry (LC-MS) techniques, a wider scope is provided by the methods applied in GC-MS. This is because a few pesticide composites make their determination by LC-MS a tedious task. The use of capillary gas chromatography coupled to MS-MS has achieved appreciable sensitivity when complex matrices are considered since they eliminate the interferences prior to the ion measurement. GC-MS is beneficial for the organochlorines, as these have low polarity and are non-ionizable and hence cannot be used in LC-MS (LeDoux 2011).

The mass spectrometers that can be commonly used in combination with gas chromatography include:

1. Triple quadrupole (QqQ) - Owing to the increased specificity, sensitivity, precision, and a wider linear range, the approaches employing triple quadrupole (QqQs) detectors have been extensively utilized in pesticide determination of food and water. Multiple-reaction monitoring (MRM) monitoring introduces an additional advantage of greater specificity by detecting both the source ion as well as any of its pre-determined fragments (Rosen 2007). These made the identification and subsequent quantification of the different pesticides possible in a single analysis. It aids in identifying isobaric species efficiently (Wołejko et al. 2016). Nonetheless, these compounds require an extensive instrument parameter optimization depending on the compounds and cannot identify the non-targeted compounds. Additionally, a set of standards have to be run to prepare the calibration curves of the standards for the identification and quantification of the standards present in the pesticides for each batch of samples (Wang et al. 2019).
2. Times of flight (TOF) - Presently accessible liquid chromatography –time of flight -mass spectrometry (LC-TOF-MS) along with LCQqTOF- MS platforms are undeveloped for pesticide residue detection (Lacorte and Fernandez-Alba 2006). Pesticide detection in food together with water widely use the TOF-MS detectors (Fernández-Alba and García-Reyes 2008; Ibáñez et al. 2008). TOF tools estimate mass-to-charge ratios relying upon distinction in the timings of the flight that is dependent on mass from the analyzer to a detector with high sensitivity, high mass resolution (more than 5000 FWHM) and high-mass accuracy for the pesticide analysis (Fernández-Alba and García-Reyes 2008). LC-TOF-MS is profitable by providing the primary configuration of the source as well as fragmented ions that might be employed for discerning or verification of target compounds. Their use can also be extended for the recognition of unspecified composites for non-target analysis. Full scan techniques like TOF-MS prove to be beneficial as compared to the targeted acquisition by MS-MS detection when an exhaustive screening of pesticide residues is being considered. The increased selectivity helps discriminate between the target pesticides by MS-MS detection that was only slightly separated by liquid chromatography (Garcia-Reyes et al. 2007).
3. Ion trap (IT) and triple quadrupole linear ion trap (QqLIT)- The ion trap (IT) mechanism by virtue of its multiple stages fragmentation (MSn) ability, can allow the continuous segregation of parent ions at variable stages, trapping, fragmentation, and mass examination as a function of time and in the same space. As a result of this, they have been widely used in pesticide residue detection for gratifying full scan sensitivity. These features cannot be availed in the triple quadrupole (QqQ) and QqTOF instruments. However, when IT-MS is operated in the multiple-reaction monitoring (MRM) mode, they display reduced sensitivity, lower linear dynamic range, along with reduced quantification ability when compared with the triple quadrupole (QqQ) complex running in the multiple-reaction monitoring (MRM) approach. The inherent sensitivity and MSn precision and spectral quality are compromised due to the plausibility of only 10–30

ions being monitored simultaneously. This can be attributed to the space charge effects (Soler and Pico 2007; Kuster et al. 2006).

4. Quadrupole time of flight (QqTOF) - When conventional TOF-MS leads up to quadrupole and a collision cell, there is an increased resolution and an enhanced sensitivity with mass accuracy in the QqTOF-MS platform. This leads to an effective determination of the pesticide residues and their altered compounds (Kuster et al. 2009; Hernández et al. 2008; Barceló and Petrovic 2007; Petrovic and Barceló 2006). The hybrid quadrupole time of flight mass spectrometry (QqTOF-MS) is being effectively employed in the detection of photo-degraded or the hydrolyzed products of pesticides found in environmental waters. These display high resolution and possess accurate mass capability (Hernández et al. 2008).
5. Orbitrap mass spectrometer- The recent development is the use of a single-stage mass spectrometer in the form of orbitrap technology for the analysis of pesticides to ensure food safety. The orbitrap hybrid mass spectrometer is an amalgam of techniques used in linear ion trap-MS (LIT-MS) together with orbitrap mass analyzer possessing sharp resolution as well as mass precision capacity (Makarov et al. 2006; Perry et al. 2008).

An enhanced selectivity, sensitivity and remarkable degree of confirmation are achieved when liquid chromatography - gas chromatography (LC-GC) is coupled. In this technology, certain specific components are prefractionated by liquid chromatography and then sent to the gas chromatography to segregate the analyte followed by its determination. The subsequent on-line transfer leads to the replacement by the liquid chromatography step of the sample formation procedures like extraction, clean-up, and concentration. The disparate techniques used for each stage in the sample preparation steps are summarized in Fig. 4.2. This helps save time and the use of voluminous toxic organic solvents which frequently lead to errors and analyte loss. Automation in the liquid chromatography - gas chromatography (LC-GC) systems is an asset when equated to the alternate procedures, especially in cases of larger sample volumes. Anyhow, the solvent utilization is again more when compared to the amount desired along with the high-priced apparatus owing to the compounded assembly (Cortés et al. 2008; Flores et al. 2008; Sanchez et al. 2006).

4.2.4 Thin-Layer Chromatography (TLC)

This technique is often used for the determination of pesticides like N-methyl carbamate pesticides (NMCs), pyrethrins or pyrethroids together with their metabolites. Owing to its lower detection limit, it is less widely used than high performance liquid chromatography (HPLC) and capillary gas chromatography. The availability of automated multiple developments of the TLC plate for analysis of pesticide residues makes even the formation of a gradient elution possible. Here, the detections are usually carried out by employing a thin-layer chromatography (TLC)

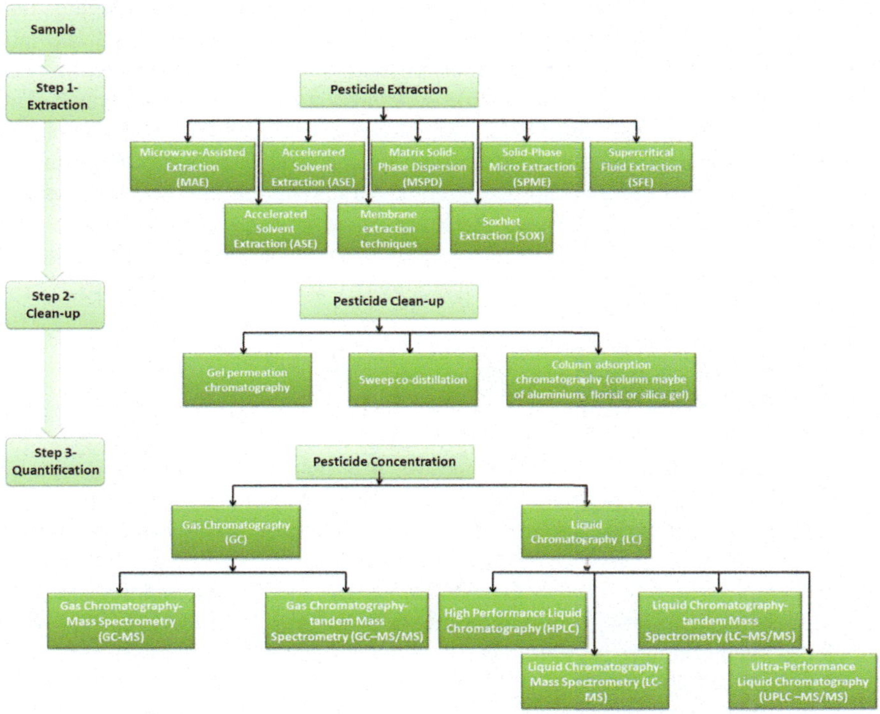

Fig. 4.2 The disparate techniques used for each stage in the sample preparation steps have been summarized in this figure. The sample preparation steps include extraction, enrichment, and concentration of the pesticides in the sample

scanner for measuring the ultraviolet (UV) absorbency for as many as six wavelengths (Ahmed 2001).

4.2.5 Capillary Electrophoresis (CE) and Capillary Electrochromatography (CEC)

This account for the recent scientific methodologies which are being employed for pesticide residues detection. The increased resolution, accelerated analysis and reduced solvent consumption are the assets of this technique. In capillary electrophoresis, an electrolytic suspension called running buffer like sodium phosphate fills up the fused silica capillary. The sodium salt of octakis (2, 3-diacetyl-6-sulfo)-Q-cyclodextrin is being employed as the resolving tool in the effective capillary electrophoresis separation of organophosphate enantiomers that possess a phosphorus-based stereogenic center. Micellar electrokinetic chromatography

(MEKC) is the familiar mode of capillary electrophoresis. Here, the surfactants that have concentrations higher than the critical micellar concentration are pushed into the background electrolyte. Subsequently, micelles are formed. On interaction with the micelles, the neutral and comparable structured compounds are separated. The surfactants used here may be anionic such as sodium dodecyl sulfate or cationic or zwitterionic. Bile acids and vancomycin may also be used. The substitute technique of MEKC is capillary electrochromatography (CEC). Herein, the micellar pseudo-stationary phase is replaced by a stationary phase of C18. This technique is a combination of the separation adequacy of capillary electrophoresis and the selectivity of high performance liquid chromatography (HPLC) (Ahmed 2001).

4.2.6 Immunoassays (IA)

Immunoassay technique is preferred because it is feasible, non-strenuous, can be performed without any high-tech instrumentation and do not require any tedious sample cleanup and extraction procedures. These methods are favored when huge sample quantities are screened simultaneously for a sole analyte in brief period. Majorly employed immunoassay in pesticide determination is immunosensors (IS), immunolabeling and enzyme-linked immunosorbent assay (ELISA). ELISA technique has been booming for pesticide measurement (Bhand et al. 2005). However, these require certain amendments to be made. When a parallel is drawn between chromatography and immunosensors, it can be seen that the immunosensors are highly selective, sensitive, can be used for in-field detection and can work with complete automation to produce results within a short period (Jiang et al. 2008). However, the application of ELISA is limited to single pesticide or some class-specific immunoassay and cannot perform multiresidue pesticide analysis. Afresh, ELISA has shown comparable results to high performance liquid chromatography (HPLC) in the analysis of carbaryl, N-methyl carbamate pesticides (NMCs), carbofuran and methiocarb in strawberries and cucumbers, involving no clean-up of the sample (Ahmed 2001).

4.2.7 Ultra-Performance Liquid Chromatography (UPLC)

A vital development in liquid chromatography field has been the increasing use of UPLC. The principle of ultra-performance liquid chromatography (UPLC) is based on the increase in the efficiency that is not affected by any increased flow rates or linear velocities when the particle size is reduced to less than 2.5 microns. Conventional high performance liquid chromatography (HPLC) columns are 10–25 cm long, have an internal diameter of 2.1–4.6 mm, particle size is 3–5 mm and run at pressures going under 6000 psi. In contrast, ultra-performance liquid chromatography (UPLC) has normal-sized columns ranging from a length of 100–150 mm, internal diameter 2.1 mm, and particle diameter within 2 mm. These require specific high performance liquid chromatography (HPLC) pumps that can apply pressures up to 1500 psi. The smaller particle size in UPLC increases the column efficiency greatly and becomes almost uninvolved with the flow rate of the mobile phase. Thus, ultra-performance liquid chromatography (UPLC) has an upper hand over HPLC due to their narrower peaks, improved confirmation of the targeted pesticides, augmented speed, enhanced accuracy owing to the elevated peak intensity and enhanced resolution (Leandro et al. 2006; Kovalczuk et al. 2006).

4.2.8 Hydrophilic Interaction Liquid Chromatography (HILIC)

An exceptional development was that of the introduction of HILIC (Hemström and Irgum 2006). It is a substitute to the ion pair liquid chromatography which is applied to the splitting of ionic compounds that cannot be retained in the reversed-phase columns. This model is similar to that of the normal-phase liquid chromatography, except the fact that they use a polar mobile phase which stands compatible with the mass spectrometry detection (Zhu et al. 2008). The polar stationary phases in hydrophilic interaction liquid chromatography (HILIC) are available to bind to polar compounds that possess a greater affinity than the non-polar compounds. The polar stationary phases might include cyanopropyl- (CN), aminopropyl- (NH2), and 2, 3-dihydroxypropyl altered (OH) together with unaltered silica. The composites are eluted in order of intensifying hydrophilic and polar character because of the employed aqueous-organic solvent gradient accompanied by a reducing organic solvent concentration. The coupling of hydrophilic interaction liquid chromatography (HILIC) with mass spectrometric detectors has proven to be effective (Kostiainen and Kauppila 2009).

4.2.9 Other Techniques

Although less informative than the capillary gas chromatography or high performance liquid chromatography (HPLC), numerous minor techniques in pesticide residues detection are employed. Since the gas chromatography and liquid chromatography are comparatively expensive, the chemiluminescence (CL) method has attracted its attention in the pesticide residues determination (Li et al. 2006). Flow injection analysis (FIA) technique offers advantages like high sample throughput, greater reproducible timing and it eliminates the need of attainment of a steady-state (Kumaran and Tran-Minh 1992; Lee et al. 2002). This technique can thus be used for the detection of pesticides. Many other methods can be alternatively used but these are time-consuming and expensive. On one hand they are very sensitive and reliable but on the other hand, they require highly trained professionals and do not hold suitable when in-field pesticides' determination is considered. In a typical case, the organochlorine pesticides analysis in plants was reported by estimating the intensity of indium monochloride's emission band at 359.9 nm that had a quantization level of 0.05 ng chlorine. The use of acetylcholinesterase for the enzyme inhibition assay helps detect carbamate as well as organophosphate pesticides. These biosensors possess a catalytic activity or inhibit numerous enzymes when in contact with pesticides. A highly pH-sensitive fluorescence probe can be used in the flow injection system to allow fluorescent monitoring of pH changes in the course of the enzymic reaction. This method provides greater sensitivity than the other reported pH-indicators (Patel et al. 2019).

4.3 Pesticide Residue Detection in Fruits and Vegetable Products

The major path of pesticide exposure is by the ingestion of food. The remnants of pesticides are generally observed in juice as well as beverages but at reduced concentrations. This might pose a risk to human health when exposed to longer time periods. And thus, maximum residue limits (MRLs) are defined by major parts of the world. Vegetable products too are exposed to pesticides to remove the insect-pests and henceforth control the pest complexes of various vegetable crops. Henceforth, the pesticides can be commonly detected in the exterior of the crops while spraying treatment or might be soaked up by the rootstock. This makes it indispensable to regularly monitor the fruits and vegetable products for pesticide residues. Ordinarily, the peel and the pulp contain higher amounts of pesticides than in their juices. And thus, the analysis becomes burdensome. The analytical methods so employed ought to be highly sensitive and specific (Gilbert-López et al. 2009).

Pesticide analysis poses some hurdles due to the complexity of the matrix and the lower concentrations of the target compounds in the desired food products. Thus, it is of utmost priority to pre-treat the sample, depending on its chemical composition, before its analysis to avoid any interference during the sample analysis. Complex multi-component samples like food require separation methodologies like gas chromatography together with liquid chromatography. When influenced by the acuteness of the chosen procedure, a cleanup stage of the raw extract becomes crucial. In general, when the matrix is more complex, the amount of desired pesticides is lower and lesser is the specificity of ultimate segregation or determination and hence, extensive clean-up becomes a pre-requisite (Gilbert-López et al. 2009).

Pesticide remnants in juice, as well as beverages, mostly use chromatographic techniques, particularly the mass spectrometry (Tadeo et al. 2004). Ultraviolet (UV) detection coupled to liquid chromatography has certain advantages like being low in cost and providing a good response for numerous compounds. However, they have a lower selectivity and thus have constraints in measuring the pesticide levels found in juices. Additionally, highly selective techniques of determination like MS-MS coupled to a triple quadrupole pave way for the reduction in the clean-up process while sample formation (Gilbert-López et al. 2009). Employing mass spectrometry together with high performance liquid chromatography (HPLC) has enhanced the responses of pesticides residues in fruits and vegetable products. Multiresidue methods are practical and direct as they ensure detection of hundreds of pesticides as well as its metabolism products in a single run. This is especially important when organophosphorus pesticides with an oxidizable sulfide group are to be detected, where the metabolites are a part of the residue content (Pihlström et al. 2007).

The quantification of pesticide residues in food can be accomplished by the use of gas chromatography-high-resolution mass spectrometry (HRMS) and liquid chromatography HRMS such as orbitrap and time of flight (TOF) instruments. These have full mass spectrometry scan capability (Wang et al. 2019). These help in the identification of compound and elucidation of its structure. Nonetheless, electro spray ionization (ESI) and atmospheric pressure chemical ionization (APCI) has led to the feasible determination of pesticide residues of distinct chemical make-up in food together with vegetables, at levels similar to those procured in GC–MS. The LC–GC technique is used in detecting the remnants of pesticides in fatty vegetable matrices such as nuts, pistachio nut, and hazelnut peanut and olive oil (Kumaran and Tran-Minh 1992).

Immunoassays and capillary electrophoresis are the other analytical methods used for the detection of fruits along with vegetable products. Enzyme-linked immunosorbent assays (ELISA) has been used mainly for screening purposes rather than for any accurate quantitative analysis in juice and beverages (Dagher et al. 1999; Urruty et al. 1997). For instance, it has been used for the quantification of the fungicide thiabendazole without the requirement of a clean-up in fruit juices and its concentrates. Photokinetic procedure for the determination of diquat in potatoes and voltammetric technique for herbicide analysis in fruits have also been described. Some others include spectrophotometric detection of organophosphorus compounds in fruits, propoxur in vegetables and grains, bendiocarb in grains, and diquat and paraquat in plants (Ahmed 2001).

4.4 Pesticide Residue Detection in Plant Material

In the contemporary world, there is a growing interest in the use of natal products, namely herbal medicines. These have been adequately used in both the developed and the developing countries. Due to the exposure of plants to chemical compounds such as pesticides during their growth, storage, and the various manipulation stages; there exists diversity in their chemical composition. Thus, herbal products possess more of the quality-related issues that are specific (Calixto 2000).

Currently, the pesticide remnants determination in plant products has been accomplished by using contemporary instrumental methodologies like gas chromatography, liquid chromatography, together with high performance liquid chromatography (HPLC). Capillary gas chromatography and high performance liquid chromatography (HPLC) by using specific detectors have been domineering for determination of pesticide surplus in the specimen that originates from plants. For screening purposes, planar chromatography (TLC), including the current substitutes like HPTLC, over-pressure layer chromatography (OPLC); are yet infrequently applied in conjunction with other specific biochemical determination techniques. However, the traditional processes have been losing usefulness (Kováč et al. 1987; Tekel 1990; Zoun and Spierenburg 1989).

Pesticides in the plant material ordinarily consist of certain heteroatoms like oxygen, phosphorus, sulfur, nitrogen, chlorine, bromine and fluorine. Several of these may often be present in a single molecule too. Subsequently, many gas chromatography procedures employed for the pesticide determination in plant extract operate with element-specific detectors. Multi-residue methods are the primarily employed mechanisms for remnant detection in plant material (Lee and Wylie 1991).

Disparate numbers of techniques and combinations of detectors have been used for the analysis of samples of plant origin. Some of them are the two-dimensional gas chromatography (Tekel and Hatrík 1996), GC-ECD-NPD system and the GC-AED (Atomic Emission Detection) system. The AED detector is a combination of the excitation of the plasma to that of optical emission spectroscopy. Conclusively, the GC-ECD-NPD system appeared to be more feasible for the analysis while screening of plant foodstuffs. Hence, the capillary gas chromatography could be used reliably when running in parallel to ECD-NPD detection (Linkerhägner and Stan 1994).

4.5 Pesticide Residue Detection in Water

Pesticides when used to combat the agricultural pests by spraying on crops or when applied to the soil are gradually released into the environment, especially water. The use of this polluted water can lead to serious contamination of the environment and may cause mishap to human health. Thus, monitoring the levels of pesticides is crucial in minimizing the hidden hazards to human well-being. The commonly found pesticide in wastewater was the insecticide diazinon, whereas the herbicide diuron was present at the greatest concentration. However, the number of pesticides found in rivers was much lower as compared to the wastewaters. The levels of pesticide residues in drinking waters are of key importance and hence they need to be evaluated by specific sensitive methods (Ballesteros and Parrado 2004).

However, contamination of water is not limited to only pesticides (Vazquez-Roig et al. 2011). The pesticide mixture itself may contain a copious number of composites that have diverse functions and a broad range of polarity as well as solubility. Whatsoever, polar contents of the mixture readily reach the surface water through process run-off, spray drifting or leaching and thus can probably contaminate the groundwater (Polati et al. 2006). Subsequently, the target compounds need to be separated from the mixture of chemical compounds present in water. Detection of pesticides in the water samples from the environment can be only evaluated by some sensitive method employing lesser extraction stages. Specimen formation protocol that is rapid, accurate and precise and the ones that consume less solvent are considered ideal to be used in the analysis. The instrumental techniques commonly used for the analysis of water samples for pesticides include gas chromatography with element-specific detectors, mass spectrometric (MS) detection and high-performance liquid chromatography (HPLC) with diode-array detection (DAD) and fluorescence detection (Jeannot et al. 2000; Rodrigues et al. 2007); Primel et al. 2010).

4.6 Pesticide Residue Detection in Air

The unrestricted use of pesticides has its aftereffects on not just the water bodies and soil, but the air also gets infected by their use. This may be attributed to the fact that there is a spray drift during its administration and volatilization after its application. An approximated 30–50% of the sprinkled pesticides infiltrate the air and can survive for several days or weeks. This happens when the pesticides adsorbed onto the soil particles enter the troposphere post the application of pesticides (Yusà et al. 2009). Post-administration, the pesticides become rapidly available in their gaseous order, dissolved in water vapor or can be adsorbed onto solid particles. However, after they are exhaled into the atmosphere, they can either be in their gaseous phase or the particles phase (Kosikowska and Biziuk 2010).

The determination of pesticides is crucial in the air since the transport of pesticides by the atmosphere is a dominant cause of environmental pollution. The varying stability of the pesticides leads to their different decomposition rates in the atmosphere. This, in turn, governs the distances that the parent and the decomposition products travel before settling. These may also be afar from the application spot of the pesticide. The trace amounts of pesticides, particularly in the areas far away from the sources; make the air specimen examination much more demanding than the scanning of water or soil samples. Ordinarily, the air samples are analyzed by extraction with soxhlet or accelerated solvent extraction. While resolving the air samples, the apparatus should be such that it can separate the gaseous phase and the solid particles. This is because of the considerably reduced levels of pesticides in the air as compared to soil and water. The principle underlying such protocols is that the pesticides are resolved from the air by pre-concentrating them onto solid sorbents when huge amounts of airflow across it. This needs to be taken care of as the pesticide quantity in the gaseous form might be in trace amounts (Kosikowska and Biziuk 2010).

The pesticides are finally resolved in both, the gaseous as well as the particulate matter, individually. Technique chosen for analysis depends on the compounds that have to be targeted and thus, the key role is played by the selection of specific detectors for a peculiar class of mixtures. The trace amounts of pesticides present in the gaseous phase together with the particulate matter demand employing analytical techniques that are of utmost selectivity and sensitivity. The standard instrumental techniques used after the isolation and pre-concentration of pesticides include the application of gas or liquid chromatography, or them being coupled with mass spectrometry (Kosikowska and Biziuk 2010).

4.7 Pesticide Residue Detection in Soil

Soil represents the predominant source containing the environmental pesticides, from which the pesticides can be released into the atmosphere, groundwater or living organisms (Luchini et al. 2000). Moreover, the first trophic level of the food chain is formed when these pesticides on direct application get deposited either through a dry gaseous or dry particle-bound or by the wet deposition onto plants and soil (Rashid et al. 2010). Owing to their persistence and hydrophobicity, the pesticides accumulate in the soils and might be withheld for years. The surface and groundwater can be contaminated when these move away from their sites, which can prove detrimental to human health as well. Soil is miscellaneous in nature which has a permeable build-up comprising of both the natural organic and inorganic components. Thus, the extent to which the pesticides can adsorb to the soil depends primarily on the type of pesticides and the soil attributes. The characteristics pertaining to soil maybe its moisture, pH, the capacity to retain, filter and degrade pesticides, texture, and the soil organic matter (SOM) quantity. Pesticides firmly adsorb to the soils that are rich in clay or organic matter (Belfroid et al. 1998).

As compared to food, the interaction between soil particles and the analytes is stronger (Rissato et al. 2005). For the reason that the organochlorine pesticides are highly lipophilic in nature, these persist in soils by binding to the soil organic matter through covalent bonds or hydrophobic interactions or diffused-controlled partitioning. Thus, they are not leached out of the soil and have the potential to travel long distances. The organochlorine pesticides are a part of the larger group of persistent organic pollutants (POPs). Thus, their detection in even the trace amounts becomes of utmost importance (Negoita et al. 2003; Zhang et al. 2011).

Pesticide residues in the soil are mainly analyzed using gas chromatography employing nitrogen-phosphorus detection (NPD) or predominantly using gas chromatography with electron-capture detection (ECD). The target compounds in the soil consist majorly of hydrophobic and the persistent organic pollutants (POPs). Thus, very little attention is paid to multiresidue analysis or the other types of pesticides. Electron impact (EI) ionization is the most widely used technique for cumulating the characteristic dissociated ions of the composites. The method is successfully implemented onto triazines, organochlorines, carbamates, organophosphorus, chloroacetamides, and fungicides for their detection (Sánchez-Brunete et al. 2002). However, since the molecular ions are sometimes invisible in electron impact spectra, the procedure supplementing the electron impact technique is the chemical ionization (CI). This helps determine the molecular weights. Commonly, the parent pesticides are detected by either gas chromatography or liquid chromatography, whereas the transformation products, due to their ionic properties and reduced volatility are analyzed only by liquid chromatography (Andreu and Picó 2004).

Due to the simultaneous extraction of humic substances along with the acidic pesticides in the soil samples makes the trace analysis of these pesticides challenging. This can be attributed to the fact that the reversed-phase liquid chromatography with ultraviolet determination (RPLC-UV) is obstructed because of the co-extraction. At low wavelengths for determination like 220 nm, the intrusion by the co-extraction is shown as a broad hump. Subsequently, the baseline deviates and the reliable quantification for the analytes at trace levels becomes demanding (Andreu and Picó 2004). Coupled-column RPLC (LC-LC) employing an analytical column filled with restricted access material (RAM) has also been used to hike the sensitivity of analyte detection in trace amounts (Hogendoorn et al. 2001). The technique of fluorescence detection has been used only with N-methyl carbamates that are broken down by sodium hydroxide at high temperatures to produce methylamine. A fluorescent isoindole is produced when this reacts with o-phthalaldehyde (OPA). This technique has been barely used, as pesticides along with its transformation products are not fluorescent and demand derivatization (Hogendoorn et al. 2001).

The hyphenated technique like gas chromatography-mass spectrometry (GC-MS) has been widely used for the multi-residue detection of pesticides in soil. The ion trap (IT) technique is an example of the tandem or MS-MS instruments coupled to gas chromatography (Andreu and Picó 2004). The foremost advantage of the ion trap is that it is capable of achieving the routine electron impact analysis of low-molecular-weight compounds along with the competence of chemical ionization and MS-MS. Although the ion trap technique can portray the aforementioned modes, these have not been fully exercised for detecting pesticide residues in soil. Liquid chromatography with mass spectrometry detection (LC-MS) is used as a substitute for pesticide detection in soil (Stipičević et al. 2003). By surging the pre-analyzer extraction (skimmer cone) voltage, the mass spectrometry fragmentation can be accomplished by the use of a single quadrupole. However, increased specific fragmentation of the protonated or deprotonated molecule is accomplished by the use of collision-induced dissociation (CID) of the tandem mass spectrometry (MS-MS) detection. Here, two tandem mass analyzers of triple quadrupole (TQ) and ion trap are used. Using specific transitions, the triple quadrupole is enforced to detect various acidic herbicides by carrying out the second fragmentation between the first and second quadrupole in a collision cell (Pozo et al. 2001).

4.8 Conclusions

It can be noted that gas chromatography and high performance liquid chromatography (HPLC) with mass spectrometry detection form the base for several detection methods. HPLC–MS-MS is the favored, growing and robust technique for multi-residue analysis (Hercegová et al. 2007). It is evident that no clean-up method is efficient enough to remove all of the components of a matrix from its source and the complete segregation of all the sample components is practically not possible for very complex samples. Thus, enhanced separation efficiency can be attained by the use of two dimensional gas chromatography (GC×GC) or the rather powerful technique of GC×GC combined with time of flight (Zrostlíková et al. 2003).

The development of ultra-performance liquid chromatography (UPLC) and fast gas chromatography in combination with different mass spectrometry platforms has led to greater detection throughput without settling for sensitivity and segregation capacity. The triple quadrupole (QqQ) platform, when operated in the multiple-reaction monitoring (MRM) system, is highly precise and sturdy for both gas chromatography as well as liquid chromatography interfaces that allow for low scan times with high speeds. They can be effectively used for the analysis of multiple residues of pesticides. The prominent intrusion in the pesticide multi-residue analysis is the use of a hybrid quadrupole linear ion trap (QqLIT) mass spectrometry platform, which is an amalgam of the advantages of both linear and QqQ ion trap mass detector (IT) geometries. The time-of-flight (TOF) and quadrupole time-of-flight (QqTOF) form successful bases for the screening, spotting along with composition determination of pesticides' transformation composites in food along with water specimens. This can be attributed to their high accuracy and greater mass resolution. In the field of pesticide residue analysis, the orbitrap mass spectrometer is a rapidly evolving and promising technology (Botitsi et al. 2011).

When compared to gas chromatography –mass spectrometry (GC-MS), single quadrupole mass spectrometers have not been much employed along with liquid chromatography –mass spectrometry (LC-MS). Some studies also indicated that numerous pesticides could be efficiently determined by electron capture detector in gas chromatography measurements rather than in the LC–MS-MS. This holds true for organochlorine compounds but additionally becomes genuine for other pesticides like dinobuton, chlozolinate, nitrofen, benfluralin, etridiazole, flumetralin or vinclozolin. This is bypassing fenchlorphos, which is more efficiently detected using gas chromatography along with nitrogen phosphorus or flame photometric detection and biphenyl that is solely detected by GC-MS.

Heretofore, only a few procedures to detect some of the organochlorine pesticides in fatty foods survived to the expectations of the method performance criteria stated by the Codex. Consequently, significant research is required for the betterment when comparing it to the detection of other pesticide analytes such as organophosphorus pesticides and carbamate pesticides.

Even for many gas chromatography-amenable pesticides, LC–MS-MS technique proves to be powerful in providing superior quality results than gas chromatography –mass spectrometry (GC-MS). Noteworthy analytical implementation of the LC– MS-MS can outweigh the gross capital of the instrument. It has abilities to save time in method development, can produce broad peaks that grants time to monitor many other co-eluting peaks without affecting the quality of results, and is a powerful technique for routine or certain special projects. Thus, hundreds of pesticide residues can be controlled in a single chromatogram of LC–MS-MS, which is not possible when using GC-MS with selected ion monitoring (SIM) or multiple stages fragmentation (MSn) techniques.

When GC-QqQ-MS was compared to GC-IT-MS, several pesticides showed intra-day precision similar in QqQ and ion trap, whereas inter-day precision was undoubtedly better in ion trap. The triple quadrupole (QqQ) analyzer showed greater sensitivity in fatty matrices owing to its greater selectivity in multiple-reaction monitoring (MRM) stage and linearity was also enhanced with the triple quadrupole (QqQ) analyzer. Four LC-MS techniques, namely, triple quadrupole (LC-QqQ-MS), ion trap (LC-IT-MS), single quadrupole (LC-Q-MS), as well as quadrupole time of flight (LC-QqTOF-MS) were correlated while determining carbosulfan along with its major alteration compounds. The results displayed better reproducibility, broader dynamic range and at least a 20-fold greater sensitivity in case of triple quadrupole (QqQ) when compared to other instruments. When checked for sensitivity, QqTOF-MS gave supreme results, succeeding IT-MS and QqQ-MS. For methodical pesticide residue determination, the amalgam of LC-QqQ-MS-MS and UHPLC-QTOF MS was found to be reasonable and efficient. Both the techniques carry the advantages of simpler and rapid sample preparations along with the ability to detect numerous compounds with varying polarities. Owing to its higher sensitivity, LC-QqQ-MS-MS can detect a greater number of target compounds.

When the median of the limits of quantification are compared, it can be seen that much higher sensitivity is portrayed when liquid chromatography and tandem mass spectrometry are in use. The median of limits found for LC–MS-MS is found to be between 0.1 and 1 ng/mL, whereas it is found to be 100 ng/mL in GC-MS. Satisfactory implementation of LC–MS-MS can be dependent on the higher injection volumes in LC–MS-MS, 20 mL vs. 1 mL, as well as the reduced fragmentation during ionization, electrospray ionization vs. electron ionization. Heedless of the above constraints, the amount of pesticides contracted by a single LC–MS-MS run has been found to be two or three times more than the total analytes studied by gas chromatography-mass spectrometry (GC-MS) in the selected ion monitoring (SIM) mode. The interference by a matrix is greatly reduced when tandem mass spectrometry is used. Accordingly, no comprehensive cleanup and refined chromatographic separation are required when LC–MS-MS methods are used.

For the analysis of air, it can be said that there is the need to develop a simultaneous gas-particulate controlling procedure to upgrade the current environmental regulations and controls of pesticides qualitatively. The need of the hour is to detect the presence of pesticides in ambient air even at trace levels. However, the detection of multi-residue pesticides in the air at low concentrations is an analytical hurdle. The methods must be efficient enough to cope up with the engaged sources, the discerning of mechanisms of transport, reactivity of atmosphere, impacts on health and the setting of environmental limits.

When compared to the conventional procedures of gas chromatography and high-performance liquid chromatography (HPLC), the enzyme-inhibition based biosensors might prove to be an upper hand. Gas chromatography and HPLC are reliable but expensive, arduous and tedious whereas the biosensors are speedy, inexpensive and require almost no sample pretreatment. Nonetheless, these biosensors are relatively short-lived that limit their use and the detections generally needs more than half an hour and several milliliters of samples (Patel et al. 2019).

The baseline is that the pesticide remnants determination by gas chromatography-mass spectrometry (GC-MS) as well as LC–MS-MS illustrated that mass spectrometry supplementing neither gas chromatography nor liquid chromatography based technique could resolve the residue analytes. The need of the hour is to develop new techniques keeping these techniques still in use. However, there are several advantages of LC–MS-MS and hence it is still the most preferred method for pesticides' residues determination. Be it the enhanced sensitivity, greater selectivity, broader scope and the paramount one is the ability to carry out most of the detections without derivatization.

Funding This work did not receive any specific grant from funding agencies in the public, commercial, or not-for-profit sectors.

References

Ahmed FE (2001) Analyses of pesticides and their metabolites in foods and drinks. TrAC Trends Anal Chem 20(11):649–661. https://doi.org/10.1016/s0165-9936(01)00121-2

Alder L, Greulich K, Kempe G, Vieth B (2006) Residue analysis of 500 high priority pesticides: better by GC–MS or LC–MS/MS. Mass Spectrom Rev 25(6):838–865. https://doi.org/10.1002/mas.20091

Andreu V, Picó Y (2004) Determination of pesticides and their degradation products in soil: critical review and comparison of methods. TrAC Trends Anal Chem 23(10–11):772–789. https://doi.org/10.1016/j.trac.2004.07.008

Arrebola FJ, Vidal JM, González-Rodrıguez MJ, Garrido-Frenich A, Morito NS (2003) Reduction of analysis time in gas chromatography: application of low-pressure gas chromatography–tandem mass spectrometry to the determination of pesticide residues in vegetables. J Chromatogr A 1005(1–2):131–141. https://doi.org/10.1016/s0021-9673(03)00887-2

Bai Y, Zhou L, Wang J (2006) Organophosphorus pesticide residues in market foods in Shaanxi area, China. Food Chem 98(2):240–242. https://doi.org/10.1016/j.foodchem.2005.05.070

Ballesteros E, Parrado MJ (2004) Continuous solid-phase extraction and gas chromatographic determination of organophosphorus pesticides in natural and drinking waters. J Chromatogr A 1029(1–2):267–273. https://doi.org/10.1016/j.chroma.2003.12.009

Barceló D, Petrovic M (2007) Challenges and achievements of LC-MS in environmental analysis: 25 years on. TrAC Trends Anal Chem 26(1):2–11. https://doi.org/10.1016/j.trac.2006.11.006

Belfroid AC, Van Drunen M, Beek MA, Schrap SM, Van Gestel CAM, Van Hattum B (1998) Relative risks of transformation products of pesticides for aquatic ecosystems. Sci Total Environ 222(3):167–183. https://doi.org/10.1016/s0048-9697(98)00298-8

Bhand S, Surugiu I, Dzgoev A, Ramanathan K, Sundaram PV, Danielsson B (2005) Immuno-arrays for multianalyte analysis of chlorotriazines. Talanta 65(2):331–336. https://doi.org/10.1016/j.talanta.2004.07.009

Botitsi HV, Garbis SD, Economou A, Tsipi DF (2011) Current mass spectrometry strategies for the analysis of pesticides and their metabolites in food and water matrices. Mass Spectrom Rev 30(5):907–939. https://doi.org/10.1002/mas.20307

Calixto JB (2000) Efficacy, safety, quality control, marketing and regulatory guidelines for herbal medicines (phytotherapeutic agents). Braz J Med Biol Res 33(2):179–189. https://doi.org/10.1590/s0100-879X2000000200004

Carro AM, Lorenzo RA (2001) Simultaneous optimization of the solid-phase extraction of organochlorine and organophosphorus pesticides using the desirability function. Analyst 126(7):1005–1010. https://doi.org/10.1039/B009910O

Chaves A, Shea D, Danehower D (2008) Analysis of chlorothalonil and degradation products in soil and water by GC/MS and LC/MS. Chemosphere 71(4):629–638. https://doi.org/10.1016/j.chemosphere.2007.11.015

Chung SW, Chen BL (2011) Determination of organochlorine pesticide residues in fatty foods: a critical review on the analytical methods and their testing capabilities. J Chromatogr A 1218(33):5555–5567. https://doi.org/10.1016/j.chroma.2011.06.066

Cortés JM, Toledano RM, Villén J, Vazquez ANA (2008) Analysis of pesticides in nuts by online reversed-phase liquid chromatography– gas chromatography using the through-oven transfer adsorption/desorption Interface. J Agric Food Chem 56(14):5544–5549. https://doi.org/10.1021/jf800773k

Dagher SM, Hawi ZK, Kawar NS (1999) Comparison of gas chromatography and immunoassay methods in quantifying fenitrothion residues in grape juice processed into alcoholic drinks. J Environ Sci Health B 34(5):849–858. https://doi.org/10.1080/03601239909373230

Dagnac T, Jeannot R, Mouvet C, Baran N (2002) Determination of oxanilic and sulfonic acid metabolites of acetochlor in soils by liquid chromatography–electrospray ionisation mass spectrometry. J Chromatogr A 957(1):69–77. https://doi.org/10.1016/s0021-9673(02)00310-2

Dasgupta S, Banerjee K, Patil SH, Ghaste M, Dhumal KN, Adsule PG (2010) Optimization of two-dimensional gas chromatography time-of-flight mass spectrometry for separation and estimation of the residues of 160 pesticides and 25 persistent organic pollutants in grape and wine. J Chromatogr A 1217(24):3881–3889. https://doi.org/10.1016/j.chroma.2015.12.076

Fernández-Alba AR, García-Reyes JF (2008) Large-scale multi-residue methods for pesticides and their degradation products in food by advanced LC-MS. TrAC Trends Anal Chem 27(11):973–990. https://doi.org/10.1016/j.trac.2008.09.009

Flores G, Díaz-Plaza EM, Cortés JM, Villén J, Herraiz M (2008) Use of absorbent materials in on-line coupled reversed-phase liquid chromatography–gas chromatography via the through oven transfer adsorption desorption interface. J Chromatogr A 1211(1–2):99–103. https://doi.org/10.1016/j.chroma.2008.09.075

Garcia-Reyes JF, Ferrer C, Gómez-Ramos MJ, Fernández-Alba AR, Molina-Díaz A (2007) Determination of pesticide residues in olive oil and olives. TrAC Trends Anal Chem 26(3):239–251. https://doi.org/10.1016/j.trac.2007.01.004

Gebara AB, Ciscato CHP, Ferreira S, Monteiro SH (2005) Pesticide residues in vegetables and fruits monitored in Sao Paulo city, Brazil, 1994–2001. Bull Environ Contam Toxicol 75(1):163–169. https://doi.org/10.1007/s00128-005-0733-9

Gilbert-López B, García-Reyes JF, Molina-Díaz A (2009) Sample treatment and determination of pesticide residues in fatty vegetable matrices: a review. Talanta 79(2):109–128. https://doi.org/10.1016/j.talanta.2009.04.022

Górecki T, Panić O, Oldridge N (2006) Recent advances in comprehensive two-dimensional gas chromatography (GC× GC). J Liq Chromatogr Relat Technol 29(7–8):1077–1104. https://doi.org/10.1080/10826070600574762

Hemström P, Irgum K (2006) Hydrophilic interaction chromatography. J Sep Sci 29(12):1784–1821. https://doi.org/10.1002/jssc.200600199

Hercegová A, Dömötörová M, Matisová E (2007) Sample preparation methods in the analysis of pesticide residues in baby food with subsequent chromatographic determination. J Chromatogr A 1153(1–2):54–73. https://doi.org/10.1016/j.chroma.2007.01.008

Hernández F, Pozo OJ, Sancho JV, Bijlsma L, Barreda M, Pitarch E (2006) Multiresidue liquid chromatography tandem mass spectrometry determination of 52 non gas chromatography-amenable pesticides and metabolites in different food commodities. J Chromatogr A 1109(2):242–252. https://doi.org/10.1016/j.chroma.2006.01.032

Hernández F, Ibáñez M, Pozo ÓJ, Sancho JV (2008) Investigating the presence of pesticide transformation products in water by using liquid chromatography-mass spectrometry with different mass analyzers. J Mass Spectrom 43(2):173–184. https://doi.org/10.1002/jms.1279

Hogendoorn EA, Huls R, Dijkman E, Hoogerbrugge R (2001) Microwave assisted solvent extraction and coupled-column reversed-phase liquid chromatography with UV detection: use of an analytical restricted-access-medium column for the efficient multi-residue analysis of acidic pesticides in soils. J Chromatogr A 938(1–2):23–33. https://doi.org/10.1016/s0021-9673(01)01202-X

Ibáñez M, Sancho JV, Hernández F, McMillan D, Rao R (2008) Rapid non-target screening of organic pollutants in water by ultraperformance liquid chromatography coupled to time-of-light mass spectrometry. TrAC Trends Anal Chem 27(5):481–489. https://doi.org/10.1016/j.trac.2008.03.007

Jeannot R, Sabik H, Sauvard E, Genin E (2000) Application of liquid chromatography with mass spectrometry combined with photodiode array detection and tandem mass spectrometry for monitoring pesticides in surface waters. J Chromatogr A 879(1):51–71. https://doi.org/10.1016/S0021-9673(00)00098-4

Jiang X, Li D, Xu X, Ying Y, Li Y, Ye Z, Wang J (2008) Immunosensors for detection of pesticide residues. Biosens Bioelectron 23(11):1577–1587. https://doi.org/10.1016/j.bios.2008.01.035

Jones KC, De Voogt P (1999) Persistent organic pollutants (POPs): state of the science. Environ Pollut 100(1–3):209–221. https://doi.org/10.1016/S0269-7491(99)00098-6

Kosikowska M, Biziuk M (2010) Review of the determination of pesticide residues in ambient air. TrAC Trends Anal Chem 29(9):1064–1072. https://doi.org/10.1016/j.trac.2010.06.008

Kostiainen R, Kauppila TJ (2009) Effect of eluent on the ionization process in liquid chromatography–mass spectrometry. J Chromatogr A 1216(4):685–699. https://doi.org/10.1016/j.chroma.2008.08.095

Kováč J, Tekel J, Kurucová M (1987) Determination of herbicide residues in agricultural crops, foods, soil and water by chronometric method. Z Lebensm Unters Forsch 184(2):96–100. https://doi.org/10.1007/BF01042081

Kovalczuk T, Jech M, Poustka J, Hajšlová J (2006) Ultra-performance liquid chromatography–tandem mass spectrometry: a novel challenge in multiresidue pesticide analysis in food. Anal Chim Acta 577(1):8–17. https://doi.org/10.1016/j.aca.2006.06.023

Kumaran S, Tran-Minh C (1992) Determination of organophosphorous and carbamate insecticides by flow injection analysis. Anal Biochem 200(1):187–194. https://doi.org/10.1016/0003-2697(92)90297-K

Kuster M, Lo'pez de Alda M, Barcelo D (2009) Liquid chromatography–tandem mass spectrometric analysis and regulatory issues of polar pesticides in natural and treated waters. J Chromatogr 1216:520–529. https://doi.org/10.1007/978-3-540-36253-1_9

Kuster M, de Alda ML, Barceló D (2006) Analysis of pesticides in water by liquid chromatography-tandem mass spectrometric techniques. Mass Spectrom Rev 25(6):900–916. https://doi.org/10.1002/mas.20093

Lacorte S, Fernandez-Alba AR (2006) Time of flight mass spectrometry applied to the liquid chromatographic analysis of pesticides in water and food. Mass Spectrom Rev 25(6):866–880. https://doi.org/10.1002/mas.20094

Leandro CC, Hancock P, Fussell RJ, Keely BJ (2006) Comparison of ultra-performance liquid chromatography and high-performance liquid chromatography for the determination of priority pesticides in baby foods by tandem quadrupole mass spectrometry. J Chromatogr A 1103(1):94–101. https://doi.org/10.1016/j.chroma.2005.10.077

LeDoux M (2011) Analytical methods applied to the determination of pesticide residues in foods of animal origin. A review of the past two decades. J Chromatogr A 1218(8):1021–1036. https://doi.org/10.1016/j.chroma.2010.12.097

Lee SM, Wylie PL (1991) Comparison of the atomic emission detector to other element-selective detectors for the gas chromatographic analysis of pesticide residues. J Agric Food Chem 39(12):2192–2199. https://doi.org/10.1021/jf00012a019

Lee HS, Kim YA, Cho YA, Lee YT (2002) Oxidation of organophosphorus pesticides for the sensitive detection by a cholinesterase-based biosensor. Chemosphere 46(4):571–576. https://doi.org/10.1016/S0045-6535(01)00005-4

Lehotay SJ (2006) Quick, easy, cheap, effective, rugged, and safe approach for determining pesticide residues. In pesticide protocols (pp. 239-261). Humana press. https://doi.org/10.1385/1-59259-929-X:239

Li XZ, Guan TT, Chi ZHOU, Yin JQ, Zhang YH (2006) Solid phase extraction chemiluminescence determination of methamidaphos on vegetables. Chem Res Chin Univ 22(1):21–24. https://doi.org/10.1016/S1005-9040(06)60037-8

Linkerhägner M, Stan HJ (1994) Screening analysis of pesticide residues in plant foodstuffs by capillary gas chromatography using the DFG multiresidue method S19: a comparison of customory detection by ECD/NPD with the novel atomic emission detector (AED). Zeitschrift fuer Lebensmittel-Untersuchung und Forschung 198(6):473–479. https://doi.org/10.1007/BF01192843

López-Blanco MC, Cancho-Grande B, Simal-Gándara J (2002) Comparison of solid-phase extraction and solid-phase microextraction for carbofuran in water analyzed by high-performance liquid chromatography–photodiode-array detection. J Chromatogr A 963(1–2):117–123. https://doi.org/10.1016/S0021-9673(02)00552-6

Luchini LC, Peres TB, De Andrea MM (2000) Monitoring of pesticide residues in a cotton crop soil. J Environ Sci Health B 35(1):51–59. https://doi.org/10.1080/03601230009373253

Makarov A, Denisov E, Kholomeev A, Balschun W, Lange O, Strupat K, Horning S (2006) Performance evaluation of a hybrid linear ion trap/orbitrap mass spectrometer. Anal Chem 78(7):2113–2120. https://doi.org/10.1021/ac0518811

Manirakiza P, Covaci A, Andries S, Schepens P (2001) Automated soxhlet extraction and single step clean-up for the determination of organochlorine pesticides in soil by GC-MS or GC-ECD. Int J Environ Anal Chem 81(1):25–39. https://doi.org/10.1080/03067310108044356

Mol HG, Van Dam RC, Steijger OM (2003) Determination of polar organophosphorus pesticides in vegetables and fruits using liquid chromatography with tandem mass spectrometry: selection of extraction solvent. J Chromatogr A 1015(1–2):119–127. https://doi.org/10.1016/S0021-9673(03)01209-3

Mondello L, Tranchida PQ, Dugo P, Dugo G (2008) Comprehensive two-dimensional gas chromatography-mass spectrometry: a review. Mass Spectrom Rev 27(2):101–124. https://doi.org/10.1002/mas.20158

Morais S, Tortajada-Genaro LA, Arnandis-Chover T, Puchades R, Maquieira A (2009) Multiplexed microimmunoassays on a digital versatile disk. Anal Chem 81(14):5646–5654. https://doi.org/10.1021/ac900359d

Negoita TG, Covaci A, Gheorghe A, Schepens P (2003) Distribution of polychlorinated biphenyls (PCBs) and organochlorine pesticides in soils from the East Antarctic coast. J Environ Monit 5(2):281–286. https://doi.org/10.1039/B300555K

Parveen Z, Khuhro MI, Rafiq N (2005) Monitoring of pesticide residues in vegetables (2000–2003) in Karachi, Pakistan. Bull Environ Contam Toxicol 74(1):170–176. https://doi.org/10.1007/s00128-004-0564-0

Patel H, Rawtani D, Agrawal YK (2019) A newly emerging trend of chitosan-based sensing platform for the organophosphate pesticide detection using acetylcholinesterase-a review. Trends Food Sci Technol. https://doi.org/10.1016/j.tifs.2019.01.007

Perry RH, Cooks RG, Noll RJ (2008) Orbitrap mass spectrometry: instrumentation, ion motion and applications. Mass Spectrom Rev 27(6):661–699. https://doi.org/10.1002/mas.20186

Petrovic M, Barceló D (2006) Application of liquid chromatography/quadrupole time-of-flight mass spectrometry (LC-QqTOF-MS) in the environmental analysis. J Mass Spectrom 41(10):1259–1267. https://doi.org/10.1016/j.chroma.2006.05.024

Picó Y, Fernández M, Ruiz MJ, Font G (2007) Current trends in solid-phase-based extraction techniques for the determination of pesticides in food and environment. J Biochem Biophys Methods 70(2):117–131. https://doi.org/10.1016/j.jbbm.2006.10.010

Pihlström T, Blomkvist G, Friman P, Pagard U, Österdahl BG (2007) Analysis of pesticide residues in fruit and vegetables with ethyl acetate extraction using gas and liquid chromatography with tandem mass spectrometric detection. Anal Bioanal Chem 389(6):1773–1789. https://doi.org/10.1007/s00216-007-1425-6

Polati S, Bottaro M, Frascarolo P, Gosetti F, Gianotti V, Gennaro MC (2006) HPLC-UV and HPLC-MSn multiresidue determination of amidosulfuron, azimsulfuron, nicosulfuron, rimsulfuron, thifensulfuron methyl, tribenuron methyl and azoxystrobin in surface waters. Anal Chim Acta 579(2):146–151. https://doi.org/10.1016/j.aca.2006.07.034

Pozo O, Pitarch E, Sancho JV, Hernandez F (2001) Determination of the herbicide 4-chloro-2-methylphenoxyacetic acid and its main metabolite, 4-chloro-2-methylphenol in water and soil by liquid chromatography–electrospray tandem mass spectrometry. J Chromatogr A 923(1–2):75–85. https://doi.org/10.1016/S0021-9673(01)01006-8

Primel EG, Milani MR, Demoliner A, Niencheski LFH, Escarrone ALV (2010) Development and application of methods using SPE, HPLC-DAD, LC-ESI-MS/MS and GFAAS for the determination of herbicides and metals in surface and drinking water. Int J Environ Anal Chem 90(14–15):1048–1062. https://doi.org/10.1080/03067310902962791

Rashid A, Nawaz S, Barker H, Ahmad I, Ashraf M (2010) Development of a simple extraction and clean-up procedure for determination of organochlorine pesticides in soil using gas chromatography–tandem mass spectrometry. J Chromatogr A 1217(17):2933–2939. https://doi.org/10.1016/j.chroma.2010.02.060

Rissato SR, Galhiane MS, Apon BM, Arruda MS (2005) Multiresidue analysis of pesticides in soil by supercritical fluid extraction/gas chromatography with electron-capture detection and confirmation by gas chromatography– mass spectrometry. J Agric Food Chem 53(1):62–69. https://doi.org/10.1021/jf048772s

Rodrigues AM, Ferreira V, Cardoso VV, Ferreira E, Benoliel MJ (2007) Determination of several pesticides in water by solid-phase extraction, liquid chromatography and electrospray tandem mass spectrometry. J Chromatogr A 1150(1–2):267–278. https://doi.org/10.1016/j.chroma.2006.09.083

Rosen R (2007) Mass spectrometry for monitoring micropollutants in water. Curr Opin Biotechnol 18(3):246–251. https://doi.org/10.1016/j.copbio.2007.03.005

Salvador IM, Frenich AG, González FE, Vidal JM (2006) Determination of organophosphorus pesticides in vegetables by GC with pulsed flame-photometric detection, and confirmation by MS. Chromatographia 64(11–12):667–672. https://doi.org/10.1365/s10337-006-0104-7

Sanchez R, Vazquez A, Villén-Altamirano J, Villén J (2006) Analysis of pesticide residues by on-line reversed-phase liquid chromatography–gas chromatography in the oil from olives grown in an experimental plot. J Sci Food Agric 86(1):129–134. https://doi.org/10.1002/jsfa.2267

Sánchez-Brunete C, Miguel E, Tadeo JL (2002) Multiresidue analysis of fungicides in soil by sonication-assisted extraction in small columns and gas chromatography. J Chromatogr A 976(1–2):319–327. https://doi.org/10.1002/jsfa.2267

Sarabia L, Maurer I, Bustos-Obregon E (2009) Melatonin prevents damage elicited by the organophosphorous pesticide diazinon on mouse sperm DNA. Ecotoxicol Environ Saf 72(2):663–668. https://doi.org/10.1016/j.ecoenv.2008.04.023

Soler C, Pico Y (2007) Recent trends in liquid chromatography-tandem mass spectrometry to determine pesticides and their metabolites in food. TrAC Trends Anal Chem 26(2):103–115. https://doi.org/10.1016/j.trac.2006.08.005

Stipičević S, Fingler S, Zupančič-Kralj L, Drevenkar V (2003) Comparison of gas and high performance liquid chromatography with selective detection for determination of triazine herbicides and their degradation products extracted ultrasonically from soil. J Sep Sci 26(14):1237–1246. https://doi.org/10.1002/jssc.200301420

Tadeo JL, Sánchez-Brunete C, Albero B, González L (2004) Analysis of pesticide residues in juice and beverages. Crit Rev Anal Chem 34(3–4):165–175. https://doi.org/10.1080/10408340490888670

Tekel J (1990) Separation of some phenylerea and s-triazine herbicides by OPLC method with a binary mobile phase. J Planar Chromatogr 3:326–330. https://doi.org/10.1093/chromsci/40.10.553

Tekel J, Hatrík Š (1996) Pesticide residue analyses in plant material by chromatographic methods: clean-up procedures and selective detectors. J Chromatogr A 754(1–2):397–410. https://doi.org/10.1016/S0021-9673(96)00489-X

Urruty L, Montury M, Braci M, Fournier J, Dournel JM (1997) Comparison of two recent solventless methods for the determination of procymidone residues in wines: SPME/GC/MS and ELISA tests. J Agric Food Chem 45(5):1519–1522. https://doi.org/10.1021/jf9609643

Uygun U, Özkara R, Özbey A, Koksel H (2007) Residue levels of malathion and fenitrothion and their metabolites in postharvest treated barley during storage and malting. Food Chem 100(3):1165–1169. https://doi.org/10.1016/j.foodchem.2005.10.063

Vazquez-Roig P, Andreu V, Onghena M, Blasco C, Picó Y (2011) Assessment of the occurrence and distribution of pharmaceuticals in a Mediterranean wetland (L'Albufera, Valencia, Spain) by LC-MS/MS. Anal Bioanal Chem 400(5):1287–1301. https://doi.org/10.1007/s00216-011-4826-5

Vidal JM, Arrebola FJ, Mateu-Sánchez M (2002) Application of gas chromatography–tandem mass spectrometry to the analysis of pesticides in fruits and vegetables. J Chromatogr A 959(1–2):203–213. https://doi.org/10.1016/S0021-9673(02)00444-2

Walorczyk S, Gnusowski B (2006) Fast and sensitive determination of pesticide residues in vegetables using low-pressure gas chromatography with a triple quadrupole mass spectrometer. J Chromatogr A 1128(1–2):236–243. https://doi.org/10.1016/j.chroma.2006.06.044

Wang J, Chow W, Wong JW, Leung D, Chang J, Li M (2019) Non-target data acquisition for target analysis (nDATA) of 845 pesticide residues in fruits and vegetables using UHPLC/ESI Q-Orbitrap. Anal Bioanal Chem:1–11. https://doi.org/10.1007/s00216-019-01581-z

Wołejko E, Łozowicka B, Kaczyński P, Jankowska M, Piekut J (2016) The influence of effective microorganisms (EM) and yeast on the degradation of strobilurins and carboxamides in leafy vegetables monitored by LC-MS/MS and health risk assessment. Environ Monit Assess 188(1):64. https://doi.org/10.1007/s10661-015-5022-4

Yusà V, Coscollà C, Mellouki W, Pastor A, De La Guardia M (2009) Sampling and analysis of pesticides in ambient air. J Chromatogr A 1216(15):2972–2983. https://doi.org/10.1016/j.chroma.2009.02.019

Zhang N, Yang Y, Tao S, Liu Y, Shi KL (2011) Sequestration of organochlorine pesticides in soils of distinct organic carbon content. Environ Pollut 159(3):700–705. https://doi.org/10.1016/j.envpol.2010.12.011

Zhu WX, Yang JZ, Wei W, Liu YF, Zhang SS (2008) Simultaneous determination of 13 aminoglycoside residues in foods of animal origin by liquid chromatography–electrospray ionization tandem mass spectrometry with two consecutive solid-phase extraction steps. J Chromatogr A 1207(1–2):29–37. https://doi.org/10.1016/j.chroma.2008.08.033

Zoun PEF, Spierenburg TJ (1989) Determination of cholinesterase-inhibiting pesticides and some of their metabolites in cases of animal poisoning using thin-layer chromatography. J Chromatogr A 462:448–453. https://doi.org/10.1016/S0021-9673(00)91376-1

Zrostlíková J, Hajšlová J, Čajka T (2003) Evaluation of two-dimensional gas chromatography–time-of-flight mass spectrometry for the determination of multiple pesticide residues in fruit. J Chromatogr A 1019(1–2):173–186. https://doi.org/10.1016/S0021-9673(03)01302-5

Chapter 5
Bioindicators of Pesticide Contaminations

Ajoy Saha, Rakesh Kumar Ghosh, P.-K. Jesna, and Partha P. Choudhury

Abstract Reducing the economic losses caused by pests, and for improving crop yields, use of pesticides in agriculture is imperative. The continuous use of pesticides in the environment is of great concern as some of them are highly persistant and causes harmful impact on non-target organisms. So, monitoring of them is utmost important for management of pesticide pollution. Among the different monitoring tools for pesticide pollution, use of bioindicator using living organism or its part or group of organisms becoming the technique of choice as this is an inexpensive, specific and easy to handle method. Moreover, number of living organisms could accumulate the pesticides and thus helps in indicating the environmental pesticides pollution.

Here, a review on biomonitoring of environmental pesticides pollution by using bioindicator is presented. Pesticides contamination in air, soil and water and their route of entry into the environmental matrices are first presented followed by a general idea have been given on well-known natural species used as environmental pesticides pollution bioindictaor. These species include microbes, animals and plants of different groups which were widely used for bioindication of environmental pesticides pollution from history to recent research which has been discussed elaborately.

Keywords Pesticides · Environmental pollution · Bioindicators · Biomonitors · Plant · Microbes · Animal

A. Saha (✉) · P.-K. Jesna
Research Centre of ICAR-Central Inland Fisheries Research Institute, Bengaluru, Karnataka, India

R. K. Ghosh
ICAR-National Institute of Natural Fibre Engineering and Technology, Kolkata, West Bengal, India

P. P. Choudhury
ICAR-Indian Institute of Horticultural Research, Bengaluru, Karnataka, India

Inamuddin et al. (eds.), *Sustainable Agriculture Reviews 48*, Sustainable Agriculture Reviews 48, https://doi.org/10.1007/978-3-030-54719-6_5

Abbreviations

GC-MS/MS	Gas Chromatography – Tandem Mass Spectrometry
LC-MS/MS	Liquid Chromatography with tandem mass spectrometry
IUCN	International Union for Conservation of Nature
OCPs	Organochlorines pesticides
OPs	Organophosphates pesticides
PCBs	Polychlorinated biphenyls
SPs	Synthetic pyrethroids

5.1 Introduction

The pest problem is a complex barrier in any agricultural production. Due to the attack of pests in major crops, including rice, wheat, maize, barley, soybeans, potatoes, cotton and coffee, farmers are losing crop yield to the extent of 20–40% (Oerke et al. 1994; Oerke 2006). Global warming makes the situation more complex as the higher temperature favors pest attack more. Deutsch et al. (2018) developed a model for the estimation of crop loss due to the changing climate. As per their estimation from the model, the crop loss of wheat, rice and maize due to the insect pest infestation would increase by 10–25% per degree celsius rise in global temperature. Since the discovery of DDT, farmers took up pesticides as the most effective tool to combat the pest problems securing crop production. The resistance development in pests against pesticides is a major concern for pest management. Farmers need to spray pesticides repeatedly in increased quantity. The increasing consumption of pesticides in crop fields and in non-crop situation leads to an environmental crisis impacting ultimately on human and animal health. Reports on effects of chronic pesticide exposure in human being and wildlife include endocrine disruption (Colborn et al. 1993, 1996; Sumpter and Johnson 2005; Anway et al. 2005), cancer (Patlak 1996; Hardell and Eriksson 1999; Schreinemachers 2000; Alavanja et al. 2004; Sanborn et al. 2004), immune suppressant acivity (Repetto and Baliga 1996), impaired neurological development (Colborn 2006), learning disorders (Guillette et al. 1998), Parkinson's disease (Brown et al. 2006), earlier onset of puberty (Guillette et al. 2006), fetal death (Bell et al. 2001), birth defects, and sex ratio changes (Garry et al. 2002), eggshell thinning in birds, liver and kidney disorder, immunodefficiency disorders, neurotoxicity, teratogenicity, delayed metamorphosis, smaller body size, reduced activity and reduced tolerance to cold or predatory stress (Murphy 1986; Colborn et al. 1993; Relyea 2005; Teplitsky et al. 2005). These are only possible when our food chain, potable water and air get contaminated heavily with pesticides. However, the toxicity of an applied pesticide depends largely on its environmental behavior.

The fate of the pesticide residues determines its impact on the environmental components. Fate of a pesticide refers to the pattern of distribution of it, its derivatives or metabolites in an organism, system, compartment or population of concern as a result of transport, partitioning, transformation or degradation (OECD 2003). A pesticide, when it is sprayed over a canopy or on soil surface, may be transported through different physical processes to various compartments of the environment, viz. soil, soil-water system, surface water, ground water, air, plants, crop produce and animals. The fate of the pesticide in the compartment is governed by various biotic and abiotic factors.

The first exposure of an applied pesticide is green vegetation and thereafter it is soil, where the pesticide deposits. A considerable part of the applied pesticide is absorbed by the waxy cuticle of leaf or root of plants and enters into the plant transport system and gets distributed into different plant parts, wherein it may be degraded to non-toxic components or it may persist for a longer period acting as a source of contamination.

In general, soil borne insect pests, diseases and weeds are managed by the direct soil application of pesticides. Also a large portion of pesticides applied on foliages ends up in soil. The pesticides present in soil gets bound to soil colloids i.e. clay and organic matter. This bound pesticide is not readily accessible to plant than the pesticide present in the soil solution. Soil colloidal particles carry a large amount of electrical charges, which help in attracting the opposite charges of pesticide molecules adsorbing them onto soil particle. As found in most cases of clay-pesticide interaction, the charges are relatively weak and thus the adsorption process is reversible. An equilibrium is attained between the amount of pesticide adsorbed to colloids and that found in solution. The ratio of adsorbed herbicide to its free state is influenced by several factors, including chemical properties of the herbicide, soil characteristics and soil water content. The opposite charges on clay surfaces strongly bind the ionic pesticides, viz. paraquat dichloride, diquat dichloride, salt of 2,4-dichlorophenoxyacetic acid, salt of glyphosate, etc. Therefore, the adsorption of ionic pesticides is more in the soil having higher clay content. However, most of the pesticides are non-ionic in nature. Soil organic matters play the major role to adsorb those non-ionic pesticides.

A heavy irrigation or rainfall causes the runoff of water from the pesticide application site. It transports water soluble pesticides or pesticides adsorbed on soil colloids downstream, finally reaching to seasonal or permanent streams, rivers, ponds, or lakes or their tributaries contaminating the surface water. Pesticide residues in surface water can cause injury to crops, livestock, or humans and the entire ecosystem. Leaching of pesticide occurs as water moves downward through the soil. Groundwater contamination is a major concern associated with the leaching of pesticides.

In these different compartments, pesticides are to face some chemical and biochemical forces, which eventually mineralize them. However, the rate of mineralization of all pesticides is not similar. Some pesticides degrade very fast; some persist for a longer time. Pesticides, which persist for a considerable period of time in the environment, may exert their intrinsic toxicity affecting the normal growth of an organism in the ecosystem. Sometimes the type of toxic effect is so specific for a set of living organismand a pesticide, we can easily apprehend from the toxic effect about the presence of that pesticide in a significant level in the ecosystem. The physical appearance of a particular living organism due to the toxic effect may indicate the presence of the pesticide. Here, the phenomenon of the toxic effect on a living organism caused by a pesticide is called 'bioindication' and the living organism is the 'bioindicator' for the pesticide.

A bioindicator is defined as a living organism or its part or group of organisms, the product of an organism (e.g., enzyme), and process of biological means which can be used to generate information on the environment quality (Killham and Staddon 2002). However, utilization of this process of using living organisms or their products as a bioindicator for pesticide contamination is limited. A number of bioindicators, including animal, plant, structure and biodiversity of microbial biomass, even enzymes may be suggested for monitoring pesticide contamination in soil or water, or in the ecosystem as a whole. Basically, biological indicators quantify the degree of stress due to the toxicological impact of pesticides on an individual organism or a group of organisms. It also indicates the extent of pesticide-related stress (Hunsaker and Carpenter 1990; Suter 1993; Dale and Beyeler 2001). These bioindicators can be utilized in developing simple and efficient method to determine and identify the ecological changes in the environment due to the pesticide contamination. They may be used as an early warning indicator of environmental changes due to pesticide pollution.

The impact of a pesticide in any ecosystem is very diverse. A pesticide is applied targeting either insect pests or weeds or disease-causing microorganisms. It kills a pest or restricts its growth by inhibiting a specific enzyme system present in the target pest. The similar enzyme may also be available in non-target organisms. Therefore, the application of the pesticide can also cause toxicity to the non-target organism. If the expression of this toxicity of pesticide is measurable, the non-target organism can be used as a bioindicator. Therefore, the selection of an organism or its product as an effective indicator is the most essential step for the overall success of any monitoring programme. On the basis of the pioneering works of Landres et al. 1988, Cairns et al. 1993, and Dale and Beyeler 2001, a set of criteria for a successful biological indicator for any ecological stress has been set. This is equally applicable for choosing a bioindicator for the stress due to pesticide. The indicator should be easily measured. It should be sensitive to the stresses. The response of the indicator to the pesticide must be in a predictable manner without any ambiguity. The changes in the affected organism, i.e. bioindicator can be averted by management action to reestablish the organism. We already have many examples of affected organisms, viz. animals, plants and microbes present in the pesticide polluted ecosystem. It is yet not possible to identify a bioindicator for the monitoring of a pesticide or a group of pesticides.

5.2 Pesticide in Environment, Air and Water

Application of plant protection chemicals has become an unavoidable practice in crop cultivation system to ensure yield. Even during storage of the harvested products, application of plant protection chemicals/pesticides has become a necessity. Not all but, there are some significant crop-pests including insects, disease causing pathogens and weeds along with some other pests like rodents etc. casue economic injury to agricultural production. Globally, crop loss due to insects is around 14%, followed by 13% each for plant pathogens and weeds. The global cultivable land area is not expanding significantly; however, the population is increasing and estimated to reach 9.2 billons within 2050. This growing population pressure will demand secured supply of food and plant protection chemicals holds promise of controlling pests and thereby ensuring crop production. It has been reported that crop loss due to non-application of pesticides may be to the tune of 32, 54 and 78% respectively, for cereals, vegetables and fruit crops (UN 2015). Thus, pesticides have become an integral part of crop cultivation system across the world. The application of pesticides started with natural compounds like sulfur (before 1870), followed by introduction of inorganic synthetic pesticides (till 1945) and then era of synthetic organic pesticides started with compounds like hexachlorocyclohexane (HCH), dichlorodiphenyltrichloroethane (DDT), 2,4-dichlorophenoxyacetic acid (2,4-D) etc. and still newer molecules are being introduced. Global pesticide production has jumped 25 times from a requirement of 0.2 million tonnes to over 5 million tonnes with an annual growth rate of 11%. The consumption of pesticides varied across globe depending on crop and agroclimatic conditions. Average pesticide usage (kg) per hactre cropland is around 3.62, 3.39, 1.67, 1.17 and 0.31 for Asia, Americas, Europe, Oceania and Africa (FAOSTAT 2017). The global trend of pesticide usage indicates around 47, 29, 18 and 6% share of insecticide, herbicides, fungicides and other pesticides, respectively. Total annual pesticide consumption wise first three countries are China (1806 million kg), followed by United States (386 million kg) and Argentina (265 million kg). However, pesticide application dose wise (kg/ha) top seven countries are Bahamas (32.22), Costa Rica (18.78), Colombia (14.57), Japan (14.18), Korea (12.74), Maldives (12.29) and Israel (11.6) (FAOSTAT 2017). Application of such huge amount of pesticides has resulted in contamination of various environmental components (EEA 2013).

Several reports have confirmed that only upto10% of applied pesticide reach the target site of crop and majority of pesticide remain in the environment and react with non-target organisms (WHO 2017). Soil acts as a major reservoir of pesticides as received through direct (soil applications) and indirect ways (soil erosion, drift, precipitation, spillage and accidental events, disposal of containers etc). After reaching soil, pesticides undergo various physico-chemical interaction with soil solid (sand, silt, clay and soil organic matter/humus) and soil water (Ghosh and Singh 2013). Pesticides which are present in soil water become available either for transport within the soil layers (leaching to ground water and horizontal transport to new soil layers) or became available to soil microbes (Ghosh and Singh 2012). Pesticides which become attached/sorbed to soil solids are prone to soil erosion, surface run-off loss and drainage loss and reach non-target areas like open surface water bodies (Carvalho et al. 2003). Again, pesticides applied directly onto soil/plant surface may undergo volatilization and transformation/photo-degradation process finally reaching the atmosphere. The volatilized pesticides/degraded pesticide molecules return back to soil and water bodies through the process of condensation and precipitation/rainfall (Simonich and Hites 1995). For example, organochlorine pesticides like chlordane, hexachlorocyclohexane and toxaphene applied in southern parts of USA were transported by the process of volatilization and precipitated in the Great Lakes of Canada (Li and Jin 2013). Not only was the organochlorines (OCPs), but also organophosphates (OPs) like chlorpyriphos, applied in the banana plantation of Central America was detected in the Arctic ice (Garbarino et al. 2002). Thus, the non-targeted and underutilized portion of applied pesticide (~70–90% of total application) moves from one component to other component of our ecosystem and in this process, a significant portion of pesticide gets accumulated in the lower trophic level organisms. This absorbed pesticide by means of bioconcentration and bioaccumulation, reach higher trophic level organisms of food chain, and by the virtue of this xenobiotic property, the bioaccumulated pesticides/metabolites affect normal physiological process of organisms, thereby putting the whole ecosystem at a stake (EEA 2013). Organochlorine pesticide residues were traced in birds, and even in human milk (EEA 2013). Unlike organochlorines (OCPs), organophosphates (OPs) and synthetic pyrethroids (SPs) are less toxic but have adverse effects on arthropods and invertebrates. Owing to molecular stability and less degradability, pesticides may persist in the environment for few days to several decades after application. Among various groups of pesticides, organochlorines are the most persistent with long half-life such as, dichlorodiphenyltrichloroethane – 10 to15 years, aldrin and dieldrin – 5 years each, endrin – 12 years, hexachlorocyclohexane – 2.7 to 22.9 years, heptachlor – 2 years, mirex – 10 years, toxaphene – 100 days to 12 years depending upon environmental conditions (Jayaraj et al. 2016). Prolonged persistence coupled with bioaccumulation of such organochlorines hold serious risk to the ecosystem and this resulted in banning of such pesticides (dichlorodiphenyltrichloroethane, hexachlorocyclohexane, aldrin, dieldrin and toxaphane) since Stockholm convention of 2002. Indiscriminate use of this persistent organochlorine has contaminated most of the rivers and water bodies across the world. Organophosphates and pyrethroids are less persistent which degrade within few days. However, frequent applications have resulted in detection of pesticide residues in river water (Gilliom 2007). Here the role of bioindicators comes for detection of pesticides pollution for taking proper management measures for this problem. Bioindicator may be an inexpensive, environment friendly approach which not only detect the pesticides but also indicates environmental quality and safety.

5.3 Bioindicators

The increased anthropogenic activities have gradually led to deterioration of environment worldwide. The environmental ecosystem is losing their biodiversity owing to human intervention. The dynamics of environment needs to be understood and monitored through comprehensive research approaches. The traditional approach of analysing the physico-chemical parameters of environment may be inadequate to evaluate the human interferences on various environmental matrices. So the use of bioindicators can aid in evaluating the ecosystem through the analysis of their response to the ecological parameters. The indices for biotic integrity are applicable especially in aquatic system, where the mere estimation of water quality may not reflect the exact state of the system due to rapid changes in hydrology. Terrestrial and aquatic fauna and flora, including invertibrates, vertibrates and microorganism, vegetation including higher and lower plants, as well as fishes can serve as bioindicators to monitor the pesticides pollution either separately or in combination (Carlisle et al. 2008).

5.3.1 Animals as Bioindicators for Pesticides Pollution

Different zooplankton, invertebrates and vertebrates are commonly used animal species as bioindicators. Several important bioindicators of aquatic pollution belongs to animal species. Changes in the populations of animal species and their diversity indicates the harmful changes caused by pesticides pollution. However, sometimes the amount of pesticides present in the animal tissue also indicates the level of contamination (Khatri and Tyagi 2015).

5.3.1.1 Zooplankton

Zooplankton are microscopic animals with feeble power of swimming, relying largely on water currents for movement and constitutes a vital food source for fish. Their growth, development and conveyance depend on various abiotic and biotic factors (Ramchandra et al. 2006). Zooplankton can be successfully utilised as important bioindicators for water pollution, eutrophication as well as primary productivity of water bodies and production of a freshwater body. Daphnids, the most sensitive zooplankton to pesticides, are proven to be a good indicator for monitoring ambient pesticide toxicity (Werner et al. 2000). Gutierrez et al. (2012) reported the response of the copepods (*Notodiaptomus conifer*, *Argyrodiaptomus falcifer*) and the cladocerans (*Daphnia magna*, *Pseudosida variabilis* and *Ceriodaphnia dubia*) as early bioindicator of endosulfan toxicity. Ecotoxicological risk to the copepods *Acartia margalefi*, *A. latisetosa* and the mysid *Siriella clausi*, can be used to assess the risk to the marine ecosystems as a precautioning signal (Fossi et al. 2001).

5.3.1.2 Invertebrates

A wide range of soil terrestrial and aquatic invertebrates have been reported as effective bioindicators. For terrestrial ecosystems invertebrate insects namely, honey bees and ants act as very important indicator for pollution transfer. Hence, these invertebrates are very good indicators to evaluate the negative impacts of pesticides particularly on non-targeted insects species. A number of invertebrate species including both terrestrial and aquatic are known to be indicator of pesticide contamination (Pelosi et al. 2014; Josephine et al. 2019).

Honey Bees and Bee Products

Pesticide consumption has increased by around 25-folds in recent decades as compared to 1950s for protecting agricultural crops and produts (Gómez-Pérez et al. 2012). However this indiscriminate use may contaminate the different environmental compartment like water, soil, air and food by pesticides. Honeybees can perform a crucial role in monitoring pesticide residues during intra and inter species pollen transport (Paradis et al. 2014). As bee communities are with large number of population and high mobility, is useful for large area monitoring programme, capable of inspection of various flowers, makes them essential component in environmental monitoring of pesticides. Similarly, due to specific morphological features, bee can carry on pesticides during the time of contact, which may unconsciously brought to the hive. Thus bee hives may also be contaminated by transported pesticide by bee or pollen. Sometimes, treatment of bee hives during honey collection may be the reason of pesticide contamination (Kujawski and Namieśnik 2011). This indicates the use of honey bees as a potential informer to determine the amount of pesticide usage and their level of contamination. Sometime without honey bees, it is difficult to prove incorrect/misuse of pesticides. Use of bees helps in monitoring the banned or restricted pesticides as reported by Porrini et al. 2003.

Furthermore, tracing of pesticides in a large geographical area is laborious and expensive. Use of honeybees for trace analysis of pesticides is non-expensive. In addition, bees can provide more valuable additional information about the pesticide pollution (Porrini et al. 2003). Though previous study mostly concentrated on presence of organochlorine pesticides (Tahboub et al. 2006) and organophosphate (Ghini et al. 2004; Chiesa et al. 2016) pesticides in honey bees, recent study indicated that their population is also affected by the new-generation pesticides like neonicotinoid and fipronil (Pisa et al. 2015). Several European countries have banned few neonicotinoid pesticides due to bee poisoning effects (Carrington 2018). As pesticides are carried by honey bee to the beehives, bee products like beeswax, beebread and honey can be the potential indicator of pesticide pollution. Due to lipophilic nature of beewax, they are good indicator of lipophilic pesticides (Calatayud-Vernich et al. 2017; Niell et al. 2015). Whereas due to hydrophilic nature, honey is less contaminated with pesticides and is a good indicator for hydrophilic pesticides (Perugini et al. 2018). Table 5.1 confirms the bioindicator potential of bees for biomonitoring of pesticides and confirms their ability to transfer pesticides which they come in contact to the hives and bee products. Though a potential indicator, still honey bees have some drawback as a bioindicator due to its irregular and unpredictable foraging behavior for long distance or duration. Their flying characters are strongly regulated by air temperature, where an air temperature of 20–25 °C is optimal for foraging (Bargańska et al. 2016). Furthermore, since there is an ample chance that bees can come into contact with multiple number of pesticides with different nature of chemistry, selection of suitable extraction and analysis techniques is also important. QuEChERS (Quick Easy Cheap Effective Rugged Safe) extraction techniques coupled with analytical techniques like Gas Chromatography-Mass Spectrometry/Mass Spectrometry, Liquid Chromatography-Mass Spectrometry/Mass Spectrometry may be the suitable option to exclude any false identification of pesticides. Hence, development of suitable extraction cum clean-up technique in order to minimize the interference from complex matix of bee and bee products with appropriate analytical technique is essential.

Ants

Ants (Hymenoptera: Formicidae) has an abundant and diverse population in almost all of the terrestrial ecosystems. Their population is highly stable irrespective of the seasons and years. Due to its high abundance and stability, they are placed in an important position in insects ecosystems. They play significant roles as predator, recycling of various nutrients, maintaining the structure of herbaceous vegetation and improving soil quality (Petal 1978). Their abundance, functional significance, comparative easiness of sampling, and quick response to the changes in the environment (Andersen 1990) make them a suitable and a preferred bioindicator for impact assessment of environmental stressors, like pesticides over many other taxa. Being tied to the same place, they are more susceptible to environmental stressors (Andersen 1988). Several reports indicates that ants are highly susceptible to the application of pesticides (Schread 1948; Su et al. 1980, Perfecto 1994) which make them a useful and important biological indicators in monitoring pesticides contamination or pollution (Peck et al. 1998; Hoffmann and Andersen 2003).

There are several contradictory findings reported about the non-target effect of pesticides on ant assemblages. Carbofuran or/and chlorpyrifos significantly reduced the ant foraging activity in maize plot (Perfecto 1990). Similarly, Pereira et al. (2005) reported herbicide/chlorpyrifos mixture and chlorpyrifos reduced the ant activity in maize plots. However, several other studies reported that certain ant species in maize field are affected by chlorpyrifos and deltamethrin, but effects were less significant on overall activity (Michereff-Filho et al. 2004; Badji et al. 2006). Ant richness or abundance was unaffected by the fenitrothion in pine forests (Kwon et al. 2005), microbial pesticides on oak-dominated forests (Wang and Xie 2000), or sparying of broad spectrum insecticides in banana plantations (Matlock Jr and de la Cruz 2003).

Table 5.1 Monitoring studies using honeybees as bioindicators for pesticide pollution

Matrix	Pesticides	Sample preparation	Analysis	Sampling			Notes	References
				Country	Location	Site and sampling		
Honey, pollen and honey bees	Neonicotinoid insecticides	QuEChERS	LC-MS/MS	Canada	City of Saskatoon, Saskatchewan	7 independent apiaries located within a 30 km radius in Central Saskatchewan	Over winter concentration estimate that dietary intake may impact bee health	Codling et al. (2016)
Honeybees	Organophosphorus and carbamates pesticides	LLE	GC-NPD and LC-MS/MS	Italy	3 townships in the province of Bologna	31 samples were analyzed	Indicates the feasibility of using honeybees for assessing pesticide exposure in agriculture settings.	Ghini et al. (2004)
Bee honey	50 multi-class pesticides	LLE	GC-NPD	Greece	Six agricultural areas of Greece (north, central, south)	Fifty representative honey samples were randomly collected from apiaries	Using honeybees as bioindicators are useful in identifying possible incorrect use of pesticides.	Balayiannis and Balayiannis (2008)
Honey and pollen	16 OCPs	Modified QuEChERS	GC-ECD	Mexico	Tapachula, located in the Soconusco region, Chiapas in southern Mexico.	36 for honey (18 for *Apis mellifera* and 18 for *Scaptotrigona mexicana*) and 36 for pollen (18 for A. mellifera and 18 for *Scaptotrigona mexicana*).	Provide a clear indication of the widespread use of OCs in the study area, confirming honey bee and beehive matrixes as appropriate sentinels for bioenvironmental monitoring.	Ruiz-Toledo et al. (2018)

Beebread, trapped pollen, brood nest wax, beeswax foundation, and adult bees and brood	Fipronil along with 171 pesticides	Modified QuEChERS	LC/MS-MS and GC/MS	The USA	Across 23 states, one Canadian province	Total of 320 beebread, 28 trapped pollen, and 2 anther samples, 238 wax (derived mainly from the brood nest of colonies) and 21 foundation samples, and 34 immature (brood) and 106 adult bee samples for pesticide residues.	Two samples of dead bees were linked by analysis to prior environmental applications of permethrin (19.6 ppm residue, LD_{50} of 1.1 ppm) and fipronil (3.1 ppm, LD_{50} 0.05 ppm).	Mullin et al. (2010)
Organic honey	11 OCs & 6 OPs along with PAHs, PBDEs and PCBs	ASE using hexane/ethyl acetate (4:1, v/v)	GC-MS/MS	Italy	North and Central Italy	95 sample from market, high anthropisation; low anthropisation; intensive farming; husbandry; intensive farming and husbandry area	OCs and OPs were found in the farming arca, confirming that the contamination could be linked to the area.	Chiesa et al. (2016)

(continued)

Table 5.1 (continued)

Matrix	Pesticides	Sample preparation	Analysis	Sampling				References
				Country	Location	Site and sampling	Notes	
Honey	Glyphosate	LLE	ELISA techniques	The USA	Hawaiian island of Kaua'i	Honey taken directly from 59 bee hives on the Hawaiian island of Kaua'i	Herbicide migration from the site of use into other areas by bees	Berg et al. (2018)

QuEChERS Quick, easy, cheap, effective, rugged, and safe, *LLE* Liquid liquid extraction, *LC-MS/MS* Liquid Chromatography with tandem mass spectrometry, *GC-NPD* GC with Nitrogen Phosphorous Detector, *GC-ECD* GC with Electron Capture Detector, *GC/MS* Gas chromatography–mass spectrometry, *ASE* Accelerated solvent extraction, *ELISA* Enzyme-linked immunosorbent assay

OC Organochlorine, *OP* Organophosphorus, *PAH* Polycyclic Aromatic Hydrocarbon, *PBDE* Polybrominated diphenyl ethers, s and *PCB* polychlorinated biphenyls

Exhaustive survey of literature by Wojcik and Porter 2001 indicate that, out of 138 references only two studies suggested that ants are susceptible to pesticide application. Both the studies carried out by DeBouge et al. (1987) and DeBouge and Thome (1989) observed unsuitability of ant as bioindicator of organochlorine pesticide pollution due to its inability to accumulate this group of pesticides.

Earthworms

Earthworms are frequently identified organisms in the soil ecosystem and play an important role in soil health and quality improvement (Spurgeon et al. 2003). Their number reduction in soil ecosytem may alter the nutrient cycling and their availability to plants (Rizhiya et al. 2007). Their living habits make them exposed to chemicals like pesticides present in their terrestrial ecosystem (Schnug et al. 2014). As soil act as a major sink of applied pesticides, these pesticides may be introduced to the part of the soil where secretions and castings of earthworm occurs, which is commonly known as drilosphere (Givaudan et al. 2014). Exposure of earthworm to the pesticides can be by direct physical contact or by ingestion of pesticide contaminated soil particles (Pisa et al. 2015). These make them as essential bioindicators of pesticide contamination particularly in soil environment (Song et al. 2009). Sometime pesticide accumulation in earthworms may not cause significant effects to it, but may have severe impact to other organisms and humans as pesticides may be transferred to higher trophic-level organisms through food chain (Uwizeyimana et al. 2017). In addition, the use of earthworm species for laboratory-based ecotoxicological tests for short time pesticides exposures makes them excellent species for biomonitoring of pesticides (Stenersen 1979). In recent years, there is an increasing interest on application of pesticides in agricultural ecosystem and its relation with eco-toxicology on earthworm along with mechanisms of eco-toxicology. Due to high fertility, inexpensive and easy availability, earthworm specie *Eisenia fetida* is increasingly used for laboratory based ecotoxicological study for pesticide exposure and biomonitoring (OECD 1984). Growth, reproduction and survivbility of the earthworms may be affected by pesticides both at individual and population levels as shown in Table 5.2. Recently, stress has been given on impact assessment of pesticides on the key enzymes of earthworm which modulate the different functions like neurotransmission, energy and amino acid metabolism and oxidative stress which makes it a more prosperous bioindicators for soil pollution (Tiwari et al. 2016). However, studies on impacts of pesticides at realistic field concentrations under actual field situation are lacking which would give more comprehensive bioindication efforts.

Aquatic Insects

Aquatic insects have a high diversity and abandaunce in most of the freshwater habitats (Hershey et al. 2010). They have an important role in understanding the patterns of aquatic ecology and environmental quality (Heino 2009; Cereghino et al.

Table 5.2 Overview of the recent reports on the effects of pesticides towards earthworm species

Pesticide applied	Comments	References
Chlorpyrifos	Chlorpyrifos strongly inhibited cholinesterase and carboxylesterase activities of earthworms	Sanchez-Hernandez et al. (2018)
Atrazine	Damage to the chloragogenous layer, epithelial tissues; glandular enlargement of the epithelial tissues, prominent vacoulations and pyknotic cells of earthworm Nsukkadrilus mbae	Oluah et al. (2010)
Imidacloprid (neonicotinoid)	Changes in body cellur changes occurred after 24 h of exposure to the lowest concentrations	Dittbrenner et al. (2011)
Chlordane (organochlorine)	Significant reduction in spermatozoa of *Lumbricus terrestris*	Cikutovic et al. (1993)
Mineral fungicides (copper and sulfur)	A slight reduction in Cholinesterase activity	Denoyelle et al. (2007)
Fenvalerate and esfenvalerate (synthetis pyrethroides)	Enantiospecific induction in oxidative stress to earthworm *Eisenia fetida*.	Ye et al. (2016)
Fipronil	R-fipronil was more toxic (almost 2 times) than S-fipronil to earthworms (*Eisenia feotida*).	Qu et al. (2014)

2003). Compared to other groups of organisms, they have been widely used to assess aquatic integrity as they possess several advantages. Compared to fish and plankton, higher ability of aquatic insect to bear pollution-induced environmental stress (Ravera 1998), make them a suitable bioindicator. As aquatic insect spends nearly whole life in a water body, they may act as an early indicator for pesticides pollution since they are exposed to the effects of physical habitat alteration, point and nonpoint contaminants over their life cycles. Therefore, they are the commonly used indicators for water quality monitoring (Morse et al. 2007). Similar to terrestrial ecosystems, aquatic ecosystems will also be adversely affected by pesticides due to its increased use (Morrissey et al. 2015). Due to surface runoff and leaching, pesticides may enter into the aquatic ecosystems in high quantity (Huseth and Groves 2014). Impacts of pesticides on aquatic insects may helps in bioindication of pesticide pollution also. In addition, sensitivity of aquatic insects towards pesticides makes it as useful indicators for overall ecosystem function (Wallace et al. 1996). Nair, 1981 reported that other than insect, water mite *Hydrachna trilobata,* is also sensitive to OCPs under laboratory condition. Reduction in number of mite species in wetlands due to organophosphorus pesticide application has been reported by Dieter et al. (1996). Imidacloprid (@ 10 kg/ha) in mesocosm rice ecosystem significantly reduced the number of aquatic insects *Crocothemis servilia mariannae* and *Lyriothemis pachygastra* nymphs (Kobashi et al. 2017). Studies should focused on use of aquatc insect as a bioindicator to identify the pesticide exposure route into the aquatic ecosystems which will definitely helps in the proper implementation of mitigation measures.

5.3.1.3 Vertebrates

Feeding habit is crucial factor for bioaccumulation of pesticides in animal. Several animal like fish, birds and mammals is placeed in an important posting of food chain and accumulation of pesticides in the tissue of animal subsequently transfer the contaminant to next higher level and finally to the human and thus they are very crucial for biomonitoring of pesticides.

Birds

Use of birds as bioindicators of environmental change with particular context of pollutants like pesticides in various systems (Carson 1962; Colborn et al. 1996; Green and Elmberg 2014) has been documented. Due to their higher position in the food chain and sensitivity towards the environmental changes make them useful for bioaccumulation study (Furness and Greenwood 1993). By ingesting pesticides contaminated prey, birds are likely to bioaccumulate pesticides (Furness and Greenwood 1993; Ortiz-Santaliestra et al. 2019) and thus can serve as a potential bioindicators for pesticides pollution. In particular, they act as a connecting bridge between terrestrial, marine and freshwater ecosystems due to their high spatial abandaunce and ability to move or spread within an environment which led them to expose maximally to function as environmental sentinels. Fish is considered as important bioindicator for pesticides, however as compared to birds, fish accumulates low concentrations of pesticides and this makes bird's bodies as background for pesticides pollution (Tanabe and Subramanian 2006). Trophic position/feeding association, taxa of species, dietary exposure, habitat, sex, age and reproductive stage of birds are important factor which influence the bioindication potential (Eulaers et al. 2013). Predator species of birds are the most appropriate bioindicator due to their abandaunce in worldwide. Due to this reason, carnivore bird's species are more often used for pollution research. Short distance migrating birds reflects the background pesticide contamination level, whereas transboundary transport of birds reflects the global pollution (Tsygankov et al. 2016).

Thinning of eggshell, reduction in reproductive success (Ratcliffe 1967), changes in normal sexual activities, smaller brain size, and neurotoxicity (Iwaniuk et al. 2006) to the birds are long back related with the pollution of OCPs. Birds are highly vulnerable to OCPs, and thus offer a potential prospect of use of them for monitoring pesticide contamination. The dangerous and disruptive effects of OCPs on birds, led to the Rachel Carson's work, *Silent Spring*, in long back 1962. Adverse effects of these OCPs on endocrine and reproductive systems of birds led to the worldwide banning and regulation of OCPs (Godduhn and Duffy 2003).

However with the advancement of technology, the method used to assess the pesticide toxicity has been changed. Previusly, mortality events, destructive sampling by analysis of internal tissues such as liver, muscle and brain (Auman et al. 1997; Rivera-Rodríguez et al. 2007), or data related to abundance and distribution change, and migratory patterns of bird were linked with pesticides pollution (Becker 2003). However, these destructive sampling is not advisable as many of the bird species is considered as endangered species by International Union for Conservation of Nature (IUCN).

Use of nestling feathers in bird's populations (Rutkowska et al. 2018) can be an alternative option of use of bird's species as a bioindicator. Non-destructive materials like blood, preen oil gland secretions can be the option for pesticides load estimation (Rutkowska et al. 2018). Among these, bird's feathers can be the best option which can proportionally represents the pesticides contamination in the body to the feathers and studies showed the significant correlation between the pesticide contamination level in seabird's food and feathers. Collection of feather is minimally invasive as they can be easily collected and stored. Moreover, they provide history of pesticides exposure towards birds' annual cycle. Wide concentration range of pesticides can be traced by feather as studies indicates that low level OCPs concentrations in bird feathers from Patagonia ($6.49 \pm 5.95\ \mu g\ g^{-1}$) (Martínez-López et al. 2015) to relatively high concentration in birds feathers in Spain (870.48 ± 614.48 ng g^{-1}) (Espín et al. 2012) and Ireland (17.94 ± 2.19 ng g^{-1}) (Acampora et al. 2018) were monitored. Therefore, feathers can be a used as bioindicator throughout the different geographic region of the world. For biomonitoring of OCPs, Penguin feathers are good tool (Metcheva et al. 2017). Feathers can represent the, regional variations in pollution as evidence from report that feathers of magpie (*Pica pica*) from rural areas contain more OCPs than industrialized/urbanized counterpart (Jaspers et al. 2009). Failed eggs are also suitable indicator for OCPs biomonitoring as it involved non-destructive sampling and good correlation between the pesticides sequestration in eggs and its contamination level in the adult breeding female (Espín et al. 2016) was reported. However, selection of life history traits of birds like migratory behavior or moult chronology is very much important, as there is a chance of false positive or negative detection of pesticides. For example pesticide load in migrant individual indicates its source population, but not the study site. Likewise, if recently grown feathers is the sampling matrix it will not true representative for local contamination in a resident bird. So, utmost precaution has to be taken during sampling of bird with source history of contamination.

Fish

Fishes are being used to monitor river ecosystem health around the globe owing to its wide range of characters, complex habitat requirements as well as being a valuable component of the aquatic ecosystems (Plafkin et al. 1989). The response of fishes to various stressors is well documented (Leonard and Orth 1986). A number of fish-based biotic indices are in application for river monitoring including multimetric indices (Karr et al. 1987). Knowledge of the indicators' taxonomy, ecological demands and physiology is a pre requisite which affects the success of bioindicator studies (Schiemer et al. 2001).

Following are the reasons for fishes being an excellent candidate species as bio indicator of aquatic environment (Barbour et al. 1999; Andreas et al. 2003).

- Fishes have relatively high longevity which makes them suitable for indicating long-term effects and broad habitat conditions, the chronic effects of different toxicants or other pollutants on ecosystem health.
- The ecological and physiological requirements of a number of fish species are well studied and recorded, which guide to easy understanding of the response to alterations in ecology.
- Fishes with diversified feeding behaviour, occupy almost every trophic level of the aquatic ecosystem from the lowest that feed on phytoplankton to the carnivorous fishes. Hence the fish population structure reflects the comprehensive environment health.
- The habitat requirements of fishes vary widely with species and their ontogenic development. Due to this specific habitat requirements, fishes are suitable indicators of the ecological status of aquatic ecosystem.
- Some fishes like Atlantic salmons being highly migratory, are suitable indicators of habitat connectivity or fragmentation.
- Fishes are having good size with well developed organs. Hence different analytical procedures such as histopathological, haematological and biochemical analysis can be carried out to evaluate toxic impacts by determining the accumulation of toxicants in tissues.
- Top-down approaches of analysing the changes in fish communities in the ecosystem, testing for sources and causes of possible problems as well as the bottom-up assessments, using laboratory data to estimate the impacts in the sympathetic ecosystems are possible when using fish as bioindicator.

The response of fishes to toxins is represented in Fig. 5.1. The low level of pollutants in the ecosystem stimulates sub cellular response without much effect on the health of the organisms. But prolonged and severe exposure induces functional and structural changes in the organism which impair vital biological functions. Prolonged exposure to moderate toxicity interferes with hormonal and biochemical processes leading to increased metabolism, immunosuppression, disturbed osmoregulation, failure of reproduction or tissue damages.

Fishes are good indicators for hydrosphere contamination for polychlorinated biphenyls contamination in water, especially the silverfish, tench accumulates polychlorinated biphenyls in tissues. Table 5.3 summarizes the use of fish species for bioindication of pesticides pollution. The skipjack tunas (*Katsuwonus pelamis*) collected from offshore waters around the countries such as Japan, Taiwan, Philippines, Indonesia, Seychelles, Brazil, and the Japan Sea, the East China Sea, the South China Sea, the Bay of Bengal, and the North Pacific Ocean, were tested for the concentrations of OCPs, such as polychlorinated biphenyls, dichlorodiphenyl trichloroethane and its metabolites (DDTs), chlordane compounds (hexachlorocyclohexane isomers and hexachlorobenzene in the liver. The

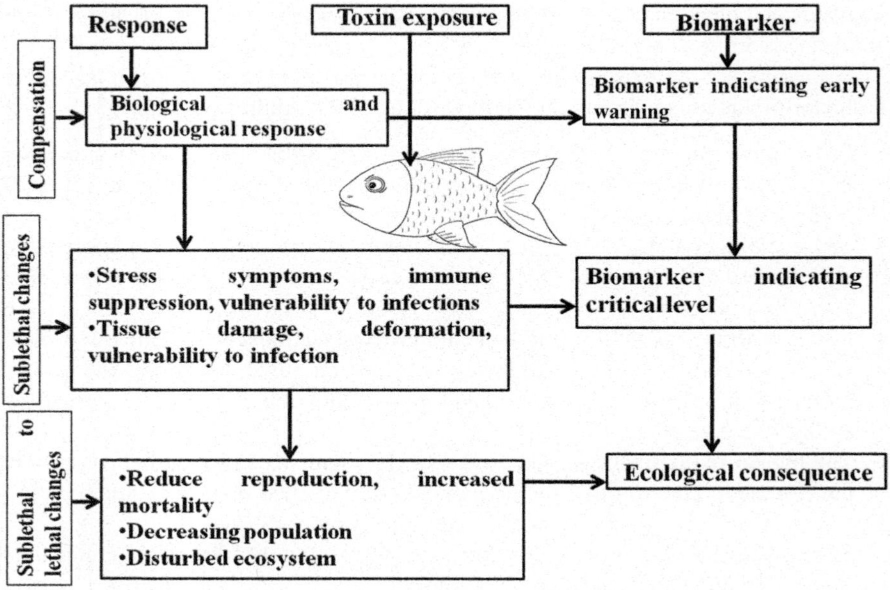

Fig. 5.1 The response of fish to toxicants and the action as a bioindicator. (Adapted from Andreas et al. 2003)

concentration of OCs in liver of skipjack tuna was comparable to the concentration estimated from surface waters, which reflect the pollution levels of these chemicals in sea water. This study results indicate that the skipkack tuna is a suitable bioindicator of seawater pollution (Ueno et al. 2003). The embryo of flat fish, *Limanda limanda* showed increased malformation indicating DDT and PCBs pollution (Cameron and Berg 1992). The fishes *Culter erythropterus* and *Protosalanx hyalocranius* reported to be sensitive to OCPs and would serve as a bioindicator for assessing contamination by OCPs in the aquatic system. *Megalobrama amblycephala* would serve as an indicator for human carcinogenic risk of OCPs especially DDT in Lake Chaohu (Liu et al. 2016). Down water pollution with pesticides may be assessed by estimating the change in reproductive efficiency (reduction in mean oocytes count, gonadal index and gonado-somatic relationship) of of*Astyanax fasciatus* (Jain et al. 2010).

Wild Animals and Small Mammals

Use of wild animals and small mammals as a bioindicator can be valuable aspect of environmental pesticides pollution. Selection of a particular species as a bioindicator should be based on type of investigation which includes species diversity and availability, pesticides to be studied and their potential toxic level, area and aim of

Table 5.3 Summary of the effects of certain pesticides on the biological and biochemical activities of the organisms, which may act as efficient bio indicators of pesticide pollution as adapted from Filimonova et al. (2016)

Pesticide	Effect on the organism	Organism	References
1,2,4-Trichlorobenzene (organochlorine pesticides) 0.245 ppm for 5 days	Alterations in morphological and fatty acid composition after longer exposure periods.	Diatom *Cyclotella meneghiniana*	Sicko-Goad et al. (1989a)
Pentachlorobenzene (organochlorine pesticides) 0.245 ppm for 5 days	Results in changes in diatom FA composition and cell structure increase the lipid volume in diatom	Diatom *Cyclotella meneghiniana*	Sicko-Goad et al. (1989b)
Substituted pyridazinone herbicide: 4-chloro-5-(dimethylamino)- 2-phenyl-3(2H)- pyridazinone (Sandoz 9785)	Inhibitor of long chain FA desaturation	Microalgae *Chroomonas salina*	Henderson et al. (1990)
Pyridazinone herbicide: norflurazon (SAN 9785) 0.2 mM	Effect on fatty acid profile	Cyanobacterium *Spirulina platensis*	Kachroo et al. (2006)
Herbicide: Diuron 0.20 mg L^{-1}	Moderately toxic to fish, affect social behavior, survival, inhibition of the nervous system and anemia; causes an increase in liver and spleen weight; fatty acid profile change.	Gilt-head sea bream *Sparus aurata*	Sanchez-Muros et al. (2013)
Triazine herbicide: Atrazyne 100 and 150 mg L^{-1}	Interferes with the photosynthesis function, affects the quality of the diatom cells	Diatom *Seminavis robusta*	De Hoop et al. (2013)
Triazine herbicide Irgarol 1051 (2-(tert-butylamino)-4-(cyclopropylamino)-6-(methylthio)-1,3,5-triazine) 0.535 mg/ml for 21 days	Significant increase of larval malformation and sperm fertilization rate, interact with calcium homoeostasis, inhibit ATP synthesis, decrease normal chemoreceptive response	Asian sea-bass *Lates calcarifer*	Ali et al. (2015)

study. Ideally, there should be proper information about the existing environmental studies, along with simultaneous compilation of other environmental parameters. Health of animals will be affected due to accumulation of pollutants if they are living in polluted environments and thus they can be good bioindicators of that particular polluted environmental. Measurement of morphological and hematological parameters of animals indicates the impacts of pollution on their health status (Velickovic 2007; Talmage and Walton 1991). Wild mammals, like wild boar (Tomza-Marciniak et al. 2014), roe deer (Naso et al. 2004), brown hare (Karner 1978) etc. are known to accumulate pesticides and thus can be used as a bioindicators for environmental pesticides pollution.

There are various opinions about the choice of species for pesticides monitoring. These also depend on the feeding habit and habitat of the species. Some studies shows that herbivorous mammals are best example to fulfill the requirements for a good bioindicator for pesticide contamination because of their adequate representative of animal species number, true representativeness of spatial and ecological niches, sufficient knowledge about the trophic level of study area as well as their dietary composition (Tataruch and Kierdorf 2003). However, other study proposes that carnivores are more suitable bioindicator particularly for lipophilic pesticides like OCPs due to their more accumulation in the fatty tissue than herbivores. Reports indicates that OCP pollution resulted in decline in population number of some species of aquatic carnivores like American mink (*Neovison vison*), American otter (*Lutra canadensis*) and Eurasian otter (*Lutra lutra*) (Giesy et al. 1994; Harding et al. 1999), whereas there number got increased when the concentration of pesticides got decreased. Their population recovery and successful reintroductions were only accomplished after declining OCPs concentration as reported by Castrillon et al. (2010). However, choice of animal indicators is comparatively more difficult than the use of aquatic habitats for pesticides monitoring. Though wild mammalian species like European hare, red deer, wild mink, and wolves are good bioindicator, but their natural habitats which includes forests or rural areas, and moreover they occupy a large territories make them diffult to represent the pesticides pollution associated with the exposure level to human being. A convenient alternative species like red fox (*Vulpes vulpes*) has been proposed by Corsolini et al. (2000) as they lives in close vicinity to humans by colonizing in suburban and urban habitats. Table 5.4 indicates the presence of the pesticides in wild mammals. In addition to use of wild animals, use of small mammals, mostly rodents as a bioindicator seems to be an important step due to their genetic and physiological similarities with humans and thus impacts of pollutants can be exploited. Rodent carcinogens are also human carcinogens, and therefore they may be a suitable model to understand the mechanism of genotoxicity of human carcinogens. Small mammal species, like Algerian mouse (*Mus spretus*) (Udroiu et al. 2008) and wood mouse (*Apodemus sylvaticus*) (Barber et al. 2003) are frequently used as environmental pollution bioindicators studies. However, more emphasis should be given on long-term studies on the use of mammals as a bioindicator along with environmental studies (Wren 1986).

5.3.2 Plants as Bioindicators

Due the availability, cost-effectiveness and simplest matrix, plant can be a suitable bioinidcator for environmental pesticide pollution. They are susceptible to alteration in chemical composition in environmental and the fact they accumulate pollutants. Higher plants are particularly important for atmospheric pesticide contamination at remote or poorly accessible locations. Aquatic plants are sensitive indicator for aquatic pesticide pollution as they are sessile and can rapidly equilibrated with natural habitat (Klemm et al. 1990). The presence or absence of particular plants

Table 5.4 The types of wild mammals, where presence of pesticides can be determined or monitored

Wild mammals	Pesticides	Sample reparation	Analysis	Country	Location	Site and sampling	Notes	References
European roe deer *Capreolus capreolus*	OCs and metabolites	LLE and SPE clean-up	GC-ECD	Italy	Emilia-Romagna region of Northern Italy	38 European roe deer	Roe deer from the Emilia Romagna region are exposed to OCPs, indicating the atmospheric transport of pesticides.	Naso et al. (2004)
Wild boar (*Sus scrofa*)	DDT, lindane, Endrin along with PCBs	LLE with florisil clean-up	GC/MS	Poland	West Pomerania, in North-Western Poland	32 no. of wild boars of various ages and sexes	Meat and offal of hunted animals from the area contains elevated concentrations of pesticides and hence may pose some risk for venison consumers.	Tomza-Marciniak et al. (2014)
Wolves (*Canis lupus* L. 1758)	OCs with PCBs	LLE	GC-ECD	Russia	Tver and Smoliensk regions of Northwest Russia	58 wolves	Organochlorine pesticides are unlikely to have significant impacts on wolf populations	Shore et al. (2001)
Chamois (*Rupicapra rupicapra*)	OCs	LLE	GC-ECD	Spain	Cadí-Moixeró Natural Park, Catalonia, Spain	19 chamois	Content of organochlorine residues stored in adipose tissue of chamois are lesser than those found in several terrestrial mammalian species	Guitart et al. (1990)
Iberian lynx (Lynx pardinus) and other 55 wild carnivores belonging to five species	OCs with PCBs	LLE	GC-ECD	Spain	Natural areas of Spain (Doñana National Park, Sierra Morena and Lozoya River)	80	Organochlorine levels have declined since the 80s in carnivores form Doñana National Park	Mateo et al. (2012)

OC organochlorine, *PCBs* polychlorinated biphenyls, *LLE* liquid liquid extraction, *SPE* solid phase extraction, *GC-ECD* GC with Electron Capture Detector, *GC/MS* gas chromatography–mass spectrometry species may provides sufficient information about environmental health and quality. Several factors like plant species, the levels of pesticide contamination in the environmental matrices, pesticide nature and many other factors (Tarricone et al. 2015) influence may the accumulation and distribution of pesticides in plant.

5.3.2.1 Higher Plant

For many decades, plants have been used as bioindicator for pesticide pollution particularly atmospheric concentration (Sharma et al. 1990). Sometimes, by reacting with contaminants, plants showes visible injuries and symptoms to express their exposure to contminants (De Temmerman et al. 2004). However, several factor like sampling site target species, selection of individual sample and target compartments, period, and size of sampling, weather and transport condition of sample effects the applicability of tree as a pesticide bioindicator. Plant leaves are excellent indicators of environmental pesticide pollution due to their capability of accumulation of pesticides present in air (Tarricone et al. 2015). Pine needles have been found to be an excellent bioindicator for local and regional ambient air pollutions monitoring (Josephine et al. 2019).

However, most of the study concentrated on use of pine needles for detection of OCPs (Romanić and Klinčić 2012; Al Dine et al. 2015). Aston and Seiber, 1996 reported the use of *Pinus ponderosa* needles as a bioindicator for pesticide like diazinon, chlorpyrifos and methidathion. They have also reported the method for chlorpyrifos and methidathion assessment in air and pine needle samples (Aston and Seiber 1997). Age of assimilation organs is also important as Romanić and Klinčić 2012 reported higher concentrations of OCPs in pine needles of two-year-old plant as compared to one-year-old plant.

However, wind or precipitation can easily removes the contamination on leaves and this method may be failed to derive any conclusive results. To overcome this, in recent years tree bark is becoming more popular for determining atmospheric pollution of pesticides (Yang et al. 2013). They are excellent in bioindication of pesticide source/sink regions, spatial and temporal trends (Schulz et al. 1999) of pesticide pollution, and their global cycling (Simonich and Hites 1995). HCHs distribution and transport in atmosphere on the national scale in China was carried out by using tree bark (Niu et al. 2014). Several studies have been using tree bark for monitoring Hexachlorocyclohexanes (Yang et al. 2013; Zhao et al. 2008; He et al. 2014). Geographic distributions and transportation of Hexachlorocyclohexanes in atmosphere has been successfully detected by using tree bark (Zhao et al. 2008; He et al. 2014).

Recently, bioassays using plants are found to be an excellent bioindication method for pesticides especially with herbicides (de Souza et al. 2016). Because, bioassays is simple compared to different environmental assessments method (Geras'kin et al. 2005). Both UNEP (United Nations Environment Programme) and IPPB (International Program on Plant Bioassays) (Ma 1999) has validated these bioassays techniques. Morover, WHO (World Health Organization) and

USEPA (the US Environmental Protection Agency) have validated the use of plant bioindicators as efficient suitable model to identify environmental genotoxicity of herbicides. Both in-vivo and in-vitro assays (Rank and Nielsen 1994) has been correlated with the data obtained using plants bioindicators and compared to the Ames assay. Plants bioindicators express a higher sensibility to toxic agents. Plant bioassays as bioindicators are ethically recommended because it is not using any animals.

5.3.2.2 Lower Plants

Phytoplankton

Various phytoplankton species are currently being used for biomonitoring of water bodies due its characteristics which ensemble to a bio indicator of water quality. Their nutrient requirement, reproduction rate and short life cycle as well as the change in species composition and cell density in accordance with the change in water quality parameters makes plankton one of the most suitable indicator species of aquatic ecosystem. Quantitative and qualitative analysis of plankton samples collected from water bodies are more useful than nutrient or chlorophyll-a concentration in providing evidence of water quality changes. A change in the density of the phytoplankton community is the first response to water quality change resulting in subsequent change in the diversity of the phytoplankton (Markert et al. 1997). Bioindicators should have a moderate tolerance to variations in the ecosystem variability compared to rare species with narrow tolerance or cosmopolite species with wide distribution pattern. This property of moderate tolerance makes them sensitive to indicate ecosystem alterations, as well as they survive withstanding some variability.

There are some properties of phytoplankton which characterize them as a reliable bioindicator (Coste et al. 2008)

- They are important functional group of the ecosystem, existing in large biomass
- The sampling procedure is easy and monitoring the phytoplankton indicator is economically viable
- The algae being sessile are unable to migrate to move away from pollution. They either tolerate the toxicants or vanish.
- The diverse algal species have different tolerance to different stressors.
- The short life span and smaller size of algae imparts rapid, sensitive and predictable response to ecosystem variations, while the algal community lives long to indicate the environmental impacts over a long run.
- The algae are densely populated over a particular area and homogenously distributed in the water in any aquatic system.
- It displays great statistical strength, a trend should be detectable even with relatively small changes, and it should respond clearly to a stressor such as nutrient addition.

- An easily identifiable species with distinct identification features is preferred.

Many species of phytoplankton are reported to be good indicators of pesticide contamination in water. Since algae are incapable of decomposing pesticides, they are main source of bioaccumulation in higher animals like fishes and mollusks which feed on them (Walsh 1978). Vance and Drummond (1969) have studied the bio concentration of the chlorinated pesticides aldrin, dieldrin, endrin and DDT in two algal species such as *Microcystis aeruginosa*, and *Anabaena cylindrica*. Tests were also made to analyse the effect of DDT, endrin, and dieldrin on *Skeletonema costatum, Dunaliella tertiolecta, Coccolithus huxleyi* and *Cyclotella nana* isolated from different oceanic environments by Menzel et al. (1970). Dunaliella was found to be insensitive to all the insecticides and hence are not recommended as an indicator species. The rate of carbon-14 uptake in Skeletonema and Coccolithus was reduced significantly at concentrations above 10 ppb of all three insecticides. Wurster 1968 studied the effects of DDT on photosynthesis by marine phytoplankton viz., *Skeletonema costatum, Coccolithus huxleyi*, Pyramimonas and *Peridinium trochoideum*. The use of alterations in FA profiles as a bioindicator in marine algal species after exposed to organic and inorganic toxicants has been reported by various researchers.

Lichen and Mosses

In terrestrial environments, pine needles and other vegetation are frequently used for monitoring of contaminants including pesticides. However, their short life, make them unsuitable regarding chronic exposure as they are poor accumulator for large amounts of organic pollutants. Contrary to this, organisms which have slow growth rate and long life expectancy are very useful in biomonitoring of contaminant especially in atmosphere (Harmens et al. 2011). In this aspect, organism like lichens which is a fungi and algae symbiosis is very much useful. Due to their persistent and robust nature, sustaining behaviour in extreme climatic conditions, make them important biomonitoring indicator (Bergamaschi et al. 2004).

Mosses have also been found to be another important indicator for biomonitoring of pesticides particularly for atmospheric pollution. Fulfilling the nutrients requirement from the air due to lack of roots, direct exposure to pollutants due to lack of waxy cutin compounds in the outermost epidermal cells, sensible and tolerance to a range of contaminants (Harmens et al. 2011), high surface area, wide habitats, slow growth rate make mosses an important bioindicator as they integrate information on long time pollutants accumulation. Though, lichens and mosses are totally different cryptogrammic organisms, but they are very effective environmental biomonitor due to their abundance, special morphological structure and physiological features. Their use as bioindicator for pesticide pollution is rising gradually (Lim et al. 2006; Yang et al. 2013; Tarcau et al. 2013). Moss and lichen was first time used by Bacci et al. (1986) for detection of OCPs Antarctic Peninsula. Successful use of lichen as a bioindicator for OCPs (Villeneuve et al. 1988) detection in Southern France and its correlation with detection of OCPs in industrial areas in Italy and Norway indicates its potential for monitoring of pesticide across the state. Even, in recently, Hanedar et al. (2019) used lichen as a bioindicator for OCPs pesticide pollution in industrial areas. Yang et al. (2013) reported that lichen shows as much potential like conifer needle for detection of DDT pesticides in Southeast Tibetan Plateau.

Sometime they are also used to identify the background levels and source of pesticides pollution. For example, ratio of α-HCB/γ-HCB and p,p′-DDE/p,p′-DDT indicates the history of arrival of these OCPs. In old formulations of hexachlorocyclohexanes insecticides predominated with α-form, whereas recent, products dominated with γ-isomer (almost 90%). Mosses collected from Antarctica in 1985 (Bacci et al. 1986) indicates the recent arrival of hexachlorocyclohexanes pesticides as it predominance with γ-form. Whereas mosses collected from Antarctica in 1988 and 1999 indicates the old arrival of hexachlorocyclohexanes as α-forms were higher than γ-form (Borghini et al. 2005; Focardi et al. 1991). Similarly ratio of DDT and its toxic microbially degradate metabolites such as p,p′-DDE, is used to indicates the sources of DDT. p,p′-DDE and p,p′-DDT ratio with value more than 1 indicates the aged DDT, wheras value less than 1 indicates the fresh inputs. p, p′-DDE/p,p′-DDT ratio is always more than 1 in the mosses collected from Antarctica which can be correlated with fact of banning of DDTs since 1980s for both production and uses (Bacci et al. 1986; Borghini et al. 2005; Cabrerizo et al. 2012). However, a value less than 1 in the mosses collected from some parts of Sinagpore indicates the continous deposits of DDTs though it has been banned in Singapore in 1985 (Lim et al. 2006). In addition, in the same study indicates that there have been no fresh hexachlorocyclohexanes inputs, since HCH concentration in the moss samples were below the limit of detection. Tarcau et al. (2013) reported that due to long-range air transport and deposition of pesticides, mosses collected from northeastern Romania showed uniform concentration of HCB, HCHs and chlordanes. The same study indicates the intense past history of use of in the southern part of Moldavia as the mosses showed high content of ΣDDT. These all suggested the important use of lichen and mosses as a potential bioindicator. However further study should be focused on correlation between the pesticide concentrations in lichen and mosses and in the atmospheric environment and its relation with human health issues.

5.3.2.3 Aquatic Macrophytes

Aquatic macrophytes plays a major role as a primary producer in oxygen production, control of water quality through nutrient cycling and provide living space for small aquaticorganisms. The aquatic plants are especially useful for indicating nutrient pollution due to its capability of reflecting the change in the nutrient levels of the system. Many cosmopolitan species have short life cycle and grow smaller in maximum size, which makes them a suitable species for analyzing different biological parameters indicating the presence of an unfavourable situation in the water body. For example fairly good, oxygenated, meso-eutrophic waters are characterised by the presence of the fresh water macrophyte *Potamogeton crispus*. Likewise *Lemna minor* has been studied widely for indicating a number of toxicants such as pesticides 2.4-D (Belgers et al. 2007), diquat and fomesafen (Gorzerino et al. 2009), glyphosate (Kielak et al. 2011), imazapyr (Cruz et al. 2015), herbicides bentazon, atrazine and clomazone and pyraclostrobin-based fungicide (Della Vechia et al. 2016). The floating aquatic plant *Azolla caroliniana* is used for indicating pollution with pesticides such as glyphosate, clomazone, oxyfluorfen, 2.4-D (Silva et al. 2012) and pyraclostrobin based fungicides. Chlorophyll-*a* fluorescence induction for estimating photosynthetic electron transport in plants as a bioindicator of toxiceffects of chemicals have been well documented in the aquatic plants *L. gibba* and *Myriophyllum spicatum*. Many aquatic plants are now better understood for its tolerance to alterations in habitat and presence of toxicants, which makes the researchers to exploit this plant community as an indicator species.

5.3.3 Microorganisms

Once pesticide enters into the soil or aquatic system, and they get bio-accumulated easily by the lower trophic level organisms which act as food for subsequent higher trophic level organisms. Thus, lower trophic level organism like microorganisms, which are abundant in soil and water, will react with these xenobiotics and may act as bioindicators for pesticide contamination. Microorganisms are an important part of both terrestrial and aquatic ecosytem; hence they can indicate the both terrestrial and aquatic pesticide pollution. Due to their fast growth and rapid expression with environmental changes even at low levels of contaminants, they are excellent bioindicator for pesticide pollution (Hans et al. 2003).

5.3.3.1 Terrestrial Microorganism

Soil provides home for enumerable numbers of microflora and fauna, and the total biomass of such microbes is over 20 times than that of human population. These microbial population mostly various prokaryots and eukaryots including bacteria, cyanobacteria, fungi, algae and actinomycetes, play various important roles including transformation and recycling of various elements. Soil microbial populations and their behaviour make a soil fertile, lively and it is an indicator of good soil health. But, presence of undesirable xenobiotics like pesticides may affect various microbial populations and thereby affecting the balanced system of soil ecology. This in other term affects soil productivity and finally, the crop yield. However, effects on pesticides on microbial populations are very specific to a particular type and species of microbe. Studies revealed an initial decrease in microbial population immediately after pesticide application with overall a small effect as microbes developed

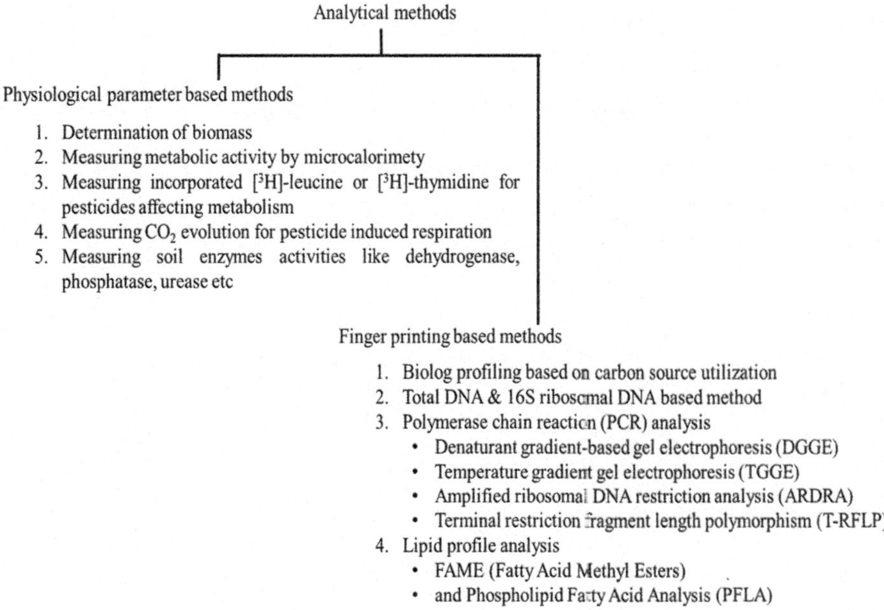

Fig. 5.2 Analytical methods for measuring bacterial bioindicators

resistance with time. Pesticides may impart change incell growth, cell division, total microbial biomass count, soil respiration, photosynthetic activity, biosynthetic pathways etc. and monitoring of changes in these parameters may indicate pesticide contamination level (DeLorenzo et al. 2001). However, the impact of pesticide is dependent on application dose of pesticide, soil parameters and lastly, on environmental conditions. The behavioural changes of bacterial based bioindicators can be measured by two categories of methods namely, physiological parameter-based methods and finger printing methods (Fig. 5.2). Most of the physiological based methods are suitable for whole soil microbial communities, rather than for a specific bacterial community. Whereas, finger printing based methods are mostly very species specific.Biologand PCR based methods arevery popular (Johnsen et al. 2001). Soil bacteria and actinomycetes have significant roles in carbon (C) and nitrogen (N) cycle. Table 5.5 indicated effects of various pesticides on soil bacterial population. Reports indicated that pesticides have negative impact on N-fixing bacteria and formation of fewer nodules. Hence, these N-fixers can be a good bioindicator for pesticide pollution in soil. Soil fungi, are comparatively more resistant toward pesticides than soil bacteria and actinomycetes. However, pesticides also have detrimental effects on soil fungi, and there by affecting various processes of transformation, mineralization and decomposition (Table 5.5). Most of fungicides and herbicides have negative impacts on soil fungi and majority of insecticide (except endosulfan) do not affect soil fungi population (Iqbal et al. 2001). Among varioussoil fungi, one of the most significant pesticide-affected population

Table 5.5 Effect of pesticides on soil microbial population

Pesticide	Effect on bacteria	References
Oxytetracycline, Captan	↓ number of cultivable bacteria, but no effect on overall soil microbial biomass	Bossuyt et al. (2001)
Endosulfan, Butachlor	↓ methanogenic bacterial population	Kumaraswamy et al. (1998)
Methyl bromide	Affected microbial communities, predominance (↑) of gram-positive bacteria	Ibekwe et al. (2001)
Cypermethrin	Affected microbial communities, predominance (↑) of gram-negative bacteria	Zhang et al. (2008)
Isofenphos, thimet, foratax	↑ total viable bacterial number, no effect on other soil microbes	Digrak and Kazanici (2001)
Organochlorines (lindane, dieldrin), organophosphates (pirimphos methyl, malathion) & carbamates (benomyl, methomyl)	Organochlorines and carbamets are more toxic than organophosphates to nitrogen mineralizing bacterias like *Nitrsomonas, Nitrobacter* and *Thiobacillus*	Odokuma and Osuagwu (2004)
Telen-17(mixture of 1,3-dichloropropene and chloropicrin), vortex (mixture of epoxiconazole, fluxapryroxad and pyraclostrobin), dazomet and fenamiphos,	↓ nitrificating bacteria, ↓ soil rhizobia and less nodule formation	Tu (1995)
Trifluralin, metribuzin	↓ nitrificating bacteria	Bertholet and Clark (1985)
	↓ nodule formation	
Thiobencarb	↓ *Azospirillum* & *Azotobacter* populations, hence less N-fixation	Jena et al. (1990)
	Effect on fungi	
Pyrazofos (fungicide)	↑ Total-N in *Trichoderma harzianum* and *Fusarium solani* & ↓ Total-N in *Aspergillus niger* and *Penicillium chrysogenum*	Moharram et al. (1994)
Bromoxynil (herbicide)	↓ Total-N in A. *niger, F. solani, T. harzianum* and *Mucor racemosus*, & ↑ Total-N in *Stachybotrys chartarum.*	
Profenfos (insecticide)	↓ Total-N in *Penicillium chrysogenum*	
Procloraz, Triazamate	↓ most of yeast population	Slavikova and Vadkertiova (2003)
Captan	↓ most of soil fungi growth	Bossuyt et al. (2001)
Endosulfan	↓ most of soil fungi growth	Iqbal et al. (2001)

(continued)

Table 5.5 (continued)

Pesticide	Effect on bacteria	References
Benomyl	↓viability and growth of arbuscular mycorrhizal fungi (*Glomus mosseae*)	Chiocchio et al. (2000)
	Effect on algae and cyanobacteria	
DDT, atrazine	↓ soil algae population	Mallavarapu (2002)
	↑ green algae population than cyanobacteria	
Atrazine	↓ cyanobacteria *Hantzschia amphioxys* (Ehrenberg); *Stauroneis obtuse* (Lagerstedt) population by inhibiting phytosysten-II pathway	Berard et al. (2004)
Monuron, linuron, diuron, chlortoluron, chloroxuron	↓ cyanobacteria population	Pipe and Cullimore, (1984)
Alachlor	22% growth ↓ in *Chlamydomonas moewusii*	Vaishampayan (1985)
	Only 3% survival of *Nostoc muscorum*	
	Effect on protozoans	
Atrazine	↓ chlorophyte and chrysophyte taxa	Downing et al. (2004)
Chlorothalonil	↓ chrysophyte and heterotrophic protists	
Endosulfan	↓ chrysophyes, cryptophytes and dinoflagellate taxa	
Thiodan and Karate	↓ protozoa population	Adebavo et al. (2007)
Pentachloronitrobenzene	Complete elimination of soil protozoa population	Ojo et al. (2007)

isglomealian arbuscularmycorrhizal fungi which plays very important role in absorption of plant nutrient like P and N, and even imparting resistance to disease. Hence, growth inhibition of arbuscularmycorrhizal fungi may act as a bioindicator for pesticide pollution in soil (Chiocchio et al. 2000). The next affected soil microbial populations are algae and cyanobacterica, the nitrogen fixers (*Anabaena* sp., *Gloeocapsa* sp., *Nostoc* sp. and *Oscillatoria* sp.). These two bottom level organism of every food chain of terrestrial and aquatic system plays very significant role by capturing gaseous nitrogen and carbon di-oxide. Application of pesticides, particularly herbicides drastically affects the photosynthetic activities of the algae and cyanobacteria by reducing the total chlorophyll content and by hindering the electron transport (Pipe and Cullimore 1984; Berard et al. 2004). Hence reduced photosynthetic activity could be a good bioindicator for soils contaminated with pesticides (Table 5.5). Similarly, soil protozoans are another group of important organisms which includes naked amoebae, ciliates and flagellates. They act as grazer and predators of soil fungi and bacteria, and hence play important role in food chain.

They are sensitive to pesticides and may act as bioindicator for pesticide pollution (Table 5.5).

5.3.3.2 Aquatic Microorganism

Reports across the world indicated that majority of water bodies including ponds, lakes, rivers, coastal areas, deltas etc. Pesticides are more common (probability >50%) in river water as compared to groundwater (Gilliom 2007; Smith et al. 2012). Once these pesticides become water soluble, they are prone to the process of bioaccumulation and biomagnification. Organochlorine insecticides like DDT, hexachlorocyclohexanes, endosulfan, endrin, toxaphene etc. can persists for long time from few days to over some years in aquatic sediments or on particulate matters. Though, reports indicated that organophosphates have lower persistence of few days, but they got absorbed by aquatic organisms during that time of existence. Owing to the lipophilic property pesticides get easily absorbed in the adipose cell/tissues of lower trophic organisms and move towards higher organisms of food web. Pesticide residues present in aquatic system impart deleterious effects on various aquatic microbes (Table 5.6). In general, herbicides and insecticides affect the phytosystem-ll of algae and cyanobactrias; thereby decrease their population (DeLorenzo et al. 2001). At the same time, they promote green algae and result in bloom of green algae as bioindicator for pesticide contamination (Pesce et al. 2006). Restricted growth/inhibition of these specific microorganisms affect the lower trophic level producers who depend on these microbes for food and nutrition. Even residues of some herbicides have been found to affect the algal symbiosis and thereby, affecting the coral reef (Smith et al. 2012).

5.3.4 Pesticides Toxicity Bioindicators for Humans

Several adverse health effects on human are associated with pesticide contamination which ranging from endocrine disruption, cancers, moderate to extreme neurotoxicity, etc. which indicates that human matrices can be suitably used as a bioindicator for pesticide pollution. Identifying the contaminants route for human exposure to pesticide i.e. air, water, soil and food is very much important. Selection of appropriate human matrices i.e. different human tissues like hair, urine, blood etc. is also important. There is an extensive use of human hair in forensic and clinical sciences for detection of drug exposure. Over the last decades, there is increasing interest in analyzing human hair for environmental contaminants biomonitoring. Compared other human matrix like blood, urine, saliva etc. which only indicates the recent exposure of pesticides, hair can serve is a bioindication of history of pollution. The reason is since there is no metabolism or excretion occurs in the human hair and pesticide will be remaining stable which makes it as an excellent bioindicato. At the same time, collection of human hair is easy, non-destructive and cheap. Several pesticides like OCPs (Covaci et al. 2008), OPs (Salquèbre et al. 2012), and SPs

Table 5.6 Effect of pesticides on aquatic microbial population

Pesticide	Effect on aquatic microbe	References
DDT	Inhibited photosynthetic	Lal and Lal (1988)
	CO_2 fixation in the green alga & ↓ population *Selenastrum capricornutum.*	
	Chlorella sp.,	
	↓ cyanobacteria population *Anabaena* sp.	
Endosulfan	↓ reproduction of green alga *Chlamydomonas reinhardtii by delaying meiosis,*	Mohapatra and Mohanty (1992)
	↓ growth of marine red alga *Champia parvula*, bluegreen algae *Anabaena* sp. and *Aulosira fertilissima.*	
Chlorpyrifos	Toxic to the marine diatoms	Butcher et al. (1977)
	Skeletonema costatum and *Minutocellus polyorphus*,	
	↓ zooplankton abundance,	
	↑ blooms of cyanobacteria (*Anabaena, Anabaenopsis*)	
Temephos	↑ growth of nitrogen-fixing bluegreen algae and the green alga *Chlamydomonas* & simultaneously	Low-Goh et al. (1981)
	↓ growth of diatom *Navicula pelliculosa*	
Atrazine	↓ growth of diatoms, diatom, dinoflagellate, chlorophyte and cryptophyte,	DeLorenzo et al. (1999)
	↑ cyanophyte biovolume	
Diquat	↓ growth of blue green algae and chrysophytes,	DeLorenzo et al. (2001)
	↑ population of green algae mainly, Oocistisspp	
Diuron	↓ phytosynthesis by affecting photosystem-ll of phototrophs,	Pesce et al. (2006)
	↓ reproduction of phytoplanktons like *Asterionella, Oocistis* and *Cyclotella*	

(Lehmann et al. 2018) have already been detected in human hairs. Hair has ability to detect the pesticides of different age group and not necessarily associated with the occupational exposure and type of pesticides. For example, maximum concenatrtion of *o,p*-DDT and *p,p*-DDT were detected in Romanian adolescents hair (Covaci et al. 2008). Whereas, in Philippines, pregnant women hair contains more of OPs (malathion and chlorpyrifos) and SPs (bioallethrin) (Ostrea Jr et al. 2009). Moreover, significant correlation of data obtained from hair analysis with other compartments of the body reaffirms the use of this matrix as a suitable bioindicator.

During the spraying of pesticide on fields, skin is the most exposed organ (Wolfe et al. 1966) and due to exposure, human skin could be critically harmed (Huang et al. 2000) which led to irritating, allergising and carcinogenic effects on skin (Zahm and Blair 1993). For example, herbicide, paraquat is very toxic to human skin and mucous membranes and it is very lethal if patients ingest (Tawatsin 2015). Human blood can also be used as a suitable matrix for evaluation of body exposure to pesticide and can provides the level of pesticide residues loading in the body. Pyrethroid residues in the blood and body fluids give an indication of level of pesticide exposure (Ramesh and Ravi 2004). Saeed et al. (2017) also reported that presence of OCPs human blood is closely associaed with exposure of pesticides. The correlation between pesticide exposure and its possible intake can suitably determine by using human blood. Human adipose tisuue could be another important indicator due to accumulation lipophilic pesticide like OCPs through the food chain and thus bioaccumulation (Daley et al. 2014; Boada et al. 2014). As the postion of Humans, is the topmost in the food chain, this accumulation will be magnified in the tissue rich in lipids (Caba et al. 2015; Boada et al. 2014). Therefore, the determination of pesticide levels in human adipose tisuue may helps in understanging the global environmental pesticide pollution (Calderón-Garcidueñas et al. 2018). Other matrices like milk (Ferronato et al. 2018), and urine (Bravo et al. 2019) also successfully used for biomonitoring of pesticides. However, impacts of pesticide in the human body are closely associated with the molecular structure and physical-chemical properties of pesticides, environmental exposure and source. Integrated study of them may give an interesting insight about the pesticides behavior in human tissues.

5.4 Conclusion and Future Research

Environmental pesticides monitoring is crucial step to manage pollution. Since use of pesticides is an unavoidable phenomenon, their excessive use may cause the environmental pollution and negative impacts on human health. So, early monitoring is essential to prevent or control the damages caused by pesticides toward humans and ecosystems. Strict regulation on the use of pesticides and its limit in different environmental matrices has been fixed throughout the world including The USA, Europe, and Asian countries like China, India, Japan and South Korea. With this, monitoring of pesticide pollution has been a regular phenomenon for safe use of pesticides for food safety and human health. Searching a suitable bioindicator may a crucial step in this endeavor for pesticide pollution management. Since, bioindicator involves the use of an organism or its part or group of organisms for pollution detection, their use will be very much helpful for assessment the pesticides impacts on human health. Considering all the results, pine needles, mosses and lichen may be best orgaism for atmospheric biomonitoring for pesticides. Fish and aquatic weeds are important indicator for water pollution by pesticides. Different specis of earthworm may an important indicator for soil conrmtination by pesticides. Honey bee and birds is the important indicator for pesticides pollution across the state. However, choice of suitable extraction and analytical method is also important for the complex matrix like honey bees and bee products and also for different human matrices. Since each environment system has different level and type of pesticides contamination, so it is highly recommended for simultaneous use of all these matrices together for bioindication.

Acknowledgments Authors thank the editors for invitation to contribute the chapter. Dr. Rakesh Kumar Ghosh, Mrs. Jesna P. K. and Dr. P. P. Choudhury acknowledge the support from Indian Council of Agricultural Research, New Delhi. Senior author is thankful to the Director, ICAR-Central Inland Fisheries Research Institute for support.

References

Acampora H, White P, Lyashevska O, O'Connor I (2018) Contrasting congener profiles for persistent organic pollutants and PAH monitoring in European storm petrels (*Hydrobates pelagicus*) breeding in Ireland: a preen oil versus feathers approach. Environ Sci Pollut Res 25:16933–16944. https://doi.org/10.1007/s11356-018-1844-2

Adebavo TA, Oio OA, Olaniran OA (2007) Effect of two insecticides Karate and Thiodan on population dynamics of four different soil microorganisms. Res J Biol Sci 2(5):557–560

Al Dine EJ, Mokbel H, Elmoll A, Massemin S, Vuilleumier S, Toufaily J, Hanieh T, Millet M (2015) Concomitant evaluation of atmospheric levels of polychlorinated biphenyls, organochlorine pesticides, and polycyclic aromatic hydrocarbons in Strasbourg (France) using pine needle passive samplers. Environ Sci Pollut Res 22:17850–17859. https://doi.org/10.1007/s11356-015-5030-5

Alavanja MCR, Dosemeci M, Samanic C, Luvin J, Lynch CF, Knott C, Barker J, Hoppin JA, Sandler DP, Cobie J, Thomas K, Blair A (2004) Pesticides and lung cancer risk in the agricultural health study cohort. Am J Epidemiol 160(9):876–885. https://doi.org/10.1093/aje/kwh290

Ali RH, Arrifin MM, Sheikh MA, Shazili NAM, Zainudin B (2015) Toxicological studies of Irgarol-1051 and its effects on fatty acid composition of Asian sea-bass, *Lates calcarifer*. Reg Stud Mar Sci 2:171–176. https://doi.org/10.1016/j.rsma.2015.09.008

Andersen AN (1988) Immediate and longer-term effects of Þre on seed predation by ants in sclerophyllous vegetation in South-Eastern Australia. Aust J Ecol 13:285–293. https://doi.org/10.1111/j.1442-9993.1988.tb00976.x

Andersen AN (1990) The use of and communities to evaluate change in Australian terrestrial ecosystems: a review and a recipe. Proc Ecol Soc Aust 16:347–357

Andreas C, Hofer R, Schiemer F (2003) Fish as bioindicators. In: Trace metals and other contaminants in the environment. Elsevier, pp 639–676

Anway MD, Cupp AS, Uzumcu J, Skinner MK (2005) Epigenetic transgenerational actions of endocrine disruptors and male fertility. Science 308(5727):1466–1469. https://doi.org/10.1126/science.1108190

Aston LS, Seiber JN (1996) Methods for the comparative analysis of organophosphate residues in four compartments of needles of *Pinus ponderosa*. J Agric Food Chem 44:2728–2735. https://doi.org/10.1021/jf950197s

Aston LS, Seiber JN (1997) Fate of summer time airborne organophosphate pesticide residues in the Sierra Nevada mountains. J Environ Qual 26:1483–1492. https://doi.org/10.2134/jeq1997.00472425002600060006x

Auman HJ, Ludwig JP, Summer CL, Verbrugge DA, Froese KL, Colborn T, Giesy JP (1997) PCBS, DDE, DDT, and TCDD-EQ in two species of albatross on Sand Island, Midway Atoll, North Pacific Ocean. Environ Toxicol Chem 16:498–504. https://doi.org/10.1002/etc.5620160315

Bacci E, Calamari D, Gaggi C, Fanelli R, Focardi S, Morosini M (1986) Chlorinated hydrocarbons in lichen and moss samples from the Antarctic Peninsula. Chemosphere 15:747–754. https://doi.org/10.1016/0045-6535(86)90041-X

Badji CA, Guedes RNC, Correa AS, Ferreira GH Nascimento IC (2006) Deltamethrin-induced impact on ant assemblages in tropical maize fields under conventional and no-tillage cultivation. Sociobiology

Balayiannis G, Balayiannis P (2008) Bee honey as an environmental bioindicator of pesticides' occurrence in six agricultural areas of Greece. Arch Environ Contam Toxicol 55(3):462. https://doi.org/10.1007/s00244-007-9126-x

Barber I, Tarrant KA, Thompson HM (2003) Exposure of small mammals, in particular the wood mouse *Apodemus sylvaticus*, to pesticide seed treatments. Environ Toxicol Chem 22(5):1134–1139. https://doi.org/10.1002/etc.5620220523

Barbour MT, Gerritsen J, Snyder BD, Stribling JB (1999) Rapid bioassessment protocols for use in streams and wadeable rivers: periphyton, benthic macroinvertebrates and fish, EPA 841-B-99-002, 2nd edn. U.S. Environmental Protection Agency, Office of Water, Washington, DC

Bargańska Ż, Ślebioda M, Namieśnik J (2016) Honey bees and their products: bioindicators of environmental contamination. Crit Rev Environ Sci Tech 46(3):235–248. https://doi.org/10.1080/10643389.2015.1078220

Becker PH (2003) Biomonitoring with birds. In: Markert BA, Breure AM, Zechmeister HG (eds) Trace metals and other contaminants in the environment. Elsevier, pp 677–736

Belgers JDM, Van Lieverloo RJ, Van der Pas JT, Van den Brink PJ (2007) Effects of the herbicide 2,4-D on the growth of the nine aquatic macrophytes. Aquat Bot 86(3):260–268. https://doi.org/10.1016/j.aquabot.2006.11.002

Bell EM, Hertz-Picciotto I, Beaumont JJ (2001) A case-control study of pesticides and fetal death due to congenital anomalies. Epidemiology 12:148–156

Berard A, Rimet F, Capowiez Y, Leboulanger C (2004) Procedures for determining the pesticide sensitivity of indigenous soil algae: a possible bioindicator of soil contamination? Arch Environ Contam Toxicol 46:24–31. https://doi.org/10.1007/s00244-003-2147-1

Berg CJ, King HP, Delenstarr G, Kumar R, Rubio F, Glaze T (2018) Glyphosate residue concentrations in honey attributed through geospatial analysis to proximity of large-scale agriculture and transfer off-site by bees. PLoS One 13(7):e0198876. https://doi.org/10.1371/journal.pone.0198876

Bergamaschi L, Rizzio E, Giaveri G, Profumo A, Loppi S, Gallorini M (2004) Determination of baseline element composition of lichens using samples from high elevations. Chemosphere 55(7):933–939. https://doi.org/10.1016/j.chemosphere.2003.12.010

Bertholet J, Clark W (1985) Effect of trifluralin and metribuzin on faba bean growth, development and symbiotic nitrogen fixation. Can J Plant Sci 65:9–20. https://doi.org/10.4141/cjps85-002

Boada LD, Sangil M, Álvarez-León EE, Hernández-Rodríguez G, Henriquez-Hernández LA, Camacho M, Zumbado M, Serra-Majem L, Luzardo OP (2014) Consumption of foods of animal origin as determinant of contamination by organochlorine pesticides and polychlorobiphenyls: results from a population-based study in Spain. Chemosphere 114:121–128. https://doi.org/10.1016/j.chemosphere.2014.03.126

Borghini F, Grimalt JO, Sanchez-Hernandez JC, Bargagli R (2005) Organochlorine pollutants in soils and mosses from Victoria Land (Antarctica). Chemosphere 58(3):271–278. https://doi.org/10.1016/j.chemosphere.2004.07.025

Bossuyt H, Denef K, Six J, Frey SD, Merckx R, Paustian K (2001) Influence of microbial populations and residue quality on aggregate stability. Appl Soil Ecol 16(3):195–208. https://doi.org/10.1016/S0929-1393(00)00116-5

Bravo N, Grimalt JO, Bocca B, Pino A, Bin M, Brumatti LV, Rosolen V, Barbone F, Ronfani L, Alimonti A, Calamandrei G (2019) Urinary metabolites of organophosphate and pyrethroid pesticides in children from an Italian cohort (PHIME, Trieste). Environ Res 176:108508. https://doi.org/10.1016/j.envres.2019.05.039

Brown TP, Rumsby PC, Capleton AC, Rushton L, Levy LS (2006) Pesticides and Parkinson's disease—is there a link? Environ Health Perspect 114(2):156–164. https://doi.org/10.1289/ehp.8095

Butcher JE, Boyer MG, Fowle CD (1977) Some changes in pondchemistry and photosynthetic activity following treatment withincreasing concentrations of chlorpyrifos. Bull Environ Contam Toxicol 17:752–758. https://doi.org/10.1007/BF01685965

Caba M, Meza E, Waliszewski SM, Martínez-Valenzuela C (2015) Inverse correlation among organochlorine pesticide levels to total lipid serum contents: a preliminary study in Veracruz, México. Environ Monit Assess 187:5613–5621. https://doi.org/10.1007/s10661-015-4694-0

Cabrerizo A, Dachs J, Barceló D, Jones KC (2012) Influence of organic matter content and human activities on the occurrence of organic pollutants in Antarctic soils, lichens, grass, and mosses. Environ Sci Technol 46(3):1396–1405. https://doi.org/10.1021/es203425b

Cairns J, McCormick PV, Niederlehner BR (1993) A proposed framework for developing indicators of ecosystem health. Hydrobiologia 236:1–44. https://doi.org/10.1007/BF00006084

Calatayud-Vernich P, Calatayud F, Simó E, Picó Y (2017) Occurrence of pesticide residues in Spanish beeswax. Sci Total Environ 605–606:745–754. https://doi.org/10.1016/j.scitotenv.2017.06.174

Calderón-Garcidueñas AL, Waliszewski SM, Ruiz-Ramos R, del Carmen Martinez-Valenzuela M (2018) Time trend tendency (1988–2014 years) of organochlorine pesticide levels in the adipose tissue of Veracruz inhabitants. Environ Monit Assess 190:206. https://doi.org/10.1007/s10661-018-6581-y

Cameron P, Berg J (1992) Morphological and chromosomal aberrations during embryonic development in dab *Limanda limanda*. Marine ecology progress series, vol 91, no 1/3. Biological effects of contaminants in the North Sea: Results of the ICES/IOC Bremerhaven Workshop, pp 163–169

Carlisle DM, Hawkins CP, Meador MR, Potapova M, Falcone JA (2008) Biological assessments of Appalachian streams based on predictive models for fish, macroinvertebrate, and diatom assemblages. J N Am Benthol Soc 27(1):16–37. https://doi.org/10.1899/06-081.1

Carrington D (2018) EU agrees total ban on bee-harming pesticides. The Guardian, 27 April 2018 (page visited on 29 June 2019)

Carson R (1962) Silent spring. Houghton Miffin, London

Carvalho FP, Montenegro-Guillén S, Villeneuve JP, Cattini C, Tolosa I, Bartocci J, Lacayo-Romero M, Cruz-Granja A (2003) Toxaphene residues from cotton fields in soils and in the coastal environment of Nicaragua. Chemosphere 53:627–636. https://doi.org/10.1016/S0045-6535(03)00451-X

Castrillon J, Gomez-Campos E, Aguilar A, Berdié L, Borrell A (2010) PCB and DDT levels do not appear to have enhanced the mortality of striped dolphins (*Stenella coeruleoalba*) in the 2007 Mediterranean epizootic. Chemosphere 81:459–463. https://doi.org/10.1016/j.chemosphere.2010.08.008

Cereghino R, Park YS, Compin A, Lek S (2003) Predicting the species richness of aquatic insects in streams using a limited number of environmental variables. J N Am Benthol Soc 22:442–456. https://doi.org/10.2307/1468273

Chiesa LM, Labella GF, Giorgi A, Panseri S, Pavlovic R, Bonacci S, Arioli F (2016) The occurrence of pesticides and persistent organic pollutants in Italian organic honeys from different productive areas in relation to potential environmental pollution. Chemosphere 154:482–490. https://doi.org/10.1016/j.chemosphere.2016.04.004

Chiocchio V, Venedikian N, Martinez AE, Menendez A, Ocampo JA, Godeas A (2000) Effect of the fungicide benomyl on spore germination and hyphal length of the arbuscular mycorrhizal fungus *Glomus mosseae*. Int Microbiol 3:73–175

Cikutovic MA, Fitzpatrick LC, Venables BJ, Goven AJ (1993) Sperm count in earthworms (*Lumbricus terrestris*) as a biomarker for environmental toxicology: effects of cadmium and chlordane. Environ Pollut 81(2):123–125. https://doi.org/10.1016/0269-7491(93)90076-Z

Codling G, Al Naggar Y, Giesy JP, Robertson AJ (2016) Concentrations of neonicotinoid insecticides in honey, pollen and honey bees (*Apis mellifera* L.) in Central Saskatchewan, Canada. Chemosphere 144:2321–2328. https://doi.org/10.1016/j.chemosphere.2015.10.135

Colborn T (2006) A case for revisiting the safety of pesticides: a closer look at neurodevelopment. Environ Health Perspect 114(1):10–17. https://doi.org/10.1289/ehp.7940

Colborn T, vom Saal S, Soto AM (1993) Developmental effects of endocrine-disrupting chemicals in wildlife and humans. Environ Health Perspect 101(5):378–384. https://doi.org/10.1289/ehp.93101378

Colborn T, Dumanoski D, Myers JP (1996) Our stolen future. Penguin, New York, p 306

Corsolini S, Burrini L, Focardi S, Lovari S (2000) How can we use the red fox as a bioindicator of organochlorines? Arch Environ Contam Toxicol 39(4):547–556. https://doi.org/10.1007/s002440010139

Coste M, Botry S, Tison-Rosebery J, Delmas F (2008) Improvements of the Biological Diatom Index (BDI): description and efficiency of the new version (BDI-2006). Ecol Indic 9:621–650. https://doi.org/10.1016/j.ecolind.2008.06.003

Covaci A, Hura C, Gheorghe A, Neels H, Dirtu AC (2008) Organochlorine contaminants in hair of adolescents from Iassy, Romania. Chemosphere 72:16–20. https://doi.org/10.1016/j.chemosphere.2008.02.058

Cruz C, Silva AF, Shiogiri NS, Garlich N, Pitelli RA (2015) Imazapyr herbicide efficacy on floating macrophytes control and ecotoxicology for non-target organisms. Planta Daninha 33:103–108. https://doi.org/10.1590/S0100-83582015000100012

Dale VH, Beyeler SC (2001) Challenges in the development and use of ecological indicators. Ecol Indic 1(1):3–10. https://doi.org/10.1016/S1470-160X(01)00003-6

Daley JM, Paterson G, Drouillard KG (2014) Bioamplification as a bioaccumulation mechanism for persistent organic pollutants (POPs) in wildlife. Rev Environ Contam Toxicol 227:107–155. https://doi.org/10.1007/978-3-319-01327-5_4

De Hoop L, De Troch M, Hendriks AJ, De Laender F (2013) Modeling toxic stress by atrazine in a marine consumer-resource system. Environ Toxicol Chem 32:1088–1095. https://doi.org/10.1002/etc.2160

de Souza CP, de Andrade Guedes T, Fontanetti CS (2016) Evaluation of herbicides action on plant bioindicators by genetic biomarkers: a review. Environ Monit Assess 188(12):694. https://doi.org/10.1007/s10661-016-5702-8

De Temmerman L, Nigel J, Garrec JP, Klumpp A, Krause GHM, Tornneijck EG (2004) Biomonitoring of air pollutants with plants—considerations for the future. In: Klumpp A, Ansel W, Klumpp G (eds) Urban air pollution bioindication and environmental awareness. Cuvillier Verlag, Göttingen

Debouge M, Thome J (1989) Dispersoin of Linade on 5 ant species IN Belgium (Wallone) and accumulation in Formica polyctena. Ann Sci Nat Zool Biol Anim 10:25–30

Debouge MH, Thomé JP, Jeuniaux C (1987) Bioaccumulation de trois insecticides organochlores (Lindane, Dieldrine et DDT) et des PCB chez plusieurs especes de fourmis [Hymenoptera-formicidae] en Belgique. Entomophaga 32(5):551–561. https://doi.org/10.1007/BF02373524

Della Vechia JF, Cruz C, Silva AF, Cerveirajr WR, Garlich N (2016) Macrophyte bioassay applications for monitoring pesticides in the aquatic environment. Planta Daninha 34(3):3. https://doi.org/10.1590/s0100-83582016340300021

DeLorenzo ME, Lauth J, Pennington PL, Scott GI, Ross PE (1999) Atrazine effects on the microbial food webin tidal creek mesocosms. Aquat Toxicol 46:241–251. https://doi.org/10.1016/S0166-445X(98)00132-5

DeLorenzo ME, Scott GI, Ross PE (2001) Toxicity of pesticides to aquatic microorganisms: a review. Environ Toxicol Chem 20:84–98. https://doi.org/10.1002/etc.5620200108

Denoyelle R, Rault M, Mazzia C, Mascle O, Capowiez Y (2007) Cholinesterase activity as a biomarker of pesticide exposure in *Allolobophora chlorotica* earthworms living in apple orchards under different management strategies. Environ Toxicol Chem 26(12):2644–2649. https://doi.org/10.1897/06-355.1

Deutsch CA, Tewksbury JJ, Tigchelaar M, Battisti DS, Merrill SC, Huey RB, Naylor RL (2018) Increase in crop losses to insect pests in a warming climate. Science 361(6405):916–919. https://doi.org/10.1126/science.aat3466

Dieter CD, Duffy WG, Flake LD (1996) The effect of phorate on wetland macroinvertebrates. Environ Toxicol 15:308–312. https://doi.org/10.1002/etc.5620150313

Digrak M, Kazanici F (2001) Effect of some organophosphorus insecticides on soil microorganisms. Turk J Biol 25:51–58

Dittbrenner N, Schmitt H, Capowiez Y, Triebskorn R (2011) Sensitivity of *Eisenia fetida* in comparison to *Aporrectodea caliginosa* and *Lumbricus terrestris* after imidacloprid exposure. Body mass change and histopathology. J Soils Sedim 11(6):1000–1010. https://doi.org/10.1007/s11368-011-0397-5

Downing HF, Delorenzo ME, Fulton MH, Scott GI, Madden CJ, Kucklick JR (2004) Effects of the agricultural pesticides atrazine, chlorothalonil and endosulfan on South Florida microbial assemblages. Ecotoxicology 13:245–260. https://doi.org/10.1023/B:ECTX.0000023569.46544.9f

EEA (2013) Late lessons from early warnings: science, precaution, innovation. European Environment Agency, report no 1/2013. EEA, Copenhagen

Espín S, Martínez-López E, María-Mojica P, García-Fernández AJ (2012) Razorbill (*Alca torda*) feathers as an alternative tool for evaluating exposure to organochlorine pesticides. Ecotoxicology 21(1):183–190. https://doi.org/10.1007/s10646-011-0777-z

Espín S, García-Fernández AJ, Herzke D, Shore RF, van Hattum B, Martínez-López E, Coeurdassier M, Eulaers I, Fritsch C, Gómez-Ramírez P, Jaspers VL (2016) Tracking pan-continental trends in environmental contamination using sentinel raptors—what types of samples should we use? Ecotoxicology 25(4):777–801. https://doi.org/10.1007/s10646-016-1636-8

Eulaers I, Jaspers VLB, Bustnes JO, Covaci A, Johnsen TV, Halley DJ, Moum T, Ims RA, Hanssen SA, Erikstad KE, Herzke D, Sonne C, Ballesteros M, Pinxten R, Eens M (2013) Ecological and spatial factors drive intra- and interspecific variation in exposure of subarctic predatory bird nestlings to persistent organic pollutants. Environ Int 57-58:25–33. https://doi.org/10.1016/j.envint.2013.03.009

FAOSTAT (2017) Pesticides use. Available online at: http://www.fao.org/faostat/en/#data/RP. Accessed Aug 2018

Ferronato G, Viera MS, Prestes OD, Adaime MB, Zanella R (2018) Determination of organochlorine pesticides (OCPs) in breast milk from Rio Grande do Sul, Brazil, using a modified QuEChERS method and gas chromatography-negative chemical ionisation-mass spectrometry. Int J Environ Anal Chem 98(11):1005–1016. https://doi.org/10.1080/03067319.2018.1518441

Filimonova V, Goncalves F, Marques JC, De Troch M, Ana Goncalves MM (2016) Fatty acid profiling as bioindicator of chemical stress in marineorganisms: a review. Ecol Indic 67:657–672. https://doi.org/10.1016/j.ecolind.2016.03.044

Focardi S, Gaggi C, Chemello G, Bacci E (1991) Organochlorine residues in moss and lichen samples from two Antarctic areas. Polar Rec 27(162):241–244. https://doi.org/10.1017/S003224740001264X

Fossi MC, Minutoli R, Guglielmo L (2001) Preliminary results of biomarker responses in zooplankton of brackish environments. Mar Pollut Bull 42(9):745–748. https://doi.org/10.1016/S0025-326X(00)00214-9

Furness RW, Greenwood JJD (1993) Birds as monitors of environmental change. Chapman and Hall, London, 356 pp

Garbarino JR, Snyder-Conn E, Leiker TJ, Hoffman GL (2002) Contaminants in arctic snow collected over northwest Alaskan sea ice. Water Air Soil Pollut 139:183–214. https://doi.org/10.1023/A:1015808008298

Garry VF, Harkins ME, Erickson LL, Long-Simpson LK, Holland SE, Burroughs BL (2002) Birth defects, season of conception, and sex of children born to pesticide applicators living in the Red River Valley of Minnesota, USA. Environ Health Perspect 110(3):441–449. https://doi.org/10.1289/ehp.02110s3441

Geras'kin SA, Kim JK, Dikarev VG, Oudalova AA, Dikareva NS, Spirin YV (2005) Cytogenetic effects of combined radioactive (^{137}Cs) and chemical (Cd, Pb, and 2,4-D herbicide)

contamination on spring barley intercalary meristem cells. Mutat Res 586:147–159. https://doi.org/10.1016/j.mrgentox.2005.06.004

Ghini S, Fernandez M, Pico Y, Marin R, Fini F, Manes J, Girotti S (2004) Occurrence and distribution of pesticides in the province of Bologna, Italy, using honeybees as bioindicators. Arch Environ Contam Toxicol 47(4):479–488. https://doi.org/10.1007/s00244-003-3219-y

Ghosh RK, Singh N (2012) Managing metolachlor and atrazine leaching losses using lignite fly ash. Ecotoxicol Environ Saf 84:243–248. https://doi.org/10.1016/j.ecoenv.2012.07.015

Ghosh RK, Singh N (2013) Adsorption–desorption of metolachlor and atrazine in Indian soils: effect of fly ash amendment. Environ Monit Assess 185:1833–1845. https://doi.org/10.1007/s10661-012-2671-4

Giesy JP, Verbrugge DA, Othout RA, Bowerman WW, Mora MA, Jones PD, Newsted JL, Vandervoort C, Heaton SN, Aulerich RJ, Bursian SJ, Ludwig JP, Dawson GA, Kubiak TJ, Best DA, Tillitt DE (1994) Contaminants in fishes from Great Lakes-influenced sections and above dams of three Michigan rivers. II: Implications for health of mink. Arch Environ Contam Toxicol 27:213–223. https://doi.org/10.1007/BF00214265

Gilliom RJ (2007) Pesticides in U.S. streams and groundwater. Environ Sci Technol 41(10):3408–3414. https://doi.org/10.1021/es072531u

Givaudan N, Binet F, Le Bot B, Wieg C (2014) Earthworm tolerance to residual agricultural pesticide contamination: field and experimental assessment of detoxification capabilities. Environ Pollut 192:9–18. https://doi.org/10.1016/j.envpol.2014.05.001

Godduhn A, Duffy LK (2003) Multi-generation health risks of persistent organic pollution in the far north: use of the precautionary approach in the Stockholm Convention. Environ Sci Pol 6(4):341–353. https://doi.org/10.1016/S1462-9011(03)00061-3

Gómez-Pérez ML, Plaza-Bolaños P, Romero-González R, Martínez-Vidal JL, Garrido-Frenich A (2012) Comprehensive qualitative and quantitative determination of pesticides and veterinary drugs in honey using liquid chromatography–Orbitrap high resolution mass spectrometry. J Chromatogr A 27(1248):130–138. https://doi.org/10.1016/j.chroma.2012.05.088

Gorzerino C, Quemeneur A, Hillenweck A, Baradat M, Delous G, Ollitrault M, Azam D, Caquet T, Lagadic L (2009) Effects of diquat and fomesafen applied alone and in combination with a nonylphenol polyethoxylate adjuvant on *Lemna minor* in aquatic indoor microcosms. Ecotoxicol Environ Saf 72(3):802–810. https://doi.org/10.1016/j.ecoenv.2008.08.001

Green AJ, Elmberg J (2014) Ecosystem services provided by waterbirds. Biol Rev 89:105–122

Guillette EA, Meza MM, Aquilar MG, Soto AD, Garcia IE (1998) An anthropological approach to the evaluation of preschool children exposed to pesticides in Mexico. Environ Health Perspect 106(6):347–353. https://doi.org/10.1289/ehp.98106347

Guillette EA, Conard C, Lares F, Aguilar MA, McLachlan J, Guillette LJ Jr (2006) Altered breast development in young girls from an agricultural environment. Environ Health Perspect 114(3):471–475. https://doi.org/10.1289/ehp.8280

Guitart R, Riu JL, Puigdemont A, Arboix M (1990) Organochlorine residues in adipose tissue of chamois from the Catalan Pyrenees, Spain. Bull Environ Contam Toxicol 44:555–560. https://doi.org/10.1007/BF01700875

Gutierrez MF, Paggi JC, Gagneten AM (2012) Microcrustaceans escape behavior as an early bioindicator of copper, chromium and endosulfan toxicity. Ecotoxicology 21:428–438. https://doi.org/10.1007/s10646-011-0803-1

Hanedar A, Güneş E, Kaykioğlu G, Çelik SÖ, Cabi E (2019) Presence and distributions of POPS in soil, atmospheric deposition, and bioindicator samples in an industrial-agricultural area in Turkey. Environ Monit Assess 191(1):42. https://doi.org/10.1007/s10661-018-7159-4

Hans W, Dyble PJ, Moisander PH, Noble RT, Piehler MF, Pinckney JL, Steppe TF, Twomey L, Valdes LM (2003) Microbial indicators of aquatic ecosystem change: current applications to eutrophication studies. FEMS Microbiol Ecol 46:233–246. https://doi.org/10.1016/S0168-6496(03)00200-9

Hardell L, Eriksson M (1999) A case–control study of non-Hodgkin lymphoma and exposure to pesticides. Cancer 85(6):1353–1360

Harding LE, Harris ML, Stephen CR, Elliott JE (1999) Reproductive and morphological condition of wild mink (*Mustela vison*) and river otters (*Lutra canadensis*) in relation to chlorinated hydrocarbon contamination. Environ Health Perspect 107:141–147. https://doi.org/10.1289/ehp.99107141

Harmens H, Foan L, Simon V, Mills G (2011) Mosses as biomonitors of atmospheric POPs pollution: a review. Report for Defra contract AQ08610

He C, Jin J, Xiang BL, Wang Y, Ma ZH (2014) Upper Yellow River air concentrations of organochlorine pesticides estimated from tree bark, and their relationship with socioeconomic indices. J Environ Sci China 26:593–600. https://doi.org/10.1016/S1001-0742(13)60455-6

Heino J (2009) Biodiversity of aquatic insects: spatial gradients and environmental correlates of assemblage-level measures at large scales. Fr Rev 2:1–29. https://doi.org/10.1608/FRJ-2.1.1

Henderson RJ, Mackinlay EE, Hodgson P, Harwood JL (1990) Differential-effects of the substituted pyridazinone herbicide Sandoz 9785 on lipid-composition and biosynthesis in photosynthetic and nonphotosynthetic marine microalgae: II. Fatty acid composition. J Exp Bot 41(227):729–736. https://doi.org/10.1093/jxb/41.6.729

Hershey AE, Lamberti GA, Chaloner DT, Northington RM (2010) Aquatic insect ecology. In: Thorp JH, Covich AP (eds) Ecology and classification of North American freshwater invertebrates. Academic, London, pp 659–694

Hoffmann BD, Andersen AN (2003) Responses of ants to disturbance in Australia, with particular reference to functional groups. Austral Ecol 28(4):444–464. https://doi.org/10.1046/j.1442-9993.2003.01301.x

Huang J, Qiao F, Zhang L, Rozelle S (2000) Farm pesticide, rice production, and human health. CCAP's Project Report 11

Hunsaker CT, Carpenter DE (eds) (1990) Environmental monitoring and assessment program: ecological indicators. Office of Research and Development, United States Environmental Protection Agency, Research Triangle Park, NC

Huseth AS, Groves RL (2014) Environmental fate of soil applied neonicotinoid insecticides in an irrigated potato agroecosystem. PLoS One 9(5):1–11. https://doi.org/10.1371/journal.pone.0097081

Ibekwe AM, Papiernik SK, Gan JY, Yates SR, Yang CH, Crowley DE (2001) Impact of fumigants on soil microbial communities. Appl Environ Microbiol 67(7):3245–3257. https://doi.org/10.1128/AEM.67.7.3245-3257.2001

Iqbal Z, Hussain A, Latif A, Asi MR, Chaudhary JA (2001) Impact of pesticide applications in cotton agroecosystem and soil bioactivity studies I: Microbial populations. J Biol Sci 1(7):640–644. https://doi.org/10.3923/jbs.2001.640.644

Iwaniuk AN, Koperski DT, Cheng KM, Elliott JE, Smith LK, Wilson LK, Wylie DR (2006) The effects of environmental exposure to DDT on the brain of a songbird: changes in structures associated with mating and song. Behav Brain Res 173:1–10. https://doi.org/10.1016/j.bbr.2006.05.026

Jain A, Singh BN, Singh SP, Singh HB, Singh S (2010) Exploring biodiversity as bioindicators for water pollution. In: National conference on Biodiversity, Development and Poverty Alleviation. Uttar Pradesh State Biodiversity Board, Lucknow

Jaspers VLB, Covaci A, Deleu P, Eens M (2009) Concentrations in bird feathers reflect regional contamination with organic pollutants. Sci Total Environ 407:1447–1451. https://doi.org/10.1016/j.scitotenv.2008.10.030

Jayaraj R, Megha P, Sreedev P (2016) Organochlorine pesticides, their toxic effects on living organisms and their fate in the environment. Interdiscip Toxicol 9(3–4):90–100. https://doi.org/10.1515/intox-2016-0012

Jena PK, Adhya TK, Rao VR (1990) Nitrogen-fixing bacterial population as influenced by butachlor and thiobencarb in rice soils. Zentralbl Mikrobiol 145:457–460. https://doi.org/10.1016/S0232-4393(11)80165-4

Johnsen K, Jacobsen CS, Torsvik V, Sørensen J (2001) Pesticide effects on bacterial diversity in agricultural soils – a review. Biol Fertil Soils 33:443–453. https://doi.org/10.1007/s003740100351

Josephine AA, Chbani A, Faljoun Z, Millet M (2019) The use of vegetation, bees, and snails as important tools for the biomonitoring of atmospheric pollution – a review. Environ Sci Pollut Res 26(10):9391–9408. https://doi.org/10.1007/s11356-019-04388-8

Kachroo D, Jolly SMS, Ramamurthy V (2006) Modulation of unsaturated fattyacids content in algae *Spirulina platensis* and *Chlorella minutissima* in response to herbicide SAN 9785. Electron J Biotechnol 9:4. https://doi.org/10.4067/S0717-34582006000400006

Karner I (1978) The brown hare as an indicator of pollution with pesticides (*Lepus europaeus* Pall.) Mohacs Island, Tompa, Fuzesgyarmat, Martonvasar, Lajta-Hansag region, Poroszlo, Dombrad; Hungary. Allattani Kozlemenyek (Hungary)

Karr JR, Yant PR, Fausch KD, Schlosser IJ (1987) Spatial and temporal variability of the index of biotic integrity in three midwestern streams. Trans Am Fish Soc 116:1–11. https://doi.org/10.1577/1548-8659(1987)116<1:SATVOT>2.0.CO;2

Khatri N, Tyagi S (2015) Influences of natural and anthropogenic factors on surface and groundwater quality in rural and urban areas. Front Life Sci 8(1):23–39. https://doi.org/10.1080/21553769.2014.933716

Kielak E, Sempruch C, Mioduszewska H, Klocek J, Leszczynski B (2011) Phytotoxicity of Roundup Ultra 360 SL in aquatic ecosystems: biochemical evaluation with duckweed (*Lemna minor* L.) as a model plant. Pestic Biochem Physiol 99:237–243. https://doi.org/10.1016/j.pestbp.2011.01.002

Killham KS, Staddon W (2002) Bioindicators and sensors of soil health and the application of geostatistics. In: Burns RG, Dick RP (eds) Enzymes in the environment. Marcel Dekker, New York, pp 391–406

Klemm DJ, Lewis PA, Fulk F, Lazorchak JM (1990) Macroinvertebrate field and laboratory methods for evaluating the biological integrity of surface waters, EPA.600/4-90/030. Environmental Monitoring Systems Laboratory, Office of Modeling, Monitoring Systems, and Quality Assurance, Office of Research and Development, U.S. Environmental Protection Agency, Cincinnati, Ohio

Kobashi K, Harada T, Adachi Y, Mori M, Ihara M, Hayasaka D (2017) Comparative ecotoxicity of imidacloprid and dinotefuran to aquatic insects in rice mesocosms. Ecotoxicol Environ Saf 138:122–129. https://doi.org/10.1016/j.ecoenv.2016.12.025

Kujawski MW, Namieśnik J (2011) Levels of 13 multi-class pesticide residues in Polish honeys determined by LC-ESI-MS/MS. Food Control 22(6):914–919. https://doi.org/10.1016/j.foodcont.2010.11.024

Kumaraswamy S, Rath AK, Satpathy SN, Ramakrishnan B, Adhya TK, Sethunathan N (1998) Influence of the carbofuran on the production and oxidation of methane in a flooded rice soil. Biol Fertil Soil 25:362–366. https://doi.org/10.1007/s003740050389

Kwon TS, Song MY, Shin SC, Park YS (2005) Effects of aerial insecticide sprays on ant communities to control pine wilt disease in Korean pine forests. Appl Entomol Zool 40(45):63–574. https://doi.org/10.1303/aez.2005.563

Lal R, Lal S (1988) Pesticides and nitrogen cycle, vol 3. CRC, Boca Raton

Landres PB, Verner J, Thomas JW (1988) Ecological uses of vertebrate indicator species: a critique. Conserv Biol 2:316–328. https://doi.org/10.1111/j.1523-1739.1988.tb00195.x

Lehmann E, Oltramare C, De Alencastro LF (2018) Development of a modified QuEChERS method for multi-class pesticide analysis in human hair by GC-MS and UPLC-MS/MS. Anal Chim Acta 999:87–98. https://doi.org/10.1016/j.aca.2017.11.009

Leonard PM, Orth DJ (1986) Application and testing of an index of biotic integrity in small, cool water streams. Trans Am Fish Soc 115:401–414. https://doi.org/10.1577/1548-8659(1986)115<401:AATOAI>2.0.CO;2

Li R, Jin J (2013) Modelling of temporal patterns and sources of atmospherically transported and deposited pesticides in ecosystems of concern: a case study of toxaphene in the Great Lakes. J Geophys Res Atmos 118:11863–11874. https://doi.org/10.1002/jgrd.50777

Lim TB, Xu R, Tan B, Obbard JP (2006) Persistent organic pollutants in moss as bioindicators of atmospheric pollution in Singapore. Chemosphere 64(4):596–602. https://doi.org/10.1016/j.chemosphere.2005.11.007

Liu WX, Wang Y, He W, Qin N, Kong XZ, He QS, Yang B, Yang C, Jiang YJ, Jorgensen SE, Xu FL (2016) Aquatic biota as potential biological indicators of the contamination, bioaccumulation and health risks caused by organochlorine pesticides in a large, shallow Chinese lake (Lake Chaohu). Ecol Indic 60:335–345. https://doi.org/10.1016/j.ecolind.2015.06.026

Low-Goh NK, Edwards PS, Frost S, Thomas MP (1981) The effect of abate on the growth of *Vibrio natriegens* and *Chlorella vulgaris*. Int J Environ Stud 17:135–139. https://doi.org/10.1080/00207238108709898

Ma TH (1999) The international program on plant bioassays and the report of the follow-up study after the hands-on workshop in China. Mutat Res 426:103–106. https://doi.org/10.1016/s0027-5107(99)00049-4

Mallavarapu M (2002) Heavy pesticide use lowers soil health. Farm Head 121:37–38

Markert B, Oehlmann J, Roth M (1997) General aspects of heavy metal monitoring by plants and animals. In: Subramanian G, Iyengar V (eds) Environmental biomonitoring – exposure assessment and specimen banking. ACS symposium series, vol 654. American Chemical Society, Washington, DC

Martínez-López E, Espín S, Barbar F, Lambertucci SA, Gómez-Ramírez P, García-Fernández AJ (2015) Contaminants in the southern tip of South America: analysis of organochlorine compounds in feathers of avian scavengers from Argentinean Patagonia. Ecotoxicol Environ Saf 115:83–92. https://doi.org/10.1016/j.ecoenv.2015.02.011

Mateo R, Millán J, Rodríguez-Estival J, Camarero PR, Palomares F, Ortiz-Santaliestra ME (2012) Levels of organochlorine pesticides and polychlorinated biphenyls in the critically endangered *Iberian lynx* and other sympatric carnivores in Spain. Chemosphere 86(7):691–700

Matlock RB Jr, de la Cruz R (2003) Ants as indicators of pesticide impacts in banana. Environ Entomol Aug 32(4):816–829. https://doi.org/10.1016/j.chemosphere.2011.10.037

Menzel D, Anderson W, Judith RA (1970) Marine phytoplankton vary in their response to chlorinated hydrocarbons. Science 167:1724–1726. https://doi.org/10.1126/science.167.3926.1724

Metcheva R, Beltcheva M, Kalinova G, Marinova M, Antonio J, Rojas H, Peneva V (2017) Organochlorine pesticides in feathers of penguins of the family pygoscelidae from Livingston and Peterman Islands, Western Antarctica. Acta Zool Bulg 8:183–188

Michereff-Filho M, Guedes RNC, Della-Lucia TMC, Michereff MFF, Cruz I (2004) Non-target impact of chlorpyrifos on soil arthropods associated with no-tillage cornfields in Brazil. Int J Pest Manag 50(2):91–99. https://doi.org/10.1080/09670870410001655885

Mohapatra PK, Mohanty RC (1992) Growth pattern changes of *Chlorella vulgaris* and *Anabaena doliolum*due to toxicity of dimethoate and endosulfan. Bull Environ Contam Toxicol 49:576–581. https://doi.org/10.1007/BF00196301

Moharram AM, Abdel-Malek AY, Abdel-Kader MIA, Omar SO (1994) Effect of three pesticides on nitrogen content of some soil fungi. J Islam Acad Sci 7(2):82–87

Morrissey CA, Mineau P, Devries JH, Sanchez-Bayo F, Liess M, Cavallaro MC, Liber K (2015) Neonicotinoid contamination of global surface waters and associated risk to aquatic invertebrates: a review. Environ Int 74:291–303. https://doi.org/10.1016/j.envint.2014.10.024

Morse JC, Bae YJ, Munkhjargal G, Sangpradub N, Tanida K, Vshivkova TS, Wang B, Yang L, Yule CM (2007) Freshwater biomonitoring with macroinvertebrates in East Asia. Front Ecol Environ 5(1):33–42. https://doi.org/10.1890/1540-9295

Mullin CA, Frazier M, Frazier JL, Ashcraft S, Simonds R, Pettis JS (2010) High levels of miticides and agrochemicals in North American apiaries: implications for honey bee health. PLoS One 5: e9754. https://doi.org/10.1371/journal.pone.0009754

Murphy SD (1986) Toxic effects of pesticides. In: Klaassen CD, Amdur MO, Doull J (eds) Casarett and Doull's toxicology: the basic science of poisons. Macmillan, New York, pp 519–581

Nair GA (1981) Toxic effects of certain biocides on a fresh water mite, *Hydrachna trilobata* Viets (*Arachnida: Hydrachnoidea*: Hydrachnidae). J Environ Biol 2:91–96

Naso B, Zaccaroni A, Perrone D, Ferrante MC, Severino L, Stracciari GL, Lucisano A (2004) Organochlorine pesticides and polychlorinated biphenyls in European roe deer Capreolus capreolus resident in a protected area in Northern Italy. Sci Tot Environ 328(1–3):83–93. https://doi.org/10.1016/j.scitotenv.2004.02.006

Niell S, Jesus F, Pérez C, Mendoza Y, Diaz R, Franco J, Cesio V, Heinzen H (2015) QuEChERS adaptability for the analysis of pesticide residues in beehive products seeking the development of an agroecosystem sustainability monitor. J Agric Food Chem 63:4484–4492. https://doi.org/10.1021/acs.jafc.5b00795

Niu L, Xu C, Xu Y, Zhang C, Liu W (2014) Hexachlorocyclohexanes in tree bark across Chinese agricultural regions: spatial distribution and enantiomeric signatures. Environ Sci Technol 48(20):12031–12038. https://doi.org/10.1021/es503372g

Odokuma LO, Osuagwu C (2004) Tolerance of chemolithotrophic bacteria to organochlorine, organophosphate and carbamate pesticides. J Agric Environ Eng Technol 1(1):7–15

OECD (2003) Descriptions of selected key generic terms used in chemical hazard/risk assessment. OECD Series on Testing and Assessment, no. 44

OECD TN (1984) Earthworm, acute toxicity tests. OECD guidelines for the testing of chemicals. Organization for Economic Co-operation and Development, Paris

Oerke E-C (2006) Crop losses to pests. J Agric Sci 144:31–43. https://doi.org/10.1017/S0021859605005708

Oerke E-C, Dehne H-W, Scho¨nbeck F, Weber A (1994) Crop production and crop protection – estimated losses in major food and cash crops. Elsevier Science, Amsterdam

Ojo OA, Adebayo TA, Olaniran OA (2007) Biological effects of four fungicides on soil microbial population. Res J Agron 1(1):33–37

Oluah MS, Obiezue RNN, Ochulor AJ, Onuoha E (2010) Toxicity and histopathological effect of atrazine (herbicide) on the earthworm *Nsukkadrilus mbae* under laboratory conditions. Anim Res Int 7(3):1287–1293

Ortiz-Santaliestra ME, Tauler-Ametller H, Lacorte S, Hernández-Matías A, Real J, Mateo R (2019) Accumulation of pollutants in nestlings of an endangered avian scavenger related to territory urbanization and physiological biomarkers. Environ Pollut Part B 252:1801–1809. https://doi.org/10.1016/j.envpol.2019.06.101

Ostrea EM Jr, Bielawski DM, Posecion NC Jr, Corrion M, Villanueva-Uy E, Bernardo RC, Jin Y, Janisse JJ, Ager JW (2009) Combined analysis of prenatal (maternal hair and blood) and neonatal (infant hair, cord blood and meconium) matrices to detect fetal exposure to environmental pesticides. Environ Res 109:116–122. https://doi.org/10.1016/j.envres.2008.09.004

Paradis D, Bérail G, Bonmatin JM, Belzunces LP (2014) Sensitive analytical methods for 22 relevant insecticides of 3 chemical families in honey by GC-MS/MS and LC-MS/MS. Anal Bioanal Chem 406(2):621–633. https://doi.org/10.1007/s00216-013-7483-z

Patlak M (1996) Estrogens may link pesticides, breast cancer. Environ Sci Technol 30(5):210A–211A. https://doi.org/10.1021/es962228l

Peck SL, Mcquaid B, Campbell CL (1998) Using ant species (Hymenoptera: Formicidae) as a biological indicator of agroecosystem condition. Environ Entomol 27(5):1102–1110. https://doi.org/10.1093/ee/27.5.1102

Pelosi C, Barot S, Capowiez Y, Hedde M, Vandenbulcke F (2014) Pesticides and earthworms. Agron Sustain Dev 34(1):199–228. https://doi.org/10.1007/s13593-013-0151-z

Pereira JL, da Silva AA, PicanÇco MC, Barros EC, Jakelaitis A (2005) Effects of herbicide and insecticide interaction on soil entomofauna under maize crop. J Environ Sci Health B 40:45–54. https://doi.org/10.1081/PFC-200034212

Perfecto I (1990) Indirect and direct effects in a tropical agroecosystem: the maize-pest-ant system in Nicaragua. Ecology 71(6):2125–2134. https://doi.org/10.2307/1938626

Perfecto I (1994) Foraging behavior as a determinant of asymmetric competitive interaction between two ant species in a tropical agroecosystem. Oecologia 98(2):184–192. https://doi.org/10.1007/BF00341471

Perugini M, Tulini SMR, Zezza D, Fenucci S, Conte A, Amorena M (2018) Occurrence of agrochemical residues in beeswax samples collected in Italy during 2013–2015. Sci Total Environ 625:470–476. https://doi.org/10.1016/j.scitotenv.2017.12.321

Pesce S, Fajon C, Bardot C, Bonnemoy F, Portelli C, Bohatier J (2006) Effects of the phenylurea herbicide diuron on natural rivierine microbial communities in an experimental study. Aquat Toxicol 78:303–314. https://doi.org/10.1016/j.aquatox.2006.03.006

Petal J (1978) The role of ants in ecosystems. In: Brian MV (ed) Production ecology of ants and termites. International biology programme, no. 13. Cambridge University Press, New York, pp 292–325

Pipe AE, Cullimore DR (1984) Influence of five phenylurea herbicides on the diatom Hantzschia in a sandy loam soil. Bull Environ Contam Toxicol 33:439–443. https://doi.org/10.1007/BF01625567

Pisa LW, Amaral-Rogers V, Belzunces LP, Bonmatin JM, Downs CA, Goulson D, Kreutzweiser DP, Krupke C, Liess M, McField M, Morrissey CA, Noome DA, Settele J, Simon-Delso N, Stark JD, Van der Sluijs JP, Van Dyck H, Wiemers M (2015) Effects of neonicotinoids and fipronil on non-target invertebrates. Environ Sci Pollut Res 22:68–102. https://doi.org/10.1007/s11356-014-3471-x

Plafkin JL, Barbour MT, Porter KD, Gross SK, Hughes RM (1989) Rapid bioassessment protocols for use in streams and rivers: benthic macroinvertebrates and fish. EPA/444/4-89-001 U.S. Environmental Protection Agency, Washington, DC

Porrini C, Sabatini AG, Girotti S, Fini F, Monaco L, Celli G, Bortolotti L, Ghini S (2003) The death of honey bees and environmental pollution by pesticides: the honey bees as biological indicators. Bull Insectol l56(1):147–152

Qu H, Wang P, Ma RX, Qiu XX, Xu P, Zhou ZQ, Liu DH (2014) Enantioselective toxicity, bioaccumulation and degradation of the chiral insecticide fipronil in earthworms (*Eisenia feotida*). Sci Total Environ 485:415–420. https://doi.org/10.1016/j.scitotenv.2014.03.054

Ramchandra TV, Rishiram R, Karthik B (2006) Zooplanktons as bioindicators: hydro biological investigation in selected Bangalore lakes. Tech Rep 115

Ramesh A, Ravi PE (2004) Electron ionization gas chromatography-mass spectrometric determination of residues of thirteen pyrethroid insecticides in whole blood. J Chromatogr B 802:371–376. https://doi.org/10.1016/j.jchromb.2003.12.016

Rank J, Nielsen MH (1994) Evaluation of the *Allium* anaphase-telophase test in relation to genotoxicity screening of industrial wastewater. Mutat Res 312(10):17–24. https://doi.org/10.1016/0165-1161(94)90004-3

Ratcliffe DA (1967) Decrease in eggshell weight in certain birds of prey. Nature 215:208–210

Ravera O (1998) Utility and limits of biological and chemical monitoring of the aquatic environment. Ann Chim 88(11–12):909–913. https://doi.org/10.1038/215208a0

Relyea RA (2005) The lethal impacts of roundup and predatory stress on six species of North American tadpoles. Arch Environ Contam Toxicol 48:351–357. https://doi.org/10.1007/s00244-004-0086-0

Repetto R, Baliga S (1996) Pesticides and the immune system: the public health risks. World Resources Institute, Washington, DC, p 100

Rivera-Rodríguez LB, Rodríguez-Estrella R, Ellington JJ, Evans JJ (2007) Quantification of low levels of organochlorine pesticides using small volumes ($\leq$100 μl) of plasma of wild birds through gas chromatography negative chemical ionization mass spectrometry. Environ Pollut 148:654–662. https://doi.org/10.1016/j.envpol.2006.11.018

Rizhiya E, Bertora C, van Vliet PC, Kuikman PJ, Faber JH, van Groenigen JW (2007) Earthworm activity as a determinant for N_2O emission from crop residue. Soil Biol Biochem 39(8):2058–2069. https://doi.org/10.1016/j.soilbio.2007.03.008

Romanić SH, Klinčić D (2012) Organochlorine compounds in pine needles from Croatia. Bull Environ Contam Toxicol 88(6):838–841. https://doi.org/10.1007/s00128-012-0575-1

Ruiz-Toledo J, Vandame R, Castro-Chan R, Penilla-Navarro R, Gómez J, Sánchez D (2018) Organochlorine pesticides in honey and pollen samples from managed colonies of the honey bee *Apis mellifera* linnaeus and the stingless bee *Scaptotrigona mexicana* guérin from Southern, Mexico. Insects 9(2):54. https://doi.org/10.3390/insects9020054

Rutkowska M, Płotka-Wasylka J, Lubinska-Szczygeł M, Różańska A, Możejko-Ciesielska J, Namieśnik J (2018) Birds' feathers–suitable samples for determination of environmental pollutants. TrAC Trend Anal Chem 109:97–115. https://doi.org/10.1016/j.trac.2018.09.022

Saeed MF, Shaheen M, Ahmad I, Zakir A, Nadeem M, Chishti AA, Shahid M, Bakhsh K, Damalas CA (2017) Pesticide exposure in the local community of Vehari District in Pakistan: an assessment of knowledge and residues in human blood. Sci Total Environ 587:137–144. https://doi.org/10.1016/j.scitotenv.2017.02.086

Salquèbre G, Schummer C, Millet M, Briand O, Appenzeller BM (2012) Multi-class pesticide analysis in human hair by gas chromatography tandem (triple quadrupole) mass spectrometry with solid phase microextraction and liquid injection. Anal Chim Acta 710:65–74. https://doi.org/10.1016/j.aca.2011.10.029

Sanborn M, Cole D, Kerr K, Vakil C, Sanin LH, Bassil K (2004) Pesticides literature review: systematic review of pesticide human health effects. The Ontario College of Family Physicians, Toronto

Sanchez-Hernandez JC, Del Pino JN, Capowiez Y, Mazzia C, Rault M (2018) Soil enzyme dynamics in chlorpyrifos-treated soils under the influence of earthworms. Sci Total Environ 612:1407–1416. https://doi.org/10.1016/j.scitotenv.2017.09.043

Sanchez-Muros MJ, Villacreces S, Miranda-de la Lama G, de Haro C, Garcia-Barroso F (2013) Effects of chemical and handling exposure on fatty acids, oxidative stress and morphological welfare indicators in gilt-head sea bream (*Sparus aurata*). Fish Physiol Biochem 39:581–591. https://doi.org/10.1007/s10695-012-9721-2

Schiemer F, Keckeis H, Flore L (2001) Ecotones and hydrology: key conditions for fish in large rivers. Ecohydrol Hydrobiol 1(1):49–55

Schnug L, Leinaas HP, Jensen J (2014) Synergistic sub-lethal effects of a biocide mixture on the springtail Folsomia finlet. Environ Pollut 64–186:158. https://doi.org/10.1016/j.envpol.2013.12.004

Schread JC (1948) Control of sail insects. J Econ Entomol 41:314–318

Schreinemachers DM (2000) Cancer mortality in four northern wheat-producing states. Environ Health Perspect 108(9):873–881. https://doi.org/10.1289/ehp.00108873

Schulz H, Popp P, Huhn G, Stärk HJ, Schüürmann G (1999) Biomonitoring of airborne inorganic and organic pollutants by means of pine tree barks. I. Temporal and spatial variations. Sci Total Environ 232:49–58. https://doi.org/10.1016/s0048-9697(99)00109-6

Sharma CB, Panneerselvam N, Ma TH (1990) Genetic toxicology of pesticides in higher plant systems. Crit Rev Plant Sci 9(5):409–442. https://doi.org/10.1080/07352689009382299

Shore RF, Casulli A, Bologov V, Wienburg CL, Afsar A, Toyne P, Dell'Omo G (2001) Organochlorine pesticide, polychlorinated biphenyl and heavy metal concentrations in wolves (*Canis lupus* L. 1758) from north-west Russia. Sci Total Environ 280(1–3):45–54. https://doi.org/10.1016/s0048-9697(01)00802-6

Sicko-Goad L, Lazinsky D, Hall J, Simmons MS (1989a) Effects of chlorinated benzenes on diatom fatty acid composition and quantitative morphology. Arch Environ Contam Toxicol 18:629–663. https://doi.org/10.1007/BF01225001

Sicko-Goad L, Hall J, Lazinsky D, Simmons MS (1989b) Effects of chlorinated benzenes on diatom fatty acid composition and quantitative morphology. II.1,3,5-trichlorobenzene. Arch Environ Contam Toxicol 18:638–646. https://doi.org/10.1007/BF01225003

Silva AF, Cruz C, Neto NA, Pitelli RA (2012) Ecotoxicity of herbicides for the aquatic macrophyte (*Azolla caroliniana*). Planta Daninha 30(3):541–546. https://doi.org/10.1590/S0100-83582012000300009

Simonich SL, Hites RA (1995) Global distribution of persistent organochlorine compounds. Science 269:1851–1854. https://doi.org/10.1126/science.7569923

Slavikova E, Vadkertiova R (2003) Effects of pesticides on yeasts isolated from agricultural soil. Z Naturforsch 58:855–859. https://doi.org/10.1515/znc-2003-11-1220

Smith R, Middlebrook R, Turner R, Huggins R, Vardy S, Warne M (2012) Large-scale pesticide monitoring across Great Barrier Reef catchments – paddock to reef integrated monitoring, modelling and reporting program. Mar Pollut Bull 65:117–127. https://doi.org/10.1016/j.marpolbul.2011.08.010

Song Y, Zhu LS, Wang J, Wang JH, Liu W, Xie H (2009) DNA damage and effects on antioxidative enzymes in earthworm (*Eisenia foetida*) induced by atrazine. Soil Biol Biochem 41(5):905–909

Spurgeon DJ, Weeks JM, Van Gestel CA (2003) A summary of eleven years progress in earthworm ecotoxicology: the 7th international symposium on earthworm ecology, Cardiff, Wales, 2002. Pedobiologia 47(5):588–606. https://doi.org/10.1078/0031-4056-00234

Stenersen J (1979) Action of pesticides on earthworms. Part I: The toxicity of cholinesterase-inhibiting insecticides to earthworms as evaluated by laboratory tests. Pestic Sci 10(1):66–74. https://doi.org/10.1002/ps.2780100109

Su TH, Beardsley JW, McEwen FL (1980) AC217, 300, a promising new insecticide for use in baits for control of the bigheaded ant in pineapple. J Econ Entomol 73:755–756

Sumpter JP, Johnson AC (2005) Lessons from endocrine disruption and their application to other issues concerning trace organics in the aquatic environment. Environ Sci Technol 39 (12):4321–4332. https://doi.org/10.1021/es048504a

Suter G (1993) Ecological risk assessment. Lewis Publishers, Ann Arbor

Tahboub YR, Zaater MF, Barri TA (2006) Simultaneous identification and quantitation of selected organochlorine pesticide residues in honey by full-scan gas chromatography–mass spectrometry. Anal Chim Acta 558(1–2):62–68. https://doi.org/10.1016/j.aca.2005.11.004

Talmage SS, Walton BT (1991) Small mammals as monitors of environmental contaminants. Rev Environ Contam Toxicol 119:47–108. https://doi.org/10.1007/978-1-4612-3078-6_2

Tanabe S, Subramanian A (2006) Bioindicators of POPs: monitoring in developing countries. Kyoto University Press/Trans Pacific Press, Kyoto/Melbourne

Tarcau D, Cucu-Man S, Boruvkova J, Klanova J, Covaci A (2013) Organochlorine pesticides in soil, moss and tree-bark from North-Eastern Romania. Sci Total Environ 456–457:317–324. https://doi.org/10.1016/j.scitotenv.2013.03.103

Tarricone K, Wagner G, Klein R (2015) Toward standardization of sample collection and preservation for the quality of results in biomonitoring with trees – a critical review. Ecol Indic 57:341–59. https://doi.org/10.1016/j.ecolind.2015.05.012

Tataruch F, Kierdorf H (2003) Mammals as biomonitors. In: Markert BA, Breure AM, Zechmeister HG (eds) Bioindicators and biomonitors. Principles, concepts and applications. Elsevier, Amsterdam, pp 737–772

Tawatsin A (2015) Pesticides used in Thailand and toxic effects to human health. Med Res Arch 1(3):1–10

Teplitsky C, Piha H, Laurila A, Merila J (2005) Common pesticide increases costs of antipredator defenses in *Rana temporaria* tadpoles. Environ Sci Technol 39(16):6079–6085. https://doi.org/10.1021/es050127u

Tiwari RK, Singh S, Pandey RS, Sharma B (2016) Enzymes of earthworm as indicators of pesticide pollution in soil. Adv Enzyme Res 4(4):113–124. https://doi.org/10.4236/aer.2016.44011

Tomza-Marciniak A, Marciniak A, Pilarczyk B, Drozd R, Ligocki M, Prokulewicz A (2014) Wild boar (*Sus scrofa*) as a bioindicator of organochlorine compound contamination in terrestrial ecosystems of West Pomerania Province, NW Poland. Environ Monit Assess 186(1):229–238. https://doi.org/10.1007/s10661-013-3368-z

Tsygankov VY, Boyarova MD, Lukyanova ON (2016) Bioaccumulation of organochlorine pesticides (OCPs) in the northern fulmar (*Fulmarus glacialis*) from the Sea of Okhotsk. Mar Pollut Bull 110(1):82–85. https://doi.org/10.1016/j.marpolbul.2016.06.084

Tu CM (1995) Effect of five insecticides on microbial and enzymatic activities in sandy soils. J Environ Sci Heal Part B 30(3):289–306

Udroiu I, Cristaldi M, Leradi LA, Tanzarella C, Moreno S (2008) Biomonitoring of Doñana National Park using the *Algerian mouse* (*Mus spretus*) as a sentinel species, vol 17, Fresen Environ Bull, pp 1519–1525

Ueno D, Takahashi S, Tanaka H, Subramanian AN, Fillmann G, Nakata H, Lam PKS, Zheng J, Muchtar M, Prudente M, Chung KH, Tanabe S (2003) Global pollution monitoring of PCBs and organochlorine pesticides using Skipjack Tuna as a bioindicator. Arch Environ Contam Toxicol 45(3):378–389. https://doi.org/10.1007/s00244-002-0131-9

UN (2015) United Nations, Department of Economic and Social Affairs, Population Division. World population prospects: the 2015 revision, key findings and advance tables. Working paper no. ESA/P/WP.241. United Nations, New York

Uwizeyimana H, Wang M, Chen W, Khan K (2017) The eco-toxic effects of pesticide and heavy metal mixtures towards earthworms in soil. Environ Toxicol Pharmacol 55:20–29. https://doi.org/10.1016/j.etap.2017.08.001

Vaishampayan A (1985) Mutagenic activity of alachlor, butachlor and carbaryl to a nitrogen-fixing cyanobacterium Nostocmuscorum. J Agric Sci 104:571–576. https://doi.org/10.1017/S0021859600044336

Vance BD, Drummond W (1969) Biological concentration of pesticides by algae. J Am Water Works Assoc:360–362

Velickovic M (2007) Measures of the developmental stability, body size and body condition in the black-striped mouse (*Apodemus agrarius*) as indicators of a disturbed environment in northern Serbia. Belg J Zool 137:147–156

Villeneuve JP, Fogelqvist E, Cattini C (1988) Lichens as bioindicators for atmospheric pollution by chlorinated hydrocarbons. Chemosphere 17(2):399–403. https://doi.org/10.1016/0045-6535(88)90230-5

Wallace JB, Grubaugh JW, Whiles MR (1996) Biotic indices and stream ecosystem processes: results from an experimental study. Ecol Appl 6:140–151. https://doi.org/10.2307/2269560

Walsh GE (1978) Toxic effects of pollutants on plankton. In: Butler GC (ed) Principles of ecotoxicology, vol 12. Wiley, New York, pp 257–274

Wang Y, Xie J (2000) Ant colony optimization for multicast routing. In: IEEE APCCAS 2000. 2000 IEEE Asia-Pacific conference on Circuits and Systems. Electronic communication systems. (Cat. No. 00EX394). IEEE, pp 54–57

Werner I, Deanovic LA, Connor V, de Vlaming V, Bailey HC, Hinton DE (2000) Insecticide-caused toxicity to *Ceriodaphnia dubia* (Cladocera) in the Sacramento–San Joaquin River delta, California, USA. Environ Toxicol Chem 19:215–227. https://doi.org/10.1002/etc.5620190126

WHO (2017) Agrochemicals, health and environment: directory of resources. Available online at: http://www.who.int/heli/risks/toxics/chemicalsdirectory/en/index1.html. Accessed June 2018

Wojcik DP, Porter SD (2001) Formis: a master bibliography of ant literature. http://cmave.usda.ufl.edu/~formis

Wolfe HR, Armstrong JF, Durham WF (1966) Pesticide exposure from concentrate spraying. Arch Environ Health 13:340–344. https://doi.org/10.1080/00039896.1966.10664567

Wren CD (1986) Mammals as biological monitors of environmental metal levels. Environ Monit Assess 6(2):127–144. https://doi.org/10.1007/BF00395625

Wurster CF Jr (1968) DDT reduces photosynthesis by marine phytoplankton. Science 159:1474–1475. https://doi.org/10.1126/science.159.3822.1474

Yang R, Zhang S, Li A, Jiang GJing C (2013) Altitudinal and spatial signature of persistent organic pollutants in soil, lichen, conifer needles, and bark of the southeast Tibetan Plateau: implications for sources and environmental cycling. Environ Sci Technol 47(22):12736–12743. https://doi.org/10.1021/es403562x

Ye X, Xiong K, Liu J (2016) Comparative toxicity and bioaccumulation of fenvalerate and esfenvalerate to earthworm *Eisenia fetida*. J Hazard Mater 310:82–88. https://doi.org/10.1016/j.jhazmat.2016.02.010

Zahm SH, Blair A (1993) Cancer among migrant and seasonal farmworkers: an epidemiologic review and research agenda. Am J Ind Med 24:753–766. https://doi.org/10.1002/ajim.4700240612

Zhang B, Zhang H, Jin B, Tang L, Yang J, Li B, Zhuang G, Bai Z (2008) Effect of cypermethrin insecticide on the microbial community in cucumber phyllosphere. J Environ Sci 20 (11):1365–1362. https://doi.org/10.1016/s1001-0742(08)62233-0

Zhao YL, Yang LM, Wang QQ (2008) Modeling persistent organic pollutant (POP) partitioning between tree bark and air and its application to spatial monitoring of atmospheric POPs in mainland China. Environ Sci Technol 42:6046–6051. https://doi.org/10.1021/es800188q

Chapter 6
Occurrence and Removal of Pesticides in Drinking Water

Anuradha and Jagvir Singh

Abstract Primarily solvents, petroleum products, and pesticides are chemicals that are considered as agents of groundwater contamination. Pesticides that belong to a chemical family have attracted our attention due to their hazardous effects on the environment. In the beginning, organochlorine insecticides were used extensively but the persistence and bio-chemulation properties of this class of pesticides make a real problem. Highly hazardous pesticides and their excessive use have led to health problems and fatal incidents world-wide. Data provided by World Health Organization and research conducted in 2016 suggest that the global impact of self-poisoning from preventable pesticide ingestion was estimated at 155,488 deaths and 7,362,493 disabilities, much higher than estimates. Since the health data available so far are limited to estimate the global health effects of such dangerous chemicals.

Pesticides have acute, carcinogenicity, and mutagenicity or causing aesthetic problems and chronic toxic risk to children. World Health Organization has set the maximum pesticides residue limit in drinking water of 0.1 micrograms/liter. This chapter reviews various removal technologies such as adsorption, ion exchange for various types of pesticides present in water. The monitoring of contaminated water and its regulation, supply is working under the United States Environmental Protection Agency and Federal Country safe drinking water act 1979.

Keywords Organochlorine pesticides · Clinical effect · Membrane filtration · Ion exchange treatment technology · Activated carbon · Electrochemical treatment · Phytoremediation · Bioaugmentation · Electrocoagulation · Pesticide transportation

Anuradha
Department of Zoology, Raghuveer Singh Government P.G. College, Lalitpur, Uttar Pradesh, India

J. Singh (✉)
Department of Chemistry, ARSD College (University of Delhi), New Delhi, India

Inamuddin et al. (eds.), *Sustainable Agriculture Reviews 48*, Sustainable Agriculture Reviews 48, https://doi.org/10.1007/978-3-030-54719-6_6

6.1 Introduction

Pesticides are being used globally as a fungal weed. These chemicals allow crops to flourish vigorously in use time, thereby increasing global production rapidly (Rosa et al. 2019; Aly et al. 2019; Yihan et al. 2018). These substances predominantly enter ground water through water sources and leaching through drift and agricultural runoff due to one or more substances being active. Diseases caused by water pollution through insect mites, weed contaminated surface water leaks, improper disposal (Arne et al. 2019; Eduard et al. 2019; Hicham et al. 2008) and even waste material injected into wells, crop be a developed nation or a developing nation, both harmful to pesticides and causing economic damage. The problem of water pollution by pesticides in India is so severe that 50% urban and 80% of rural people were getting much affected. Insecticides may contain weeds, fish, birds, molluscs, nematodes, plant pathogens, and a pesticide chemical substance that may be a biological agent, disinfectant, or any device that forces economic power (Xiaochu et al. 2019; Tawfeek et al. 2019; Gangesh et al. 2006; Wangyang 2017) by using them.

The use of pesticide chemicals started after 1940 to protect crops, increase yields. Along with protecting crops, plants and increasing crop production, their use is also increasing in defense of organisms. Their use is constantly increasing in every production filed. Organochlorine pesticide (Adgate et al. 2000; Ankley et al. 2010) usage has increased substantially since World War II. The article by Rachel Carson in "Silent Spring" was published in 1962, in which shows the excessive use of pesticides in agriculture and due this sources of water are at peak of dangerous point, these dangerous chemicals cause eating disorders and at the same time their negative effects on the environment can be described in detail. Research records of the researcher obtained suggest that people in ancient Mesopotamia had already used pesticides to protect their crops about 4500 years ago, and primary sulfur dusting has been the first known used pesticide (Arena and Sgolastra 2014; Arias et al. 2008; Atkins et al. 1986).

If we talk about water, then after air, has been like the most important part of nature here for the creatures on earth. Only 3% of the water as fresh-water resources exists on earth. These water resources are very important for daily use water, ecosystem, energy industry, health of animals and other water-dependent subjects. Developing countries are still struggling with this kind of problem compared to other developed countries (Epstein and Bassein 2003; Epstein and Zhang 2014). Industrialization and increased population are fully responsible for this. Due to the quality of its illiteracy, they are also causing financial problems in the environment along with the problem of water pollution. If we talk about the last 20 years, mostly seen that the use of some pesticides has increased to a very high percentage, which makes water (Zijian and Aaron 2018; Arisekar et al. 2019) resource of polluted. Due to their poisonous effects, these chemicals are poisonous for the life of our earth.

The pesticide-rich water source is a major concern all over the world because many deadly diseases start with such water pollution. 18,000 or more people die every day due to pesticide pollutants. If we compare this kind of problem (Mourabit and Boulaid 2019; Rani and Kumaraguru 2014; Rasul and Thapa 2003; Abarikwu and Adesiyan 2009) with the developed countries, then we find that pesticide pollution is a very terrible problem in developing countries. In addition to pesticides, water pollution can also be natural in which storms, earthquakes, volcanoes, algae blooms and ecosystems can occur. Molluscicides, plant growth regulators, fungicides, herbicides, nematicides, rodents are all types of pesticides. Before 1900, organochlorine, a pesticide, was used to control many deadly diseases such as malaria and typhus, which was later banned in most countries (Williamson 2007; Westbom et al. 2008).

The arrival of organophosphates, carbamates, herbs and pyrethroids in the decade of 1960–1970 gave a new direction to agriculture and its production. Although a pesticide should be lethal only for its target and has been observed that pesticides has proved to be very deadly for non-target species, even humans. This has created a controversy over the use and non-use of pesticides (Baker and Wilkenson 1990; Bohmont 1997). In 1952, with the production of Benzene Hexachloride, the use of pesticides took birth in India and its use started on a large scale. After India, China is the second largest agrochemical producing country in Asia and ranked twelfth globally. These pesticides in drinking water remain a health concern even after years. If very small part or quantity of pesticides that reach drinking water from various sources which enters the human body with drinking water and causes adverse health effects.

Every country uses pesticides in agriculture, residential gardens and commercial areas. Due to their exposure, these natural waters reach the human body through source, breath and dermal contact. Organophosphorus insecticides (Salam et al. 2015) have been used in agriculture since the ban on organochlorine pesticides, which has created dangerous health problems. Pollution of ground water by pesticides has become a national problem because more than 50% of the people here depend on ground water. In countries or places where these chemicals are used more or say that agricultural areas are more affected because their life depends on ground water. Before 1970 this was thought that the soil, like a filter, prevents pesticides from going into this ground water (Sattar and Rahman 1987; Satyavardhan 2013). New research tells us that these dreaded chemicals do not get filtered and go to ground water through leaching from agricultural land, which is a huge source of such type water pollution. The problem of pesticide pollution also arises through the disposal of waste material, improper disposal, leakage of contaminated surface water, accidental spills and leakage.

6.2 Dispersal of Water on Earth

So far drinking water has been possible to see life only on Earth, the reason for this is the presence of water only on this planet. Which covers 71% of it. Table 6.1 shows the different sources and availability of water on the earth.

Table 6.1 Distribution of water on our earth planet

Sources of water	Quantity in ppm	Percentage	Area in km
Fresh ground water	7600	0.76	10,530,000
Saline ground water	9400	0.94	12,870,000
Ice form	17,400	1.74	24,064,000
Biological water	1	0.0001	1120
Atmospheric	10	0.001	12,900
Swamp water	8	0.0008	11,470
Rivers	2	0.0002	21,200
Saline lakes	60	0.006	85,400
Oceans like water	965,000	96.5	1,338,000,000
Ground ice and permafrost	220	0.22	300,000
Soil moisture	10	0.001	16,500
Fresh lakes	70	0.007	91,000

Quantity and their percentage indicate that all most water quantity is useless for household purposes with as drinking (Gross 2015)

6.3 Historical Perspectives

The history of pesticides is very old, the whole world has been using such chemicals in agriculture for centuries. People started using them in the nineteenth to the twentieth century by using synthetic pesticides. By mixing the right proportion of pesticides, we got rid of undesirable pests, weeds and gave the agricultural world a chance to flourish (Sharmin et al. 2015; Bolognesi 2003).

In the agriculture farm, and other places, these chemicals pesticide named 'Paris Green' was the first chemical pesticides have been starting to use in United States Nation since the year of beginning 1867. The rise of synthetic pesticides in the nineteenth century led to Paris Green being discharged with pesticides containing calcium arsenate, nicotine sulfate and sulfur (Bolognesi and Merlo 2011; Cerejeira et al. 2003). Initially, the Romans used to control weeds with nonmetallic salt and sulfur which is have many anti properties against pets, fungus, bacteria, used to control them and both are the same work as done by pesticides or other herbicides, fungicides. In the 1600 century, the mixture of honey and arsenic compounds in a fixed proportion has been used to control ants and such others. Pyrethrin compound containing pesticides that very hazardous chemicals were used to control various types of insects and hellebore to control plant lice. Dichlorodiphenyltrichloroethane as shown in Fig. 6.1 was the first synthetic pesticide of the time to establish a new identity with its use in the whole world (WWF 1998). The reason for its popularity was its use against mosquitoes and conquered the disease like malaria. Along with utility and also gave birth to its derivatives.

These chemicals created a revolution in the world by giving birth to a new era with the availability of agriculture and domestic applications. Along with the utility, these pesticide chemicals have also given rise to their dangerous and deadly derivatives. Dichlorodiphenyltrichloroethane became very famous due to its multiple

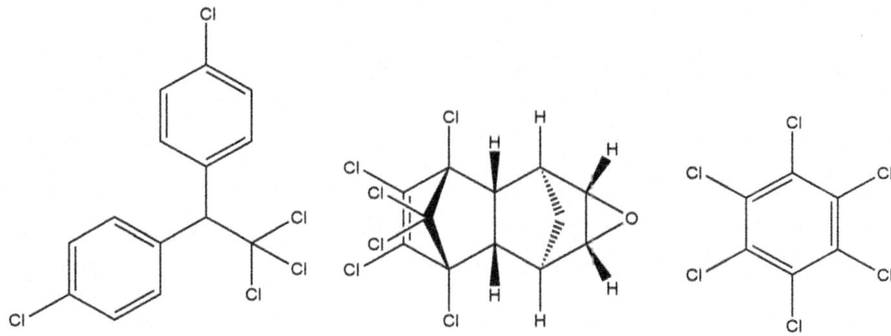

A. B. C.

Fig. 6.1 Dichlorodiphenyltrichloroethane (**a**), Dieldrin (**b**) and Benzene-hexachloride (**c**) was the starting the chemicals that used as pesticides but now-a-days banned. These belong to highly persistent organochlorine insecticide. (WWF 1998)

utility in agriculture but unfortunately the persistence of its properties made pesticide almost non-existent after World War II. After this, some more pesticides started to be used in agriculture, in which Aldrin, Benzene-hexachloride, Andrine, Dyedrin are some of the major chemicals (Shi et al. 2011; Siddika et al. 2012). These pesticides became popular with being effective, cheap. Rachel Carson, through her research, in her book 'Silent Spring in 1962', presented a harsh picture of people's understanding of the environmental risks of pesticides.

6.4 Pesticide Classification

The term pesticide is a group of some dangerous, contaminated and negatively impacted chemicals (Jing et al. 2016; Schipper and Marc 2008). They are mainly divided into two classes natural and artificial or manmade. The classification has also been made by the World Health Organization based on the potential health risks and toxicity of pesticide chemicals as following-.

Class 1: Extremely dangerous pesticides. Example- Aldicarb, Brodifacoum, Bromadiolone, Calcium cyanide, Ethoprophos, Flocoumafen, Hexachlorobenzene, Phenylmercury acetate, Phosphamidon, Sodium fluoroacetate, Sulfotep, Tebupirimfos, Terbufos.

Class 2: Highly dangerous pesticides. Example- Acrolein, Allyl alcohol, Azinphosethy, Blasticidin-S, Butocarboxim, Butoxycarboxim, Carbofuran, 3-Chloro-1,2-propanedio, Coumaphos, Zeta-cypermethrin, Demeton-S-methyl, Dicrotophos, Fluoroacetamide, Heptenophos, Oxydemeton-methy, Pentachloropheno, Thallium sulfate, Zinc phosphide.

Class 3: Moderately dangerous pesticides. Example- Alanycarb, Anilofos, Azaconazole, Azocyclotin, Benfuracarb, Bensulide, Bifenthrin, Bioallethrin, Butylamine, Carbosulfan, Chlorfenapyr, Cyfluthrin, Beta-cyfluthrin, Cypermethrin, Alpha-cypermethrin, Difenzoquat, Esfenvalerate, Fentin hydroxide, Fuberidazole, Imidacloprid, Methyl isothiocyanate, Phoxim, Pyrethrins, Quizalofop-p-tefuryl, Thiacloprid, Xylylcarb.

Class 4: Slightly dangerous pesticides. Example- Acephate, Acetochlor, Acifluorfen, Alachlor, Allethrin, Ametryn, Amitraz, Azamethiphos, Bensultap, Bentazone, Butralin, Butroxydim, Chinomethionat, Chlormequat, Chloroacetic acid, Copper oxychloride, Cycloate, Cyhexatin, Cymoxanil, Cyproconazole, Dichlorobenzene, Dichlorophen, Diethyltoluamide, Difenoconazole, Dimepiperate, Dimethylarsinic acid, Etridiazole, Fenothiocarb, Fluazifop-p-butyl, Glufosinate, Hexazinone, 2-Napthyloxyacetic acid, N-octylbicycloheptene dicarboximide, Pirimiphos-methyl, Quinoclamine, Resmethrin, Tebuconazole, Undecan-2-one, Ziram.

These pesticides are based on functions such as orchards, wood preservatives, domestic germ-ants, fungicides, herbicides and rodents. Many pesticides are increasing their amounts in the environment based on their special properties and their specific targets and water solubility (Alavanja et al. 2004; Arinaitwe et al. 2016). The main basis of classification is based on their chemical composition, properties, size, target and the overall classification have been shown in Fig. 6.2.

Manmade chemical pollutants can be classified into subclasses based on their varying physicochemical properties.

6.4.1 Mode of Action

Mode of action is the class of pesticides in which the chemical or pesticide affects its target. Such chemicals or insecticides fall under the mode of action. After this, they are classified into non-systemic and systemic category. Systemic insecticides exhibit their effects by moving into the target's vascular system while externally displaying their effects by not entering non-systemic target tissues (Bertolote et al. 2006).

6.4.2 Target Insect

The target insect class includes insecticides or chemicals that are classified based on the organism of the target insect. This is the most familiar category. In this category, pesticides are named because of their unique properties (Devleesschauwer et al. 2014). At the end of their names, the Latin word cide mean dyeing is used which is used as a suffix as a victim. For example, insecticides target plants with pests and

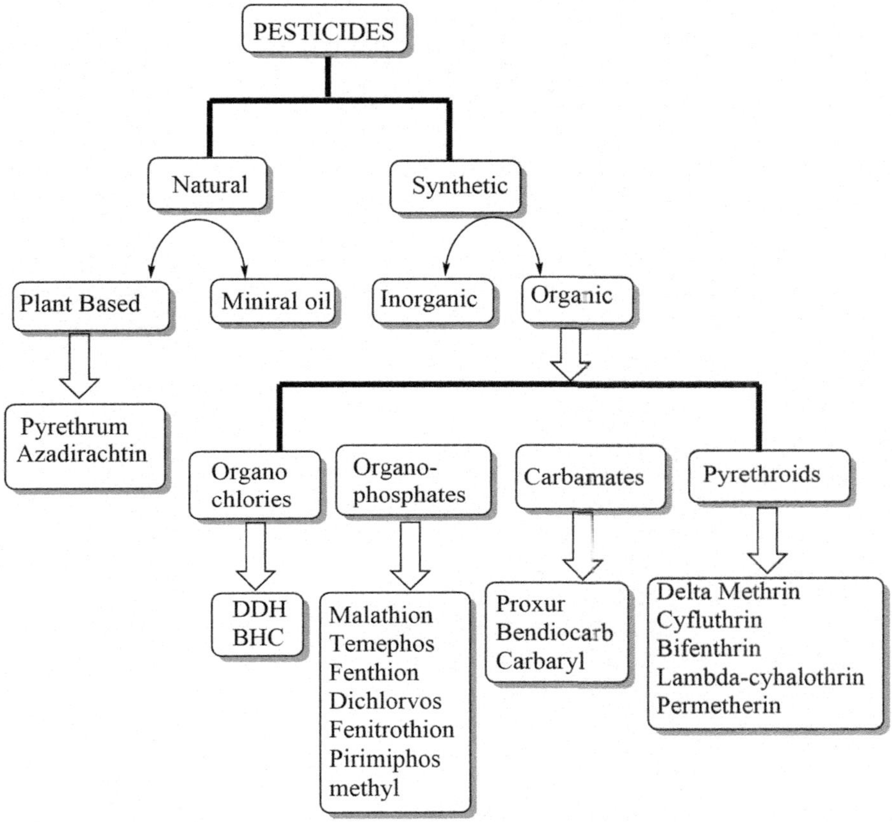

Fig. 6.2 This diagram shows the general classification of pesticides. The proposed classification is mainly based on the nature, character of insecticides and their specific targets. (Jing et al. 2016)

herbicides, while mollicides for rodents, fungicides for fungus, bactericides for bacteria, insecticides for insects are affected.

6.4.3 *Chemical Composition*

As we know that insecticide is a type of chemical compound. Chemical composition class as pesticide is known by the name that they are classified on the basis of their chemical nature and active elements. This category is also very important from the point of view of research in the field of pesticides and environment (Uddin et al. 2012; Ullah et al. 2014; Van den Brink 2013).

Depending on the chemical properties, pesticides can then be divided into seven subclasses, organochlorines, organophosphorus, carbamates, pyrethroids, organochlorins, organophosphorus, agotic heterocyclics, amides, and anilines. Organochlorine pesticides have five or more chlorine atoms. This class was the first synthetic organic pesticide to be used in agriculture, in which the chemical composition is stable and keeps accumulating in the atmosphere. The first target of this pesticide is the nervous system, which, under control, leads to the goal of paralysis or death convulsions (Drechsel et al. 2010; Dufour et al. 2012). They cause severe endocrine disorders in aquatic animals and birds and was banned for the use of agriculture.

Organophosphates containing phosphate group are very toxic pesticides. From the point of view of the user, it was occupied by 48.6%. Organophosphates were used as a war material in World War II. This pesticide has also established records of its utility in industries, cosmetics, medicine, agriculture. In humans and some other species this compound prevents acetylcholine from being formed in the nervous system by the acetylcholinesterase enzyme formed in the hydrolyze process. The residue of this pesticide is dangerous for both ecosystems and food industries due to acute toxicity (Eerkes et al. 2015; Epstein 2015). Based on annual reports, three million cases of its acute poisoning have led to the death of 250 to 370,000 people annually, so this pesticide was also banned worldwide.

Carbamate is an organic compound and this compound's derived from carbamic acid (NH_2COOH) which is highly toxic pesticide. This pesticide is completely different from reversible, cholinesterase inhibition and organophosphates and inactivates the acetylcholinesterase enzyme. Pyrethroids, aniline, amides and abiotic heterocyclic insecticides are toxic compared to the above.

6.5 Sources of Pesticide Pollutants

Pesticides reach different sources of water and contaminate it. Examples of mediums such as air flow, water flow, soil like through leaching can be found, even directly, the evidence of contamination of water is fair. These pollutants originate from agriculture itself, which are the source of pesticide polluted water (Accra et al. 2008; WWF 1998). Excessive use of agrochemicals has created competition for their food, habitat for vegetation risks as well as wildlife living there.

When these agrochemicals are used in the fields, as a result of improper formatting, the emission of these pesticides pollutes ground water through surface water or air flow and drinking water sources on the way of leaching (Agricultural Chemical Regulations 2006).

If some amount of these pesticides gets deposited on the tree plants through the air, then by damaging them, they get deposited through rain water in the source of these water, where it also becomes a target for humans and animals as well. They give birth to terrible diseases and end of life. Pesticides persist for a long time in water bodies due to their physical and chemical properties such as deposition rate, partition coefficient, degradation rate (Erlanger et al. 2005). According to a study,

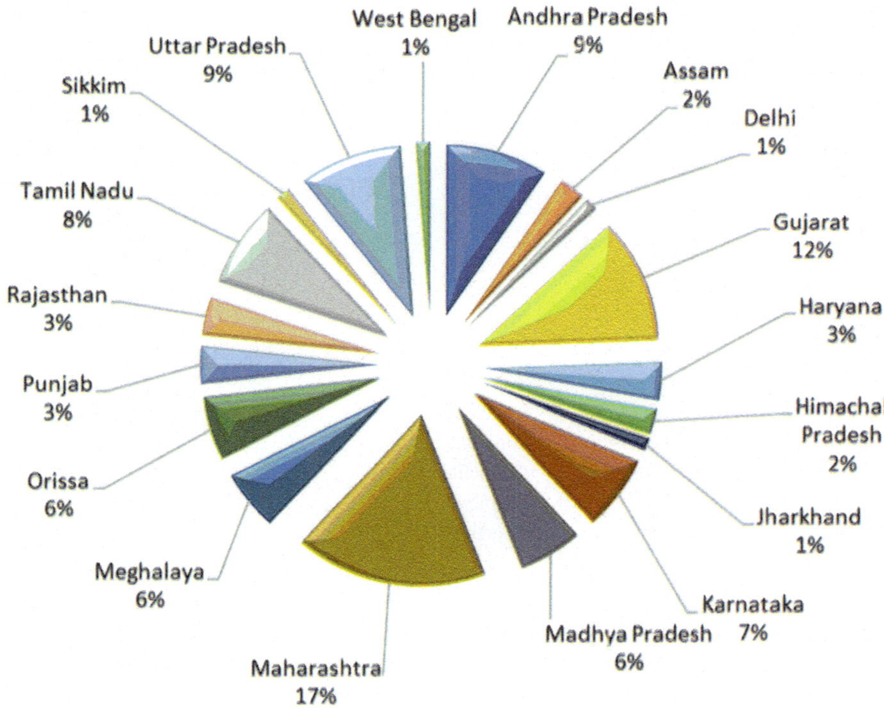

Fig. 6.3 This picture represents the pesticides pollution in water different percentage in Indian states. This graph showed that Andhra Pradesh has higher these sites. (WWF 1998)

the percentage of pesticides were different in different regions of India and the data percentage as graphically shown in the following Fig. 6.3.

Due to the physical and chemical properties of the pesticide such as partition coefficient, erosion rate, deposition rate persists for a long time in water bodies and in the environment. Due to having these properties, their importance has also been seen in climate change.

6.5.1 Pesticides Water Pollution

In the American Geographic Survey has been found that out of 76 pesticides, seven pesticides were found which were broken down and converted into other products. In a subsequent research, researchers found in 90% water streams and 50% wells research findings that every water source had at least one pesticide presence (WHO 2009). Carbamate pesticide is found most often in ground water or drinking water. Aldicarb pesticides have been found in more than two thousand wells in some states

Table 6.2 The nature and polluted concentrations of various chemicals that behave like pesticides are tabulated. (WHO 2009)

Pesticides	Nature	Fouling concentration in µg/L
Carbofuran	Nematicide	40
Dalapon	Herbicide	200
Dibromochloropropane	Nematocide	0.2
Dinoseb	Insecticide	7
Dioxin	Herbicide	0.0003
Diquat	Herbicide	20
Endothall	Algicide	100
Ethylene dibromide	Insecticide	0.5
Glyphosate	Herbicide	700
Methoxychlor	Insecticide	40
Oxamyl	Insecticide	200
Pentachlorophenol	Fungicide	1
Picloram	Herbicide	500
Simazine	Herbicid	3
Toxaphene	Insecticide	4

along the island of India. Due to the different nature of these chemicals, their quantity in drinking water is different and the various data are given in Table 6.2.

"Our water is our world" is a motto for public awareness of water pollution. Given the risks of pesticides, water pollution prevention becomes very important. The spread of these chemicals can be prevented through awareness campaigns such as schools, colleges, newspapers and by informing people about the ways of using pesticides and poisoning of pesticides.

In this way, by managing their, control strategies, information, scientific studies and tests, you can save the environment and its organism. Pollard et al. (1992) in their research paper low-cost alternatives like biomass such as wood, coconut shell, bio-sorbents like chitosan, fiber have been used to reduce the limit of pesticides from drinking water. In milligram per liter to microgram per liter amounts of pesticides, they are very effective in disturbing the environment and life to a large extent.

Along with pesticides to pollute our drinking water, we are also responsible to a great extent because the population of the earth has grown very fast in the last few decades (Richa and Shilpi 2018), due to which the use of these pesticides in agriculture for food reimbursement is very high. The following graphic data shown in Fig. 6.4 strongly support it.

As a result, the amount of these pesticide pollutants in the water has also increased, which has serious consequences for the environment as well as human health.

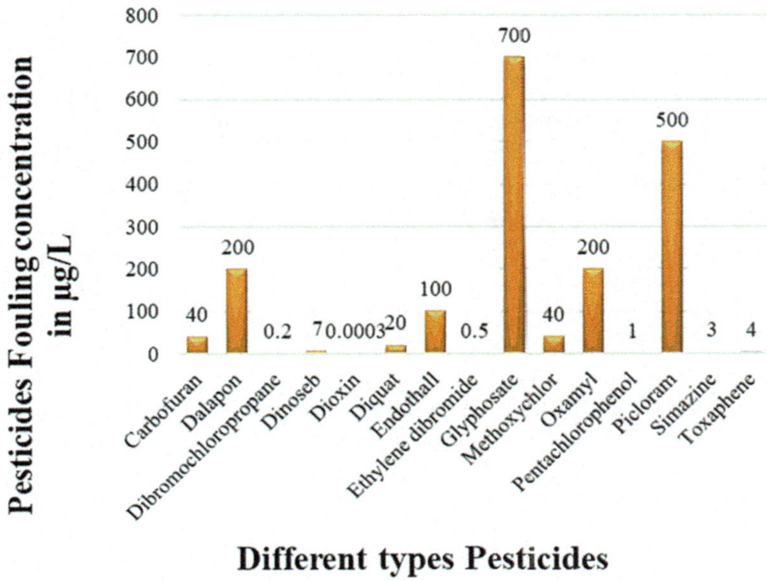

Fig. 6.4 This bar graph plotted between the different types of pesticides chemicals versus their fouling concentration in µg/L. Glyphosate is herbicide showed the maximum concentration to contaminant. (WHO 2009)

6.6 Pesticide Properties

Adsorption, evaporation, firmness and solubility are the properties of pesticides that reflect the eccentricity of leach or runoff. Through these processes, these chemicals get adjusted to ground water.

6.6.1 Adsorption

In this process, the pesticide gets stuck by the particles on the upper surface of the agricultural land while using chemicals, this is called the process of adsorption. Etrazine, paraquat which is an herbicide and the adsorption strength of paraquat is higher than atrazine. This action also depends on the adsorption properties of the soil.

6.6.2 Solubility

This property depends on the solubility characteristic of the pesticide chemical and can be written in milligrams per liter or in parts per million. If this characteristic is too much of a chemical, then the amount of a pesticide chemical is more than 1.0 parts per million and Finally, they reach ground water as well as various sources of drinking water very rapidly through the above medium.

6.6.3 Firmness

The term literally means how often a pesticide chemical is decomposed. Chemically pesticides have specific property that can control the environment stagnation and its loss or for how long. Some other factors such as microbial action, chemical reactions, half-life and photodegradation also have their effect on this specificity.

6.6.4 Evaporation

As we all know that in the evaporation process, the pesticide may be solid or liquid differ at different temperatures which is converted into a gaseous state. Pesticides, at this stage are less polluting the environment. The risk of pesticides increasing in the water environment increases due to rain. In this manner, mainly the amount of pesticides is very low.

6.7 Pesticide Transportation

The transportation of pesticide may mainly consist of air and water. Air is a type of transport method in which picks up pesticides and reached other sources such as fields, water reserves. If the agriculture fields absorbed these chemicals, the surface of the land, then air contaminates both ground water and surface water.

6.7.1 Surface Runoff and Erosion

Runoff is a type of action that involves water, soil and pesticide like chemicals. This type is found mainly where staircase is cultivated. When water is given in the fields, the overflows along with soil erosion. Pesticides and other chemicals along with that goes through these steps and goes to these water sources. The amount of runoff pesticides depends on land's slope, soil texture and moisture.

6.7.2 *Diffuse Source*

This pollution is caused by various activities without any particular point. This type of transport brings a series of such pollutants that water alone does not have any effect on the environment, but these insecticides by capturing these have specific effects. Diffuse is also governed by the spread of pesticides in agriculture. Apart from this, forestry, rural housing, atmospheric deposition, urban land can also be a source of diffuse.

6.7.3 *Spray Drift*

In this, the inadvertent spread of pesticides and their negative effects are mainly described. In this transport, the spray of these pesticides with spray, runoff from the soil is also polluted off-target, which severely affects property, environment and human health. Including off-target contamination due to spray runoff as well as runoff from plants or soil.

6.7.4 *Leaching Processes*

This transportation provides a medium for pesticides to reach ground water through such means. This is called leaching. Leaching solubility, soil types such as sandy, rain after some time of spraying succeeded in increasing leaching. Such moments help to maintain its properties to a great extent on pesticides and soil physicochemical properties, vapor pressure, dependence, solubility.

6.7.5 *Overland Flow*

In this, water flows on earth from rain, snow or other sources, which also plays a major role in the water cycle. The surface runoff before reaching a source is called surface runoff/overland flow. Overland also corrupts soil.

$$\textbf{Total runoff} = \textbf{Surface runoff} + \textbf{groundwater Baseflow}$$

After the ice melts, the river merges with water from various sources. Which then overflows and causes great harm to the atmosphere. The overflow flows downstream from the top surface to the hydrographic network due to gravity. Another way of overflow can also be that this works to uplift the source of water, after that when this level exceeds the maximum level, then pesticide starts to flow from the water. Which also takes pesticide contaminants along with them and mix them in the source of clean water. When this water is used, the presence of pesticides shows its opposite effect.

6.8 Pesticide Poisoning

Pesticide chemicals are very toxic in nature. Those act like sweet poison in our food and which are constantly being served in front of us. These are being included in grains, vegetables and fruits through irrigation water and pesticides or in every food item of agricultural production. Continuous use of pesticide chemicals has increased fertility, quality and toxicity of agricultural land (Atreya 2006; Austin 1999).

The sources of water like ponds, rivers make the water there poisonous, causing great harm to humans as well as animals, birds and all living environment. Excessive spraying of pesticide chemicals leads to very serious fatal diseases. Their toxicity results in increased incidence of diseases such as ingestion, inhalation or cutaneous absorption, neurological, psychological and behavioral disorders, hormonal imbalances, leading to infertility, breast pain.

6.8.1 Organophosphates

Organophosphate pesticides were created in the year 1940 after the tetraethyleprophosphate pesticide was synthesized. Organophosphate poisoning is a type of pollution caused by pesticides containing organophosphate. The chemical species contain of organophosphate, they affect the nervous system directly. Some fatal diseases of such pesticides like increase in saliva, tears, diarrhea, vomiting, red marks in the eyes, sweating, body swelling and confusion occur. The initial symptoms of which may take a few minutes, 7 weeks or even months.

In the developing nations, it has been observed that patients with organophosphate poisoning suffer from these pesticides in such a way that they are forced to commit suicide attempts. This toxicity is also increased by drinking water, breathing and skin when touch such chemicals. Its basic process is similar to inhibition of acetylcholinesterase which later becomes cetylcholine. Its confirmation is based on its symptoms. Butyryl cholinesterase has its activity in the blood after confirmation.

6.8.2 *Chlorpyrifos*

Chlorpyrifos is a versatile insecticide. Since its use has been banned in today's date. Chlorpyrifos is a type of organophosphate insecticide. This pesticide affects termites, grubs, flies, lice, fire ants, cockroaches and corn root worms. Chlorpyrifos acts primarily as a poison (Jing et al. 2016). Its presence was found in some fruits specimen responsible for dangerous damage to the liver, lungs, kidneys, eyes and central nervous system. Its presence in some of the well-known brands of milk stored here also damages carcinogenic and sensory systems.

6.8.3 *Dichlorodiphenyltrichloroethane*

Dichlorodiphenyltrichloroethane is a well-known pesticide with a low soluble tendency in water whereas fats and organic oils dissolve easily. After their use, mostly seen that this type of chemical kills the surrounding aquatic and flora. Their decaying dead parts accumulate in living creatures and flora which are then added to the food chain by organisms and flora. Since solubility in dichlorodiphenyltrichloroethane vasa and dissolves in the vasa of the body which increases both concentration and toxicity (WHO 2003).

The molecules of these pesticides do not decompose over a long period of time and lying in the ecosystem their effects all the ecosystem factors. Once used, they are present in the soil and environment for a long time like dichlorodiphenyltrichloroethane for 5 years, lindane for 7 years and arsenic molecules for an indefinite period of time (Michael et al. 2018). Exist in nature. Thus, insecticides are active all the time to destroy our nature.

6.8.4 *Aldrin and Dieldrin*

Since the 1940s, aldrin and dieldrin have started to use pesticide chemicals. These pesticides fall under the organochlorine class until 1970. They were used in crops, and in 1987 they started being used for termite control. Their high concentration in water affects dopaminergic and monoaminergic neurons, increasing the likelihood of diseases like Parkinson's and Alzheimer's (Honeycutt and Shirley 2014). Neurotoxicity is the result of these pesticides that aim to block the central nervous system of vertebrates. Headache, nausea, vomiting, spasms and coma are some of such physical onset symptoms.

The World Health Organization (WHO 1995) has determined body weight and is 100 nanogram per kilograms. A research has been shown by the United States that 0.5 micrograms of dieldrin contains our food directly from various sources including food, water, and air, daily. (Pengfei et al. 2011; Alan 2014).

In the body of mankind, these pesticide species can live a half-life of 300 days. Dieldrin pesticide in tap water has a maximum limit of 0.001 μg/liter and river water maximum are 0.002 μg/liter.

6.9 Pesticide Treatment Technologies

Pesticide chemicals have been used continuously in botany for the past several decades. They have been used in the prevention of pests, weeds and plant diseases. The use of these pesticides by the agricultural world has opened up new dimensions in the economic world. Along with this, these chemicals have created a crisis in front of his life by contaminating the environment to a great extent. If we talk about our drinking water in the environment, then these chemicals have also contaminated the groundwater with these sources.

The presence of these chemicals has become a concern in human society. In view of the above problems, this is absolutely necessary to remove these chemicals from the source of drinking water or to negligible its effect.

6.9.1 Membrane Technology

In the early 1990s, this way was in the form of membrane filtration that began to remove a wide spectrum of contaminants from the growing wastes. The membrane separation process is the basis on its pore size and the range from 0.03–10 μm, where1 μm is equal to 0.0001 millimeter. The two class of this technology, material science and membrane fabrication technology as well as increasing demand due to inferior quality have increased the usefulness of this technology (Shinichi 1994; Lilane et al. 2019). This technology has emerged as a very useful option due to clean and safe drinking water.

Some of favorable features are as follows.

(i) This is technically applicable at any ambient temperature without changing the state.
(ii) When used, there is no accumulation of substances in it which makes volatile nature. For example, in the ion-exchange resin process.
(iii) This does not include any chemical additives when removing the membrane which proves helpful in enhancing the environment and water quality.
(iv) In the science of technologies, membrane filtration is the very simple and traditional technology for the process of water separation.
(v) These membranes made with the use of polymeric chemistry or at low pressure also prove to be very suitable. Like new technology, these membranes or very less foam.

(vi) Membrane filtration also gives very important features like low energy, low cost operation, cleanliness and long life.

6.9.1.1 Reverse Osmosis

The methodology of this technique is based on high pressure. Reverse osmosis is used in wastewater treatment technology (Haidari et al. 2019). Reverse osmosis is a very popular technique throughout the world (Endre 2019; Simon 2019). The contaminated water is passed through an impermeable membrane. By this, almost all inorganic ions, turbidity, bacteria, solids and organic matter present in water can be separated from the water.

Advantages

1. Removes contaminants and most soluble chemicals.
2. This technology works without break-in period.
3. Low flow concentrations possible.
4. In addition to pesticide, reverse osmosis is helpful in removing bacteria and other contaminated particles.
5. Along with this small reverse osmosis system, this technique can be operational simplicity and automation its applications.
6. This technic has high capital and high cost of starting
7. Along with pollutants, present technic is also suitable in removing salinity of water.
8. Produces the most waste water like 25–50% of feed water.

6.9.1.2 Microfiltration

Microfiltration technique can be used to separate particles and bacteria of size up to 0.04–1.0 micrometer from water. Microfiltration is available as a filter cartridge in the form of a tubular, disc plate, spiral or hollow fiber. This technique is not applicable to remove the smaller particles that ultrafiltration and nanofiltration but microfiltration is effective in eliminating bacteria, colloids, pyroxenes, suspended solids with high atomic weight 0.05 to 0.10 micrometers and used in domestic water recycling systems.

6.9.1.3 Ultrafiltration

The ultrafiltration technique mainly uses this method in the removal of high-molecular-weight compounds of 0.5 to 0.10 micrometers, colloids, pyroxenes, bacteria and suspended solids by ultra-filtration. An ultrafilter is used as a membrane or filter tubular, disc plate, spiral and hollow fiber.

Advantages

1. The biggest advantage of this technique is that this technique does not require any chemicals.
2. Useful in silt filtration of its cavity.
3. Important role in the removal of contaminated particles and microbials of certain size.
4. Simple automation and,
5. Process and plant compactness.

6.9.1.4 Nanofiltration

Nanofiltration technology uses nano filters with holes up to 0.6–5 nanometers. Nanofiltration technology is multifunctional, clean water that separates contaminants from ions, microorganisms, and suspended solids which is free from these contaminants. In nanofiltration, monovalent ions cannot be separated from the water. Nanofiltration technology is also used to remove membrane alkalinity, hardness.

6.9.2 Ion Exchange Resins

Ion exchange resin, a polymeric material, is known to be one of the oldest techniques used to filter pesticide containing water. Along with this, the hardness of water also goes away by this means. This method (Alexandratos 2009) is also used for the separation of dyes, heavy metals and other organic pollutants present in the drinking water.

This contaminated water from various industries is treated according to its nature. Combined contaminated water treatment plants can be set up to treat this water from industrial units installed in industrial areas or clusters. The resin is also possible to treat such types contaminated water that discharge of two different forms by mixing them together. For example, contaminated water of acidic and alkaline nature can be mixed and neutralized. Similarly, many metallic pollutants are precipitated by mixing them together.

After proper treatment of polluted water (Deepti et al. 2016; Liang et al. 2018; Irina et al. 2018) they can be recycled under process. The resin is also possible to be used in plantations as required. The classification of ion exchange resins is as follows.

1. **Strong acid cation-** R-S(=O)$_2$-OH group containing species are strong acid cation.
2. **Strong base ion-** That have negative ions, called anion exchange resins.
3. **Weak acid cation-** H-ions lover species are weak acid cation.

4. **Weak base ion-** These species are more stable than strong base as used for removal of mineral acids, organic acids and other organic materials.

6.9.3 Activated Carbon

Activated carbon is a technique in which pesticides separate pollutants from polluted water, resulting in clean and disease free water. Along with the elimination of pesticides from water, this technique is a very effective method in the pesticide manufacturing industry (Atkins 1972; Kobylinski et al. 1984). Activation carbon is used as a cleaning method in these factories. The carbon adsorption treatment method consists of a prefilter or alum flocculation chamber made of sand when the contaminated water containing these effluents passes through these chambers or these profiles, the carbon has strong adsorption so that pesticides are absorbed. As a result, clean, sterile water is obtained by this mechanism.

Some successful results have been found in some researches that the presence of some pesticides such as chlorpyrifos, malathione, dioxanone, dimethoate and propoxur insecticides have been observed in the rinsing water. Thus, this method of adsorption has efficiently absorbed the waste and pollutants arising from the water. (Dennis and Kobylinski 1983; Nye 1984) Many researchers have used this technique for remove or separate of pesticides and these techniques have also proved very useful.

6.9.4 Phytoremediation

Phytoremediation technique is an environmentally friendly technique. In this technique, plants are mainly used to remove pollutants such as pesticides from ground water. In this way, money, time can be avoided with any other water cleaning technique (Nele et al. 2017; Prabakaran et al. 2019). Aquatic vegetation can also be prevented from getting dirty by absorbing pollutants of this type from its source. Coat and Anderson, in their research study and found a reduction of about 45% in the presence of pesticides such as atrazine, metolachlor, and trifluene, indicating that phytoremediation is a very effective emerging technique. But the above aquatic plants recorded the presence of these pollutants up to 2–50%. Phytoremediation techniques have proved very helpful in removing pesticide pollutants.

6.9.5 Bioaugmentation

Bioaugmentation is a technique in which pesticide pollutants are isolated with the help of bacteria-microbes and all the process they have play an important role. These germs convert these dangerous lethal pesticide chemicals into low toxicity products through aerobic and anaerobic processes through metabolic activity and pure water is obtained. Some bacterial strains that are the only nutrient or carbon source of these are degraded (Mariusz et al. 2017). As the carbon monoxide is released during metabolism, pesticide is released during this process with a large amount of carbon monoxide, which causes these reactions to be known as minerals.

Bacterial microorganisms in their metabolic process, secretes some enzymes in which these pesticides are consumed as food, which reduces the such pollutants present in the aquatic environment. They can be easily used as a result of bacterial microbe found everywhere. They always exist in the environment and serve as nature and maintain life cycle while serving nature (Do et al. 2015).

Microorganisms use this energy to complete this process, from the energy generated by the breakdown of pesticides to the energy used to increase their number or biomass. The breakdown of the bond and the taking of electrons from the contaminants receive energy and more and more new cells are produced.

Now bioaugmentation comes that depending on which cars these pesticides are destroyed by the environment or water source. This type is as follows.

1. Types of bacteria or microbes
2. Pesticide Chemical Type
3. Geological composition and presence of chemicals on contaminated water and water bodies

These factors greatly influence the result of this technique. Overall, this is concluded that this technology or organisms used in this technology converts harmful chemical compounds, ions present in water and environment into other such products which are reduce into less harmful.

6.9.6 Electrocoagulation

Nowadays, electrocoagulation technology is being used on a large scale in wastewater treatment. Among the classical physicochemical treatment processes. The principle of this technique is a reaction based on an advanced electrochemical technique. In this process, water is treated as a medium and is used to remove hazardous pesticide chemicals by electrifying them in various types of anodes and cathodes. These can be in the form of anode and cathode with rods, fluidized bed shells, plates, tubes, balls and wire mesh. In this, metal salts polymer and polyelectrolyte can be used to break the emulsion and suspension (Nuno et al. 2019; Miguel et al. 2019; Rabiatuladawiyah et al. 2019). Polymers which are highly charged used to remove water contamination through pesticides or other such as colloidal solids, soluble inorganic and metals from aqueous media.

6.10 Conclusion

From the above study concluded that pesticides which is a deadly chemical and affected by both the fauna and flora. Its bigger source is mainly agricultural where these pesticides are used the most. Some pesticides form as complex with organic compounds and these compounds have carcinogenic properties. If these chemicals sprayed on the surface of plants by spraying methods, during rainy days when the plants are watered, these chemicals dissolve in water, or form colloidal solution with them. In both the states, they come out in the drinking water source and make the drinking water harmful to all lives.

The most pesticides or biocides are chlorinated hydrocarbons. These pesticides are non-biodegradable i.e. why their extreme effects are on water bodies and aquatic life. So, these type chemicals are need to remove them from our water system. Many modern technologies are using to remove them. The bioaugmentation method is the best method for treatment of pesticides contaminants. Bioaugmentation is proved the green and very significant method.

Acknowledgement The authors are grateful to Dr. HS Bhargav, former coordinator-UGC, India for critical reading of the chapter.

References

Abarikwu SO, Adesiyan AC (2009) Changes in sperm characteristics and induction of oxidative stress in the testis and epididymis of experimental rats by an herbicide, atrazine. Arch Environ Contam Toxicol 58(3):874–882. https://doi.org/10.1007/s00244-009-9371-2

Accra G, Williamson S, Ball A, Pretty J (2008) Trends in pesticide use and drivers for safer pest management in four African countries. Crop Prod 27:1327–1334. https://doi.org/10.1016/j.cropro.2008.04.006

Adgate JL, Kukowski A, Stroebel C (2000) Pesticide storage and use patterns in Minnesota households with children. J Expo Anal Environ Epidemiol 10:159–167. https://doi.org/10.1038/sj.jea.7500078

Agricultural Chemical Regulations (2006) The Control of Agricultural Chemicals (Registration and Control) Regulations (Under Section 15 of the Act 2006). The Republic of Uganda, Entebbe, Uganda, pp 22–23

Alan H (2014) Hazard identification of the potential for dieldrin carcinogenicity to humans. Environ Res 131:188–214. https://doi.org/10.1016/j.envres.2014.02.007

Alavanja MC, Hoppin JA, Kamel F (2004) Health effects of chronic pesticide ex-posure: cancer and neurotoxicity. Annu Rev Publ Health 25:155–197. https://doi.org/10.1146/annurev.publhealth.25.101802.123020

Alexandratos SD (2009) Ion-exchange resins: a retrospective from industrial and engineering chemistry research. Ind Eng Chem Res 48(1):388–398. https://doi.org/10.1021/ie801242v

Aly D, Russel C, Waqar J, Hiroshi ST (2019) Trends in organophosphorus pesticides use and concentrations in river water in Japan, and risk assessment. J Environ Sci 79:135–152. https://doi.org/10.1016/j.jes.2018.11.019

Ankley GT, Bennett RS, Erickson RJ (2010) Adverse outcome pathways: a conceptual framework to support ecotoxicology research and risk assessment. Environ Toxicol Chem 29:730–741. https://doi.org/10.1002/etc.34

Arena M, Sgolastra F (2014) A meta-analysis comparing the sensitivity of bees to pesticides. Ecotoxicology 23:324–334. https://doi.org/10.1007/s10646-014-1190-1

Arias ME, Lopez EP, Martinez EC (2008) The mobility and degradation of pesticides in soils and the pollution of groundwater resources. Agric Ecosyst Environ 123:247–260. https://doi.org/10.1016/j.agee.2007.07.011

Arinaitwe K, Kiremire BT, Muir DCG, Fellin P, Li H, Teixeira C, Mubiru DN. (2016) Legacy and currently used pesticides in the atmospheric environment of Lake Victoria, East Africa. Sci Total Environ. 543:9–18. https://doi.org/10.1016/j.scitotenv.2015.10.146

Arisekar U, Robinson JS, Shalini R, Palani K, Asha HM, Rani V (2019) Accumulation of organochlorine and pyrethroid pesticide residues in fish, water, and sediments in the Thamirabarani river system of southern peninsular India. Environ Nanotechnol Monit Manag 11:100194. https://doi.org/10.1016/j.enmm.2018.11.003

Arne D, Niels DT, Michael H, Luis DG, Indira N, Wout VE, Marie AEF, Pieter S, Peter G (2019) Distribution of agricultural pesticides in the freshwater environment of the Guayas river basin (Ecuador). Sci Total Environ 646:996–1008. https://doi.org/10.1016/j.scitotenv.2018.07.185

Atkins PR (1972) The pesticide manufacturing industry-current waste treatment and disposal practices. Water pollution control research series. US Environmental Protection Agency, Washington, DC

Atkins EL, Kellum D (1986) Comparative morphogenic and toxicity studies on the effect of pesticides on honeybee brood. J Appl Res 25(4):242–255. https://doi.org/10.1080/00218839.1986.11100725

Atreya N (2006) Chemophobia pesticide residues in food. Outlooks Pest Manag 17:242. https://doi.org/10.1564/17dec01

Austin RB (1999) Yield of wheat in the United Kingdom: recent advances and prospects. Crop Sci 39(6):1604–1610. https://doi.org/10.2135/cropsci1999.3961604x

Baker SR, Wilkenson CF (1990) The effects of pesticides on human health. Princeton Sci. Pub., Princeton

Bertolote JM, Fleischmann A, Eddleston M, Gunnell D (2006) Deaths from the pesticide poisoning global response. Br J Psychiatry 189:201–203. https://doi.org/10.1192/bjp.bp.105.020834

Bohmont BL (1997) Educational and informational strategies to reduce pesticide risks. Prev Med 26(2):191–200. https://doi.org/10.1006/pmed.1996.0122

Bolognesi C (2003) Genotoxicity of pesticides: a review of human biomonitoring studies. Mutat Res 543(3):251–272. https://doi.org/10.1016/S1383-5742(03)00015-2

Bolognesi C, Merlo FD (2011) Pesticides: human health effects. In: Nriagu JO (ed) Encyclopedia of environmental health. Elsevier, Burlington. https://doi.org/10.1016/B978-0-444-52272-6.00592-4

Cerejeira MJ, Viana P, Batista S, Pereira T, Silva E, Valerio MJ (2003) Pesticides in Portuguese surface and ground waters. Water Res 37(5):1055–1063. https://doi.org/10.1016/S0043-1354(01)00462-6

Deepti SP, Sanjay MC, John UKO (2016) A review of technologies for manganese removal from wastewaters. J Environ Chem Eng 4(1):468–487. https://doi.org/10.1016/j.jece.2015.11.028

Dennis WH, Kobylinski EA (1983) Pesticide-laden wastewater treatment for small waste generators. J Environ Sci Health B 18(30):317–331. https://doi.org/10.1080/03601238309372372

Devleesschauwer B, Havelaar AH, Maertens C, Haagsma JA, Praet N, Dorny P, Duchateau L, Torgerson PR, Van OH, Speybroeck N (2014) Calculating disability-adjusted life years to quantify burden of disease. Int J Public Health 59(3):565–569. https://doi.org/10.1007/s00038-014-0552-z

Do GL, Kun CC, Kung HC (2015) Removal of triclosan in nitrifying activated sludge: effects of ammonia amendment and bioaugmentation. Chemosphere 125:9–15. https://doi.org/10.1016/j.chemosphere.2014.12.085

Drechsel P, Scott CA, Raschid-Sally L, Redwood M, Bahri A (2010) Wastewater irrigation and health: assessing and mitigating risk in low-income countries. Earthscan, London, p 404

Dufour A, Bartram J, Bos R, Gannon V (2012) Animal waste, water quality and human health. IWA Publishing, London. https://doi.org/10.2166/9781780401249

Eduard F, Valenzuela HC, Menezes ZL (2019) New passive sampling device for effective monitoring of pesticides in water. Anal Chim Acta 1054:26–37. https://doi.org/10.1016/j.aca.2018.12.017

Eerkes MD, Thompson RC, Aldridge DC (2015) Microplastics in fresh water systems: a review of the emerging threats, identification of knowledge gaps and prioritization of research needs. Water Res 75:63–82. https://doi.org/10.1016/j.watres.2015.02.012

Endre N (2019) Reverse Osmosis. In: Basic equations of mass transport through a membrane layer, 2nd edn. Elsevier, Amsterdam, pp 497–503. https://doi.org/10.1016/B978-0-12-813722-2.00020-0

Epstein E (2015) Disposal and management of solid waste: pathogens and diseases. CRC Press, Boca Raton. https://doi.org/10.1201/b18070

Epstein L, Bassein S (2003) Patterns of pesticide use in California and the implications for strategies for reduction of pesticides. Annu Rev Phytopathol 41:351–375. https://doi.org/10.1146/annurev.phyto.41.052002.095612

Epstein L, Zhang M. (2014) The impact of integrated pest management programs on pesticide use in California, USA. doi: https://doi.org/10.1007/978-94-007-7802-3_7

Erlanger TE, Keiser J, Caldas M, Bos R, Singer BH, Tanner M, Utzinger J (2005) Effect of water resource development and management on lymphatic filariasis and estimates of populations at risk. Am J Trop Med Hyg 73:523–533. https://doi.org/10.4269/ajtmh.2005.73.523

Gangesh S, Anoop K, Mayank B, Joseph PE, Ajay T (2006) Organochlorine pesticide contamination of ground water in the city of Hyderabad. Environ Int 32(2):244–247. https://doi.org/10.1016/j.envint.2005.08.027

Gross M (2015) How life shaped earth. Curr Biol 25(19):847–850. https://doi.org/10.1016/j.cub.2015.09.011

Haidari AH, Gonzalez O, Heijman SGJ (2019) Scaling after remineralisation of reverse osmosis permeate. Desalination 467:57–63. https://doi.org/10.1016/j.desal.2019.06.002

Hicham EB, Abdelhamid O, Jose M, Jose U (2008) Pesticides in ground water beneath Loukkos perimeter, Northwest Morocco. J Hydrol 348(3–4):270–278. https://doi.org/10.1016/j.jhydrol.2007.10.002

Honeycutt M, Shirley S (2014) Aldrin. In: Encyclopedia of toxicology, 3rd edn. Elsevier, Amsterdam, pp 126–129. https://doi.org/10.1016/B978-0-12-386454-3.00094-4

Irina L, Juan J, Rueda M, Mika S (2018) Removal of natural organic matter (NOM) from water by ion exchange – a review. Chemosphere 192:90–104. https://doi.org/10.1016/j.chemosphere.2017.10.101

Jing G, Mengxiao L, Donglan W, Zhiyong Z, Xiangyang Y (2016) Dissipation and distribution of chlorpyrifos in selected vegetables through foliage and root uptake. Chemosphere 144:201–206. https://doi.org/10.1016/j.chemosphere.2015.08.072

Kobylinski EA, Dennis WH, Rosencrance AB (1984) Treatment of pesticide-laden waste water by recirculation through activated carbon. Treatment of pesticide wastes. ACS Symp Ser 259:125–154. https://doi.org/10.1021/bk-1984-0259.ch008

Liang W, Lu L, Zhaohui Z, Bin Z, Junjing L, Bingjie D, Nian L (2018) 17α-Ethinylestradiol removal from water by magnetic ion exchange resin. Chin J Chem Eng 26(4):864–869. https://doi.org/10.1016/j.cjche.2017.08.006

Lilane A, Saifaoui D, Hariss S, Jenkal H, Chouiekh M (2019) Modeling and simulation of the performances of the reverse osmosis membrane. Mater Today Proc. https://doi.org/10.1016/j.matpr.2019.07.694

Mariusz C, Agnieszka M, Zofia PS (2017) Bioaugmentation as a strategy for the remediation of pesticide-polluted soil: a review. Chemosphere 172:52–71. https://doi.org/10.1016/j.chemosphere.2016.12.129

Michael K, Denk N, Milutinovic S (2018) Reductive dehalogenation of DDT with folate models: formation of the DDT metabolite spectrum under biomimetic conditions. Chemosphere 191:408–411. https://doi.org/10.1016/j.chemosphere.2017.10.055

Miguel AS, Rosalba F, Jose LN, Oscar C, Yanmei L, Jesus HH (2019) Simultaneous removal of fluoride and arsenic from groundwater by electrocoagulation using a filter-press flow reactor with a three-cell stack. Sep Purif Technol 208:208–216. https://doi.org/10.1016/j.seppur.2018.02.018

Mourabit F, Boulaid M (2019) Pesticide removal in drinking water treatment using biodegradable polymers. Mater Today Proc 13(3):1033–1038. https://doi.org/10.1016/j.matpr.2019.04.068

Nele E, Jason CW, Jaco V, Nele W (2017) Bio- and phytoremediation of pesticide-contaminated environments: a review. Adv Bot Res 83:277–318. https://doi.org/10.1016/bs.abr.2017.01.001

Nuno SG, Ana MR, Alirio ER (2019) Modeling the electrocoagulation process for the treatment of contaminated water. Chem Eng Sci 197:379–385. https://doi.org/10.1016/j.ces.2018.12.038

Nye JC (1984) Treating pesticide-contaminated wastewater, development and evaluation of a system. Treatment and disposal of pesticide wastes. ACS Symp Ser 259:153–160. https://doi.org/10.1021/bk-1984-0259.ch009

Pengfei X, Toshio M, Ichiro K (2011) Novel metabolic pathways of organochlorine pesticides dieldrin and aldrin by the white rot fungi of the genus Phlebia. Chemosphere 85:218–224. https://doi.org/10.1016/j.chemosphere.2011.06.028

Prabakaran K, Jian L, Anandakumar A, Zhanrui L, Chris B, Zou DD (2019) Managing environmental contamination through phytoremediation by invasive plants: a review. Ecol Eng 138:28–37. https://doi.org/10.1016/j.ecoleng.2019.07.002

Rabiatuladawiyah D, Shafreeza S, Luqman CA, Mohsen NM (2019) FTIR, CHNS and XRD analyses define mechanism of glyphosate herbicide removal by electrocoagulation. Chemosphere 233:559–569. https://doi.org/10.1016/j.chemosphere.2019.06.010

Rani GI, Kumaraguru AK (2014) Behavioural responses and acute toxicity of Clariasbatrachusto synthetic pyrethroid insecticide, ʎ-cyhalothrin. J Environ Appl Biores 2(1):19–24

Rasul G, Thapa G (2003) Sustainability analysis of ecological and conventional agricultural systems in Bangladesh. World Dev 31(10):1721–1741. https://doi.org/10.1016/S0305-750X(03)00137-2

Richa K, Shilpi G (2018) Agrochemicals as a potential cause of ground water pollution: a review. Int J Chem Stud 6(3):985–990

Rosa MS, Pascal JF, Arnautvan LA, Annemarie PV, Van W (2019) Occurrence of pesticides in Dutch drinking water sources. Chemosphere 235:510–518. https://doi.org/10.1016/j.chemosphere.2019.06.207

Salam MA, Shahjahan M, Sharmin S, Haque F, Rahman MK (2015) Effects of sub-lethal doses of an organophosphorous insecticide sumithion on some hematological parameters in common carp. Pak J Zool 47:1487–1491

Sattar MA, Rahman MM (1987) Techniques of soil analysis. Bangladesh Agricultural University, Mymensingh, p 120

Satyavardhan K (2013) A comparative toxicity evaluation and behavioral observations of fresh water fishes to Fenvalerate. Middle-East J Sci Res 13(2):133–136

Schipper PNM, Marc V (2008) Pesticides in groundwater and drinking water wells: overview of the situation in the Netherlands. Water Sci Technol 57(8):1277–1286. https://doi.org/10.2166/wst.2008.255

Sharmin S, Shahjahan M, Hossain MA, Haque MA, Rashid H (2015) Histopathological changes in liver and kidney of common carp exposed to sub-lethal doses of malathion. Pak J Zool 47(5):1495–1498

Shi W, Zhang FX, Zhang XW, Su GY, Wei S, Liu HL, Cheng SP, Yu HX (2011) Identification of trace organic pollutants in freshwater sources in Eastern China and estimation of their associated human health risks. Ecotoxicology 20(5):1099–1106. https://doi.org/10.1007/s10646-011-0671-8

Shin-ichi N (1994) Determination of pore size and pore size distribution: 3. Filtration membranes. J Membr Sci 96(1–2):131–165. https://doi.org/10.1016/0376-7388(94)00128-6

Siddika F, Shahjahan M, Rahman MS (2012) Abundance of plankton population densities in relation to bottom soil textural types in aquaculture ponds. Int J Agric Res Innov Technol 2(1):56–61. https://doi.org/10.3329/ijarit.v2i1.14006

Simon A (2019) PVDF optimises UF membranes for reverse osmosis pre-treatment. Membr Technol 2019(6):5. https://doi.org/10.1016/S0958-2118(19)30114-4

Tawfeek K, Ali AH, Ashraf ME, Ekramy H, Mostafa S (2019) Degradation of some organophosphorus pesticides in aqueous solution by gamma irradiation. J Hazard Mater 373:23–28. https://doi.org/10.1016/j.jhazmat.2019.03.011

Uddin MN, Shahjahan M, Haque MM (2012) Manipulation of species composition in small scale carp polyculture to enhance fish production. Bangladesh J Prog Sci Technol 10:9–12

Ullah R, Zuberi A, Tariq M, Ullah S (2014) Toxicity to hematology and morphology of liver, brain and gills during acute exposure of mahseer (Tor putitora) to cypermethrin. Int J Agric Biol 17(1):199–204

Van den Brink PJ (2013) Assessing aquatic population and community level risks of pesticides. Environ Toxicol Chem 32:972–973. https://doi.org/10.1002/etc.2210

Wangyang L (2017) Pesticide use and health outcomes: evidence from agricultural water pollution in China. J Environ Econ Manag 86:93–120. https://doi.org/10.1016/j.jeem.2017.05.006

Westbom R, Hussen A, Megersa N, Retta N, Mathiasson L, Bjorklund E (2008) Assessment of organochlorine pesticide pollution in Upper Awash Ethiopian state farm soils using selective pressurized liquid extraction. Chemosphere 72:1181–1187. https://doi.org/10.1016/j.chemosphere.2008.03.041

WHO (1995) Pesticide residues in food-1994. Report of the joint meeting of the FAO Panel of Experts on Pesticide Residues in Food and the Environment and WHO Toxicological and Environmental Core Assessment Groups. Food and Agriculture Organization of the United Nations, Rome

WHO (2003) Aldrin and dieldrin in drinking-water. Background document for preparation of WHO Guidelines for drinking-water quality. World Health Organization, Geneva

WHO (2009) Guns, knives, and pesticides. Reducing access to lethal means. World Health Organization, Geneva, pp 1–16

Williamson S (2007) The hidden costs of pesticide use in Africa. Pesticides Management in West Africa, ECOWAS Newsletter no.6

WWF (1998) Resolving the DDT dilemma: protecting biodiversity and human health. World Wildlife Fund Canada, Toronto, pp 1–8

Xiaochu CQ, Zhou FL, Qingrong PPT (2019) Removal of nine pesticide residues from water and soil by biosorption coupled with degradation on biosorbent immobilized laccase. Chemosphere 233:49–56. https://doi.org/10.1016/j.chemosphere.2019.05.144

Yihan C, Kaifeng Y, Muhammad H, Cong X, Bo Z, Karina YHG, Yiliang H (2018) Occurrence, distribution and risk assessment of pesticides in a river-reservoir system. Ecotoxicol Environ Saf 166:320–327. https://doi.org/10.1016/j.ecoenv.2018.09.107

Zijian L, Aaron J (2018) Global variations in pesticide regulations and health risk assessment of maximum concentration levels in drinking water. J Environ Manag 214:426

Chapter 7
Degradation of Pesticides Residue by Engineered Nanomaterials

Manviri Rani, Uma Shanker, Jyoti Yadav, and Keshu

Abstract The extensively used pesticides have severely impacted on environment in view of their toxicity and persistence. Among them, organochlorines are highly toxic with half-lives of many years followed by organophosphates. Being banned in many countries, most of pesticides are still persisting into the environment. Traditional methods are not enough for removal of those structurally stable and naturally persistent toxicants. Sustainable technologies based on engineered nanomaterials have better treatment efficiency credited to enhanced multidimensional properties including surface functionalization and porosity, specific targeting capabilities, increased surface-area, and catalysis. Further, improvisation in engineered nanomaterials for potential adsorption of pollutants could be achieved by application of organic ligands, enzyme immobilization, inorganic moieties and surface polymerization.

In this chapter, we have described present status of pesticides removal by engineered nanomaterials. Moreover, categorization of pesticides along with various health effects is also discussed. Titanium dioxide, zinc oxide and iron based nanomaterials are highly used for treatment of organochlorine and organophosphorus pesticides. Fabrication of types of engineered nanomaterials also added enhanced properties using templates of zinc, titanium, and tungsten as semiconductor photocatalyst; iron and carbon as nanoadsorbents and polymer like polyaniline and polythiophene have also described. The chapter concluded with major gaps in implementing the hybrid technologies in water treatment plants, industries and their future scope and perspectives of engineered nanomaterials.

Keywords Engineered nanomaterials · Environmental contaminants · Pesticides · Photocatalyst · Degradation

M. Rani · J. Yadav
Malaviya National Institute of Technology, Jaipur, Rajasthan, India
e-mail: manviri.chy@mnit.ac.in

U. Shanker (✉) · Keshu
Dr B R Ambedkar National Institute of Technology, Jalandhar, Punjab, India
e-mail: shankeru@nitj.ac.in

Inamuddin et al. (eds.), *Sustainable Agriculture Reviews 48*, Sustainable Agriculture Reviews 48, https://doi.org/10.1007/978-3-030-54719-6_7

7.1 Introduction

Recently, water contamination by toxic and recalcitrant compounds is becoming a grave problem to its proper consumption. These micro-pollutants and persistent organic pollutants are discharged from various direct and indirect sources that include industries and domestic discharges as well as agricultural runoff (Rani and Shanker 2018a, b, c, d). They are hard to detect and degrade even at trace levels concentrations. Pesticides have been used for boosting agricultural produce worldwide. Moreover, the runoff from agricultural fields carries the risk of surface water contamination which may directly affect the consumers. Drinking water may have pesticides with permissible limit of 0.1–0.5 μgL^{-1} for total dissimilar constituents of water used for human ingestion as per Water Directive (98/83/EC) (Karabelas et al. 2009). Dichlrodiphenyletrichlorethane and benzene hexachloride were the most primitive organochlorine pesticide that created havoc in environment (Gupta 2004). Pesticides are commercially synthesized from toxic constituent majority of them being phenols and cyanide derivatives. The inhibitory effect of phenols on biological microorganisms makes them recalcitrant towards biological treatment techniques. The leakage of methyl isocyanate from a carbamate pesticide production plant in Bhopal resulted in a deadly disaster (Broughton 2005). Pesticides have extensive production and consumption as well as prevalence resulted in toxicity caused by themselves and their metabolites. In view of these challenges, pesticides are classified into united state-environmental protection agency priority pollutants list (Pillai and Gupta 2015a). Moreover, strict regulations have been applied for their marketing and use of pesticides derivatives by The European Union (Rani et al. 2017a, b). Nevertheless, there is alarming rise in pesticides poisoning cases reported from Asian countries.

Being chemically hydrophobic, pesticides showed bioaccumulation over a long range of periods even in years. They affect the biological system including central nervous and endocrine system (McKinlay et al. 2008). Pesticides like atrazine, parathion, lindane, endosulfan, dichlorodiphenylaceticacid, endrin and oxychlordane are used in agricultural fields and domestic places (Devi and Raha 2013). The recalcitrant nature of the contaminants headed to development of various advanced efficient techniques based on engineered nanomaterials that are cheap, accessable and can able to complete removal or degradation of pesticides (Morasch et al. 2010; Ormad et al. 2011; Pitarch et al. 2010). Engineered nanomaterials are promoted to industrial scale due to their enhanced characteristic, wide band gap and improved output of applications. There are several review available on cheap traditional adsorbents (Ahmed et al. 2010), advanced oxidation processes (Burrows et al. 2002) and doped photocatalysts (Devipriya and Yesodharan 2005; Sud and Kaur 2012). Konstantinou and Albanis (2003) have reported photocatalytic-degradation of organophosphates under aqueous suspension of titanium dioxide under the exposure of artificial and solar light.

Engineered nanomaterials also showed potentials as reactive adsorbents in eradication of organophosphates pesticides and several warfare agents (Amin et al.

2014). Rani et al. (2017a, b) presented a comprehensive review on degradation of organochlorines by nanomaterials. It has been conclued that use of engineered nanomaterials like titanium dioxide and zerovalent iron are basically used. Limited discussion on employment of nanocomposites and nano-biocomposite for pesticides are reported (Sahithya et al. 2015). In this light of discussion, a whole packet on pernicious effects of several categories of pesticides along with the major methods for their removal by different types of engineered nanomaterials from environmental matrices is imperative. Present status of pesticides, their classification, consumption and toxicity are also discussed. Engineered nanomaterials as such or modified forms are categoriezed into several groups and specific engineered nanomaterials is highlighted for their application towards traditional and new emerging pesticides that are removed through adsorption, redox, rapid reactive adsorption (SN^2 reaction) and photocatalytic mechanism (Herrmann 1999). In view of environmental concern, importance of green synthesized engineered nanomaterials is also deliberated (Maurino et al. 1999; Gaya and Abdullah 2008).

7.1.1 Wordwide Consumption of Pesticides

Pesticides are cheap and easily available with great potential of action that led to the rapid increase in call for pesticides and analogues. Statistics from food and agricultural organization in 2015 indicated that approximately 2.8 metric tons of pesticides are spent all over the world. Europe was the highest annual consumer of pesticides followed by Asia and United States of America worldwide. Asian countries are the major producers of these chemicals (Gupta 2004; Zhang et al. 2011). Singh et al. (2014) reported worldwide pesticides consumption in the order: Taiwan with the highest (17 Kg ha^{-1}) followed by China (13 Kg ha^{-1}), Japan (12 Kg ha^{-1}), Korea (7 Kg ha^{-1}), France (5 Kg ha^{-1}) and United Kingdom (5 Kg ha^{-1}). A report of European Crop Protection Association stated that pesticide usage is more widespread in Europian, American and Asian countries with total of 80% of yearly consumption in the year 2001 (Tadeo 2008). It has also been observed that the cost of life and quality that is being degraded by pesticide is being tolerated majorly by the developing countries. Global estimates report that number of deaths and diseases that these toxic substances cause can be estimated more than 1 million per year (Aktar et al. 2009).

7.1.2 Chemical Sketch and Grouping of Pesticides

Pesticides are categorized either on their chemical structures or based on control-action towards pests. With different functional moieties and unique structure, they are placed as organochlorines, organophosphorus, carbamates and pyrethroids pesticides (Konstantinou et al. 2006). While on based of action, they can be grouped into insecticides, fungicides, and herbicides. World health organization also categorizes pesticides based on toxicity (oral and dermal) dependent lethal dose that is capable of killing 50% of the target usually represented as lethal dose 50 (mg kg^{-1}) (World Health Organization 2002, 2010).

The most accepted classification is based on chemical compositions is discussed below:-

Organochlorines The chlorinated ethane derivatives, mainly used as insecticides and known as persistent organic pollutants. Examples of organochlorines are dichlrodiphenyletrichlorethane, benzenhexachloride and chlordane. They are mostly banned for use and still found in environmental segments due to their long half-lives.

Organophosphorus They are derivatives of organophosphoric acid and extensively used worldwide after strict regulations on organochlorines. They are not as such toxic and persistent like organochlorines but cannot be evaluated as harmless (Ecobichon 1991; Padilla et al. 1994). They are also treated as poisoning and homicidal agents. Organophosphates and its metabolic products are proved to be toxic to organisms (Fernandez-Casalderrey 1992; Ray and Ghosh 2006). Examples of this category are malathion, parathion and chlorpyriphos.

Substituted Ureas These herbicides with free iminohydrogen have high selectivity and lower persistence (half-life ranges from days to weeks) (Wang et al. 2009). Photosynthesis efficiency of plants can be reduced even by application of 1.5 μg L^{-1} of diuron (Ricart et al. 2010).

Carbamate They are the derivatives of carbamic acid and cheaper than the organochlorines and organophosphate pesticides e.g., carbofuran, diazinon carbaryl, thiram and mancozeb. Carbamates with reproductive toxicities are found to occurrence in atmosphere (air and fog) (Pesticide news 2000) and their byproducts have shown more toxicities than parent ones (Chapalamadugu and Chaudhry 1992).

Biopesticides These are safer alternatives obtained from either derivatives of plants or micro-organism extracts. Many of them are in use and some are coming into the market. Examples of this group are dipel (*bacillus thuringiensis*, bacteria), pyrethrum (*chrysanthemum cineraiaefolium*) and neem (*azadirachta indica*) (Isman 1994, 1997).

Besides the categories discussed, a number of miscellenous pestides with different or modified functional groups are also used e.g., pyridazinone, neonicotinoids, triazines and anilides. *They are newely introduced and less used, but still showing emerging concerns to environment due to caused toxicity and persistence.*

Various pesticides structures are presented as Fig. 7.1.

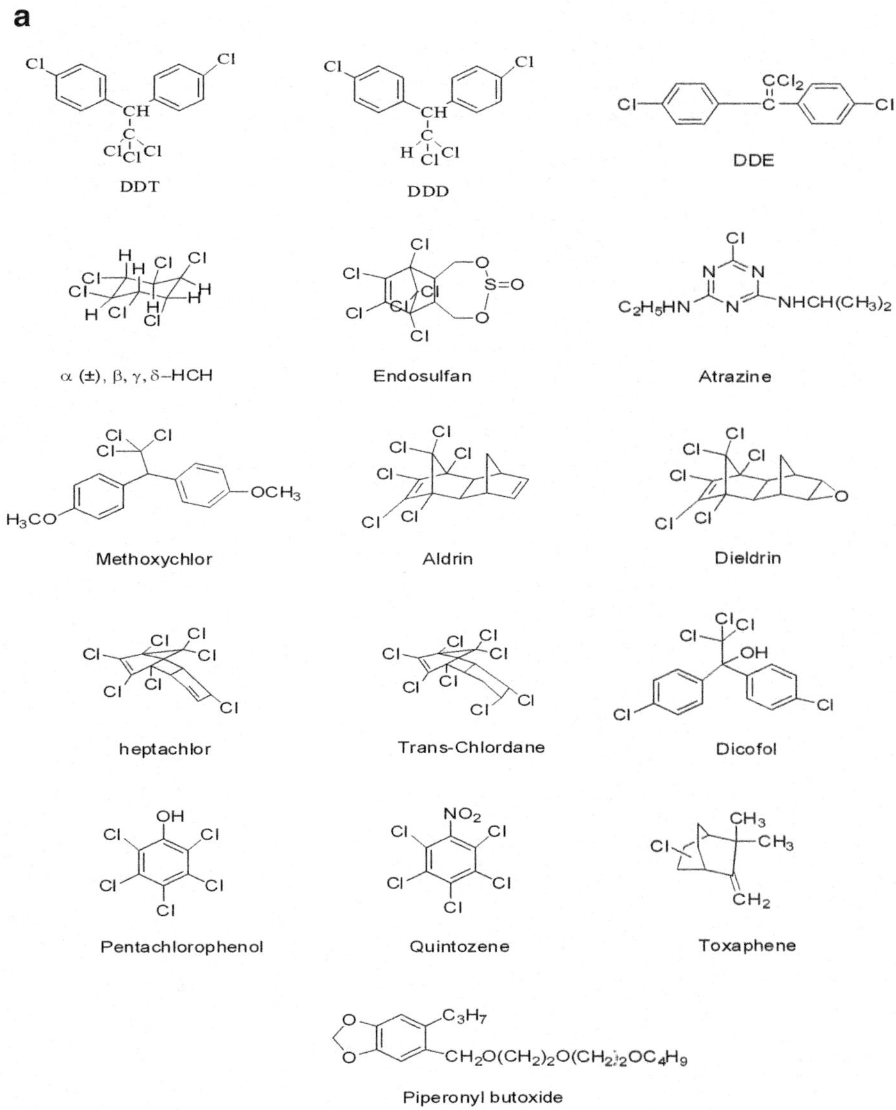

Fig. 7.1 (**a**) Structures of various organochlorine pesticides. (**b**) Structures of various organophosphate pesticides. (**c**). Structures of various carbamate pesticides

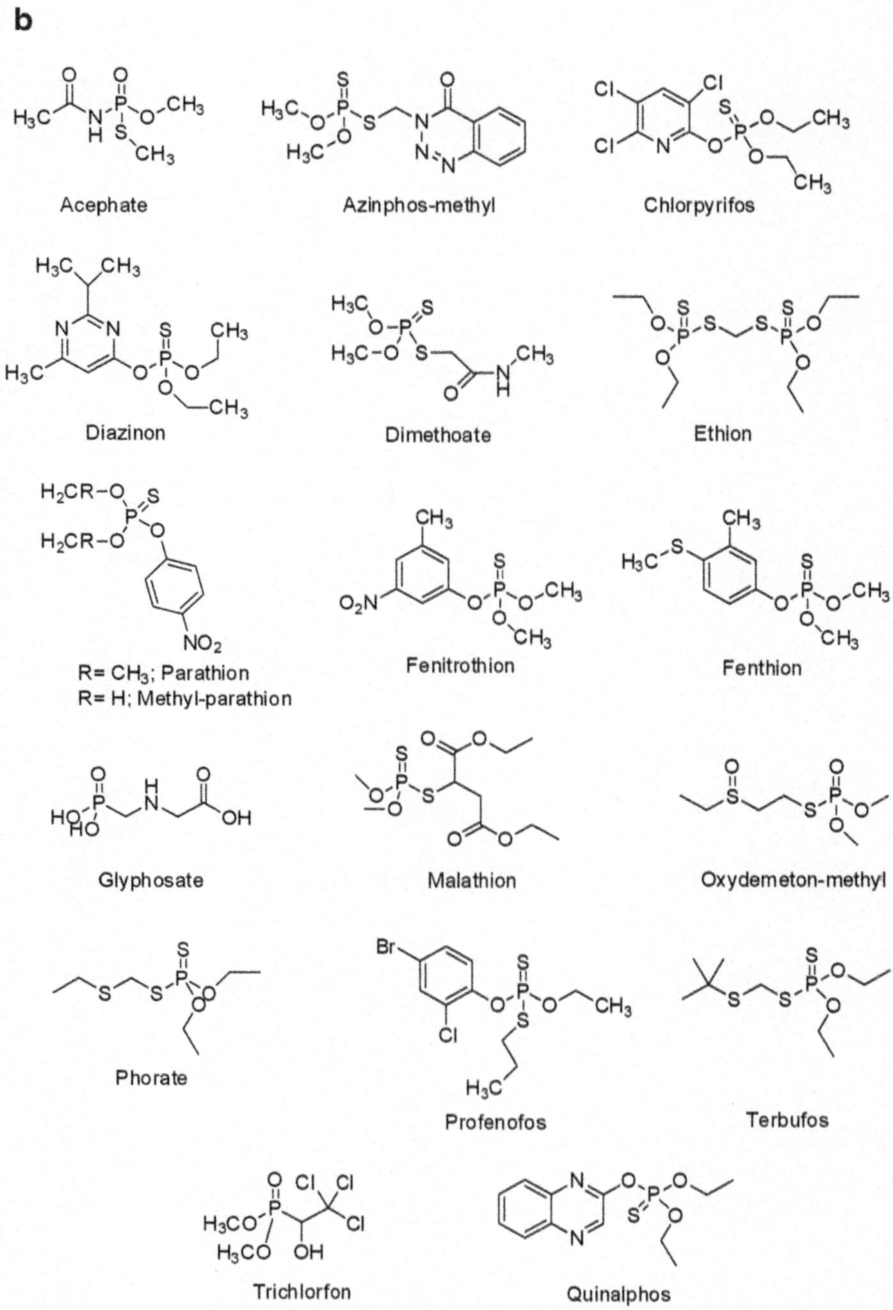

Fig. 7.1 (continued)

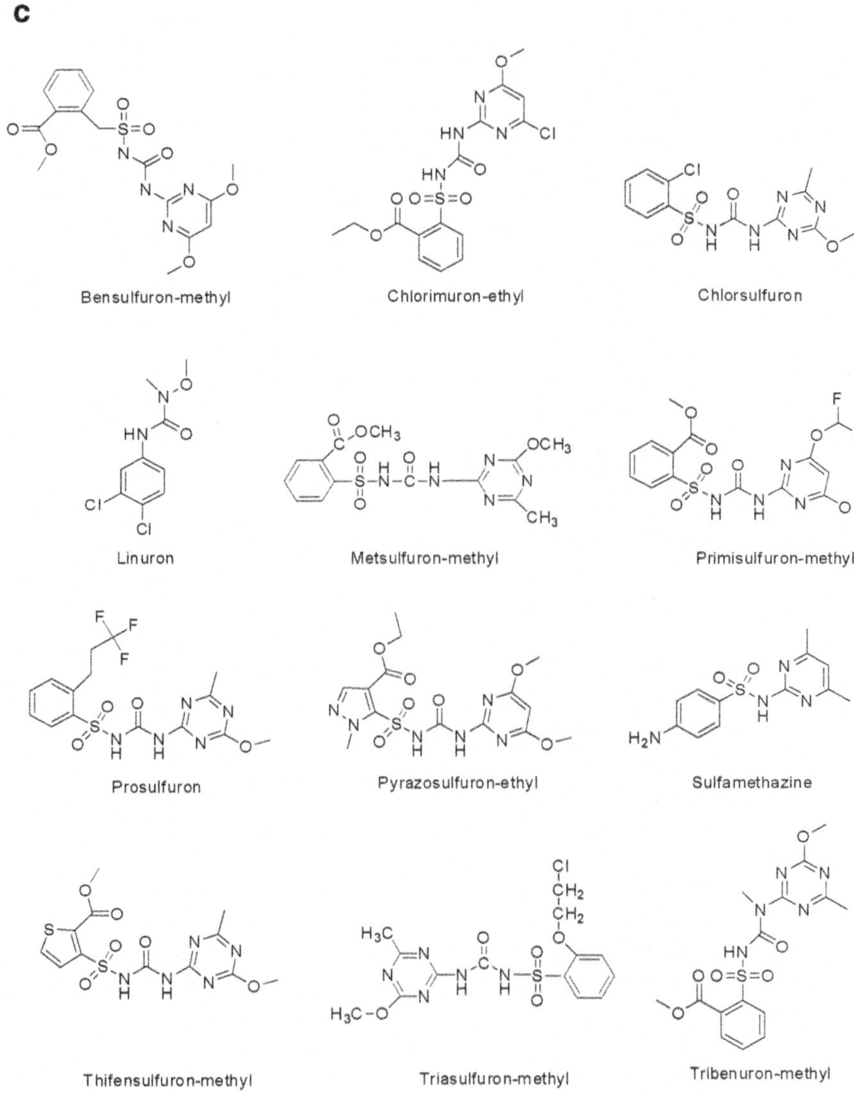

Fig. 7.1 (continued)

7.1.3 Environmental Concerns of Harmful Pesticides

Pesticides have a numerous effects not only on human beings but also on the entire biosphere. The extensively employed pesticides are prevalent in developed and developing countries. Oganophosphates and carbamate pesticides are included into federal economics poison act in view of their various kinds of harmful effects

(Landrigan et al. 1999; Pradeep and Anshup 2009). Long term exposure of pesticides is also resulted in damage of central nervous system by obstructive sodium-potassium gates and inhibition of cholinesterase and chitin synthesis (McKinlay et al. 2008). Fungicides inhibit ergosterol synthesis which is a potential precursor of vitamin D2. Heptachlor epoxides is highly toxic due to strong adsorption to organic matter and recalcitrant nature (Devi and Raha 2013; Grundlingh et al. 2011). Pyrethroids are also seen with carcinogenicity, reproductive toxicity and endocrine disruptors (Hrelia et al. 1996; Ishihara et al. 2003; Slotkin 1999). Moreover, it was noted that a very minute quantity of pesticide could even cause a potential threat due to their longer persistence (Yadav et al. 2013). They also inhibit the acetylcholinesterase activity of nerve cells (Tomer et al. 2015; McKinlay et al. 2008). Pesticides have caused the reproductive failure in some fish species and reduction of the thickness of birds-eggshells (Al Hattab and Ghaly 2012). The pesticides related accidents are summarized as Table 7.1

7.2 Classes of Engineered Nanomaterials

7.2.1 Carbon Based Engineered Nanomaterials

The backbone of this type of engineered nanomaterials mainly made up of Carbon, for example, uni and multi-walled carbon-nanotubes, graphene and fullerenes along with derivatives, nanofibers of carbon, and amorphous carbonaceous composites (Iijima 1991a, b; Chen et al. 2011a, b; Zhu et al. 2010; Gupta and Saleh 2013). Carbon based engineered nanomaterials are inherently hydrophobic materials and promising materials for removal of pollutants due to exceptionally high surface-area. These materials can only adsorb the contaminants physically and not able to transform or completely degrade the targets. Moreover, surface of engineered nanomaterials can be modified by functionalization (introduction of functional group on surface) or by making composites (aligned or incorporated with fibre or polymer) through chemical and electrical interactions (Wang et al. 2013; Mostafavi et al. 2009; Han et al. 2013; Gupta and Saleh 2013; Ren et al. 2011).

7.2.2 Metal and Metal Oxides Based Engineered Nanomaterials

Metal and metal oxide engineered nanomaterials are extensively used for their useful applications such as catalysis, sensor, electrochemical cells, and optical. They are mainly transition metal compounds that may consist of one, two, or, less commonly, three metals and/or their oxides (Li et al. 2008; Xu et al. 2012; Pradeep and Anshup 2009; Stoimenov et al. 2002). Titanium dioxide (Soni et al. 2008; Lu et al. 2013;

Table 7.1 Some important pesticide related accidents worldwide

Pesticide	Place	Year	Causes	References
Parathion	India	1958	Contaminated food due to leakage	Rani (2012)
	India	1962	Inhalation in manufacturing plant	
Hexachlorohexane	India	1963	Contaminated rice	
Endrin	India	1964	Contaminated food	
Hexachlorohexane	India	1963	Contaminated rice	
Endrin	India	1964	Contaminated food	
Dichlrodiphenyltrichloroethane	India	1965	Contaminated chutney	
Diazinon	India	1968	Contaminated food	
Hexachlorohexane	India	1976	Mixed with wheat	
Endrin	India	1977	Contaminated crabs in rice field	
Aluminum Phosphide	India	1983	Contaminated food grain	
Methyl Isocyanate	India	1984	Storage tank leakage	
Cartap hydrochloride	India	1988	Factory workers	
Endosulfan	India	1997	Contamination due to aerial spray	
Phorate	India	2001	Spray drift from banana field	
Endosulfan	India	2002	Contaminated wheat flour	
2,3,7,8-tetrachlorobenzo-10-dioxin	Italy	1976	Air pollution due to poisonous gas	De (2010)
Sarin	Japan	1985–95	Mass-poisoning	Nagami (2010)
Pesticides	USA[a]	1968–78	Contaminated food	Laseter (1978)
Phenol	USA	1974	Accidental spillage	Baker et al. (1978)
Phenol	India	1999	Accidental over-dose of phenol	Gupta et al. (2009)
Phenol	New Zealand	1980	Absorption of phe-nol through skin	Lewin and Cleary (1982)
2,4-dinitrophenol	China	2009	Non-oral exposure to workers in a chemical factory	Lu et al. (2011)
2,4-dinitrophenol	USA	1933–38	Poisoning due to weight loss pill	Bartlett et al. (2010)

(continued)

Table 7.1 (continued)

Pesticide	Place	Year	Causes	References
Phenol	USA	1974	Accidental spillage of 37,900 1	Baker et al. (1978)
PolyAromatic Hydocarbons (Lakeview Gusher)	USA	1910	1200 tonnes of crude oil released	Rani and Shanker (2018a, b, c, d)
PolyAromatic Hydocarbons (Kuwaiti oil lakes)	Kuwait	1991	Kuwaiti oil lakes accidental spillage	
PolyAromatic Hydocarbons (Kuwaiti oil fires)	Kuwait	1991	136,000 tonnes of crude oil released	
PolyAromatic Hydocarbons	S. Korea	2007	MT *Hebei Spirit* oil spill	
PolyAromatic Hydocarbons	USA	2010	*Deep Water Horizon* oil spill)	
PolyAromatic Hydocarbons (Sundarbans oil spill)	Bangladesh	2014	Accidental spillage	
PolyAromatic Hydocarbons (Ennore oil spill)	Chennai	2017	Accidental spillage	
PolyAromatic Hydocarbons (Lakeview Gusher)	USA	1910	1200 tonnes of crude oil released	

[a]*USA* United States of America

Bennett and Keller 2011; Yang et al. 2002, 2013), zinc oxide (El-Kemary et al. 2010, Alikhani et al. 2012), nanoscale zero-valent iron (Mueller et al. 2012; Karn et al. 2009; Lacinova et al. 2012), and Ag (Dankovich and Gray 2011; Ren et al. 2013; Yang et al. 2013; Georgekutty et al. 2008; Zodrow et al. 2009; Ehdaie et al. 2014) are some commonly used metal and metal oxides based engineered nanomaterials for treatment of wastewater and environmental clean-up. Those engineered nanomaterials work on basis of chemical degradation, adsorption, chemical disinfection and photodegradation mechanism.

7.2.3 Engineered Nanomaterials of Magnetic-Core Based Composites

This category includes of basically magnetic-core nanoparticles that consists of magnetic elements (iron, cobalt, nickel) or their oxides and alloys with ferromagnetic or superparamagnetic properties, and shells coatings (Ngomsik et al. 2005; Ambashta and Sillanpää 2010; Liu et al. 2008; Girginova et al. 2010; Clark and Keller 2012a, b; Huang and Keller 2015). Shells can have inorganic (e.g., silica or alumina) (Bagheri et al. 2012; Mashhadizadeh and Amoli-Diva 2013; Clark and Keller 2012a, b; Huang and Keller 2013; Wang et al. 2008; Karimi et al. 2012), organic (polymers) (Ashtari et al. 2005; Li et al. 2010) or surfactants components (Faraji et al. 2010; Afkhami et al. 2010). The major advantage of fabrication of magnetic core based engineered nanomaterials is that superparamagnetism allows quick separation of catalyst form water (Huang and Keller 2013; Wang et al. 2008).

7.2.4 Engineered Nanomaterials Based on Composites

These composite nanomaterials can be decorated on existing grapheme, polymers, and clay materails to make their use on industrial scale. Presently, composite nanomaterials are being showed applications in sustainable chemistry, electrials and sensor, catalysis and environmental remediation (Mohmood et al. 2013; Qu et al. 2013). Advantages of engineered nanomaterials lie in improvement of specific surface- area, surface defects, electron mobility and wider range of light absorption along with greater adsorption capacity (Reddy and Kim 2015; Sarkar et al. 2007). Recent work of modifying nanomaterials to make composite materials has led to development of versatile materials which can bridge the loop holes of simple nanomaterials.

Several research articles showed the application of nanocomposites in adsorption, photocatalysis, membrane and modification. Broadly, composite nanomaterials utilize a basic architecture of a stable nanomaterial that has been modified by many substances. Doping with metals or nonmetals, complexation of surface, sensitization with dye, polymer-nanocomposites, and dip-coating on solid substrates are some of examples of composites engineered nanomaterials. Majorly of engineered nanomaterials are based on carbon (carbon nanotubes, graphene), oxides of metals (zinc oxide, hematite, manganese oxides and tungusten dioixde) and magnetic-core (magnetite) as discussed above. They are extremely useful and efficient in control of water pollution caused by pesticides (Ahmed et al. 2011; Al-Hamdi et al. 2016; Khajeh et al. 2013). These aforementioned materials have been reported to be efficient in targeting a wide range of contaminants (adsorption and oxidation) by their surface modification and reducing the recombination tendency of electron and hole pairs (Adachi et al. 2004). Illustration of various engineered nanomaterials used for pesticides degradation is presented as Fig. 7.2.

7.2.5 Working Mechanisms of Engineered Nanomaterials

For removal of pesticides from water, there are several working mechanisms on which engineered nanomaterials act. Initially, such materials adsorb the organic contaminants on their porous-surface via the solid-liquid interface followed by degradation by semiconduction (photochemical), redox or reactive adsorbents approaches (Pillai et al. 2015; Reddy et al. 2016; Sahoo et al. 2012). Engineered nanomaterials based on metals; transition metals and oxides of other metals have

Fig. 7.2 Illustration of use of various types of engineered nanomaterials like metals, metal oxides, polymeric materials, nanocomposites for pesticides degradation from water and soil

tendency for absorption of a precise spectrum of sunlight (Debnath and Gupta 2017). More the band gap of engineered nanomaterials, more is energy required for generation of electron-hole (e and h) pair. Various engineered nanomaterials with the mechanism of degradation of pesticides is presented in Fig. 7.3. The absorbed photonic energy facilitates the movement of charge carriers from valence band to conduction band, which, in turn, generates active radicals. Moreover, the objective of charge separation can be easily achieved by mixing/doping such single constituent engineered nanomaterials with metal and non-metal complexes or with species of polymer and plasmons and the detailed phenomenon has been represented in Fig. 7.3.

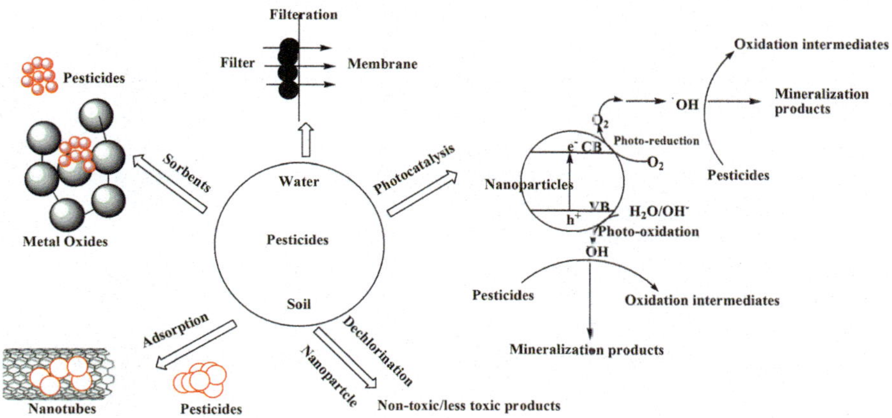

Fig. 7.3 Various techniques used for pesticides degradation based on engineered nanomaterials like adsorption, photocatalysis, membrane and filteration based. These techniques are found highly effective during recent times

7.3 Degradation of Pesticides by Engineered Nanomaterials

Engineered nanomaterials based on titanium dioxide, zinc oxide and zerovalent iron are most economic, rapid and effective heterogeneous photocatalyst for abundant removal of pesticides under ultra-violet irradiation and sunlight (Kralj et al. 2007; Moctezumaa et al. 2007; Zhang et al. 2006). They not only adsorbed contaminants but also degrade them via photocatalytic, redox and reactive mechanism. Hydrogen peroxide further improved their oxidative efficiency with more hydroxyl radical produced. Zinc oxide-clay, iron-zeolite and titanium dioxide-zeolite (bionanocomposites with enhanced activity and biocompatibility) are also used. No much attention has been paid to emerging contaminants like carbamates and substituted urea.

7.3.1 Titanium Based Engineered Nanomaterials

In the last decades and among the metal oxides, titanium dioxide is one of the most extensively used in view of better light absorption in the ultraviolet spectrum. Titanium dioxide has advantages of cheapness, easy availability and wide band gap (3.2 eV) which overlaps with the energy range of ultraviolet light). The high conduction of electron resulted in easy formation of electron and hole pairs in valence and conduction band, respectively.

$$TiO_2 + hv \rightarrow e_{CB}^{-} + h_{VB}^{-}$$

The interaction of those e− and h+ pairs with atmospheric oxygen and water produced superoxide and •OH on the catalyst surface. Furthermore, Cl• and sulfide radicals are also formed by the trigger mechanism of electron on available chlorine or sulfur. Due to these specific properties, it was a promising technology to degrade carbon-based contaminants (Lopez-Ayala et al. 2015; Sivagami et al. 2015). Titanium dioxide is found in three phases (anatase, rutile and brookite) where rutile phase being thermodynamically stable. The high photo activity of the material has been instrumental in degrading hydrophilic contaminants (Chitose et al. 2003; Follut and Vel Leitner 2007). However, low visible-light activity and rapid recombination of charge carriers limit the proper and effective usage of titanium (Gupta et al. 2006; Sahoo and Gupta 2013). Those cause the requirement of modification of titanium dioxide, through doping, functionalization or by making composites with materials having good absorption capacity so that they can span the visible range of light.

Infusing metals like silver and gold, non-metals, or attaching engineered nanomaterials with mesoporous materials, graphene, carbon nanotubes, polymer materials and glass surfaces are some process of engineered nanomaterials modifications (Lee et al. 2015; Linsebigler et al. 1995). Hybrid engineered nanomaterials promoted charge separation and visible light absorption via introduction of newly energy states (Cao et al. 2008). The respective bandgap and corresponding light wavelength is presented as Fig. 7.4.

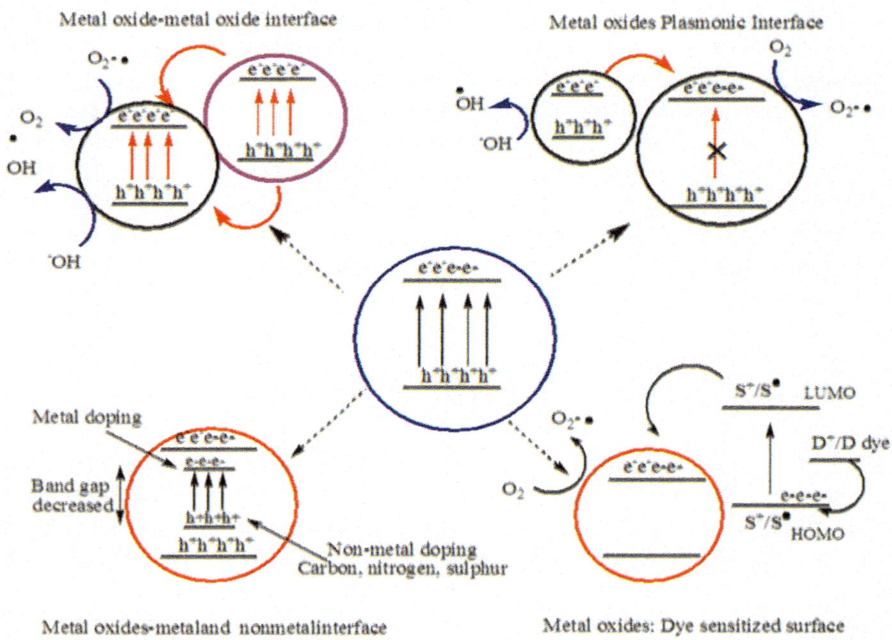

Fig. 7.4 Schematic representation of bandgap of different types of engineered nanomaterials and corresponding light wavelength. This figure revelas the efficiency of various engineered nanomaterials under the influence of electromagnetic irradiation

The extensive usage of silver and gold nanoparticles as novel catalysts having improved efficiency credited to their versatile behavior. Using this property of silver, Tian and co-workers (2014) prepared a visible-light active photocatalyst having sandwiched structures. Here, silver chloridel act as negative surface which help in production of active Cl• and entrapment of h+ (Dong et al. 2013; Naik et al. 2012; Wang et al. 2012). Zhang et al. (2012) also observed that silver/titanium dioxide (80%) composites is more effective than bared nanoparticles (<5%) in degradation of pentachlorophenol under visible light conditions.

Various other metals like copper and zirconium were also used to form a p-n heterojunction with titanium dioxide. Electropositive metals facilitate interfacial charge- transfer which leads to the formation of zerovalent metals (Hernández-Gordillo and González 2015; Naraginti et al. 2015). Additionally surface defects caused by doping metals promoted charge separation for trapping of excess of light energy (Hossaini et al. 2014; Sahoo et al. 2005; Sahoo and Gupta 2015; Linsebigler et al. 1995). Zhu and co-workers (2007) modified titanium dioxide with cetyltrimethylammonium bromide surfactant for better interaction and rapid degradation of organic pollutants via improved stability of free radicals. Co-adsorptive tendency of photocatalyst-surfactant bilayer and better charge transfer on the tailored hybrid materials facilitated degradation process (Senthilnathan and Philip 2010). Titanium dioxide is also being modified with silicon or iron nanoparticles for quick separation via magnetic process. Titanium dioxide immobilized on silica beads potentially degraded organophosphate pesticides in 7-h (Shifu and Gengyu 2005). Iron/titanium dioxide nanocomposite prepared for removal of 2,4-dinitrophenol via dechlorination mechanism (Liu et al. 2012). Iron nanoparticles able to inhibit charge-recombination through transfer of e− to iron and h^+ to semiconductor. Overall, the active ionic and radical species ($O_2^{\cdot}$, $^{\cdot}OH$, Cl^-) promoted degradation of contaminants via following equations.

$$TiO_2 + hv \rightarrow e_{CB}{}^- + h_{VB}{}^+ \tag{7.1}$$

$$MNP + hv \rightarrow MNP^+ + MNP^- \tag{7.2}$$

$$MNP^+ + Cl^- \rightarrow Cl^0 \tag{7.3}$$

$$MNP^- + OH^- \rightarrow O_2- \tag{7.4}$$

$$Cl^0 + OH^- \rightarrow Cl^- + OH^+ \tag{7.5}$$

Lindane was degraded successfully with N-doped titanium dioxide (Fig. 7.5) (Senthilnathan and Philip 2010) and bimetallic Fe-Pd within 8 h (Joo and Zhao 2008).

Under solar irradiation, malathion was all-out removed by photocatalyt made of 2% tungusten trioxide/titanium dioxide while 47% mineralization was obtained with bare titanium dioxide (Ramos-Delgado et al. 2013). Doping with two or more transition metals further improved the efficiency or quantum yield of surface process due to increase in number of energy levels between CB and VB edge of titanium dioxide as well increase in surface area (malathion removal: 172% with gold-palladium titanium dioxide nanotube film) (Yu et al. 2010).

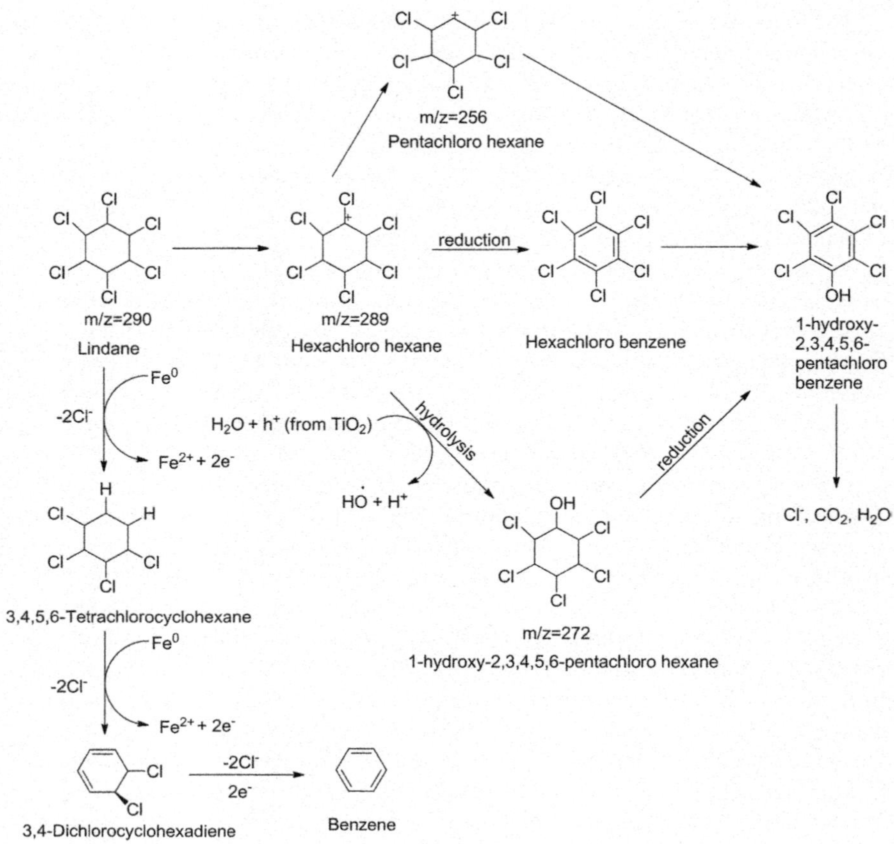

Fig. 7.5 Degradation pathway of lindane under photocatalytic conditions by nitrogen-doped titanium dioxide and nano zerovalent iron. The pathway suggest that degradation followed oxidation/reduction mechanism

Moreover, metal ions facilitate charge parting of e− and h^+ and acted as charge carrier trappers. On the base of doping identity and concentration, a number of Ti^{3+} ion with more surface defects are produced and enable effectual adsorption of atmospheric or moisured oxygen on titania surface. Charge transfer transition of electron occurs between energy levels of dopants and titanium dioxide (Ramacharyulu et al. 2015). Titanium dioxide functionalized with silver and gold are effectively employed for the degradation of organophosphate containing phenols at pH~ 3 (Shet and Shetty 2016). Monocrotophos and quinalphos were degraded to maximum extent by zinc oxide/ titanium dioxide (7:3 or 8:2) (Kaur et al. 2013). Monocrotophos with dichlorvos and phoxim were ereadicated by doped nanocomposites titanium dioxide-silicon dioxide beads, titanium dioxide supported on zeolite (Anandan et al. 2007a, b; Shifu and Gengyu 2005; Gomez et al. 2015; Dai et al. 2009). Henych et al. (2016) suggested reactive degradation of toxic organophosphate like malathion and parathion methyl with mixed metals and their oxides in organic medium.

Both surface area and porosity were enhanced for organophosphate due to solid collaboration of titanium with Cerium that resulted in high production of Ce^{3+} and formation of Ti^{4+} states. Titanium dioxide functionalized with polyacrylonitrilenanofiber and iron supported on zeolite was used for complete oxidation of aldicarb and methomyl under ultra-violet irradiation, respectively (Dixit et al. 2009; Tomasevic et al. 2010). Gold doped to zinc oxide and titanium dioxide increased the degradation percentage of chloridazone within shorter time due to prevention of the electron hole recombination (Fouad and Mohamed 2012).

The application of engineered nanomaterials based on titanium dioxide for degradation of pesticides and derivatives have been provided in Table 7.2 (Fig. 7.6).

7.3.2 *Zinc Oxide Based Composite Nanomaterial*

Another engineered nanomaterials used for treatment of organophosphates (dichlorvos, monocrotophos, malathion) are zinc oxide – zeolite, copper oxide-chitosan, copper oxide -montmorillonite chitosan and copper oxide -montmorillonite gumghatti bionanocomposites (Tomasevic et al. 2010; Jaiswal et al. 2012; Sahithya et al. 2015).

The development of photocatalytic treatment methods led to the demand of novel materials with enhanced properties and ability to reuse it for repeated reduction of organic compounds. Zinc oxide have been found efficient in the degradation of pesticides owing to several enhanced properties such as fast movement of electron-hole pair, minor harmfulness, cheap, effortlessly modified exterior properties, and little scattering of sunlight. Hindrance in electron-hole pair, quantum entrapment, and release of heat are the main factors for an efficient photocatalysis process. Researchers observed that the fast electron movement in zinc oxide conduction band generates extra reductive species compared to titanium dioxide and oxidative potential of hydroxyl radicals is superior in zinc oxide with band gap ~3.2 eV than titanium dioxide (2.7 eV) (Kumar and Rao 2015). Furthermore, the thermal and mechanical stability of zinc oxide remarkably enhances its stable efficiency even after repeated use makes zinc oxide an efficient photocatalyst and antimicrobial agent (Bechambi et al. 2015; Navarro et al. 2009). Though the significant contributions of zinc oxide make it an outstanding option for wastewater treatment, few issues of photo-corrosion, fast recombination of electron-hole pair, photo-dissolution and surface passivation generation at basic pH hinder the flexibility of zinc oxide. The reactions of the process are given below in Eq. (7.6) (Kumar and Rao 2015)

$$ZnO + 2h^{+} n\,H_2O \rightarrow Zn(OH)n^{(2-n)+} + 1/2\,O_2 + n\,H^{+} \qquad (7.6)$$

Table 7.2 Application of nanocomposites of titanium dioxide in degradation of pesticides in water

Material	Contaminants	Brief summary	Mechanism	References
AgCl/Ag/ TiO_2 60 mg	2,4-dichloro-phenol 40 ppm	Removal: 94% in 60 min	500 W halogen tungsten lamp	Tian et al. (2014)
Ag/TiO_2 2.5 mg/20 mL	Pentachlorophenol 20 ppm	100% in 160 min	UV-A 365 nm	Zhang et al. (2012)
Fe/TiO_2 1g/L	2,4-dinitrophenol 15 ppm	97% in120 min TOC removal: 71%	UV-20 W	Liu et al. (2012)
Ag/Zr/TiO_2 0.1 g/ 50 mL	4-nitrophenol 0.1 mmol/L	5 mL of 0.05 M NaBH4; 100% in 8 min	200 W Philips tungsten lamp	Naraginti et al. (2015)
Ag/Cu/ TiO_2 0.15 g L^{-1}	4-nitrophenol 8.5 ppm	0.5 M N_2H_4 96% in 30 min	UV (254 nm: 4400 $\mu W/cm^2$)	Hernández-Gordillo and González (2015)
Ag/TiO_2 0.4 g L^{-1}	Acetamiprid 20 ppm	40 min at 25 °C, pH: 6	250 W high pressure lamp	Cao et al. (2008)
Porous TiO_2	Atrazine Thiobencarb 5 ppm	Removal upto 95.7% Removal upto 96.7% 120 min	UV light	Xing et al. (2014)
FeFNS/ TiO_2 0.1 g L^{-1}	Diazinon 1.3 ppm	pH: 7 87.6% in 100 min	LED 385 nm	Hossaini et al. (2014)
TiO_2/Fe_2O_3 0.1 g L^{-1}	Diazinon 10 ppm	95.07% in 45 min	14 W/cm^2	Mirmasoomi et al. (2016)
Ni_2O_3/ TiO_{2-x} B_x (1 g L^{-1})	Trichloro-phenol (1×10^{-4}M)	Removal upto 80% Removal upto 70% 150 min		Zhao et al. (2004)
CTAB-TiO_2 1.5 g L^{-1}	Pyridaben C0= 2.63×10^{-5}M	pH: 10 100% in 560 min	350 W Hg-lamp	Zhu et al. (2007)
TiO_2-SiO_2 8 gdm^{-3}	Dichlorovos, (0.65×10^{-4} mol/dm^3)	pH: 5 Air flow: 0.02 m^3h^{-1} 100% 420 min	UV Light intensity: 2860 $\mu W/cm^2$	Shifu and Gengyu (2005)
Fe_2O_3/ CuO/TiO_2 0.5 g L^{-1}	2,4 D C0= 100 ppm	pH: 6 1190 ppm H_2O_2 UV-A: 40 W/m^2 97 % 300 min		Lopez-Ayala et al. (2015)
Fe/TiO_2	Thiacloprid	Dose: 0.5 g L^{-1} C0= 0.3 mM pH: 2.8 45 mM H_2O_2 125 W high pressure lamp 96 % 240 min		Banić et al. (2011)

(continued)

Table 7.2 (continued)

Material	Contaminants	Brief summary	Mechanism	References
Fe_2O_3-TiO_2	Propachlor	Dose: 0.2 g L^{-1} C0= 5 ppm 1500 W Xenon lamp – 50 min		Belessi et al. (2009)
Au-Pd-TiO_2	Malathion 10 ppm	450Whigh-pressure mercury Lamp 98.2% TOC removal: 50.7% 240 min		Yu et al. (2010)

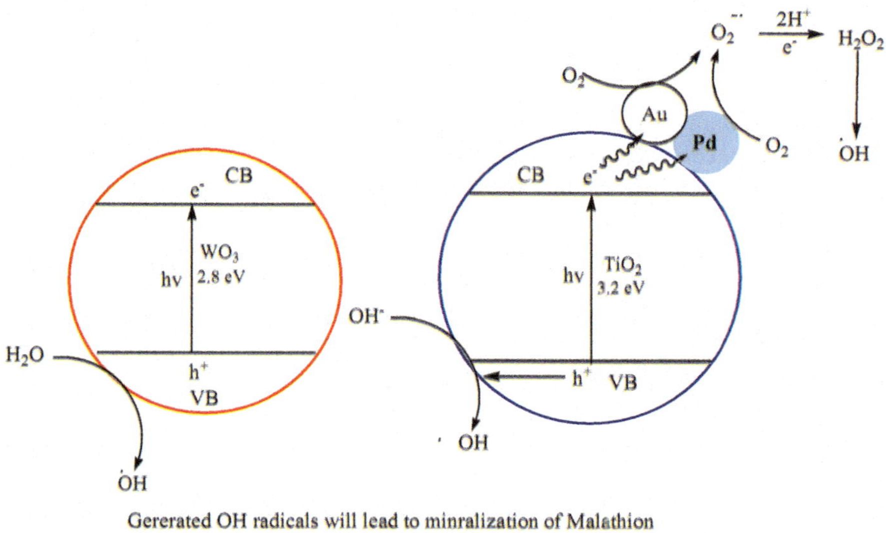

Fig. 7.6 Photocatalytic degradation of Malathion by tungusten trioxide/zinc oxide nanocomposites. This shows generation of massive hydroxyl radicals is responsible for the degradation of Malathion. Here, synergistic effect of zinc oxide with tungusten trioxide playing vital role towards efficient degradation of malathion

The composite of chitosan with zinc oxide was studied for active substituent of permethrin pesticide (Comparelli et al. 2005; Moradi Dehaghi et al. 2014). These investigations showed that at a dose of 0.5gL^{-1} polymer- zinc oxide composite 99% of pesticide was degraded and also reported a reuse efficiency of 56% even after three cycles. Furthermore, dissolution of zinc oxide in acidic solution inhibited due to the fact that biopolymers can be protonated under pH less than 7 it will further enables the binding of negatively charged pollutants strongly (Divband et al. 2013; Khatamian et al. 2012; Qamar et al. 2015).

Many reports reveal that encapsulation of zinc oxide with lanthanum, niobium decrease rate of combination of electron –hole pair (Anandan et al. 2007a, b; Lam et al. 2014; Xie et al. 2015). The hydroxyl radicals formed by the oxidation of hydroxide by valence band holes while superoxide obtained by conduction band electron in Eqs. (7.7), (7.8) and (7.9).

$$O_2 + e^- \rightarrow O^-{}_2 \quad (7.7)$$

$$h^+ + OH^- \rightarrow OH^{\cdot} \quad (7.8)$$

$$OH^{\cdot} + OH^{\cdot} \rightarrow H_2O \quad (7.9)$$

Composite of zinc oxide with silver used for effective removal of phosphamidon an organophosphorus pesticide reported by Korake et al. (2012). However, the extensive use of zinc oxide has been restricted due to outward diffusion of electron resulted in low quantum yield. Few studies have reported p-n hetero-junctions to be an effective way to inhibit recombination of electron-hole pair (Pirhashemi and Habibi-Yangjeh 2017; Darwish et al. 2017; Debnath and Gupta 2017). Superiority of zinc oxide for photocatalytic ability among other contemporary material is due to absorption of multiple ranges of wavelength and great photo-stability. Insight in to sunlight assisted degradation of bisphenol A by zinc oxide @zinc hexacyanoferrate nanocomposites has been explored and found that such engineered nanomaterials ae highly effective in complete removal of variety of contaminants like polycyclic aromatic hydrocarbons, and phenols (Fig. 7.7). The synthesis of zinc oxide based enginerred nanomaterials composites and their photocatalytic efficiency have been presented in Table 7.3.

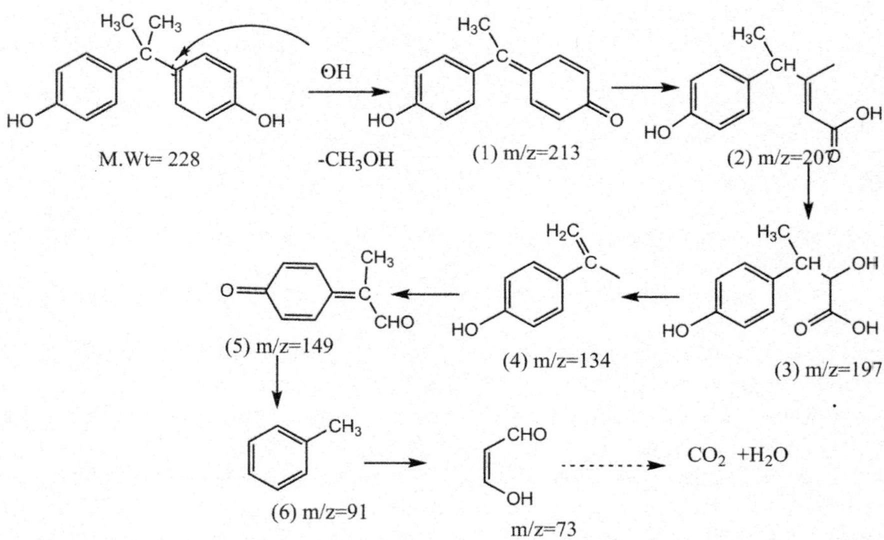

Fig. 7.7 Degradation pathways of Bisphenol A over the surface of zinc oxide @zinc hexacyanoferrate nanocubes. The best conditions for efficient removal were; Concentration of bisphenol A; 2 mgL^{-1}; pH:~7; catalyst amount: 25 mg under direct sunlight for 12 h

Table 7.3 Zinc based nanocomposites mediated degradation of pesticides in water

Nanocomposites	Contaminants	Reaction conditions	Degradation /adsorption capacity	Time (minutes)	References
Chitosan-ZnO 0.5 gL^{-1}	Permethrin	pH: 7 Concentration: 0.1 mg L^{-1}	99%	90	Moradi Dehaghi et al. (2014)
ZnO-bentonite 0.3 g L^{-1}	Phenol	pH: 12 Concentration: 200 mg L^{-1}	80%	–	
Ag/ZnO 1 gdm^{-3}	Phosphamidon	pH: 7 Visible light (520 nm) Concentration: 5×10^{-2} mol L^{-1}	100%	150	Korake et al. (2012)
ZnO/TiO_2 0.5 g L^{-1}	Diazinon	pH: 7 UV irradiation Concentration: 5 mg L^{-1}	99.99%	60	Jonidi-Jafari et al. (2015)
ZnO/Fe_2O_3 0.5 g L^{-1}	Dicamba	pH: 7; Concentration:10 mg L^{-1}	100%	300	Maya-Treviño et al. (2014)
Ag/ZnO 0.1 g mL^{-1}	4-nitrophenol	UV irradiation Concentration: 10 mg L^{-1}	100%	180	Divband et al. (2013)
Ln/ZnO 0.1 g L^{-1}	4-nitrophenol	UV irradiation Concentration: 10 mg L^{-1}	83.4%	200	Khatamian et al. (2012)
CuO/ZnO 1 g L^{-1}	4-nitrophenol	Sunlight Concentration: 30 mg L^{-1}	99% 83%	180	Qamar et al. (2015)
Graphene/ZnO 50 mg L^{-1}	Nitrophenol	30 W UV irradiation 0.53 mW/cm^2 Concentration: 10 mg L^{-1}	98%	150	Darwish et al. (2017)
La/ZnO 2 gdm^{-3}	Metasystox	pH: 8; UV irradiation Concentration: 1.4×10^{-3} M	90%	150	Korake et al. (2014)
La/ZnO 1 g L^{-1}	2,4,6 trichlorophenol	pH: 10; 8 W low pressure Mercury lamp Concentration:40 ppm	100%	120	Anandan et al. (2007a)
La/ZnO 1 g L^{-1}	Monocrotophos	pH: 10; 8 W mercury lamp Concentration: 40 ppm	100%	120	Anandan et al. (2007b)

(continued)

Table 7.3 (continued)

Nanocomposites	Contaminants	Reaction conditions	Degradation /adsorption capacity	Time (minutes)	References
Fe_2O_3/ZnO 0.15 g L^{-1}	Pentachlorophenol	500 W solar simulator Concentration: 10 ppm	98%	240	Xie et al. (2015)
W/ZnO 100 mg L^{-1}	Chlorophenol	Sunlight: $1000 \pm 100 \times 10^2$ lux Concentration:50 ppm	100% 93%	150	Aslam et al. (2015)
Sm/ZnO 1 g L^{-1}	Phenol	pH: 5; 55 W fluorescent lamp; Concentration:20 ppm	89.5%	480	Sin et al. (2013)
Nb_2O_5/ZnO 1 g L^{-1}	Phenol	Sunlight: 8.9 ×104 lux, UV 15 W Concentration:20 mg L^{-1}	100% 76%	40	Lam et al. (2014)

7.3.3 Tungsten Based Composite Nanomaterial

Ultra-violet active photocatalysts having wide band gap and hence suffers major limitations due to their incapability to absorb visible light which is the main constituent of solar spectrum. This limitation was overcome by synthesizing visible light active tungsten oxides with a band gap of 2.3–2.5 eV. Photocatalysis has been considered as one of the sustainable options for wastewater treatment. Recently, tungsten trioxide combined to several semiconducting materials with high surface to volume ration for mineralization of contaminants. Tungsten trioxide exhibits two different phases monoclinic I and monoclinic II (γ and ε- Tungsten trioxide) (Tahir et al. 2017).

Because of low gap between valence and conduction band of tungsten trioxide combination of electron-hole pair occurs and thereby decreasing its efficiency in degrading organic contaminants.

Modifications in past reported the complexation of Tungsten trioxide with titanium dioxide, zinc selenide or metals like palladium or on carbon derivatives like graphene, carbon nanotubes. Some studies showed tungsten trioxide – titanium dioxide mediated degradation of various contaminants like phenolic compounds more than >98%, malathion (>76%) and imazapyr (100%) on mesoporous with removal efficacy (Aslam et al. 2014; Ismail et al. 2016). The oxidatative degradation enhanced due to facile relocation of h+ from tungsten trioxide to titanium dioxide brings about formation of large number of active species (Mkhalid 2016). WO_3 coated graphite electrodes have been found efficient materials for the removal of 2-nitrophenol with the efficacy of 82% (Umukoro et al. 2017). The catalyst was tested for the degradation of recalcitrant pollutant 2,4-Dichlorophenoxyacetic acid. Earlier studies illustrating the applications of modified tungsten trioxide have been given as Table 7.4.

Table 7.4 List of nanocomposites of tungsten for degradation of pesticides in water

Nanocomposites	Contaminants/ concentration	Reaction conditions	% Removal / adsorption capacity	Time (minutes)	References
Tungsten trioxide	2-chloro-phenol 30 mg L^{-1}	Sunlight: 1000 ± 100×10^2 lux	>98	180	Aslam et al. (2014)
Tungsten trioxide –Titanium dioxide	Malathion 12 mg L^{-1}		99	300	Ramos-Delgado et al. (2013)
Tungsten trioxide –Titanium dioxide	Imazapyr 0.08 mmol L^{-1}	pH: 4 Visible light: 2 mW/cm^2	100	120	Ismail et al. (2016)
Zinc selenide-tungsten trioxide	Bisphenol A 10 ppm	pH: 7 Sunlight: 30×103±100 lux	99	90	Kumar et al. (2017)
Palladium-tungsten trioxide	2,4-D 40 ppm	300 W Xenon lamp	100	50	Mkhalid (2016)

7.3.4 *Composite Nanomaterials Using Adsorption and Reduction Approaches*

The surface area of composite nanomaterials offers more active sites to interact the solid phase with the liquid phase in the heterogeneous system which results in phase separation by mass transfer phenomenon (Pillai and Gupta 2015). The present section explores and identifies the exceptional properties of composite materials that enhance their adsorptive and reductive properties. The development of nanomaterials with high surface area to volume ratio has proved to be of significant importance in their use as water treatment materials.

The fabrication of composite nanomaterials attempts to overcome the potential loopholes of nanomaterials like instability due to high agglomeration, passivation of surface and high surface energy. The process of adsorption serves to be an energy efficient simple method to remove pesticides from aqueous solution without any additional by products. Moreover, the high reductive properties of iron based materials facilitate simultaneous degradation and adsorption of the pesticides.

7.3.5 *Iron Based Composite Nanomaterial*

Iron is the third most abundant element in Earth crust as mineral ores. Due to geological abundance, it exists in various forms like magnetite, maghemite, hematite, iron oxyhydroxide, nano zerovalent iron each of which has a unique a specific application in removal of water contaminants. Many studies have reported the use of iron to imitate its natural potential in degrading and removing emerging contaminants. The properties like enhanced surface area, potential to act as both reducing and oxidizing agent and easy separation of particles from water have made Fe based nanocomposites as effective materials for treatment of wastewater.

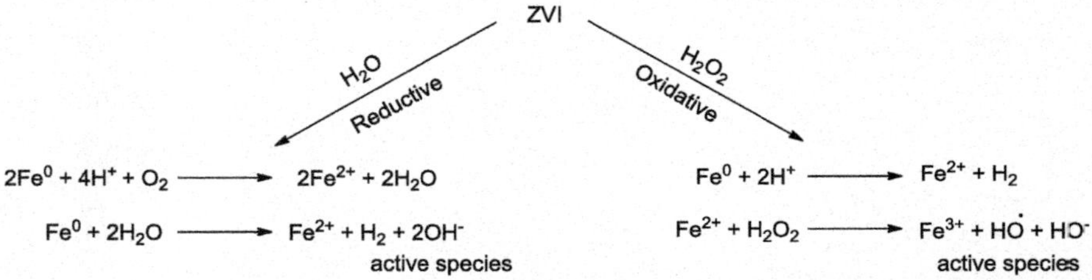

Mechanism for the working of zero-valent iron under different conditions such as oxidative and reductive generation of active species utilized for removal of presticides

The efficient de-chlorination of chlorinated organic compounds such as polychlorinated biphenyls, hexachlorocyclohexane and hydrocarbons by nickel and palladium based compounds has been reported (Cao et al. 2005; Elliott et al. 2009; Tian et al. 2015). The degradation profile of Dichlorodimethyltrichloroethane with iron-nickel nanocomposites has been illustrated as Fig. 7.8.

Traditionally, iron(II) sulfate and hydrogen peroxide were used as Fenton's reagent for the effective degradation of organic contaminants. Fe nanoparticles were reported highly efficient in removal of pesticides but their implication was restricted in view of agglomeration of particles in water. The experimental details using iron based composite nanomaterials have been presented in Table 7.5. This catalyzed the emergence of various nanocomposites of iron that fueled its use in water treatment. Mineralization 4-nitrophenol using copper-maghemite nanocomposite has been reported by Feng et al. (2013). The composites not only provide stability to iron nanoparticles but also increase the efficiency utilizing the

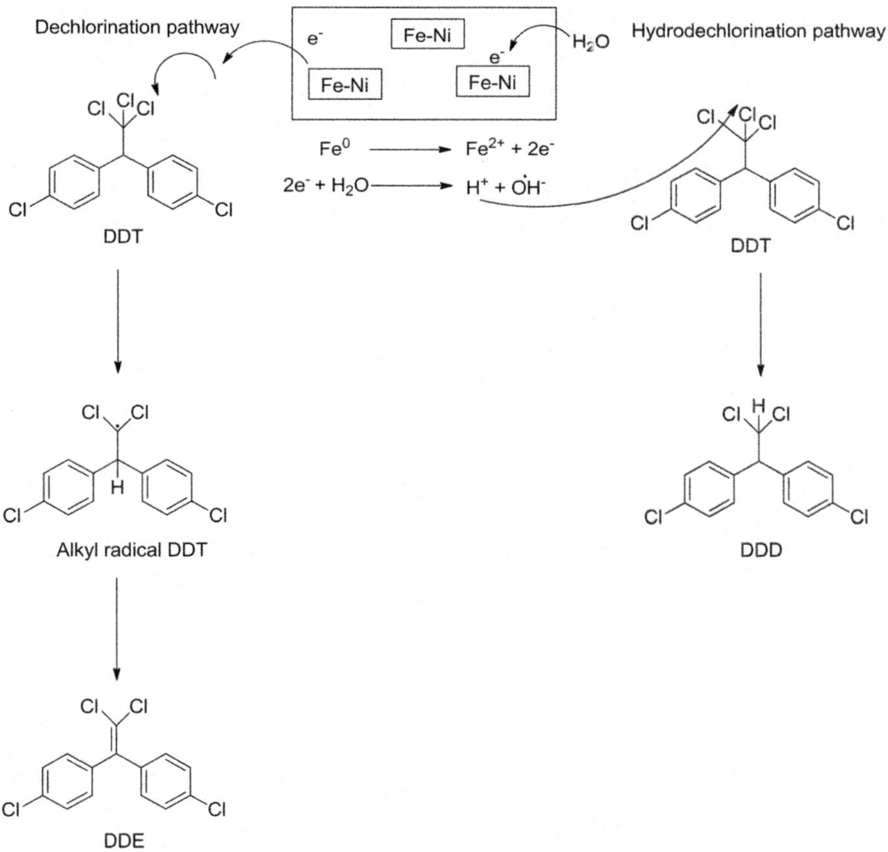

Fig. 7.8 Degradation profile of Dichlorodiphenyltrichloroethane by iron-nickel bimetallic nanocomposites via dechlorination and hydrodechlorination mechanism. This degradation ended with less toxic metabolite as compared to DDT

Table 7.5 List of nanocomposites of iron for removal of pesticides in water

Nanocomposite	Contaminants	Reaction conditions	Removal efficiency	Time minutes	References
Copper ferrate	4-nitrophenol	0.005 mol L^{-1}; 0.1 mol L^{-1} $NaBH_4$, 2 mL H_2O_2, in	95%	4	Feng et al. (2013)
Iron/Palladium	Lindane	5 ppm; pH: 6.5	89	5	Nagpal et al. (2010)
Hematite/multiwalled carbon nanotubes	Atrazine	10 ppm	81.4	120	Yu et al. (2015)
Starched Iron/palladium	Trichloroethylene	25 ppm; Dose: 0.1 g L^{-1}	98	60	He and Zhao (2005)
Silver/iron oxide	Nitrophenol	0.1 mM; Dose: 0.02 g L^{-1}	–	30	Chiou et al. (2013)
Nickel/iron-megnetite	2,4 dichlorophenol	20 ppm; Dose: 3 g L^{-1}, pH: 6	100	180	Xu et al. (2016a)
Iron oxides	Atrazine	30 μg L^{-1}; Dose: 2.5 g L^{-1}, pH: 7	95	30	Ali et al. (2016)
Silver/iron microbox	Nitrophenol	0.1 mM; Dose: 7 mg, 0.2 M $NaBH_4$	97	150	Jiang et al. (2015)
Manganese ferrate-Graphene	Glyphosate	20 ppm; Dose: 1 g L^{-1}	97	480	Yamaguchi et al. (2016)
Iron zinc indium sulphide	Tribromophenol	0.12 mmol/L; Dose: 0.5 g L^{-1}, 20 W low pressure Mercury lamp	95	60	Gao et al. (2013)
Cetyl trimethylammonium bromide-iron oxide	Bisphenol A	Dose: 0.125 g L^{-1}	98	–	Zhao et al. (2008)
Copper ferrate	Tetrabromobisphenol A	10 ppm; Dose: 0.1 g L^{-1}, 0.2 mmol L^{-1}, permonosulfate	99	30	Ding et al. (2013)
Graphene oxide-magnetite	2,4-D, 25 ppm	pH: 3; 67.2 mgg^{-1}		–	Nethaji and Sivasamy (2017)
Iron oxide@n-silica oxide@ m silica oxide	Dichlorodiphenyltrichloroethane	1.5 μgm L^{-1}	97	60	Tian et al. (2015)

Cobaltferrate@Titanium dioxide-Graphene oxide, 1.5 g L^{-1}	Chloropyrifos	5 ppm; pH: 5.8	–	60	Gupta et al. (2015)
CexFe1-xO_2 0.5 g L^{-1}	Chlorophenol, Dichloro Phenol, 2,4-D	500 ppm; 250 ppm; 250 ppm	58.75; 42.44 45.78	90 45 60	Kurian et al. (2017)
Iron oxide/activated carbon	Chlordecone	2.2 μg L^{-1} ; pH: 7, 79.8 μgmg^{-1}		1400	Rana et al. (2017)
Palladium/iron, oxide@Multiwalled carbon nanotubes	2,4-dichloro Phenol	0.12 mM; pH: 6.5	92.3	300	Xu et al. (2016b)

optimum amount of catalyst making the treatment an economic option (Ayad et al. 2017).

Recently, the efficient use of nickel/iron-manganetite for the de-chlorination of a phenolic compound (dichlorophenol) rapidly in 3 h has been reported (Xu et al. 2016a, b). The basic fundamental of iron nanocomposites relies on the ability of iron to reduce as well as provide a porous architecture for rapid adsorption of organic species. Being mostly hydrophobic, pesticides and phenolic residues can adsorb onto to the iron surface. Since iron based composites having high permeability and porosity hence, operative diffusion of metabolites of pesticides was observed. The active mass-transfer amid solid-liquid phase offers adequate time for quick degradation. Moreover, the existence of oxygen onto iron interface resulted oxidation the secondary metabolites. Nagpal et al. (2010) reported effective degradation of lindane in the presence of iron-palladium bimetallic nanocomposite (concentration: 5 mgL^{-1}).

In this study comparison between of iron nanoparticles with surface area to bulk iron particles showed the effectiveness of iron-palladium nanocomposite was recognized to quick breaking of Carbon-Chlorine bond in organochlorine pesticides due to superior electronegativity (Joo and Zhao 2008). Moreover, it was also established that efficiency mainly depends upon the optimization of palladium quantity in the nanocomposite that competently controls aggregation (Xu et al. 2005). Development of dendrimer-like structures with starch coated magnetic nanoparticles were found to remove 98% of trichloroethylene in 1 hour (He and Zhao 2005) while some studies showed that starch itself may be used for pesticide removal.

Removal of glyphosate by manganese ferrate-graphene composite was reported depended of positive surface charge and surface area of magnetic nanocomposite (Yamaguchi et al. 2016; Yu et al. 2015). It was concluded that that fast dealkylation, alkylic oxidation and/or dechlorination as well as more adsorption sites at the carbon nnotubes surface enhances the removal of pesticide (Graymore et al. 2001). Bimetallic iron-nickel nanostructures were found effetive for removal of Dichlorodiphenyltrichloroethane under mild acidic and alkaline conditions (Tian et al. 2014). Aldrin, endrin and lindane was massively adsorbed by magnetite-polystyrene (Jing et al. 2015). On the other hand, carbon-zinc oxide-cadmium disulphide combination worked on photocatalysis for 98% removal of 4-chloro phenol (Lavand and Malghe 2015).

On similar way, zerovalent iron/magnetite and magnetite@gold composite worked for thiamethoxam and imidacloprid at acidic pH from water under irradiation (De Urzedo et al. 2009). Depending on absence or presence of hydrogen peroxide, zerovalent iron/magnetite behaved as a reducing or an oxidizing catalysts (Fouad and Mohamed 2012; Rodriguez et al. 2011) (Fig. 7.9).

Ruthenium doped titanium dioxide and Ag/chitosan nano-biocomposite used for 80–98% removal of metsulfuron-methyl and Atrazine (Senthilnathan and Philip 2010; Saifuddin et al. 2011) while only 40% achieved with bare metal oxides.

$$\text{Fe(III)-OH} + H_2O_2 \longrightarrow (H_2O_2)_{ads}$$

$$(H_2O_2)_{ads} \longrightarrow \text{Fe(II)} + H_2O + HO_2^{\bullet}$$

$$HO_2^{\bullet} \rightleftharpoons O_2^{\bullet -} + H^+$$

$$\text{Fe(III)-OH} + HO_2^{\bullet} \longrightarrow \text{Fe(II)} + H_2O/OH^- + O_2$$

$$\text{Fe(II)} + H_2O_2 \longrightarrow \text{Fe(II)-OH} + HO^{\bullet}$$

$$^{\bullet}OH + \text{R (pesticide)} \longrightarrow \text{Products}$$

Fig. 7.9 Reaction of oxidative degradation pesticides based on hydroxyl radicals generated with the help of magnetite under the electromagnetic irradiation

7.3.6 Carbon Based Composite Nanomaterials

Carbon-based material, such as activated carbon has already been commercialized for water filtration. The various physical properties such as electrical conduction, high mechanical and thermal of carbon are found enhanced once the particle size decreased to the nanoscale dimension. Carbon nanoparticles exists as carbon nanotubes, graphene, carbon quantum dots, carbon-based composites with high surface area and has been found excellent adsorbent for removal of pesticides and their derivatives (Yue and Economy 2005). The hydrophobic nature of carbon nanoparticles due to π-π interactions have a great attraction for hydrophobic pesticides and phenols but due to deficiency of chemical functional group carbon nanoparticles generally take long time for adsorption. To improve performance of carbonous nanomaterial's their surface was altered with several other materials. Activated carbon with improved surface properties has been found effective for low cost removal technique for pesticides from water. Furthermore, carbon based nanomaterial's showed substantial stability at high temperature; this helps their easy recovery after their use as an adsorbent by heating. Activated carbon generated from waste vehicular tires employed for extraction of many persistent pesticides such as atrazine, methoxychlor, and methyl parathion (Gupta et al. 2011). Silver encapsulated reduced graphene oxide with high surface-area reported effective for removal of various pesticides like endosulfan, chlorpyrifos (Koushik et al. 2016). Graphene is a single later carbon sheet exfoliated from graphite (Cai et al. 2005; Peng et al. 2003; Zhou et al. 2006). The carbon nanotubes are of more advantage than the activated carbon due to their hydrophobicity, specific grooves, π-π interactions, surface interactive sites and can be easily modified by metals to impart a catalytic effect to the material. The extraordinary characteristics resulted from the geometrical structure, purity, and degree of curl, tube-diameter, and the synthetic routes (Peralta-Hernandez et al. 2017; Chen et al. 2011a, b; Pan and Xing 2008).

Table 7.6 List of carbon-based functional material for removal of pesticdes from water

Nanocomposites	Contamiants	Reaction conditions	Mechanism	References
Multiwalled carbon nanotubes/Titanium dioxide 8 g L^{-1}	Dinitrophenol; 38 ppm	pH: 6, 99% removal in 150 min	Sunlight	Wang et al. (2009)
Titanium/ Multiwalled carbon nanotubes 0.2 mg L^{-1}	Atrazine 4.6×10^{-4}- 9.3×10^{-2}, mmol/L	3 min		Chen et al. (2011a, b)
Ag/AgBr/ carbon nanotubes 30 mg/50 mL;	Tribromophenol; 100 μmol/L	pH: 10; 100% removal in 50 min	250 W metal halide lamp	Shi et al. (2013)
Au/Mesoporous Carbon, 75 mg	p-Nitrophenol; 200 ppm	pH: 10.1, 0.06 M, $NaBH_4$, 87% removal in 30 min		Guo et al. (2016)
Ag/Graphene-Dopamine	4-nitrophenol	$NaBH_4$	8 min	Jeon et al. (2013)
Multiwalled carbon nanotubes (0.85%), 5 mg/25 mL	Atrazine; 4.2 ppm	0.01 M $NaNO_3$ + 0.1g L^{-1} NaN_3, pH: 6, capacity:17.3 mgg^{-1}		Chen et al. (2009)
RGO-ZnS	Nitrophenol	87% removal in 70 min	Solar light	Ibrahim et al. (2017)
Reduced Graphene Oxide-Silver	Lindane, 2 ppm	99% removal in 15 min		Sen Gupta et al. (2015)
Magnetite/graphene 0.5 g L^{-1}	Ametryn, 10 ppm	93.6% removal in 70 min		Boruah et al. (2016)

The emergence of complex priority pollutants with highly stabilized aromatic structure need the development of co-removal systems which can degrade such toxic components as well as adsorb it from aqueous systems. Quick removal of dinitrophenol within 180 min by multiwalled carbon nanotubes/titanium dioxide nanocomposite was reported (Wang et al. 2009). The effective use of silver/silver chloride/carbon nanotubes nanocomposites for the eradication of tribromophenol explored by Shi et al. (2013) using ultrasonic-deposition-precipitation. List of studies explored the applications of carbon nased nonocomposites have been given as Table 7.6.

7.3.7 Polymeric Composite Nanomaterials

Recently, researchers have reported the effective removal of pesticides and derivatives using composite polymeric materials (Hu et al. 2015; Lu et al. 2011). Polymers like chitosan, polyacrylamide, polyvinylchloride, polypyrrole, polythiophene and clay obtained from green sources and easily extractable from low cost sources for pesticides degradation. Numerous polymers were employed for fabrication of membranes and polymeric resins to target contaminants in wastewater. Furthermore, such polymers showed high porosity and stability and their surface can be easily functionalized to improve its interaction with pesticides.

Efficient removal of 2,4-Dinotrophenol of concentration (70 ppm) by carbon coated polyacrylonitrile have been reported (Zhao et al. 2017; Sawicki and Mercier 2006; Zhao et al. 2017; Pradeep and Anshup 2009). Likewise, these materials have been mixed with montmorillonite, a natural clay material to mimic the sediment sorption process of non-polar contaminants (Sahithya et al. 2015, 2016; Shabtai and Mishael 2017; Milojević-Rakić et al. 2013; Sarkar and Das 2017). Table 7.7 showed list of composites of polymers for degradation of pesticides. The natural abundance of polymers can be exploited to develop solid substrates for wastewater treatment applications.

7.4 Importance of Green Synthesized Nanomaterials

In order to avoid use of toxic solvents such as dicholomethane, dimethylformamide, hydrazine, sodium borohydride, green routes employing environmentally benign and renewable materials are preferred. It is more reliable, sustainable and bioinspired bottom-up approach (Makarov et al. 2014; Virkutyte et al. 2013). Green synthesis of engineered nanomaterials has several benefits over traditional methods such as cheaper, ecofriendly, safe, reusable, easy couture of the size, morphology, shape, and surface functionality, generation of relative stability, remarkable biocompatibility and biodegradability (Makarov et al. 2014; Li et al. 2010; Virkutyte and Varma 2013). Green synthesis employing sunlight, or nontoxic, environmentally benign and renewable materials such as water, ionic liquids, microbes, plant-based surfactants or combination of those can be used on large-scale. Intracellular or inner surface biosynthesis modulates the crystal growth and nucleation processes microbial synthesis of nanomateirals (Narayanan and Sakthivel 2010). However, the method is time consuming (~30 days) and expensive too.

Figure 7.10 represents the various green methodologies used for synthesis of nanomaterials. Engineered nanomaterials synthesized via green methods are low-cost and effective catalysts for remediation of environmental pollutants. Despite this, they are not probed for much of contaminants. Reduced graphene oxide-based plasmonic nanohybrids synthesized using extracts of Baker's yeast showed exceptional photothermal transformation adeptness for waste desalination and purification under visible-light (Attia et al. 2016). Moreover, reduced graphene oxide-silver synthesized using Psidium guajava leaves extracts showed detection capability for methylene blue even at very low concentration (Chettri et al. 2017). The nanocomposite palladium-gold produced by *Shewanella oneidensis* was found in effect for dechlorination of many carbon-based impurities in pharmaceutical drugs (Corte et al. 2012). Compared to bared silver or gold, engineered nanomaterials like

Table 7.7 List of polymeric nanocomposites for eradication of pesticides from water

Polymeric nanocomposites	Contamiants	Brief summary	Mechanism	References
Chitosan- β cyclodextrin-hematite	Bisphenol A (200 ppm)	pH: 6, capacity 133 mgg^{-1} in 60 min	Adsorption	Huang et al. (2017)
Polyvinyl-pyridineco-Styrenemont morillonite clay composite (0.5 $g L^{-1}$)	Diazinon (0.13 mM)	pH: 3.5; 100% removal in 240 min		Shabtai and Mishael (2017)
B-cyclodextrin 1 $g L^{-1}$	Bisphenol A 0.1 $mmol L^{-1}$	113 mgg^{-1}	Adsorption	Wang et al. (2017)
Carbon coated polyacrylonitrile 50 mg/80 mL	2,4 D (70 ppm)	pH: 3; 61.02 mg/g 10 h	Adsorption	Zhao et al. (2017)
Montmorillonite- CuO-Chitosan 1.5 $g L^{-1}$	Dichlorovos 80 ppm	pH: 8 93% in 300 min		Sahithya et al. (2015)
Montmorillonite-CuO- polylactic acid 15 $g L^{-1}$	Monocrotophos100 ppm	pH: 5 80% in 6 h		Sahithya et al. (2016)
Starch polymerlaponite Clay 30 mg/25 mL	Dicamba 500 ppm	251 mgg^{-1} at pH: 7	Adsorption	Pinto et al. (2016)
Polyaniline-zeolite 50 mg/5 cm^3	Glyphosate 400 ppm	98.5 mgg^{-1}	Adsorption	Milojević-Rakić et al. (2013)
Polyaniline-silica gel 1 g/L	2,6dichlorophenol 80 ppm	31.9 mgg^{-1} in 2 h	Adsorption	Pan et al. (2011)
Polyvinylpyrrolidonemagnetic nanoparticles 50 $\mu g L^{-1}$	Bisphenol-A	pH: 7; 90 mgg^{-1}; 20 min	Adsorption	Alizadeh Fard et al. (2017)

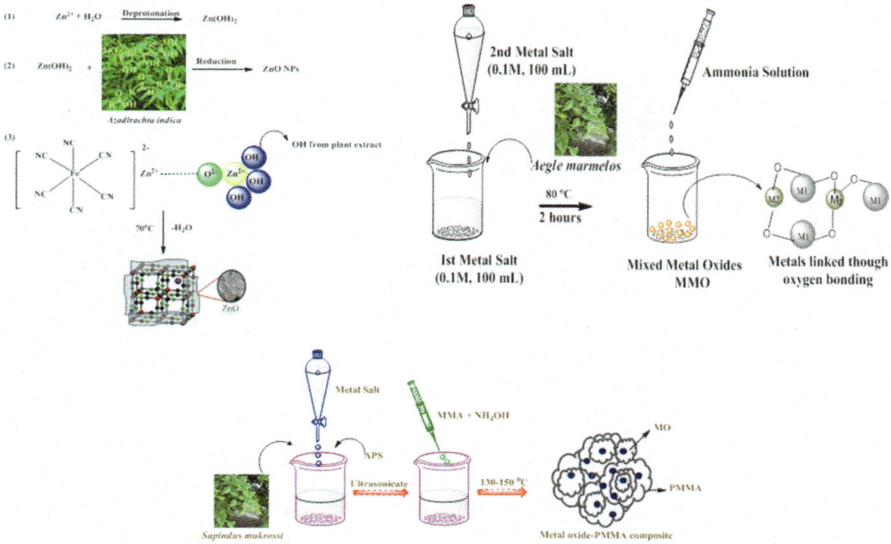

Fig. 7.10 Green strategy for synthesis of engineered nanomaterials

silver-copper prepared by use of leaf extracts of *Prosopis cineraria* has advantages of superior antibacterial activity and cytotoxicity against human breast cancer cell line (Jinu et al. 2017). Bio-functionalization of silver and graphene oxide with phenylalanine peptide i.e., graphene oxide -peptide-silver nanohybrids showed electrochemical detection of small molecules like hydrogen peroxide (Wang et al. 2017). Nanocomposites of graphene oxide with crystalline cellulose displayed astonishing adsorption capacity for methelene blue (2630 mgg^{-1}) (Wei et al. 2017). Besides, cellulose-silver nanocomposites have unlimited talent for commercial applications (catalysis, antibacterial, sensor and environment) (Cao et al. 2008). Silver nanoparticles with *amaranthus gangeticus* could able to remove more than 50% of congo red dye in 15 min (Kolya et al. 2015).

Polycyclic aromatic hydrocarbons and dyes could be photocatalytic removed by biosynthesized nanocomposites of zinc oxide and copper oxide with graphene oxide (Hassan et al. 2015; Lellala et al. 2016). Fardood and research group biofabricated several nanocomopites of metal oxides (zinc, copper and nickel) and ferrites nickel-copper-zinc, nickel–copper–magnesium and copper-magnetite) by use of gel, gum or plants. They used them as catalyst for organic synthesis and photo-catalyst for removal of organic pollutants (Fardood et al. 2017a, b; Sorbiun et al. 2018). In addition, supermagnetic magnese based engineered nanomaterials were used for remediation of several reactive dyes (Fardood et al. 2017; Atrak et al. 2018). Green tea or starch fabricated iron based engineered nanomaterials have been estimated for photocatalysis (Hoag et al. 2009; Huang and Keller 2013). Compared to bared ones, green synthesized engineered nanomaterials showed superior activities: bimetallic iron/palladium> iron and hematite@zinc heaxcyanoferrate nanocubes> zinc heaxcyanoferrate > hematite naoparticles (Rachna et al. 2018; Smuleac et al. 2011). *Better catalytic efficiency of* green tea-iron iron nanoparticles than iron nanoparticles fabricated by borohydride reduction further supported this fact (Shahwan et al. 2004).

Shanker and his research group applied several green methods based on sunlight assisted synthesis and plant extracts (*A. marmelos, S. mukorossi, A. indica)* for fabrication of bimettalic oxides, metal hexacyanoderrates, cobaltates nanostructures for photocatalytic degradation of several organic pollutants (Jassal et al. 2015, 2016a, b; Shanker et al., 2016a, b, 2017a, b,c; Rani and Shanker 2017; Rani et al., 2017a; Rani and Rachna 2018; Rani and Shanker 2018a, b, c, d).

Green synthesized highly crystalline sharp metal hexacyanoferrates and transition metal oxide nanostructures obtained via use of sunlight and natural surfactant (*A. marmelos*) have were able to photodegrade quantitative (more than 90%) amount of dyes (Jassal et al. 2015, 2016a, b, c; Shanker et al. 2016).

Metal hexacyanoferrate nanoparticles of particle zise approximate 10–100 nm were more effective for toxic polycyclic aromatic hydrocarbons under sunlight than ultra-violet and dark exposure. Depending upon size of rings, molecular weight and aromaticity, anthracene was initially adsorbed followed by degradation to maximum content over the surface of photocatalyst than phenanthrene, fluorine, chrysene, benzopyrene in water (70–93%) along with soil (68–84%) at neutral pH under daylight contact. The high potential is further seen by finding of minor and lesser-toxic metabolites led to mineralization.

Crystalline nanocubes of hematite@zinc heaxcyanoferrate and zinc oxide@zinc heaxcyanoferrate nanocomposite were synthesized using water and plant extract of *A. indica*, locally available plants. Phytochemicals like benzoquinones, saponin, polyphenols have tendency to regulate the particle growth by minimizing interfacial tension. Functionalization with doping improved the surface-to-volume ration area from 343 $m^2\ g^{-1}$ to 114 $m^2 g^{-1}$ and band energy from 2.18 eV to 2.2 eV) as well because of synergestic effect of both semiconductors that is metal oxides and zinc heaxcyanoferrate. This fact is supported by 3–6 times better adsorption of chrysene (X_m = 45.45 mg g^{-1}) by doped nanocubes than bared zinc heaxcyanoferrate (X_m = 16.22 mg g^{-1}) and Fe_2O_3 (X_m = 7.348 mg g^{-1}) (Rachna et al. 2018). For bisphenol A, degradation trend of zinc oxide – zinc heaxcyanoferrate was highest (97%) followed by zinc heaxcyanoferrate (88%) and zinc oxide (75%). Finally, bisphenol A was converted to minor and small non-toxic oxidative chemicals like malealdehyde, propionic acid and but-2-ene-1,2,4-triol by countless hydroxyl radical generated by conduction of wraped-nanocomposite. Moreover, both the functionalized catalysts are easily regenerated and used upto ten-cycles emphasizing their high potential for industrial applications.

Functionalization further improved charge separation and prevent charge carriers recombination because of movement of photogenerated electrons from photocatalysts to the photosensitizer. Green synthesis is commonly used for metal oxides but limited for bimetaalic and higher forms. Farood and research group (2003) carried out green funtionalized of zinc oxide with cerium ion or silver for photocatalytic degradation of organic pollutants. Rani et al. 2019, fabricated bimetallic oxides (below 50 nm) such as nickel copper oxide nanorods, copper ferrate nanoflowers and nickel chromate nanospheres via use of *A. marmelos* leaf extract consisting of terpenoids, alkaloids and phenylpropanoids as phytochemicals for controlling the morphology of engineered nanomaterials. Degradation efficiency of bimetallic oxide of Nickel- oxide, Nickel-Chromium oxide and Copper -Chromium oxide are in order of zeta potential, surface area and uniform distribution of nanopartciles of both oxides involved. The by product obtained from degradation (oxidation and cleavage of benzene-rings) of phenols were oxalic acid, benzoquinone, (Z)-hex-3-enedioic acid, (Z)-but-2-enal, (Z)-4-oxobut-2-enoic acid. Moreover, like engineered nanomaterials of metal oxide with zinc hexacyanoferrate, mixtures of metal oxides are less expensive as synthesized from cheap process and reusable (n=10) with enhancement of charge separation.

Diverse potentials were observed by different reasearcher for engineered nanomaterials fabricated from green source containing different phytochemicals surfactants. Due to those, engineered nanomaterials of different size and morphology were obtained. Hence, in view of this, commercialization of green synthesized engineered nanomaterials should be promoted.

7.5 Future Scope and Perspectives

In view of recalcitrant nature and serious health hazard of the pesticides, their complete removal from contaminated wastewater is now the need of hour. Though the pesticides concentration in environment found in ng L^{-1} to mg L^{-1}, but because of their recalcitrant behavious they may expressively increase their impression to the extended period of time. Due to incomplete removal and generation of secondary pollutants using conventional treatment methods pesticides removal needs advanced treatment methods that not only adsorb it but also completely mineralize. In this perspective, the generation of functional nanomaterial's can be considered as potential candidates in view of advanced properties to solve the removal of pesticides. These contaminants are not only noxious in its primary form but also generate highly toxic secondary metabolites that are even more challenging to eradicate than parental compound. Therefore, comprehensive details of possible subordinate metabolites must be documented for all the pesticides. Furthermore, an advanced-eradication technology must be recognized subsequently an exhaustive study is required of the secondry metabolites made during the degradation reaction. Extensive research for synthesis of engineered nanomaterials requires enhancing their competence, economic, and technical practicability. Removal of pesticides by the engineered nanomaterials was found dependent of several environmental parameters. The combinations of the engineered materials exhibiting both adsorption and photocatalytic effect found better in removal of pesticides compared to individual ones. The reusability of the functional materials will surely achieve a sustainable treatment method for pesticide removal.

7.6 Conclusion

The present chapter summarizes the progress in development of various functional materials and their application in pesticides removal. Functionalization of nanomaterials (silica, carbonaceous, and metal-based; (zinc oxide, titanium dioxide, iron, platinum and palladium and nickel oxide) for pesticides degradation have been overviewed. Various types of functionalization methods especially the chemical approaches for its diversity and ability to decorate nanomaterials with large number of functionalities were discussed. High production and wide use of innumerable perilous man-made organic is observed the prominent sources of water pollution in Asia and worldwide. Engineered nanomaterials may overcome the principal limitations that hinder wide-scale applications of nanoparticles. Due to finalization, their working ability enhanced by manifolds at neutral pH under ultra-violet or solar radiation engineered nanomaterials are used in wide-range to exclude or remove their destructive impacts via several processes including oxidation, photocatalysis (generation of electron-hole pairs) and adsorption. Degradation resulted into more or less complete removal through mineralization or transformation to less-toxic chemicals in short span. Green synthesis produced engineered nanomaterials with variable morphology and size depending on different biogenic sources (plant surfactants, microorganisms). Those engineered nanomaterials obtained could have better efficacy than that of chemically synthesized ones because of role of surfactant that enter in lattice or sometimes found to dispersed on it. However, the way is still long and more focus is needed to employ those as a mile stone in the environmental and other commercial nanotechnology.

Furthermore, the role of operating parameters on efficiency of composite materials is reported. Hybrid functional materials involve templates of nanoadsorbents (graphene, carbon nanotubes, nano zerovalent iron, hematite, magnetite), (titanium dioxide, zinc oxide, tungusten trioxide) and polymeric materials have been used. The functionalization of composite materials helps in overcoming the limitations of simple materials such as less surface area and adsorption capacity, electron-hole recombination, visible light insensitivity. The influence of crucial factors like pH, material dose, initial contaminant concentration, temperature, and the presence of other oxidants in removing the pernicious contaminants are of paramount importance and have been addressed in brief.

Finally, chapter end with some points on suggested future scope:-

1. Green synthesis may result in cheaper, effective and sustainable engineered nanomaterials mediated degradation process. As well, biopolymers-based functioalized composites should be encouraged.
2. Low-cost, easy available, green, and highly efficient, titanium dioxide and nano zerovalent have been reported as an encouraging pesticide eradication process.
3. More research is needed to explore different nanomaterials on traditional or upcoming organophosphates, carbamates and substitute urea to ban or restrict them. Engineered nanomaterials with virtue of high potential should also be used to treat still persisting organochlorine. For some of pollutants like Quintozene, pentachlorophenol and toxaphene, they have never used.

Acknowledgement One of the authors Dr Manviri Rani is thankful to TEQIP-III, MNIT Jaipur, India and DST SERB (File Number: SRG/2019/000114), New Delhi, India for financial support. Dr Uma Shanker is also thankful to TEQIP-III, Dr B R Ambedkar National Institue of Technology Jalandhar, Punjab, India for financial support.

References

Adachi M, Murata Y, Takao J, Jiu J, Sakamoto M, Wang F (2004) Highly efficient dye sensitized solar cells with a titania thin-film electrode composed of a network structure of single crystal-like TiO_2 nanowires made by the "oriented attachment" mechanism. J Am Chem Soc 126 (45):14943–14949. https://doi.org/10.1021/ja048068s

Afkhami A, Moosavi R, Madrakian T (2010) Preconcentration and spectrophotometric determination of low concentrations of malachite green and leuco-malachite green in water samples by high performance solid phase extraction using maghemite nanoparticles. Talanta 82:785–789. https://doi.org/10.1016/j.talanta.2010.05.054

Ahmed S, Rasul MG, Martens WN, Brown R, Hashib MA (2010) Heterogenous photocatalytic degradation of phenols in wastewater: a review on current status and developments. Desalination 261:3–18. https://doi.org/10.1016/j.desal.2010.04.062

Ahmed S, Rasul MG, Brown R, Hashib MA (2011) Influence of parameters on the heterogeneous photocatalytic degradation of pesticides and phenolic contaminants in wastewater: a short review. J Environ Manag 92(3):311–330. https://doi.org/10.1016/j.jenvman.2010.08.028

Aktar MW, Sengupta D, Chowdhury A (2009) Impact of pesticides use in agriculture: their benefits and hazards. Interdiscip Toxicol 2(1):1–12. https://doi.org/10.2478/v10102-009-0001-7

Al Hattab MT, Ghaly AE (2012) Disposal and treatment methods for pesticide containing wastewaters: critical review and comparative analysis. J Environ Prot. 3(5):431–453. https://doi.org/10.4236/jep.2012.35054

Al-Hamdi AM, Sillanpää M, Bora T, Dutta J (2016) Efficient photocatalytic degradation of phenol in aqueous solution by SnO_2:Sb nanoparticles. Appl Surf Sci 370:229–236. https://doi.org/10.1016/j.apsusc.2016.02.123

Ali I, AlOthman ZA, Al-Warthan A (2016) Sorption, kinetics and thermodynamics studiesof atrazine herbicide removal from water using iron nano-composite material. Int J EnvironSci Technol 13(2):733–742. https://doi.org/10.1007/s13762-015-0919-6

Alikhani M-Y, Lee S-M et al (2012) Samarghandi, photocatalytic removal of escherichia coli from aquatic solutions using synthesized ZnO nanoparticles: a kinetic study. Water Sci Technol 67:557–563. https://doi.org/10.2166/wst.2012.593

Alizadeh Fard M, Vosoogh A, Barkdoll B, Aminzadeh B (2017) Using polymer coated nanoparticles for adsorption of micropollutants from water. Colloids Surf A Physicochem Eng Asp 531:189–197. https://doi.org/10.1016/j.colsurfa.2017.08.008

Ambashta RD, Sillanpää M (2010) Water purification using magnetic assistance: a review. J Hazard Mater 180:38–49. https://doi.org/10.1016/j.jhazmat.2010.04.105

Amin MT, Alazba AA, Manzoor U (2014) A review of removal of pollutants from water/wastewater using different types of nanomaterials. Adv Mat Sci Eng 2014:1–24. https://doi.org/10.1155/2014/825910

Anandan S, Vinu A, Mori T, Gokulakrishnan N, Srinivasu P, Murugesan V, Ariga K (2007a) Photocatalytic degradation of 2,4,6-trichlorophenol using lanthanum doped ZnO in aqueous suspension. Catal Commun 8:1377–1382. https://doi.org/10.1016/j.catcom.2006.12.001

Anandan S, Vinu A, Sheeja Lovely KLP, Gokulakrishnan N, Srinivasu P, Mori T, Murugesan V, Sivamurugan V, Ariga K (2007b) Photocatalytic activity of La-doped ZnO for the degradation of monocrotophos in aqueous suspension. J Mol Catal A Chem 266:149–157. https://doi.org/10.1016/j.molcata.2006.11.008

Ashtari P, Wang KM, Yang XH, Huang SS, Yamini Y (2005) Novel separation and preconcentration of trace amounts of copper(II) in water samples based on neocuproine modified magnetic microparticles. Anal Chim Acta 550:18–23. https://doi.org/10.1016/j.aca.2005.06.048

Aslam M, Ismail IMI, Chandrasekaran S, Hameed A (2014) Morphology controlled bulk synthesis of disc-shaped WO_3 powder and evaluation of its photocatalytic activity for the degradation of phenols. J Hazard Mater 276:120–128. https://doi.org/10.1016/j.jhazmat.2014.05.022

Aslam M, Soomro MT, Ismail IMI, Salah N, Gondal MA, Hameed A (2015) Sunlight mediated removal of chlorophenols over tungsten supported ZnO: electrochemical and photocatalytic studies. J Environ Chem Eng 3:1901–1911. https://doi.org/10.1016/j.jece.2015.07.004

Atrak K, Ramazani A, Fardood ST (2018) A novel sol–gel synthesis and characterization of MgFe 2 O 4@ γ–Al 2 O 3 magnetic nanoparticles using tragacanth gel and its application as a magnetically separable photocatalyst for degradation of organic dyes under visible light. J Mater Sci Mater Electron 29:6702–6710. https://doi.org/10.1007/s10854-018-8656-5

Attia YA, Yasser MA, Mohamed YMA, Altalhi TA (2016) Photobiosynthesis of metal/graphene nanocomposites: new materials for water desalination and purification. Desalin Water Treat 57

Ayad MM, Amer WA, Kotp MG (2017) Magnetic polyaniline-chitosan nanocomposite decorated with palladium nanoparticles for enhanced catalytic reduction of 4-nitrophenol. Mol Catal 439:72–80. https://doi.org/10.1016/j.mcat.2017.06.0232468-8231

Bagheri A, Taghizadeh M et al (2012) Synthesis and characterization of magnetic metal–organic framework (MOF) as a novel sorbent, and its optimization by experimental design methodology for determination of palladium in environmental samples. Talanta 99:132–139. https://doi.org/10.1016/j.talanta.2012.05.030

Baker J, Laflen J, Johnson H (1978) Effect of tillage systems on runoff losses of pesticides, a rainfall simulation study. Trans ASAE 21:886–0892. https://doi.org/10.13031/2013.35408

Banić N, Abramović B, Krstić J, Šojić D, Lončarević D, Cherkezova-Zheleva Z, Guzsvány V (2011) Photodegradation of thiacloprid using Fe/TiO_2 as a heterogeneous photo- Fenton catalyst. Appl Catal B Environ 107:363–371. https://doi.org/10.1016/j.apcatb.2011.07.037

Bartlett J, Brunner M, Gough K (2010) Deliberate poisoning with dinitrophenol (DNP): an unlicensed weight loss pill. Emerg Med J 27(2):159–160

Bechambi O, Chalbi M, Najjar W, Sayadi S (2015) Photocatalytic activity of ZnO doped with Ag on the degradation of endocrine disrupting under UV irradiation and the investigation of its antibacterial activity. Appl Surf Sci 347:414–420. https://doi.org/10.1016/j.apsusc.2015.03.049

Belessi V, Lambropoulou D et al (2009) Structure and photocatalytic performance of magnetically separable titania photocatalysts for the degradation of propachlor. Appl Catal B Environ 87:181–189. https://doi.org/10.1016/j.apcatb.2008.09.012

Bennett SW, Keller AA (2011) Comparative photoactivity of CeO_2, gamma-Fe_2O_3, TiO_2 and ZnO in various aqueous systems. Appl Catal B 102:600–607. https://doi.org/10.1016/j.apcatb.2010.12.045

Boruah PK, Sharma B, Hussain N, Das MR (2016) Magnetically recoverable Fe_3O_4/graphene nanocomposite towards efficient removal of triazine pesticides from aqueous solution: investigation of the adsorption phenomenon and specific ion effect. Chemosphere 168:1–10. https://doi.org/10.1016/j.chemosphere.2016.10.103

Broughton E (2005) The Bhopal disaster and its aftermath: a review. Environ Health 4:6. https://doi.org/10.1186/1476-069X-4-6

Burrows HD, Canle LM, Santaballa JA, Steenken S (2002) Invited review: reaction pathways and mechanisms of photodegradation of pesticides. J Photochem Photobiol B Biol 67:71–108. https://doi.org/10.1016/S1011-1344(02)00277-4

Cai YQ, Cai YE, Mou SF, Lu YQ (2005) Multi-walled carbon nanotubes as a solid-phase extraction adsorbent for the determination of chlorophenols in environmental water samples. JChromatogr A 1081:245–247. https://doi.org/10.1021/ac0263566

Cao J, Elliott D, Zhang WX (2005) Perchlorate reduction by nanoscale iron particles. J Nanoparticle Res 7:499–506. https://doi.org/10.1007/s11051-005-4412-x

Cao Y, Tan H, Shi T, Tang T, Li J (2008) Preparation of Ag-doped TiO_2 nanoparticles for photocatalytic degradation of acetamiprid in water. J Chem Technol Biotechnol 83:546–552. https://doi.org/10.1002/jctb.1831

Chapalamadugu S, Chaudhry GR (1992) Microbiological and biotechnological aspects of metabolism of carbamates and organophosphates. Crit Rev Biotechnol 12:357. https://doi.org/10.3109/07388559209114232

Chen GC, Shan XQ, Zhou YQ, Shen X e, Huang HL, Khan SU (2009) Adsorptionkinetics, isotherms and thermodynamics of atrazine on surface oxidized multiwalled carbon nanotubes. J Hazard Mater 169:912–918. https://doi.org/10.1016/j.jhazmat.2009.04.034

Chen Q, Wang J, Li F (2011a) Formation of carbon nanofibers from supported Pt catalysts through catalytic chemical vapor deposition from acetylene. Ind Eng Chem Res 50:9034–9042. https://doi.org/10.1021/ie200345f

Chen H, Yang S, Yu K, Ju Y, Sun C (2011b) Effective photocatalytic degradation of atrazine over titania-coated carbon nanotubes (CNTs) coupled with microwave energy. J Phys Chem A 115:3034–3041. https://doi.org/10.1021/jp109948n

Chettri P, Vendamani V, Tripathi A, Singh MK, Pathak AP, Tiwari A (2017) Green synthesis of silver nanoparticle-reduced graphene oxide using Psidium guajava and its application in SERS for the detection of methylene blue. Appl Surf Sci 406:312–318

Chiou JR, Lai BH, Hsu KC, Chen DH (2013) One-pot green synthesis of silver/iron oxide composite nanoparticles for 4-nitrophenol reduction. J Hazard Mater 248–249:394–400. https://doi.org/10.1016/j.jhazmat.2013.01.030

Chitose N, Ueta S, Seino S, Yamamoto TA (2003) Radiolysis of aqueous phenol solutions with nanoparticles. 1. Phenol degradation and TOC removal in solutions containing TiO2 induced by UV, γ-ray and electron beams. Chemosphere 50:1007–1013. https://doi.org/10.1016/S0045-6535(02)00642-2

Clark K, Keller A (2012a) Investigation of two magnetic permanently confined micelle array sorbents using nonionic and cationic surfactants for the removal of PAHs and pesticides from aqueous media. Water Air Soil Pollut 223:3647–3655

Clark KK, Keller AA (2012b) Adsorption of perchlorate and other oxyanions onto magnetic permanently confined micelle arrays (Mag-PCMAs). Water Res 46:635–644. https://doi.org/10.1016/j.watres.2011.11.025

Comparelli R, Fanizza E, Curri ML, Cozzoli PD, Mascolo G, Agostiano A (2005) UV induced photocatalytic degradation of azo dyes by organic-capped ZnO nanocrystals immobilized onto substrates. Appl Catal B Environ 60:1–11. https://doi.org/10.1016/j.apcatb.2005.02.013

Corte SD, Hennebel T, Gusseme BD (2012) Bio-palladium: from metal recovery to catalytic applications. Microb Biotechnol 5(1):5–17

Dai K, Peng T, Ke D, Wei B (2009) Photocatalytic hydrogen generation using a nanocomposite of multi-walled carbon nanotubes and TiO2 nanoparticles under visible light irradiation. Nanotechnology 20:125603. https://doi.org/10.1088/0957-4484/20/12/125603

Dankovich TA, Gray DG (2011) Bactericidal paper impregnated with silver nanoparticles for point-of-use water treatment. Environ Sci Technol 45:1992–1998. https://doi.org/10.1021/es103302t

Darwish M, Mohammadi A, Assi N, Manuchehri QS, Alahmad Y, Abuzerr S (2017) Shape-controlled ZnO nanocrystals synthesized via auto combustion method and enhancement of the visible light catalytic activity by decoration on graphene. J Alloys Compd 703:396–406. https://doi.org/10.1016/j.jallcom.2017.01.309

De AK (2010) Environmental chemistry, 7th edn. New Age International Publisher, New Delhi

Debnath D, Gupta AK (2017) Optimizing the fabrication of nano-plasmonic silver-nitrogen co-doped zinc oxide (AgxZn(1-x)NyO(1-y)) mediated by ammonia template: Insight into its enhanced physiochemical and photocatalytic behavior. J Mol Liq 249:334–345. https://doi.org/10.1016/j.molliq.2017.11.050

Devi NL, Raha P (2013) Contamination of Organochlorine pesticides (OCPs) in India. Bull Environ Sci Res 2:9–14 de Urzedo AP, Nascentes CC, Augusti R (2009) Degradation of the insecticides Thiamethoxam and Imidacloprid in aqueous solution as promoted by an innovative Fe°/Fe3O4 composite. J Brazil Chem Soc 20:51–56. https://doi.org/10.1590/S0103-50532009000100010

Devipriya S, Yesodharan S (2005) Photocatalytic degradation of pesticide contaminants in water. Sol Energy Mater Sol Cells 86:309–348

Ding Y, Zhu L, Wang N, Tang H (2013) Sulfate radicals induced degradation of tetrabromobisphenol A with nanoscaled magnetic $CuFe_2O_4$ as a heterogeneous catalyst of peroxymonosulfate. Appl Catal B Environ 129:153–162. https://doi.org/10.1016/j.apcatb.2012.09.015

Divband B, Khatamian M, Eslamian GRK, Darbandi M (2013) Synthesis of Ag/ZnO nanostructures by different methods and investigation of their photocatalytic efficiency for 4-nitrophenol degradation. Appl Surf Sci 284:80–86. https://doi.org/10.1016/j.apsusc.2013.07.015

Dixit V, Tewari JC, Obendorf SK (2009) Identification of degraded products of aldicarb due to the catalytic behavior of titanium dioxide/polyacrylonitrile nanofiber. J Chromatogr A 1216:6394–6399

Dong R, Tian B, Zeng C, Li T, Wang T, Zhang J (2013) Ecofriendly synthesis and photocatalytic activity of uniform cubic Ag@AgCl plasmonic photocatalyst. J Phys Chem C 117:213–220. https://doi.org/10.1021/jp311970k

Ecobichon DJ (1991) Pesticides. In: Amdur MO, Doull J, Klaassen CD (eds) Casarett and Doull's toxicology: the basic science of poisons, 4th edn. Pergamon Press, New York, p 580

Ehdaie B, Krause C, Smith JA (2014) Porous ceramic tablet embedded with silver nanopatches for low-cost point-of-use water purification. Environ Sci Technol 48:13901–13908. https://doi.org/10.1021/es503534c

El-Kemary M, El-Shamy H, El-Mehasseb I (2010) Photocatalytic degradation of ciprofloxacin drug in water using ZnO nanoparticles. J Lumin 130:2327–2331. https://doi.org/10.1016/j.jlumin.2010.07.013

Elliott DW, Lien H-L, Zhang W-X (2009) Degradation of lindane by zero-valent iron nanoparticles. J Environ Eng 135:317–324. https://doi.org/10.1061/(ASCE)0733-9372(2009)135:5(317)

Faraji M, Yamini Y, Rezaee M (2010) Extraction of trace amounts of mercury with sodium dodecyl sulphate-coated magnetite nanoparticles and its determination by flow injection inductively coupled plasma-optical emission spectrometry. Talanta 81:831–836. https://doi.org/10.1016/j.talanta.2010.01.023

Fardood ST, Ramazani A, Moradi S (2017a) Green synthesis of Ni–cu–mg ferrite nanoparticles using tragacanth gum and their use as an efficient catalyst for the synthesis of polyhydroquinoline derivatives. J Solgel Sci Technol 82:432–439. https://doi.org/10.1007/s10971-017-4310-6

Fardood ST, Ramazani A, Moradi S, Asiabi PA (2017b) Green synthesis of zinc oxide nanoparticles using arabic gum and photocatalytic degradation of direct blue 129 dye under visible light. J Mater Sci Mater Electron 28:13596–13601. https://doi.org/10.1007/s10854-017-7199-5

Feng J, Su L, Ma Y, Ren C, Guo Q, Chen X (2013) $CuFe_2O_4$ magnetic nanoparticles: a simple and efficient catalyst for the reduction of nitrophenol. Chem Eng J 221:16–24. https://doi.org/10.1016/j.cej.2013.02.009

Fernandez-Casalderrey A, Ferrando MD, Andreu-Moliner E (1992) Endosulfan and diazinon toxicity to the freshwater rotifier Brachionus calyciflorus. J Environ Sci Health Part B 27:155–164. https://doi.org/10.1080/03601239209372772

Follut F, Vel Leitner NK (2007) Radiolysis of aqueous 4-nitrophenol solution with Al_2O_3 or TiO_2 nanoparticles. Chemosphere 66:2114–2119. https://doi.org/10.1016/j.chemosphere.2006.09.031

Fouad DM, Mohamed MB (2012) Comparative study of the photocatalytic activity of semiconductor nanostructures and their hybrid metal nanocomposites on the photodegradation of malathion. J Nanomater 2012. https://doi.org/10.1155/2012/524123

Gao B, Liu L, Liu J, Yang F (2013) Photocatalytic degradation of 2,4,6-tribromophenol over Fe-doped $ZnIn_2S_4$: stable activity and enhanced debromination. Appl Catal B Environ 129:89–97. https://doi.org/10.1016/j.proenv.2013.04.068

Gaya UI, Abdullah AH (2008) Heterogeneous photocatalytic degradation of organic contaminants over titanium dioxide: a review of fundamentals, progress and problems. J Photochem Photobiol C Photochem Rev 9:1–12. https://doi.org/10.1016/j.jphotochemrev.2007.12.003

Georgekutty R, Seery MK, Pillai SC (2008) A highly efficient Ag–ZnO photocatalyst: synthesis, properties, and mechanism. J Phys Chem C 112:13563–13570. https://doi.org/10.1021/jp802729a

Girginova PI, Daniel-da-Silva AL et al (2010) Silica coated magnetite particles for magnetic removal of Hg2+ from water. J Colloid Interface Sci 345:234–240. https://doi.org/10.1016/j.jcis.2010.01.087

Gomez S, Marchena CL, Renzini MS, Pizzio L, Pierella L (2015) In situ generated TiO2 over zeolitic supports as reusable photocatalysts for the degradation of dichlorvos. Appl Catal B Environ 162:167–173

Graymore M, Stagnitti F, Allinson G (2001) Impacts of atrazine in aquatic ecosystems. Environ Int 26:483–495

Grundlingh J, Dargan PI, El-Zanfaly M, Wood DM (2011) 2,4-Dinitrophenol (DNP): A weight loss agent with significant acute toxicity and risk of death. J Med Toxicol 7:205–212. https://doi.org/10.1007/s13181-011-0162-6

Guo P, Tang L et al (2016) Catalytic reduction–adsorption for removal of p-nitrophenol and its conversion p-aminophenol from water by gold nanoparticles supported on oxidized mesoporous carbon. J Colloid Interface Sci 469:78–85. https://doi.org/10.1016/j.jcis.2016.01.063

Gupta PK (2004) Pesticide exposure – Indian scene. Toxicology 198:83–90

Gupta VK, Saleh T (2013) Sorption of pollutants by porous carbon, carbon nanotubes and fullerene – an overview. Environ Sci Pollut Res 20:2828–2843. https://doi.org/10.1007/s11356-013-1524-1

Gupta AK, Pal A, Sahoo C (2006) Photocatalytic degradation of a mixture of crystal violet (basic violet 3) and methyl red dye in aqueous suspensions using Ag^+ doped TiO_2. Dye Pigment 69:224–232. https://doi.org/10.1016/j.dyepig.2005.04.001

Gupta S, Ashrith G, Chandra D, Gupta AK, Finkel KW, Guntupalli JS (2009) Acute phenol poisoning: a life-threatening hazard of chronic pain relief. Clinical Toxicol 46:250–253

Gupta VK, Gupta B, Rastogi A, Agarwal S, Nayak A (2011) Pesticides removal from waste water by activated carbon prepared from waste rubber tire. Water Res 45:4047–4055. https://doi.org/10.1016/j.watres.2011.05.016

Gupta VK, Eren T, Atar N, Yola ML, Parlak C, Karimi-Maleh H (2015) $CoFe_2O_4$@TiO_2 decorated reduced graphene oxide nanocomposite for photocatalytic degradation of chlorpyrifos. J Mol Liq 208:122–129. https://doi.org/10.1016/j.molliq.2015.04.032

Han Y, Xu Z, Gao C (2013) Ultrathin graphene nanofiltration membrane for water purification. Adv Funct Mater 23:3693–3700. https://doi.org/10.1002/adfm.201202601

Hassan KA, Elbourne LD, Li L, Hewawasam Gamage HK, Liu Q, Jackson SM, Sharples D, Kolstø A-B, Henderson PJ, Paulsen IT (2015) An ace up their sleeve: a transcriptomic approach exposes the AceI efflux protein of Acinetobacter baumannii and reveals the drug efflux potential hidden in many microbial pathogens. Front Microbiol 6:333. https://doi.org/10.3389/fmicb.2015.00333

He F, Zhao D (2005) Preparation and characterization of a new class of starch-stabilized bimetallic nanoparticles for degradation of chlorinated hydrocarbons in water. Environ Sci Technol 39:3314–3320. https://doi.org/10.1021/es048743y

Henych J, Janos P, Kormunda M, Tolasz J, Stengl V (2016) Reactive adsorption of toxic organophosphatesparathion methyl and DMMP on nanostructuredTi/Ce oxides and their composites. Arab J Chem. https://doi.org/10.1016/j.arabjc.2016.06.002

Hernández-Gordillo A, González VR (2015) Silver nanoparticles loaded on Cu-doped TiO_2 for the effective reduction of nitro-aromatic contaminants. Chem Eng J 261:53–59. https://doi.org/10.1016/j.cej.2014.05.148

Herrmann JM (1999) Heterogeneous photocatalysis: Fundamentals and applications to the removal of various types of aqueous pollutants. Catal Today 53:115–129. https://doi.org/10.1016/S0920-5861(99)00107-8

Hoag GE, Collins JB, Holcomb JL, Hoag JR, Nadagouda MN, Varma RS (2009) Degradation of bromothymol blue by 'greener' nano-scale zero-valent iron synthesized using tea polyphenols. J Mater Chem 19:8671–8677. https://doi.org/10.1039/b909148c

Hossaini H, Moussavi G, Farrokhi M (2014) The investigation of the LED-activated FeFNS TiO_2 nanocatalyst for photocatalytic degradation and mineralization of organophosphate pesticides in water. Water Res 59:130–144. https://doi.org/10.1016/j.watres.2014.04.009

Hrelia P, Fimognari C et al (1996) The genetic and non-genetic toxicity of the fungicide Vinclozolin. Mutagenesis 11:445–453

Hu R, Dai S, Shao D et al (2015) Efficient removal of phenol and aniline from aqueous solutions using graphene oxide/polypyrrole composites. J Mol Liq 203:80–89. https://doi.org/10.1016/j.molliq.2014.12.0460167-7322

Huang Y, Keller AA (2013) Magnetic nanoparticle adsorbents for emerging organic contaminants. ACS Sustain Chem Eng 1:731–736. https://doi.org/10.1021/sc400047q

Huang Y, Keller AA (2015) EDTA functionalized magnetic nanoparticle sorbents for cadmium and lead contaminated water treatment. Water Res 80:159–168. https://doi.org/10.1016/j.watres.2015.05.011

Huang D, Tang Z et al (2017) Fabrication of water-compatible 1014 molecularly imprinted polymer based on β- cyclodextrin modified magnetic chitosan and its application for selective removal of bisphenol A from aqueous solution. J Taiwan Inst Chem Eng 77:113–121. https://doi.org/10.1016/j.jtice.2017.04.030

Ibrahim S, Chakrabarty S, Ghosh S, Pal T (2017) Reduced graphene oxide–zinc sulfide composite for solar light responsive photo current generation and photocatalytic 4- nitrophenol reduction. Chem Select 2:537–545. https://doi.org/10.1002/slct.201601999

Iijima S (1991a) Helical microtubules of graphitic carbon. Nature 354:56–58

Iijima S (1991b) Helical microtubules of graphitic carbon. Nature 354:56–58

Ishihara A, Nishiyama N, Sugiyama S, Yamauchi K (2003) The effects of endocrine disrupting chemicals on thyroid hormone binding to Japanese quail transthyretin and thyroid hormone receptor. Gen Comp Endocrinol 134:36–43

Ismail AA, Abdelfattah I et al (2016) Ease synthesis of mesoporous WO_3–TiO_2 nanocomposites with enhanced photocatalytic performance for photodegradation of herbicide imazapyr under visible light and UV illumination. J Hazard Mater 307:43–54. https://doi.org/10.1016/j.jhazmat.2015.12.041

Isman MB (1994) Botanical insecticides. Pestic Outlook 5:26–31

Isman MB (1997) Neem and other botanical insecticides: barriers to commercialization. Phytoparasitica 25:339. https://doi.org/10.1007/BF02981099

Jaiswal M, Chauhan D, Sankararamakrishnan N (2012) Copper chitosan nanocomposites: synthesis, characterization, and application in removal of organophosphorous pesticide from agricultural runoff. Environ Sci Pollut R 19:2005–2062. https://doi.org/10.1007/s11356-011-0699-6

Jassal V, Shanker U, Kaith BS, Shankar S (2015) Green synthesis of potassium zinc hexacyanoferrate nanocubes and their potential application in photocatalytic degradation of organic dyes. RSC Adv 5:26141–26150

Jassal V, Shanker U, Gahlot S (2016a) Green synthesis of some iron oxide nanoparticles and their interaction with 2-Amino, 3-Amino and 4-Aminopyridines. Mater Today Proc 3:1874–1880

Jassal V, Shanker U, Kaith BS (2016b) Aegle marmelos mediated green synthesis of different nanostructured metal hexacyanoferrates: activity against photodegradation of harmful organic dyes. Scientifica 2016:1–17

Jassal V, Shanker U, Kaith BS, Gahlot S, Kamaluddin IMA, Samuel P (2016c) Sapindus mukorossi mediated green synthesis of some manganese oxide nanoparticles interaction with aromatic amines. Appl Phys A 122:271–280

Jeon EK, Seo E, Lee E, Lee W, Um MK, Kim BS (2013) Mussel-inspired green synthesis of silver nanoparticles on graphene oxide nanosheets for enhanced catalytic applications. Chem Commun 49:3392. https://doi.org/10.1039/C3CC00115F

Jiang Z, Jiang D, Hossain AMS, Qian K, Xie J (2015) In situ synthesis of silver supported nanoporous iron oxide microbox hybrids from metal organic p -nitrophenol reduction. Phys Chem Chem. Phys 17:2550–2559. https://doi.org/10.1039/C4CP04594G

Jinu U, Gomathi M, Saiqa I, Geetha N, Benelli G, Venkatachalam P (2017) Green engineered biomolecule-capped silver and copper nanohybrids using Prosopis cineraria leaf extract: enhanced antibacterial activity against microbial pathogens of public health relevance and cytotoxicity on human breast cancer cells (MCF-7). Microb Pathog 105:86–95. https://doi.org/10.1016/j.micpath.2017.02.019

Jonidi-Jafari A, Shirzad-Siboni M, Yang JK, Naimi-Joubani M, Farrokhi M (2015) Photocatalytic degradation of diazinon with illuminated ZnO-TiO_2 composite. J Taiwan Inst Chem Eng 50:100–107. https://doi.org/10.1016/j.jtice.2014.12.020

Joo SH, Zhao D (2008) Destruction of lindane and atrazine using stabilized iron nanoparticles under aerobic and anaerobic conditions: effects of catalyst and stabilizer. Chemosphere 70:418–425. https://doi.org/10.1016/j.chemosphere.2007.06.070

Karabelas AJ, Plakas KV et al (2009) Impact of European legislation on marketed pesticides – a view from the standpoint of health impact assessment studies. Environ Int 35:1096–1107. https://doi.org/10.1016/j.envint.2009.06.011

Karimi MA, Hatefi-Mehrjardi A et al (2012) Solid phase extraction of trace amounts of silver (I) using dithizone-immobilized alumina-coated magnetite nanoparticles prior to determination by flame atomic absorption spectrometry. Int J Environ Anal Chem 92:1325–1340. https://doi.org/10.12691/wjac-1-4-1

Karn B, Kuiken T, Otto M (2009) Nanotechnology and in situ remediation: a review of the benefits and potential risks. Environ Health Perspect 117:1823–1831. https://doi.org/10.1289/ehp.0900793

Kaur P, Bansal P, Sud D (2013) Heterostructured nanophotocatalysts for degradation of organophosphate pesticides from aqueous streams. J Korean Chem Soc 57(3):382–388. https://doi.org/10.5012/jkcs.2013.57.3.382

Khajeh M, Laurent S, Dastafkan K (2013) Nanoadsorbents: classification, preparation, and applications (with emphasis on aqueous media). Chem Rev 113(10):7728–7768. https://doi.org/10.1021/cr400086v

Khatamian M, Khandar AA et al (2012) Heterogeneous photocatalytic degradation of 4-nitrophenol in aqueous suspension by Ln (La^{3+} , Nd^{3+} or Sm^{3+}) doped ZnO nanoparticles. J Mol Catal A Chem 365:120–127. https://doi.org/10.1016/j.molcata.2012.08.018

Kolya H, Maiti P, Pandey A, Tripathy T (2015) Green synthesis of silver nanoparticles with antimicrobial and azo dye (Congo red) degradation properties using Amaranthus gangeticus Linn leaf extract. J Anal Sci Technol 6:33. https://doi.org/10.1186/s40543-015-0074-1

Konstantinou IK, Albanis TA (2003) Photocatalytic transformation of pesticides in aqueous titanium dioxide suspensions using artificial and solar light: intermediates and degradation pathways. Appl Catal B Environ 42:319–335

Konstantinou IK, Hela DG, Albanis TA (2006) The status of pesticide pollution in surface waters (rivers and lakes) of Greece. Part I. Review on occurrence and levels. Environ Pollut 141:555–570. https://doi.org/10.1016/j.envpol.2005.07.024

Korake PV, Sridharkrishna R, Hankare PP, Garadkar KM (2012) Photocatalytic degradation of phosphamidon using Ag-doped ZnO nanorods. Toxicol Environ Chem 94:1075–1085. https://doi.org/10.1080/02772248.2012.693495

Korake PV, Dhabbe RS et al (2014) Highly active lanthanum doped ZnO nanorods for photodegradation of metasystox. J Photochem Photobiol B Biol 130:11–19. https://doi.org/10.1016/j.jphotobiol.2013.10.012

Koushik D, Sen GS, Maliyekkal SM, Pradeep T (2016) Rapid dehalogenation of pesticides and organics at the interface of reduced graphene oxide-silver nanocomposite. J Hazard Mater 308:192–198. https://doi.org/10.1016/j.jhazmat.2016.01.004

Kralj MB, Cernigoj U, Franko M (2007) Comparison of photocatalysis and photolysis of malathion, isomalathion, malaoxon, and commercial malathion: products and toxicity studies. Water Res 41:4504–4514. https://doi.org/10.1016/j.watres.2007.06.016

Kumar SG, Rao KSRK (2015) Zinc oxide based photocatalysis: tailoring surface-bulk structure and related interfacial charge carrier dynamics for better environmental applications. RSC Adv 5:3306–3351. https://doi.org/10.1039/C4RA13299H

Kumar A, Naushad M et al (2017) ZnSe-WO_3 nano-hetero-assembly stacked on gum ghatti for photo degradative removal of bisphenol a: symbiose of adsorption and photocatalysis. Int J Biol Macromol 104:1172–1184. https://doi.org/10.1016/j.ijbiomac.2017.06.116

Kurian M, Kunjachan C, Sreevalsan A (2017) Catalytic degradation of chlorinated organic pollutants over CexFe1-xO2 (x: 0, 0.25, 0.5, 0.75, 1) nanocomposites at mild conditions. Chem Eng J 308:67–77. https://doi.org/10.1016/j.cej.2016.09.039

Lacinova L, Kvapil P, Cernik M (2012) A field comparison of two reductive dechlorination (zero-valent iron and lactate) methods. Environ Technol 33:741–749. https://doi.org/10.1080/09593330.2011.592225

Lam S, Sin J, Satoshi I, Abdullah AZ, Mohamed AR (2014) Enhanced sunlight photocatalytic performance over Nb_2O_5/ZnO nanorod composites and the mechanism study. Appl Catal A Gen 471:126–135. https://doi.org/10.1016/j.apcata.2013.12.001

Landrigan PJ, Claudio L et al (1999) Pesticides and inner city children: exposures, risks, and prevention. Environ Health Perspect 107:431–437

Laseter JL (1978) Environmental contamination of food in environmental contaminants in food, chapter 2, Washington, DC

Lavand AB, Malghe YS (2015) Visible light photocatalytic degradation of 4-chlorophenol using C/ZnO/CdS nanocomposite. JSaudi Chem Soc 19:471–478. https://doi.org/10.1016/j.jscs.2015.07.001

Lee HG, Sai-Anand G et al (2015) Efficient visible-light-driven photocatalytic degradation of nitrophenol by using graphene-encapsulated TiO_2 nanowires. J Hazard Mater 283:400–409. https://doi.org/10.1016/j.jhazmat.2014.09.014

Lellala K, Namratha K, Sudhakar K, Byrappa K (2016) Hydrothermal synthesis and characterization of hybrid Al/ZnO-GO composite for significant photodegrdation of dyes. AIP p. 020627. https://doi.org/10.1063/1.4946678

Lewin JF, Cleary WT (1982) An accidental death caused by the absorption of phenol through skin. A case report. Forensic Sci Int 19(2):177–179

Li Q, Mahendra S, Lyon DY et al (2008) Antimicrobial nanomaterials for water disinfection and microbial control: potential applications and implications. Water Res 42:4591–4602. https://doi.org/10.1016/j.watres.2008.08.015

Li QL, Lam MHW, Wu RSS, Jiang BW (2010) Rapid magnetic-mediated solidphase extraction and pre-concentration of selected endocrine disrupting chemicals in natural waters by poly (divinylbenzene-co-methacrylic acid) coated Fe_3O_4 core-shell magnetite microspheres for their liquid chromatography-tandem mass spectrometry determination. J Chromatogr A 1217:1219–1226. https://doi.org/10.1016/j.chroma.2009.12.035

Linsebigler AL, Yates JT Jr, Lu G, Lu G (1995) Photocatalysis on TiO_2 surfaces: principles, mechanisms, and selected results. Chem Rev 95:735–758. https://doi.org/10.1021/cr00035a013

Liu J-F, Zhao Z-S, Jiang G-B (2008) Coating Fe_3O_4 magnetic nanoparticles with humic acid for high efficient removal of heavy metals in water. Environ Sci Technol 42:6949–6954. https://doi.org/10.1021/es800924c

Liu L, Chen F, Yang F, Chen Y, Crittenden J (2012) Photocatalytic degradation of 2,4-dichlorophenol using nanoscale Fe/TiO_2. Chem Eng J 181–182:189–195. https://doi.org/10.1016/j.cej.2011.11.060

Lopez-Ayala S, Rincon ME et al (2015) Nanocrystalline titania xerogels doped by metal precursors in the photocatalytic degradation of 2,4-D sodium salts. J Photochem Photobiol A Chem 311:166–175. https://doi.org/10.1016/j.jphotochem.2015.06.019

Lu LC, Wang CI, Sye WF (2011) Applications of chitosan beads and porous crab shell powder for the removal of 17 organochlorine pesticides (OCPs) in water solution. Carbohydr Polym 83:1984–1989. https://doi.org/10.1016/j.carbpol.2010.11.003

Lu SY, Wang QL, Wu D, Li XD, Yan JH (2013) Photocatalytic decomposition of hexachlorobenzene on nano-titanium dioxide films – experimental study and mechanistic considerations. Environ Prog Sustain Energy 32:458–464. https://doi.org/10.1002/ep.11639

Makarov VV, Love J, Sinitsyna OV (2014) Green nanotechnologies: synthesis of metal nanoparticles using plants. Acta Nat 6(1):35–44. https://doi.org/10.32607/20758251-2014-6-1-35-44

Mashhadizadeh MH, Amoli-Diva M (2013) Atomic absorption spectrometric determination of Al^{3+} and Cr^{3+} after preconcentration and separation on 3- mercaptopropionic acid modified silica coated-Fe_3O_4 nanoparticles. J Anal At Spectrom 28:251–258. https://doi.org/10.1039/C2JA30286A

Maurino V, Minero C, Pelizzetti E, Vincenti M (1999) Photocatalytic transformation of sulfonylurea herbicides over irradiated titanium dioxide particles. Colloids Surf A Physicochem Eng Asp 151:329–338. https://doi.org/10.5772/34522

Maya-Treviño ML, Guzmán-Mar JL et al (2014) Activity of the ZnO-Fe_2O_3 catalyst on the degradation of Dicamba and 2,4-D herbicides using simulated solar light. Ceram Int 40:8701–8708

McKinlay R, Plant JA, Bell JNB, Voulvoulis N (2008) Endocrine disrupting pesticides: implications for risk assessment. Environ Int 34:168–183. https://doi.org/10.1016/j.envint.2007.07.013

Milojević-Rakić M, Janošević A et al (2013) Polyaniline and its composites with zeolite ZSM-5 for efficient removal of glyphosate from aqueous solution. Microporous Mesoporous Mater 180:141–155. https://doi.org/10.1021/jp057528g

Mirmasoomi SR, Ghazi MM, Galedari M (2016) Photocatalytic degradation of diazinon under visible light using TiO_2/Fe_2O_3 nanocomposite synthesized by ultrasonic-assisted impregnation method. Sep Purif Technol 175:18–21

Mkhalid IA (2016) Photocatalytic degradation of herbicides under visible light using Pd-WO_3 nanorods. Ceram Int 42:15975–15980. https://doi.org/10.1016/j.ceramint.2016.07.100

Moctezumaa E, Leyva E, Palestino G, de Lasa H (2007) Photocatalytic degradation of methyl parathion: reaction pathways and intermediate reaction products. J Photochem Photobiol A Chem 186:71–84. https://doi.org/10.1016/j.jphotochem.2006.07.014

Mohmood I, Lopes CB et al (2013) Nanoscale materials and their use in water contaminants removal-a review. Environ Sci Pollut Res 20:1239–1260. https://doi.org/10.1007/s11356-012-1415-x

Moradi Dehaghi S, Rahmanifar B, Moradi AM, Azar PA (2014) Removal of permethrin pesticide from water by chitosan-zinc oxide nanoparticles composite as an adsorbent. J Saudi Chem Soc 18:348–355. https://doi.org/10.1016/j.jscs.2014.01.004

Morasch B, Bonvin F et al (2010) Occurrence and fate of micropollutants in the vidy bay of Lake Geneva, Switzerland. Part II: micropollutant removal between wastewater and raw drinking water. Environ Toxicol Chem 29:1658–1668. https://doi.org/10.1002/etc.222

Mostafavi ST, Mehrnia MR, Rashidi AM (2009) Preparation of nanofilter from carbon nanotubes for application in virus removal from water. Desalination 238:271–280. https://doi.org/10.1016/j.desal.2008.02.018

Mueller N, Braun J et al (2012) Application of nanoscale zero valent iron (NZVI) for groundwater remediation in Europe. Environ Sci Pollut Res 19:550–558 https://doi.org/10.1007/s11356-011-0576-3

Nagami H (2010) Historical perspective of pesticide poisoning in Japan and measures taken by the Japanese Association of Rural Medicine. J Rural Med 5(1):129–133

Nagpal V, Bokare AD et al (2010) Reductive dechlorination of gamma-hexachlorocyclohexane using Fe-Pd bimetallic nanoparticles. J Hazard Mater 175:680–687. https://doi.org/10.1016/j.jhazmat.2009.10.063

Naik B, Prasad VS, Ghosh NN (2012) Preparation of Ag nanoparticle loaded mesoporous γ-alumina catalyst and its catalytic activity for reduction of 4-nitrophenol. Powder Technol 232:1–6. https://doi.org/10.1016/j.powtec.2012.07.052

Naraginti S, Stephen FB, Radhakrishnan A, Sivakumar A (2015) Zirconium and silver co doped TiO_2 nanoparticles as visible light catalyst for reduction of 4-nitrophenol, degradation of methyl orange and methylene blue. Spectrochim Acta A Mol Biomol Spectrosc 135:814–819. https://doi.org/10.1016/j.saa.2014.07.070

Narayanan KB, Sakthivel N (2010) Biological synthesis of metal nanoparticles by microbes. Adv Coll Int Sci 156(1–2):1–13

Navarro S, Fenoll J, Vela N, Ruiz E, Navarro G (2009) Photocatalytic degradation of eight pesticides in leaching water by use of ZnO under natural sunlight. J Hazard Mater 172:1303–1310. https://doi.org/10.1016/j.jhazmat.2009.07.137

Nethaji S, Sivasamy A (2017) Graphene oxide coated with porous iron oxide ribbons for 2, 4-Dichlorophenoxyacetic acid (2,4-D) removal. Ecotoxicol Environ Saf 138:292–297. https://doi.org/10.1016/j.ecoenv.2017.01.001

Ngomsik AF, Bee A, Draye M, Cote G, Cabuil V (2005) Magnetic nano- and microparticles for metal removal and environmental applications: a review. C R Chim 8:963–970. https://doi.org/10.1016/j.crci.2005.01.001

Ormad M, Miguel N, Mosteo R, Rodríguez J, Ovelleiro JL (2011) Study of the presence of pesticides in treated urban wastewaters. In: Stoytcheva M (ed) Pesticides in the modern world-risks and benefits, vol 13. InTech, Croatia, pp 453–470. https://doi.org/10.5772/18363

Padilla S, Wilson VZ, Bushnell PJ (1994) Studies on the correlation between blood cholinesterase inhibition and 'target tissue' inhibition inpesticide-treated rats. Toxicology 92:11–25. https://doi.org/10.1016/0300-483X(94)90164-3

Pan B, Xing B (2008) Adsorption mechanisms of organic chemicals on carbon nanotubes. Environ Sci Technol 42:9005–9013. https://doi.org/10.1021/es801777n

Pan J, Yao H, Guan W et al (2011) Selective adsorption of 2,6-dichlorophenol by surface imprinted polymers using polyaniline/silica gel composites as functional support: equilibrium, kinetics, thermodynamics modeling. Chem Eng J 172:847–855. https://doi.org/10.1016/j.cej.2011.06.072

Peng X, Li Y, Luan Z, Di Z, Wang H, Tian B, Jia Z (2003) Adsorption of 1,2- dichlorobenzene from water to carbon nanotubes. Chem Phys Lett 376:154–158. https://doi.org/10.1016/S0009-2614(03)00960-6

Peralta-Hernandez OMR, Goonetilleke NJM, Bandala ER (2017) Treatment technologies for emerging contaminants in water: a review. Chem Eng J 323:361–380. https://doi.org/10.1016/j.cej.2017.04.106

Pesticides News No. 49 (2000) Diazinon. The Journal of Pesticide Action Network UK. Pesticides News No. 49, September 2000, p 20. Available at: http://www.pan-uk.org/pestnews/Actives/diazinon.htm. Accessed March 2016

Pillai IMS, Gupta AK (2015) Potentiostatic electrodeposition of a novel cost effective PbO2 electrode: degradation study with emphasis on current efficiency and energy consumption. J Electroanal Chem 749:16–25. https://doi.org/10.1016/j.jelechem.2015.04.020

Pillai IMS, Gupta AK (2015a) Batch and continuous flow anodic oxidation of 2, 4-dinitrophenol: modeling, degradation pathway and toxicity. J Electroanal Chem 756:108–117. https://doi.org/10.1016/j.jelechem.2015.08.020

Pillai IMS, Gupta AK, Tiwari MK (2015) Multivariate optimization for electrochemical oxidation of methyl orange: pathway identification and toxicity analysis. J Environ Sci Health A Tox Hazard Subst Environ Eng 50:301–310. https://doi.org/10.1080/10934529.2015.981119

Pinto MDCE, Gonçalves RGL et al (2016) Mesoporous carbon derived from a biopolymer and a clay: preparation, characterization and application for an organochlorine pesticide adsorption. Microporous Mesoporous Mater 225:342–354. https://doi.org/10.1016/j.micromeso.2016.01.012

Pirhashemi M, Habibi-Yangjeh A (2017) Preparation of novel nanocomposites by deposition of Ag2WO4 and AgI over ZnO particles: efficient plasmonic visible-light-driven photocatalysts through a cascade mechanism. Ceram Int 43:13447–13460. https://doi.org/10.1016/j.ceramint.2017.07.049

Pitarch E, Portoles T, Marin JM, Ibanez M, Albarran F, Hernandez F (2010) Analytical strategy based on the use of liquid chromatography and gas chromatography with triplequadrupole and time-of-flight MS analyzers for investigating organic contaminants in wastewater. Anal Bioanal Chem 397:2763–2776. https://doi.org/10.1007/s00216-010-3692-x

Pradeep T, Anshup A (2009) Noble metal nanoparticles for water purification: a critical review. Thin Solid Films 517:6441–6478. https://doi.org/10.1016/j.tsf.2009.03.195

Qamar MT, Aslam M, Ismail IMI, Salah N, Hameed A (2015) Synthesis, characterization, and sunlight mediated photocatalytic activity of CuO coated ZnO for the removal of nitrophenols. ACS Appl Mater Interfaces 7:8757–8769. https://doi.org/10.1021/acsami.5b01273

Qu X, Alvarez PJJ, Li Q (2013) Applications of nanotechnology in water and wastewater treatment. Water Res 47:3931–3946. https://doi.org/10.1016/j.watres.2012.09.058

Rachna Rani M, Shanker U (2018) Enhanced photocatalytic degradation of chrysene by Fe2O3@ZnHCF nanocubes. Chem Eng J 348:754–764

Rachna Rani M, Shanker U (2019) Mineralization of carcinogenic anthracene and phenanthrene by sunlight active bimetallic oxides nanocomposites. J Colloid Interface Sci 555:676–688

Ramacharyulu PVRK, Praveen Kumar J, Prasad GK, Srivastava AR (2015) Synthesis, characterization and photocatalytic activity of Ag–TiO_2 nanoparticulate film. RSC Adv 5:1309–1314. https://doi.org/10.1039/C4RA10249E

Ramos-Delgado N, Gracia-Pinilla M et al (2013) Solar photocatalytic activity of TiO_2 modified with WO_3 on the degradation of an organophosphorus pesticide. J Hazard Mater 263:36–44. https://doi.org/10.1016/j.jhazmat.2013.07.058

Rana VK, Kissner R, Jauregui-Haza U, Gaspard S, Levalois-Grutzmacher J (2017) Enhanced chlordecone (Kepone) removal by FeO-nanoparticles loaded on activated carbon. J Environ Chem Eng 5:1608–1617. https://doi.org/10.1016/j.jece.2017.01.039

Rani M (2012) Studies on decay profiles of quinalphos and thiram pesticides. Ph.D Thesis. Indian Institute of Technology Roorkee, Roorkee, Uttarakhand, India, Chapter 1–5

Rani M, Rachna SU (2018) Metal hexacyanoferrates nanoparticles mediated degradation of carcinogenic aromatic amines. Environ Nanotechnol Monit Manag 10:36–45

Rani M, Shanker U (2017) Removal of carcinogenic aromatic amines by metal hexacyanoferrates nanocubes synthesized via green process. J Environ Chem Eng 55:298–5309

Rani M, Shanker U (2018a) Degradation of traditional and new emerging pesticides in water by nanomaterials: recent trends and future recommendations. Int J Environ Sci Technol 15:1347–1380. https://doi.org/10.1007/s13762-017-1512-y

Rani M, Shanker U (2018b) Removal of chlorpyrifos, thiamethoxam, and tebuconazole from water using green synthesized metal hexacyanoferrate nanoparticles. Environ Sci Pollut Res 25:10878–10888

Rani M, Shanker U (2018c) Effective adsorption and enhanced degradation of various pesticides from aqueous solution by Prussian blue nanorods. J Environ Chem Eng 6:1512–1519

Rani M, Shanker U (2018d) Promoting sunlight-induced photocatalytic degradation of toxic phenols by efficient and stable double metal cyanide nanocubes. Environ Sci Pollut R 25:23764–23779

Rani M, Shanker U, Jassal V (2017a) Recent strategies for removal and degradation of persistent & toxic organochlorine pesticides using nanoparticles: a review. J Environ Manag 190:208–222. https://doi.org/10.1016/j.jenvman.2016.12.068

Rani M, Shanker U, Chaurasia A (2017b) Catalytic potential of laccase immobilized on transition metal oxides nanomaterials: degradation of alizarin red S dye. J Environ Chem Eng 5:2730–2739

Ray AK, Ghosh MC (2006) Aquatic toxicity of carbamates and organophosphates. Toxicol Organophosphate Carbamate Compounds 5:657–672. https://doi.org/10.1016/B978-012088523-7/50046-6

Reddy PVL, Kim KH (2015) A review of photochemical approaches for the treatment of a wide range of pesticides. J Hazard Mater 285:325–335. https://doi.org/10.1016/j.jhazmat.2014.11.036

Reddy PAK, Reddy PVL, Kwon E, Kim KH, Akter T, Kalagara S (2016) Recent advances in photocatalytic treatment of pollutants in aqueous media. Environ Int 91:94–103. https://doi.org/10.1016/j.envint.2016.02.012

Ren X, Chen C, Nagatsu M, Wang X (2011) Carbon nanotubes as adsorbents in environmental pollution management: a review. Chem Eng J 170:395–410. https://doi.org/10.1016/j.cej.2010.08.045

Ren D, Colosi LM, Smith JA (2013) Evaluating the sustainability of ceramic filters for point-of-use drinking water treatment. Environ Sci Technol 47:11206–11213. https://doi.org/10.1016/j.apcatb.2005.02.013

Ricart M, Guasch H, Barcelo D et al (2010) Primary and complex stressors in polluted mediterranean rivers: pesticide effects on biological communities. J Hydrol 383:52–61. https://doi.org/10.1016/j.jhydrol.2009.08.014

Rodriguez EM, Fernandez G, Alvarez PM, Hernandez R, Beltran FJ (2011) Photocatalytic degradation of organics in water in the presence of iron oxides: effect of pH and light source. Appl Catal B Environ 102:572–583

Sahithya K, Das D, Das N (2015) Effective removal of dichlorvos from aqueous solution using biopolymer modified MMT-CuO composites: equilibrium, kinetic and thermodynamic studies. J Mol Liq 211:821–830. https://doi.org/10.1016/j.molliq.2015.08.013

Sahithya K, Das D, Das N (2016) Adsorptive removal of monocrotophos from aqueous solution using biopolymer modified montmorillonite-CuO composites: equilibrium, kinetic and thermodynamic studies. Process Saf Environ Prot 99:43–54. https://doi.org/10.1016/j.psep.2015.10.009

Sahoo C, Gupta AK (2013) Application of statistical experimental design to optimize the photocatalytic degradation of a thiazin dye using silver ion-doped titanium dioxide. J Environ Sci Health Part A 48:694–705. https://doi.org/10.1080/10934529.2013.744598

Sahoo C, Gupta AK (2015) Characterization and photocatalytic performance evaluation of various metal ion-doped microstructured TiO_2 under UV and visible light. J Environ Sci Health Part A 50:659–668. https://doi.org/10.1080/10934529.2015.1011958

Sahoo C, Gupta AK, Pal A (2005) Photocatalytic degradation of Methyl Red dye in aqueous solutions under UV irradiation using Ag^+ doped TiO_2. Desalination 181:91–100. https://doi.org/10.1016/j.desal.2005.02.014

Sahoo C, Gupta AK, Pillai IMS (2012) Heterogeneous photocatalysis of real textile wastewater: evaluation of reaction kinetics and characterization. J Environ Sci Health A Tox Hazard Subst Environ Eng 47:2109–2119. https://doi.org/10.1080/10934529.2012.695996

Saifuddin N, Nian C, Zhan L, Ning K (2011) Chitosan-silver nanoparticles composite as point-of-use drinking water filtration system for household to remove pesticides in water. Asian J Chem 6:142–159. https://doi.org/10.3923/ajb.2011.142.159

Sarkar S, Das R (2017) PVP capped silver nanocubes assisted removal of glyphosate from water-a photoluminescence study. J Hazard Mater 339:54–62. https://doi.org/10.1016/j.jhazmat.2017.06.014

Sarkar B, Venkateswralu N, Rao RN, Bhattacharjee C, Kale V (2007) Treatment of pesticide contaminated surface water for production of potable water by a coagulation adsorption-nanofiltration approach. Desalination 212:129–140. https://doi.org/10.1016/j.desal.2006.09.021

Sawicki R, Mercier L (2006) Evaluation of mesoporous cyclodextrin-silica nanocomposites for the removal of pesticides from aqueous media. Environ Sci Technol 40:1978–1983. https://doi.org/10.1021/es051441r

Senthilnathan J, Philip L (2010) Removal of mixed pesticides from drinking water system using surfactant-assisted nano-TiO_2. Water Air Soil Pollut 210:143–154. https://doi.org/10.1007/s11270-009-0230-6

Shabtai IA, Mishael YG (2017) Catalytic polymer-clay composite for enhanced removal and degradation of diazinon. J Hazard Mater 335:135–142. https://doi.org/10.1016/j.jhazmat.2017.04.017

Shahwan T, Erten H (2004) Temperature effects in barium sorption on natural kaolinite and chlorite-illite clays. J Radioanal Nucl Chem 260:43–48. https://doi.org/10.1023/B:JRNC.0000027059.66424.b4

Shanker U, Jassal V, Rani M, Kaith BS (2016a) Towards green synthesis of nanoparticles: from bio-assisted sources to benign solvents. A review. Int J Environ Anal Chem 96:801–835

Shanker U, Jassal V, Rani M (2016b) Catalytic removal of organic colorants from water using some transition metal oxide nanoparticles synthesized under sunlight. RSC Adv 6:94989–94999

Shanker U, Jassal V, Rani M (2017a) Catalytic removal of organic colorants from water using some transition metal oxide nanoparticles synthesized under sunlight. RSc Adv 69:4989–9499

Shanker U, Jassal V, Rani M (2017b) Degradation of toxic PAHs in water and soil using potassium zinc hexacyanoferrate nanocubes. J Environ Manag 204:337–345

Shanker U, Jassal V, Rani M (2017c) Green synthesis of iron hexacyanoferrate nanoparticles: potential candidate for the degradation of toxic PAHs. J Environ Chem Eng 5:4108–4117

Shet A, Shetty VK (2016) Photocatalytic degradation of phenol using Ag core-TiO_2 shell (Ag@TiO_2) nanoparticles under UV light irradiation. Environ Sci Pollut Res 23:20055–20064. https://doi.org/10.1007/s11356-015-5579-z

Shi H, Chen J, Li G, Nie X, Zhao H, Wong PK, An T (2013) Synthesis and characterization of novel plasmonic Ag/AgX-CNTs (X = Cl, Br, I) nanocomposite photocatalysts and synergetic degradation of organic pollutant under visible light. ACS Appl Mater Interfaces 5:6959–6967. https://doi.org/10.1021/am401459c

Shifu C, Gengyu C (2005) Photocatalytic degradation of organophosphorus pesticides using floating photocatalyst TiO_2.SiO_2/beads by sunlight. Sol Energy 79:1–9. https://doi.org/10.1016/j.solener.2004.10.006

Sin JC, Lam SM, Lee KT, Mohamed AR (2013) Preparation and photocatalytic properties of visible light-driven samarium-doped ZnO nanorods. Ceram Int 39:5833–5843. https://doi.org/10.1016/j.ceramint.2013.01.004

Sivagami K, Vikraman B, Krishna RR, Swaminathan T (2015) Chlorpyrifos and Endosulfan degradation studies in an annular slurry photo reactor. Ecotoxicol Environ Saf 5:1–5. https://doi.org/10.1016/j.ecoenv.2015.08.015

Slotkin TA (1999) Developmental cholinotoxicants: nicotine and chlorpyrifos. Environ Health Perspect 107:71–80. https://doi.org/10.1289/ehp.99107s171

Smuleac V, Varma R, Baruwati B, Sikdar S, Bhattacharyya D (2011) Nanostructured membranes for enzyme catalysis and green synthesis of nanoparticles. ChemSusChem 4:1773–1777. https://doi.org/10.1002/cssc.201100211

Soni SS, Henderson MJ, Bardeau JF, Gibaud A (2008) Visible-light photocatalysis in titania-based mesoporous thin films. Adv Mater 20:1493–1505. https://doi.org/10.1002/adma.200701066

Sorbiun M, Mehr ES, Ramazani A, Fardood ST (2018) Green synthesis of zinc oxide and copper oxide nanoparticles using aqueous extract of oak fruit hull (jaft) and comparing their photocatalytic degradation of basic violet 3. Int J Environ Res 12:29–37. https://doi.org/10.1007/s41742-018-0064-4

Stoimenov PK, Klinger RL, Marchin GL, Klabunde KJ (2002) Metal oxide nanoparticles as bactericidal agents. Langmuir 18:6679–6686. https://doi.org/10.1021/la0202374

Sud D, Kaur P (2012) Heterogeneous photocatalytic degradation of selected organophosphate pesticides: a review. Crit Rev Environ Sci Technol 42:2365–2407. https://doi.org/10.1080/10643389.2011.574184

Tadeo JL (2008) Analysis of pesticides in food and environmental samples. CRC Press/Taylor & Francis, Boca Raton

Tahir MB, Nabi G, Khalid NR (2017) Nanostructured-based WO_3 photocatalysts: recent development , activity enhancement, perspectives and applications for wastewater treatment. Int J Environ Sci Technol 14:2519. https://doi.org/10.1007/s13762-017-1394-z

Tian B, Dong R, Zhang J, Bao S, Yang F, Zhang J (2014) Sandwich-structured AgCl@Ag@TiO_2 with excellent visible-light photocatalytic activity for organic pollutant degradation and E. coli K12 inactivation. Appl Catal B Environ 158–159:76–84. https://doi.org/10.1016/j.apcatb.2014.04.008

Tian H, Chen J, He J, Liu F (2015) Pd-loaded magnetic mesoporous nanocomposites: a magnetically recoverable catalyst with effective enrichment and high activity for DDT and DDE removal under mild conditions. J Colloid Interface Sci 457:195–202. https://doi.org/10.1016/j.jcis.2015.07.024

Tomasevic A, Kiss E, Petovic S, Mijin D (2010) Study on the photocatalytic degradation of insecticide methomyl in water. Desalination 262:228–234. https://doi.org/10.1016/j.desal.2010.06.019

Tomer V, Sangha JK, Ramya HG (2015) Pesticide: an appraisal on human health implications. Proc Natl Acad Sci India Sect B Biol Sci 85:451–463

Umukoro EH, Peleyeju MG, Ngila JC, Arotiba OA (2017) Towards wastewater treatment: photo-assisted electrochemical degradation of 2-nitrophenol and orange II dye at a tungsten trioxide-exfoliated graphite composite electrode. Chem Eng J 317:290–301. https://doi.org/10.1016/j.cej.2017.02.084

Virkutyte J, Varma RS (2013) Chapter 17, nanotechnology for water and wastewater treatment. In: Green synthesis of nanoparticles and Nanocatalysts. IWA Publishing, London, pp 375–408. isbn: 9781780404585

Wang P, Shi Q, Shi Y, Clark KK, Stucky GD, Keller AA (2008) Magnetic permanently confined micelle arrays for treating hydrophobic organic compound contamination. J Am Chem Soc 131:182–188. https://doi.org/10.1021/ja806556a

Wang H, Wang HL, Jiang WF, Li ZQ (2009) Photocatalytic degradation of 2,4- dinitrophenol (DNP) by multi-walled carbon nanotubes (MWCNTs)/TiO_2 composite in aqueous solution under solar irradiation. Water Res 43:204–210. https://doi.org/10.1016/j.watres.2008.10.003

Wang P, Huang B, Dai Y, Whangbo M-H (2012) Plasmonic photocatalysts: harvesting visible light with noble metal nanoparticles. Phys Chem Chem Phys 14:9813–9825. https://doi.org/10.1039/c2cp40823f

Wang S, Sun H, Ang HM, Tadé MO (2013) Adsorptive remediation of environmental pollutants using novel graphene-based nanomaterials. Chem Eng J 226:336–347. https://doi.org/10.1016/j.cej.2013.04.070

Wang Z, Zhang P, Hu F, Zhao Y, Zhu L (2017) A cross linked β-cyclodextrin polymer used for rapid removal of a broad-spectrum of organic micropollutants from water. Carbohydr Polym. https://doi.org/10.1016/j.carbpol.2017.08.059

Wei X, Huang T, Yang JH, Zhang N, Wang Y, Zhou ZW (2017) Green synthesis of hybrid graphene oxide/microcrystalline cellulose aerogels and their use as superabsorbents. J Hazard Mater 335(1):28–38

WHO (2002) Guidelines to classification 2000-2002. Geneva, Switzerland.

WHO (2010) The WHO recommended classification of pesticides by hazard and guidelines to classification 2009. World Health Organization, Geneva

Xie J, Zhang L, Li M, Hao Y, Lian Y, Li Z, Wei Y (2015) α-Fe_2O_3 modified ZnO flower-like microstructures with enhanced photocatalytic performance for pentachlorophenol degradation. Ceram Int 41:9420–9425. https://doi.org/10.1016/j.ceramint.2015.03.320

Xing Z, Zhou W, Du F, Zhang L, Li Z, Zhang H, Li W (2014) Facile synthesis of hierarchical porous TiO_2 ceramics with enhanced photocatalytic performance for micropolluted pesticide degradation. ACS Appl Mater Interfaces 6:16653–16660. https://doi.org/10.1021/am5034236

Xu X, Zhou M, He P, Hao Z (2005) Catalytic reduction of chlorinated and recalcitrant compounds in contaminated water. J Hazard Mater 123:89–93. https://doi.org/10.1016/j.jhazmat.2005.04.002

Xu P, Zeng GM, Huang DL et al (2012) Use of iron oxide nanomaterials in wastewater treatment: a review. Sci Total Environ 424:1–10. https://doi.org/10.1016/j.scitotenv.2012.02.023

Xu C, Liu R, Chen L, Tang J (2016a) Enhanced dechlorination of 2,4-dichlorophenol by recoverable Ni/Fe–Fe_3O_4 nanocomposites. J Hazard Mater 48:92–101. https://doi.org/10.1016/j.jes.2015.10.033

Xu J, Cao Z et al (2016b) Preparation of functionalized Pd/Fe-Fe_3O_4@ MWCNTs nanomaterials for aqueous 2, 4-dichlorophenol removal: Interactions, influence factors, and kinetics. J Hazard Mater 317:656–666. https://doi.org/10.1016/j.jhazmat.2016.04.063

Yadav IC, Devi NL et al (2013) Current status of persistent organic pesticides residues in air, water, and soil, and their possible effect on neighboring countries: a comprehensive review of India. Sci Total Environ 511:123–137

Yamaguchi NU, Bergamasco R, Hamoudi S (2016) Magnetic $MnFe_2O_4$–graphene hybrid composite for efficient removal of glyphosate from water. Chem Eng J 295:391–402. https://doi.org/10.1016/j.cej.2016.03.051

Yang J, Li D, Wang X, Yang XJ, Lu LD (2002) Rapid synthesis of nanocrystalline TiO_2/SnO_2 binary oxides and their photoinduced decomposition of methyl orange. J Solid State Chem 165:193–198. https://doi.org/10.1006/jssc.2001.9526

Yang XH, Fu HT, Wong K, Jiang XC, Yu AB (2013) Hybrid Ag@ TiO_2 core–shell nanostructures with highly enhanced photocatalytic performance. Nanotechnology 24:756–776

Yu H, Wang X, Sun H, Huo M (2010) Photocatalytic degradation of malathion in aqueous solution using an Au-Pd-TiO_2 nanotube film. J Hazard Mater 184:753–758. https://doi.org/10.1016/j.jhazmat.2010.08.103

Yu L, Yang X, Ye Y, Wang D (2015) Efficient removal of atrazine in water with a Fe_3O_4/MWCNTs nanocomposite as a heterogeneous Fenton-like catalyst. RSC Adv 5:46059–46066. https://doi.org/10.1039/C5RA04249F

Yue Z, Economy J (2005) Nanoparticle and nanoporous carbon adsorbents for removal of trace organic contaminants from water. J Nanoparticle Res 7:477–487. https://doi.org/10.1007/s11051-005-4719-7

Zhang L, Yan F, Wang Y (2006) Photocatalytic degradation of methamidophos by UV irradiation in the presence of nano-TiO_2. J Inorg Mater 42:1379–1387. https://doi.org/10.1134/S002016850612017X

Zhang WJ, Jiang FB, Ou JF (2011) Global pesticide consumption and pollution: with China as a focus. Proc Int Acad Ecol Environ Sci 1:125–144

Zhang H, Liang C, Liu J, Tian Z, Wang G, Cai W (2012) Defect-mediated formation of Ag cluster-doped TiO_2 nanoparticles for efficient photodegradation of pentachlorophenoli. Langmuir 28:3938–3944. https://doi.org/10.1021/la2043526

Zhao W, Ma W, Chen C, Zhao J, Shuai Z (2004) Efficient degradation of toxic organic pollutants with Ni_2O_3/TiO2-x Bx under visible irradiation. J Am Chem Soc 126:4782–4783. https://doi.org/10.1021/ja0396753

Zhao X, Shi Y, Cai Y, Mou S (2008) Cetyltrimethylammonium bromide-coated magnetic nanoparticles for the preconcentration of phenolic compounds from environmental water samples. Environ Sci Technol 42:1201–1206. https://doi.org/10.1021/es071817w

Zhao R, Li X, Sun B, Ji H, Wang C (2017) Diethylenetriamine-assisted synthesis of amino rich hydrothermal carbon-coated electrospun polyacrylonitrile fiber adsorbents for the removal of Cr (VI) and 2,4-dichlorophenoxyacetic acid. J Colloid Interface Sci 487:297–309. https://doi.org/10.1016/j.jcis.2016.10.057

Zhou Q, Xiao J, Wang W (2006) Using multi-walled carbon nanotubes as solid phase extraction adsorbents to determine dichlorodiphenyltrichloroethane and its metabolites at trace level in water samples by high performance liquid chromatography with UV detection. J Chromatogr A 1125:152–158. https://doi.org/10.1016/j.chroma.2006.05.047

Zhu X, Yuan C, Chen H (2007) Photocatalytic degradation of pesticide pyridaben 3. In surfactant/ TiO_2 aqueous dispersions. Environ Sci Technol 41:263–269. https://doi.org/10.1021/es061178+

Zhu Y, Murali S, Cai W, Li X, Suk JW, Potts JR, Ruoff RS (2010) Graphene and graphene oxide: synthesis, properties, and applications. Adv Mater 22:3906–3924. https://doi.org/10.1002/adma.201001068

Zodrow K, Brunet L, Mahendra S, Li D, Zhang A, Li Q, Alvarez PJJ (2009) Polysulfone ultrafiltration membranes impregnated with silver nanoparticles show improved biofouling resistance and virus removal. Water Res 43:715–723. https://doi.org/10.1016/j.watres.2008.11.014

Chapter 8
Environmental and Health Effects of Pesticide Residues

Sajjad Ali, Muhammad Irfan Ullah, Asif Sajjad, Qaiser Shakeel, and Azhar Hussain

Abstract Now a days, agroecosystem is largely dependent on pesticides to meet the ever-increasing demand of food and fiber. Tons of synthetic pesticides are deposited to increase the yield. This scenario poses great threat to non-target organisms because 98% sprayed pesticides directly or indirectly affect them. Several studies estimated that 80% of sprayed pesticides directly contaminate the environment. This chapter reviewed the pesticides effects on environment, natural biodiversity, pollinators, food chains and outcome health issues. It was reviewed that soil and water qualities are deteriorated because of pesticide depositions. Due to microbial life threat, soil respiration is decreased by 35% and almost 90% water sources in agro-lands are polluted with pesticides. The aquatic and terrestrial food chains are being endangered because of bioaccumulation and biomagnification of loads of pesticides. Studies revealed that biodiversity and species are endangered because of pesticide exposures. During last few decades, 70% decline in insect biomass and 50% decline in farmland birds is reported in European nations. Similarly, 42% reduction in species richness was also noted in Europe, Australia and North America. Pesticides residues have injurious effects on bees which ultimately decrease their ecological service. United Nations warned that 40% of invertebrate pollinator, particularly bees

S. Ali (✉) · A. Sajjad
Department of Entomology, Faculty of Agriculture & Environmental Sciences, The Islamia University of Bahawalpur, Bahawalpur, Pakistan
e-mail: sajjad.ali@iub.edu.pk

M. I. Ullah
Department of Entomology, College of Agriculture, University of Sargodha, Sargodha, Pakistan
e-mail: mirfanullah@uos.edu.pk

Q. Shakeel
Department of Plant Pathology, Faculty of Agriculture & Environmental Sciences, The Islamia University of Bahawalpur, Bahawalpur, Pakistan
e-mail: qaiser.shakeel@iub.edu.pk

A. Hussain
Department of Soil Science, Faculty of Agriculture & Environmental Sciences, The Islamia University of Bahawalpur, Bahawalpur, Pakistan

Inamuddin et al. (eds.), *Sustainable Agriculture Reviews 48*, Sustainable Agriculture Reviews 48, https://doi.org/10.1007/978-3-030-54719-6_8

and butterflies, are at risk to global extermination. Almost, 30% honey bee population decline in American and European countries is reported due to extreme use of pesticides. Through food chain contamination, adverse effects reach to human life. These ingestions have resulted serious carcinogenic, endocrine, neurological, reproductive and other ailments. Even, many casualties have been reported due to exposure to poisonous pesticides. Bundle of research indicated that cancer risk and mental health problems are enlarged by 25–30% after exposure to pesticides. Similarly, 50% elevated risk of leukaemia, lymphoma and brain cancer in children is linked with paternal exposure to pesticides. Now, it is time to find out some alternatives focusing the environmental protection and ultimately human health. Integrated pest management can be the one and only option to minimize the use of pesticides.

Keywords Air pollution · Biodiversity · Carcinogenic · Diseases · Food chain contamination · Pollinator · Pesticide pollution · Water pollution

8.1 Introduction

Application of pesticides to prevent, kill, repel or mitigate the harmful and invasive organisms, has become an integral input in modern agriculture system to meet the ever-increasing demands of food and fiber. Tons of synthetic chemicals, mainly herbicides, insecticides and fungicides, are being deposited into the natural and agroecosystems every year to protect the crop produces from the weeds, insect and pathogens. By 2050, we need to feed 10 billion population of the world (Eddleston 2000). Since 1940, many agricultural practices including pesticides would have become a new trend in modern farming for improved crop yields; eight times better than the previous one. Without pesticide application, losses of fruits, vegetables and cereals due to various pest injuries can reach up to 78, 54 and 32%, respectively (Cai 2008).

Around 2 million tons of pesticides (45% Europe, 25% United States and 25% other countries of the world) are consumed every year worldwide for the crop protection. Worldwide, 40% of the pesticides is confined to herbicides, 17% to insecticides, 10% to fungicides and rest of the pesticides fall into other small usages (Fig. 8.1) (Alavanja 2009; De et al. 2014).

Almost, 98% of the sprayed pesticide chemicals also influence the non-target organisms. Commonly, the pesticide residues through agro-ecosystem, domestic, garden use and spray drifts are accumulating into soil, air, water and food. The workers, who have direct exposure, are at more risks to carcinogenic and mutagenic illnesses but the common people are also not excluded from this danger being exposed to pesticide residues in nature. Besides the use of smart technology in agriculture and increased focus on organic farming, still around 25 million workers of agricultural community are diseased by exposure to pesticides residues exclusive to infection of farm land animals. Environmental scientists are trying to reduce

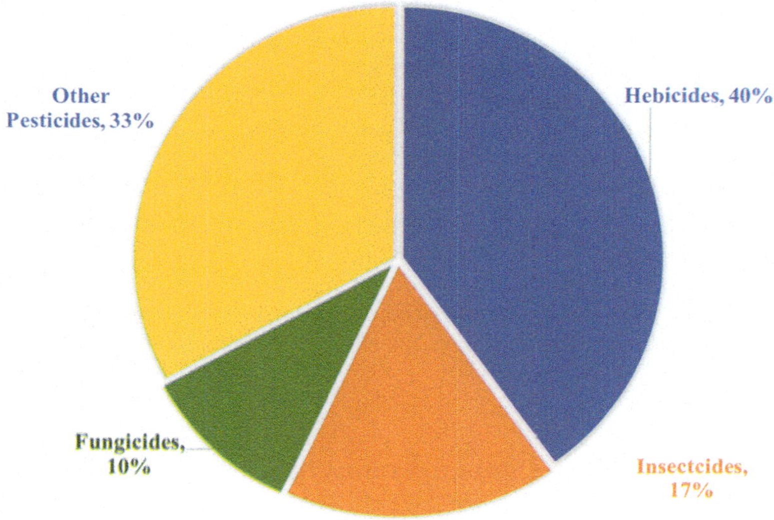

Fig. 8.1 Global share of different kinds of pesticides application

pesticide application to minimize the poisonous residues exposure (Alavanja et al. 2001; Alavanja 2009).

The monitoring of pesticide residues in food stuff was started in the European Union since 1996. Due to food uncleanness, most of the deaths are noted in hospital reviews (Eddleston 2000). In India, the first pesticides poisoning report was from Kerala in 1958, where over 100 people expired after eating wheat flour with parathion contaminations (Karunakaran 1958). The land can become contaminated during each pesticide application. Likely, 80–90% of pesticides are volatilized into environment within few hours of application (Damalas and Eleftherohorinos 2011). The drift of pesticide causes, not only, the loss of 2–25% of chemical but also causes spread of pesticide residues over longer distances. Large number of pesticide transformation products, from a widespread range of chemicals, were also documented in the past which are entering into our environment every year (Hernández et al. 2008).

Similarly, application of pesticide on plants, contaminate the surface and ground water by leaching down of pesticide residues from soil. A complete survey of at least 43 states in United States was reported by United State Department of Geo Sciences; in which 143 pesticides and 21 transformation products from all major pesticide classes were noticed in ground water (Waskom 1995). Approximately, 90% water samples from all rivers, stream and fish contain one or several pesticides in a study conducted in India. Additionally, 58% water of human consumption was polluted with Organo-Chlorine insecticides above the Environmental Protection Agency values (Kole et al. 2001). If ground water is polluted with toxic chemicals once, naturally it may take several years for purification from these toxic chemicals. In most of the cases, cleanup is very costly with complex processes, if not impossible (Waskom 1995).

The use of pesticides is increasing day by day. For example, during 1990–2006, the total area treated with pesticides and herbicides increased by 30 and 38% in the United Kingdom (Fera 2009). The excessive use of pesticides may induce different health problems to human after they are exposed to these toxins at various levels. The main problem is the duration of the exposure to pesticide residues. Some people like children, pregnant women, sick or aged persons are more sensitive to the residual effect of pesticide. Now-a-days, pesticides have passed into natural food chains and tend to be bioaccumulated in the higher tropic levels (Mostafalou and Abdollahi 2013). So, there is utmost need for scientists to make strategies to reduce the pesticide residue exposure to save the environment and living beings.

8.2 Soil and Water Contaminations by Pesticide Residues

Agro-ecosystem is being contaminated directly or indirectly through plenteous use of pesticides. In addition to reducing the harmful insects, pathogen or weeds; pesticides can be lethal to a range of non-target organisms including beneficial insects, birds, fish, and non-target plants. Insecticides are generally the most acutely toxic group of pesticides, but herbicides can also exert hazards to non-target living organisms. Agricultural systems are disturbed because of pesticide residues after they enter into ecosystem through surface water, ground water and contaminate the soil (Fig. 8.2) (Wasim et al. 2009).

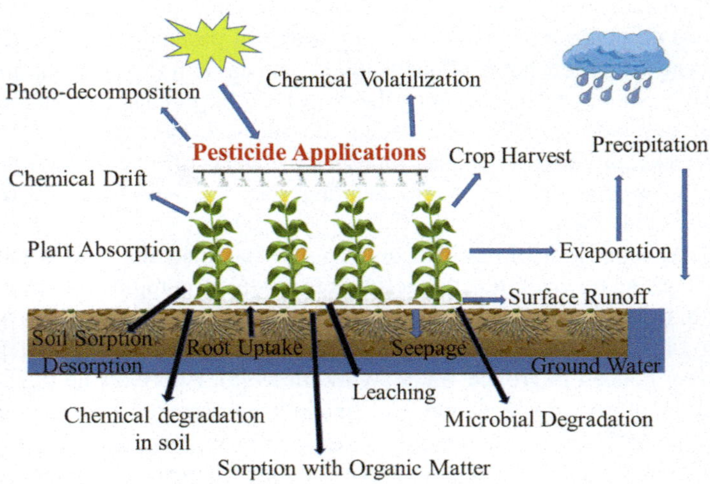

Fig. 8.2 Pesticide application, decomposition and residues entry into ecosystem through soil, surface and ground water

Similarly, pesticides transformation products are also accumulated in the soils and may disturb the soil microbial life benefitting our crop system. The beneficial micro fauna is declining because of heavy deposits of pesticides in the agricultural soils and their surrounding areas. For example, nitrogen fixing bacteria are disturbed by the heavy applications of herbicides and insecticides into the soil. Similar effects have also been noted in case of mycorrhizal fungi (Lu and Lu 2018). Due to declined populations of beneficial microbes after long exposures of pesticides; the crop productivity might be lowered in future.

A larger fraction of pesticides amasses in the soil due to unselective and frequent use of pesticides. Afterwards, properties and microbiome of soil regulate the destiny of the applied pesticides and their residues, through many degradation processes, transformation products, and adsorption-desorption courses (Weber et al. 2004; Hussain et al. 2009). Degrading pesticides tend to change the soil textures and microorganism activities by changing its microbial biodiversity, biochemical and enzyme activities in the soil (Hussain et al. 2009) which may lead to worrying soil-ecosystem and soil fertility loss. Additionally, repeated applications of pesticides pose impact on the function of helpful root-colonizing soil microorganisms like bacteria, fungi, arbuscular mycorrhiza and algae by manipulating their growth, colonization ability and metabolic events (Tien and Chen 2012). For example, herbicides like chlorsulfuron, sulfonyl-urea, metsulfuron and thifensulfuron methyl decrease the growth of *Pseudomonas* strains in agricultural soil which is key player in soil fertility (Boldt and Jacobsen 1998). Likewise, benomyl, chlorothalonil and captan tend to lower the soil respiration process (a microbiota indicator) by 30–50% (Tien and Chen 2012). Pesticides residual masses also disturb the soil biochemical reactions i.e. nitrogen-fixation, ammonification and nitrification by manipulating certain soil microbial organisms or their enzyme activities (Hussain et al. 2009). Pesticides may influence soil mineralization and organic matter, key soil property to regulate soil productivity. For example, herbicides application including atrazine, glyphosate, primeextra, and paraquat significantly reduce the soil organic matter (Sebiomo et al. 2011). It is vital to assess the responses of soil microbes and numerous enzyme activities to pesticide depositions so their deleterious effects can be minimized.

Pesticide runoff from treated plants and soil can pollute the surface water bodies which may disturb the water life in the long run. According to United States geological survey; urban water resources are polluted with pesticides than the agricultural water bodies. Pesticide leaching from the soil surfaces is accumulated into ground water system (Kole et al. 2001). In the United States, 90% of the wells water samples were found contaminated with pesticides during a study by the US Geological Survey (Gillion et al. 2006).

Accidental spillages, waste effluents, surface run-off and shifting from pesticide applied soils, cleaning of spray machinery after spraying, leaching into water bodies and aerial sprays to kill water-dwelling pests can be ways for pesticides to get into water resources. Pesticides not only affect the fish fauna, but also disturb the food-webs afterwards. Many persistent pesticides like organochlorines and polychlorinated biphenyls were noticed in the chief Arctic Ocean food-webs. The residual effects of pesticides, on aquatic environment, are governed by their solubility in water and penetration into an organism. The hydrophilic or lipophobic nature of the pesticide residual mass makes it less available in the fatty tissues of aquatic organisms which mat later its toxicity (Pereira et al. 2013). If pesticide application remains continued it will be a prodigious threat for agro-ecosystem so, now it is right time to be aware all about this issue and find alternative and environmentally friendly strategies.

8.3 Pesticide Bioaccumulation and Bio-magnification in Living Systems

Pesticide chemicals can enter into living systems from the environment directly or through agricultural runoff. They pass through different trophic levels in food web from bottom (algae, oysters and fish) to top (eagle, bears and humans) level and tend to accumulate in living organisms' tissues at each trophic level because most of the chemicals have half-lives between 1 and 4 years (Mostafalou and Abdollahi 2012). At each following food chain level, pesticide residues increase in the tissues of living organisms due to their non-biodegradable nature which is known as biomagnification. This fact was described by studying many food chains and it was noted that higher trophic level possesses elevated levels of the toxins as compared to primary exposures. For example, Fig. 8.3 is depicting the biomagnification ability of Dichloro-Diphenyl-Trichloroethane (DDT) insecticide at different trophic levels (Deribe et al. 2013; Katagi and Tanaka 2016).

This phenomenon poses greater threat to higher trophic levels. The hostile effects of pesticides residues on non-target organisms like aquatic fauna and humans, occurred due to biomagnifications of these toxins. For instance, population declines

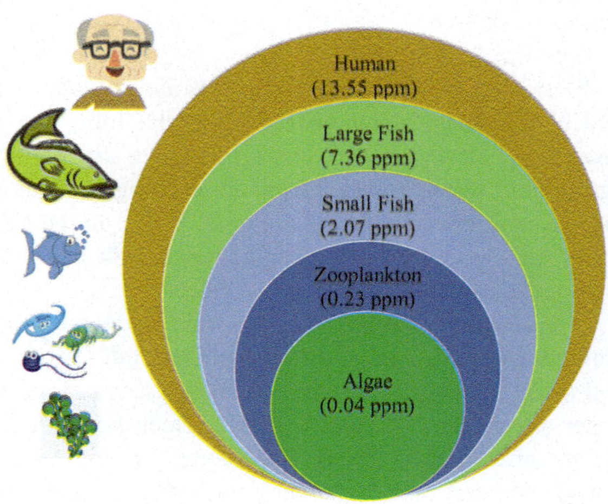

Fig. 8.3 Biomagnification of pesticides (DDT) at different trophic levels in a food chain

and reproductive disabilities in many fish-eating birds like gulls, terns, and herons etc. were observed as a consequence of Dichloro-diphenyl-dichloroethylene induced eggshell weakening (Grasman et al. 1998). It is observed that biomagnifications increase with increasing lipophilic features and persistence of the pesticide residues in living organisms. Due to this reason, organochlorines are ranked as chemicals with higher biomagnification rates and are persistent in a broader array of living organisms as compared to organophosphates. That's why, they are banned with time (Favari et al. 2002).

Pesticides can affect living organisms, in two ways. Firstly, they become the source of sudden death of directly exposed organisms or secondly, these toxicants accumulate in the living organisms or in sediments of aquatic environments and cause death chronically afterwards. Usually the bio-degradation process of pesticides and polychlorinated biphenyls is very slow as the residues in sediments below water surface cannot degraded by UV light or microorganism activity. Fish species can be exposed to these sediments through skin or gills contact or by direct consumption (Walczak and Reichert 2016). In a study, Striped weakfish (*Cynoscion guatucupa*) was used as a biomonitoring agent of environment. Nineteen pesticides standards were used to detect the pesticides and isomers of endosulfan, chlordanes and hexachlorocyclohexane were found in fish tissues and food contents of *C. guatucupa* showing extreme bioaccumulation and biomagnification of these chemicals at different trophic levels (Lanfranchi et al. 2006). Similarly, sulfotep, demeton-O, dimethoate, disulfoton, fenitrothion and malathion pesticide residues were also detected from coconut oil at higher concentrations. It was concluded that pesticides were bioaccumulated in commercial coconut trees through the rough use of agrochemicals in the fields (dos Anjos and de Andrade 2014). As most of the pesticides are non-biodegradable in nature so, their residual effect cannot be eliminated easily. Even UV light is unable to degrade the pesticide residues in sediments.

8.4 Impact of Pesticide Residues on Biodiversity

Pesticides are regarded as one of main constituent of 'Green Revolution'. Mainly, the function of a pesticide is to help the users to avoid pest losses. In addition to controlling pests, the declining biodiversity in nature is obvious now. Pesticide residues are often persistent in nature, remain stable in environment and are causing a serious threat to non-target and non-pest organisms. (Venter et al. 2006). Many of these compounds are extremely toxic to exposed insects, birds, mammals, amphibians and fish. During last few decades, 70% decline in insect biomass in Germany, 50% decline in farmland birds in Europe was recorded due to pesticide applications. Removal of weed can modify the habitat type by discarding of vegetation and eventually leading to insect population reduction. Elimination of some arable weeds was caused by the usage of fungicides which allows farmers to stop 'break crops' such as grass and roots (Storkey et al. 2012). Arial spraying was a major problem prevailing in Canada related to biodiversity loss and 62% of species decline was associated directly or indirectly with pesticide usage. A study investigated that average specie loss was 10% in common bird species from 1980 to 2006 but the decline was recorded up-to 50% in common farmland bird species in 2006 in United Kingdom leaving no option for recovery and suggesting the harmful effect of pesticides in environment (Gibbs et al. 2009). On the same lines, the currently used pesticides in Europe and Australia, are major reason for 42% regional diversity losses of invertebrates in streams and rivers. Similarly, in Europe, 42% species richness was reduced due to pesticide exposures even at environmentally safe doses (Beketov et al. 2013). It is estimated that yearly 72 million birds are killed due to pesticides applications in the United States (Fimrite 2011).

Collectively, 1211 species of birds was regarded as threatened species and among these 86% are endangered due to demolition and deprivation of habitat because of discriminating usage of pesticides and other synthetic compounds (Mitra et al. 2011). Pesticides can accumulate in the tissues of prey and can cause toxic effect to top predators. Particular rodenticides are very toxic and can accumulate in the body of rodents. These can cause secondary poisoning to predators like dogs, foxes, non-target mammals and raptors by eating pesticide exposed prey (Brakes and Smith 2005). Herbicides damage the vegetation and can affect the life of common shrew, wood mouse and badger due to the shortage of food and alteration in microclimate (Hole et al. 2005).

Organochlorine belongs to the most important group of insecticide which is the most persistent and stable compound in the environment and its application causes a huge loss to biodiversity. According to a study by (Newton 1976), it was estimated that the usage of compounds such as Dichlorodiphenyltrichloroethane and dieldrin from organochlorines group of insecticides resulted a heavy loss to many bird species like fish eating birds and Peregrine Falcon *(Falco peregrinus)* in some regions of Europe and Americas. It was concluded that applications of granular formulation of carbofuran has an effectual impact on the populations of songbird exposed to insecticide when they were breeding alongside the edge of contaminated fields (Stinson et al. 1994). Carbofuran in liquid form was also reported for its lethal effects on Burrowing Owl (*Speotyto cunicularia*), a bird species with was recalled as endangered (Gervais et al. 2000). The negative impact of diazinon applied on the grasslands showed maximum mortality in the population of Brant geese (*Branta bernicla*) which used to harbour their nests at that place to lay eggs (Stinson et al. 1994).

The negative impact of pesticides on fish have been widely documented. According to a study, the insecticidal residues of Hexachlorocyclohexane and Dichlorodiphenyltrichloroethane was found in four different species of fish in freshwater in China. Residual scale of Hexachlorocyclohexane contents was found comparatively higher in grass carp and chub while that of Dichlorodiphenyltrichloroethane was examined higher in snakehead species of fish (Wu et al. 2013).

According to a research, the predator-avoidance behaviour of guppy fish (*Poecilia reticulata*) in response to the effect of pentachlorophenol in the presence of a predator largemouth bass (*Micropterous salmoides*). The findings of this study suggested that the groups of guppies treated with higher level of chemical exhibit sluggish response and could not maintain an optimal flux of speed to get escaped after the attack of predator (Brown et al. 2009).

There are approximately 6000 species of amphibians documented in literature to date world widely. One-third of this amount is endangered due to various reasons like overexploitation, introduction of predator species and habitat destruction by using pesticides. The prominent factor in this context is the water pollution caused by runoff and leaching of pesticide residues (Brühl et al. 2013). In addition to the insecticides, herbicide like diclofop-methyl and fungicide like difenoconazole also indicated toxic effects of these chemicals on albino rats. These compounds had altered the enzymatic and metabolic activities of tested rats. These chemicals possess higher potential to cause toxic effect to humans as well as environment (Abd-Alrahman et al. 2014).

Cereal yield loss was widely documented and inspected that it was affected due to one-half reduction in plants, one-third loss in insects and about one-fifth decline in bird species in the United Kingdom (Robinson 2016). Residues of Dichlorodiphenyltrichloroethane, chlordane and Hexachlorobenzene was found in blubber of franciscana dolphins in Brazil from 1994 to 2004 (Leonel et al. 2010). A study was arranged from 1992 to 2006 determining the concentration of pesticides (Dichlorodiphenyltrichloroethane, Hexachlorocyclohexane and Dieldrin) residues in blubber of bycaught female common dolphins (Law et al. 2013).

Hazardous impact of pesticides residues to non-target organisms were documented by several specialists. Yadav (2017) highlighted the risk of biodiversity to the toxicants of pesticides. In this study, it was described that the physiological activity of pesticides has a similarity between both pest and non-target species. For example, Carbofuran, Chlorpyrifos and Terbufos are very efficient and well-known pesticides to control the corn rootworm immatures present in the soil but these insecticides impose extremely lethal effects to populations of earthworms.

Aerial application of some pesticides was responsible for total extinction of arthropods in different crops such as cotton. Systemic insecticides caused secondary poisoning in predator species like *Chrysoperla carnea* which fed on pesticide exposed insects which are threatened by agrochemicals in farmland ecosystems (Mansoor et al. 2015). Similarly, lethal and sublethal effects of some insecticides were tested against a predatory bug *Orius insidiosus*. According to results of studies, it was derived that insecticides such as abamectin, cartap, imidacloprid, and flubendiamid was seemed to be safe and those pyriproxyfen and rynaxypyr were characterized as non-injurious and pymetrozine was categorized as somewhat damaging the bugs (Moscardini et al. 2013).

Pesticides exert very devastating effects to honey bees. Impact of imidacloprid, diafenthiuron and ethofenprox was observed in response to metabolic changes in larvae and adults of wild honey bees (*Apis dorsata)*. Haemocytes of bees were badly affected resulted in the loss of immunity against diseases and other abnormalities such as agglutination, denucleation and cell shape distortion. It was concluded that pesticide exposure was the key factor in the destruction of immune system in honey bees (Perveen and Ahmad 2017).

The impact of neonicotinoids was observed affecting the immunocompetence of honey bees (*Apis mellifera*). Thiacloprid and imidacloprid reduced the number of haemocytes, encapsulation response, and antimicrobial activity. Clothianidin was responsible for carrying out these abnormalities at somehow large concentrations (Brandt et al. 2016). Sublethal effects of clothianidin, imidacloprid and thiamethoxam against the foraging behaviour of honey bees were also recorded. Neonicotinoids were associated with multisensory disruptors and had a damaging role counter to feeding activity of pollinators (Démares et al. 2018).

After green revolution, agrochemicals became an integral part of modern crop production technologies to fulfil the dietary demands of rapidly increasing populations but in current scenario, their toxic and deadly effects to living organisms demand restricted use of these agrochemicals. Organic farming is much important to specie richness and abundance of living organisms. Though, more quickly degradable chemicals are invented which exchange the persistent ones, but their residues are still capable of putting an organism's health at risk by contaminating food, environment and water. Increment in food production in present and future point of view, must oblige with production of good class of food and with less lethal contaminants (Carvalho 2017). Data recorded from various countries showed that due to pesticide toxic effect biodiversity is declining day by day.

8.5 Impact of Pesticides Residues on Pollinators and Bees

Pollinators facilitate the transfer of pollen between flowers. Almost, 90% wild plants are reliant on insect pollination (Ollerton 2017). One third of global food production is materialized due to insect pollinators, mainly because of bees which have more than 20,000 species worldwide. Honey bees add about US $200 billion to the world economy each year through crop pollination (Eilers et al. 2011).

In 2016, the United Nations warned that 40% of invertebrate pollinator, particularly bees and butterflies, are at risk to global extermination. During past few years, decline of bee species and downfall of honeybee colonies, due to continuous pesticides applications to agricultural crops, have worried the apiculturists and ecologists About, 25–30% honey bee population decline in American and European countries is due to extreme use of pesticides. More than half of native bee species are in endangered in North American regions (Sanchez-Bayo and Goka 2014).

These beneficial insects are exaggerated by sub-lethal effects of pesticides which are often ignored. Due to repeated and long-term exposure to pesticides, physiology, neurobiology, behavior and other mechanisms of survival of pollinator insects can be changed. Pesticides residues may be carried back to hive and fed to the young ones thus imposing an extra risk and pressure to bee populations which are on continuous decline in agricultural lands (Connolly 2013).

Toxicity of pesticides may vary to honey bees. Most of the miticides, fungicides and herbicides are nontoxic to the honey bee. The biological insecticide *Bacillus thuringenesis* shows very low toxicity to bees. Different formulations of the same pesticides vary significantly in their toxicity to bees. Dust formulations are more dangerous for bees than others because they can stick to their hairs. However, granular formulations are less dangerous to bees. Residual characteristic of a pesticide is an important factor in defining its safety for pollinators. An insecticide which has low residual activity and degrades within few hours can be applied with less risk when bees are not foraging actively on plants (Sponsler et al. 2019).

Bee killing, due to pesticide use, have been known since the late nineteenth century but the discovery huge declines in wild and managed pollinators indicated reduction in crop productivity and honey yields (Berenbaum 2016). The pyrethroid insecticides can kill more bees when used in combination with fungicides and their toxicity is increased by 10 to 100 times. In Europe, almost 9% bee and butterfly species are endangered to be lost due to pesticides. A decline of 37% of bee species and 31% of butterfly species is reported in European regions. Similarly, beekeepers in the United States also testified 50% hive losses yearly with some grief losses up-to 100%. Since 2006, about 10 million bee hives have been reported to be lost, with a value of about $200 per hive. These sufferings are magnified due to neonicotinoids application at large scale (Whitehorn et al. 2012).

Unfortunately, mix of pesticides is a common practice among the farmers to save time and labour. Many pesticides like pyrethroids, imidacloprid and clothianidin produce a great reduction in foraging and navigation abilities of bees and other pollinators (Schneider et al. 2012; Feltham et al. 2014). In addition to behavioral degradation, colony collapse disorder in honeybees is also reported frequently, worldwide. Several studies narrated that the declining of bee colonies is a consequence of exposure to pesticides or their residues in cropping areas.

Organophosphate, Pyrethroid and systemic neonicotinoid insecticides like thiamethoxam and imidacloprid are mainly responsible for bee declining in exposed areas due to reduced colony initiation, lowered colony growth, and lesser reproductive outcome after exposure to certain pesticides (Fig. 8.4). Imidacloprid trace dietary residues through nectar have no lethal effects, however, can cause 6–20% reduction in the expected performance of honey bees. Overall, up-to 50% colony collapse disorders are occurring due to pesticide applications (Baron et al. 2017). Many neonicotinoids were present in plant nectar and pollen of insecticide treated crop area and nearby wild flowers also and therefore, taken into bee colony food-storage which pose serious danger to the immatures (Alaux et al. 2010).

In addition to bee damages, pesticide residues have also been reported in honey and bee-wax samples around the globe which is threatening for the end consumers (Noori et al. 2012). Eventually, agricultural production and source of revenue of small farmers will be greatly affected due to loss of beneficial insects like pollinators and specially honey bees. This challenge can be overcome by formulating the strict rules and regulations for pesticides use (Connolly 2013). Due to repeatitive application of pesticides, the physiology and other mechanisms of pollinaters become modified or altered. So, it is need of an hour to protect the pollinaters and other lives either by limiting the use of pesticides or by making strict rules and regulations about its application.

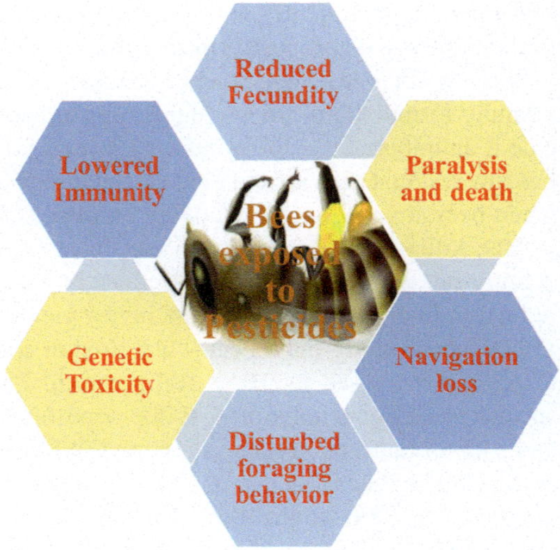

Fig. 8.4 Effect of pesticide exposures on honeybees and their life activities

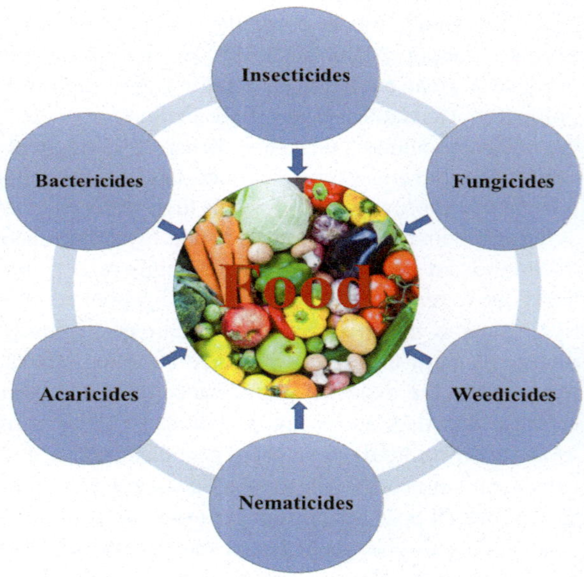

Fig. 8.5 Application various kinds of pesticides for food production leading to residual accumulation in food

8.6 Pesticides Residues and Food Chain Contamination

Pesticide contamination is a worldwide food safety issue. The magnitude and effect of pesticide exposures through food is of great concern for food consumers. Environmental contaminants like pesticides possess potential to be accumulated in both terrestrial and aquatic food chains (Fig. 8.5).

Many studies have reported pesticide buildups in soils, terrestrial and aquatic frameworks, and their lethal consequences on human and non-human biota. Innovations are utilized first and their undesired impacts appear later on. Pesticides are one of the best examples in this case (Clarkson 1995). Synthetic pesticides chemicals were presented to improve crop yields by shielding crops from pests. Because of crop yield security new pesticides products are utilized for higher crop production. The Pesticide utilization remains a typical practice particularly in Tropical and South nations (Carvalho 2017).

As the persistent synthetic chemicals have been eliminated and limited by progressively biodegradable synthetic compounds, contamination by pesticide deposits and late buildups still affect the natural human nourishment, for example, dichlorodiphenyltrichloroethane, hexachlorocyclohexane and lindane, are long-lasting chemicals. Now a days, these pesticides are prohibited for farming use in most nations. As an outcome, residues of these synthetic compounds cause contamination to the environment by scattering in the ecosystem (Reeves et al. 2019). Agriculture needs to additionally create better practices to secure general well-being, which requires progressively safe utilization of pesticides through earlier testing, cautious hazards, and additionally through training of farmers and clients, measures for better insurance of biological systems, and great practices for the reasonable improvement of agribusiness, fisheries, and aquaculture (Green et al. 2016).

Pesticide chemicals enter into the natural food chains, and lastly are ingested by people consuming such food and water. Due to pesticide resistance development, pesticide companies persistently develop new synthetic compounds. In the European Union, there are more than 800 synthetics are enlisted as pesticides. We know slightly about the ecological effects of these synthetic compounds and their negative impacts on living organisms (Köhler and Triebskorn 2013). As they are toxic and deliberately mixed in the environment, their production, distribution, and use necessitate purposeful and regular monitoring of their residues in food. The acceptable daily intakes are being established and used by governments and international risk managers, to set maximum residue limits for pesticides in food items to facilitate safer food (FAO 2010).

Numerous instances of pesticide intoxication in farmers, workers, and their families occurred during pesticide application. Pesticides application in agriculture is derived with the assistance of a few systems, from the manual services by laborers by walking to truck-and plane-based application methods. It is assumed that accidental pesticide poisoning kills an estimated 0.35 million people every year, globally. Such, poisonings are strongly related to unnecessary exposures and inappropriate use of toxic chemicals (Carvalho 2017). Dispersion of pesticide residues in the environment and large killings of human and non-human biotas, such as amphibians, bees, birds, fish and minor mammals were reported in the literature (Köhler and Triebskorn 2013).

Throughout the years, extensive research was conducted, additionally, to know the behavior of these synthetic compounds in nature to understand their decomposing cycles and fate with respect to their lethality to biota. The dichlorodiphenyltrichloroethane used to save crops was regularly noticed to be transported to the water bodies where it is quickly processed to dichlorodiphenyldichloroethylene and tend to be bio-accumulated in the oceanic food-chains and returned to end consumers in the long run (Carvalho 2006).

The organochlorine pesticides, long before were revealed as steady products, stay longer times in soils and silt, accumulating in non-human creatures with devastating impacts at the population level. Ultimately, they move in food chains with adverse effects on human health (Carvalho 2006; Köhler and Triebskorn 2013). Organochlorine synthetic compounds were queried and replaced by less persistent synthetic substances, like organophosphate, carbamate and pyrethroids based on research with respect to their degradation rates in soil and aquatic situations (Carvalho 2006). For instance, toxaphene is not used in cotton fields anymore in Nicaragua but years after end of its applications, the deposits in agricultural soils is still a contaminant source transported to marine environment with a huge danger to food farming in coastal areas (Carvalho 2006).

Soils are serving as the fundamental source of constant organochlorines. Soil erosion, surface runoff, and river bodies convey a large number of detrimental organochlorines to the ecosystem and contaminating natural food chains. Pesticide residues, conveyed to the ocean, pose higher risks to enormous marine biological systems, for example, coral reefs and other biota in the deep sea (Jamieson et al. 2017). Poisonous remains of pesticides in water systems may eradicate marine species, decrease biodiversity, and disturb the natural ecosystem functioning. Massive research in aquatic toxicology helped to comprehend pesticide bioaccumulation mechanisms to set toxicity levels for certain representative species (crustacean, fish, plants) and intricate strategies to control pesticide pollution under tolerated limits (Carvalho 2017).

The pesticides residues in food, on ingestion interact with the human gastrointestinal microbiota consisting of certain bacteria, fungi, viruses and protozoa. These pesticide residues cause disruption in the composition of microbiome, which could have lethal effects on intestinal homeostasis and overall systemic immunity (FAO 2010). The consumption and ingesting contaminated food weaken the human immunity. So, the food at MRL (maximum residual level) should be consumed that will be tolerant by the human body.

8.7 Pesticide Resdues and Health Concerns

Exposure to agrochemicals is posing serious concerns to human health. The dermal exposures account for almost 90% of the pesticide exposures (Ross et al. 2001). Toxicities by pesticide exposure are expressed in a variety of ways, ranging from mild symptoms, like slight skin irritation or allergic indications to severe symptoms, like strong headache, dizziness, or nausea. Longer exposures can cause chronic abnormalities in human ranging from cancer to other serious diseases (Fig. 8.6). The most important ailments raised by pesticide exposures are discussed here in this section.

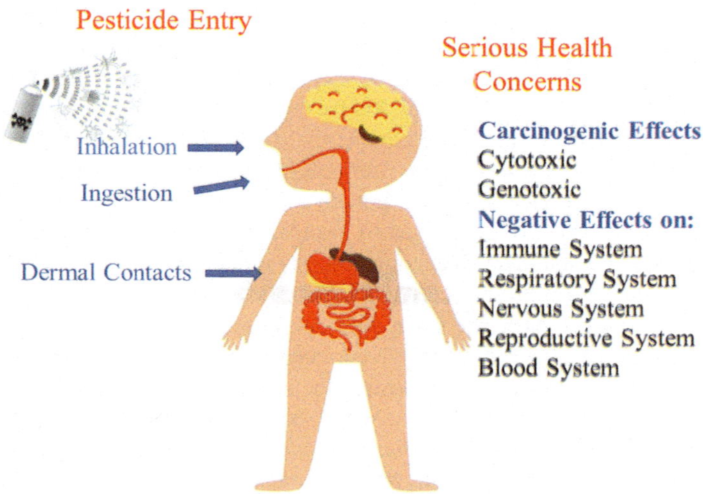

Fig. 8.6 Exposure routs and effects of pesticide exposure on human health

8.7.1 Carcinogenic Effects of Pesticide Residues

Many kinds of cancer have been produced in human due to pesticide residues exposures. Even, if a pregnant woman is exposed to the pesticide residues then there is a possibility that her child may be sensitive towards the brain cancer, leukemia, and Wilms' tumor. Pesticides exposures from outside is mainly related to the causes of blood cancer in children (Srivastava and Kesavachandran 2019). In 2010, a review found that 19 out of 32 pesticides were linked with at-least one type of cancer, including brain, bladder, colon and rectal, leukaemia, lung, melanoma, multiple myeloma, non-hodgkin lymphoma, and pancreatic cancer (Weichenthal et al. 2010).

Carcinogenic effects are still underdiscussing and there is an urgent need of research to explore the association of organochlorines and cancer, but the danger cannot be underestimated (Calle et al. 2002; Witczak and Abdel-Gawad 2014). It was documented that glyphosate was determined in endocrine disruptor (Thongprakaisang et al. 2013). Data was analyzed from ten different Districts of Spain to investigate the association of pesticide with cancer from occupational and non-occupational papulations. It was observed that cancer rated were maximum in areas belonging to intensive usage of agrochemicals compared to those of minimized consumption of such carcinogenic compounds. The study strengthened and supported the evidences of involvement of pesticides to threat of cancer (Parrón et al. 2014). Organochlorine residues, in blood stream from adipose tissues, might be mixed up with breast milk and injurious to infants (Mrema et al. 2013).

Several case-control studies analyzed rates of cancer among youngsters exposed to pesticides and inflated rates of all kinds of cancer were found higher in folks exposed to pesticides within the garden and on indoor plants and whose mothers had been exposed during pregnanancy. An outstanding study exhibited that inflated rates of childhood cancer are linked with exposure to pesticides (Srivastava and Kesavachandran 2019). In USA, a study expressed 19 and 28% excess prostate cancer risk among the farmers and commercial pesticide applicators (Koutros et al. 2010). Certain studies, from Canada, Europe and the United States, investigating the association between toxic chemical exposure and brain cancer also disclosed inflated cancer risks among the folks exposed to pesticides. Almost all organochlorines are lipophilic in nature and their bioaccumulation is very high remaining in the adipose tissues for decades (Mrema et al. 2013).

Six studies investigated the links between chemical exposure and carcinoma. Most of these, supported the supposed association. A study of feminine greenhouse employees in Kriti found that exposure to pesticides for quite 4 h daily for a minimum of 10 years inflated the chance of benign breast unwellness (Calle et al. 2002). In the same line, several studies evaluated the connection between chemical exposure and urinary organ cancer, and they found positive associations. The link was found not solely in directly exposed populations, however conjointly in youngsters of exposed folks, and was most consistent once folks had prolonged exposure (Xie et al. 2016). The relation between abdomen cancer and nitrates and some weed killer chemicals indicated that the rate of abdomen cancer was found increased in areas with high levels of weed killer contamination within the water (Sabarwal et al. 2018).

Four studies examined the association between carcinoma and chemical exposure. Results of those studies are somewhat tough to interpret as solely two collected data relating to smoking standing. In the end, pesticide associated elevated risk of carcinoma among girls exposed to pesticides at work was found (Parrón et al. 2014). Few studies were found on chemical exposure and gonad cancer. Gonad cancers are considered as a health impact of interest and evaluated its association with exposure to certain pesticides was found (Alavanja et al. 2004). During the evaluation of the connection between carcinoma and chemical exposure of pesticide residues, and it was found positive associations between pesticides exposures and this sort of cancer (Bassil et al. 2007).

8.7.2 Neurological Effects of Pesticide Residues

Experimental evidences link the hazardous exposure to higher neurologic outcomes. The residual effects of pesticide may result in each general and typical symptom. Direct exposure, as an example, will cause typical indications like contraction of the pupils, fuzzy vision, a supercilium headache, reddening and irritation in the eyes. Symptoms of general poisoning are entirely because of the buildup of neurotransmitter at the nerve ending. Under advanced poisoning, the target is pale with higher sweating, and frothing from the mouth. Alternative symptoms embrace changes in vital signs like muscle weakness, convulsions, disarray with or without coma. The victim might be dead if treatment is not given timely. In-vitro and animal-based studies showed that nanomolar concentrations of organochlorines such as endusulfan and lindane were involved in the blockage of calcium channels resulting in the deficiency of calcium ions in pheochromocytoma cells leading to mental disruptions (Heusinkveld et al. 2010).

Yokoyama (2007) highlighted Tokyo subway sarin poisoning cases along with the pesticide users (tobacco farmers) in association with Green Tobacco Sickness in Malaysia. It was determined that after 6–8 months exposure to pesticides leads to neurobehavioral and neurophysiological abnormalities. Additionally, organophosphates and dithiocarbamate disrupt peripheral nerve conduction and some related disorders (Yokoyama 2007). Similarly, mental health of Brazilian public was assessed through a questionnaire and clinched that pesticides were found in blood cholinesterase and found associated with mental disorders in all participated individuals (Buralli et al. 2019). In Mexico, farmers exposed to pesticides showed 25% elevation in depression and depression–anxiety while 24% inhibited enzymatic activity with generalized anxiety (Serrano-Medina et al. 2019).

8.7.3 Endocrine Disruptions and Pesticide Residues

Endocrine hormones play an imperative role in different body functioning particularly in growth and reproductive function. There were significant evidences of involving of Dichlorodiphenyltrichloroethane and its isomers like Dichlorodiphenyldichloroethylene in disruption of endocrine glands (Turusov et al. 2002). In general, about all classes of organochlorines were found to be associated with abnormal functioning of endocrine with very small concentration of residual contents (Lemaire et al. 2004). Organophosphates like malathion and parathion were also found active in disrupting and malfunctioning of endocrine glands (Gasnier et al. 2009).

Risk association between specific pesticide and incident polygenic disorder ranged from 20 to 200%. Risks were larger once users of specific pesticides were compared with applicators of United Nations agency who had never applied that chemical. Synthetic pesticides can also extinguish cell structure and cell functioning. Previous studies highlighted that at micromolar level of organochlorines like dieldrin, cell propagation and cell capability was declined up to a considerable extent (Slotkin and Seidler 2007). Another investigation revealed that heptachlor and dieldrin are the causal agents of raising advance apoptosis in cell cultures producing mitochondrial damage and oxidative stress (Culbreth et al. 2012).

8.7.4 *Effect of Pesticide Residues on Reproduction and Fertility*

Pesticide chemical exposures are also linked to birth imperfections, craniate death and changed fetal growth. Figure 8.7 is explaining the mode of action and effects of pesticide exposure on reproductive system on human (Bretveld et al. 2006). A weed killer, a 50:50 mixtures of 2,4,5-Trichlorophenoxyacetic Acid and 2,4-Dichlorophenoxyacetic Acid has been related to dangerous health and hereditary effects in Malaya and Vietnam. Additionally, it was found that descendants who were exposed to pesticides had birth weight and biological process defects (Srivastava and Kesavachandran 2019). A number of pesticides and a couple of 4-D compounds have been reported for impaired male fertility. Toxic chemical exposures resulted in decreased male fertility, genetic variations and reduced variety in spermatozoon, injuries to the germinal animal tissues and disturbed endocrine performance (Srivastava and Kesavachandran 2019). Sabarwal et al. (2018) emphasized that the long-term exposure of pesticides are accountable as causal agents of Parkinson's and Alzheimer's diseases in addition to various other reproductive and respiratory disorders.

8.7.5 *Pesticides Residues and Child Health*

Children are very prone and susceptible of being affected by neurotoxins of pesticides and exposure rates were very high. Scientists suggest that rapidly developing

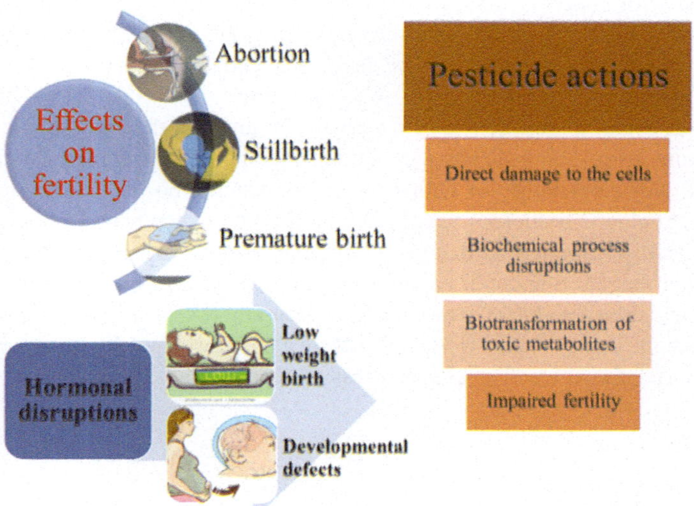

Fig. 8.7 Pesticide actions and their effects on reproductive function of human

brain tissues and cells become most vulnerable. Nevertheless, the results of pesticides were seen not as much linked with pre and postnatal exposure about neurophysiological effects. It was concluded from a study that parental exposure of pesticides resulted in deprived mental performance level and this factor was observed more in boys than in girls. It was also suggested that postnatal impact of pesticides can adversely affect child's neuropsychological behavior and parental exposure was less associated with neurodevelopmental damages (González-Alzaga et al. 2015). In 2011, a meta-analysis of 40 studies concluded that maternal pesticide exposures before birth is positively associated with 48% elevated risk of leukaemia and 53% increased risk of lymphoma in childhood. While, paternal exposures before or after the birth is linked with 50–65% higher risk of brain cancer in children (Vinson et al. 2011). Such findings are discussed in many other studies also.

8.7.6 General Health Aspects and Pesticide Residues

It is growing belief that carcinogenic activity of pesticides is the main matter of concern with public health. In addition to cancer, 4 non-cancer attributes such as dermatologic, neurologic, reproductive, and genotoxic effects and their relationships with pesticides were investigated by Sanborn et al. (2007). They concluded the positive linkage between 4 above mentioned non-cancer human health issues with pesticides. A comparative study was established to observe the effects of synthetic pesticides in contrast to naturally occurring plant chemicals. The study demonstrated that the toxicological effects of both synthetic and naturally occurring chemicals was equally harmful in context to human cancer risks. It was also observed that with the application of low doses the exposure rate of pesticides residues was insignificant (Ames et al. 1990). The key signs of pesticides boozing can be distinguished into syndromes like muscarinic syndrome; in which acetylcholine the exploitation, heart disturbances and exocrine glands with raised bronchial secretions, sweating and tearing disrupt the gastrointestinal tone. The conditions may cause diarrhea, nausea, vomiting, bronchospasm, bradycardia and elicits urinary incontinence. Some studies have reported accrued risks of eczema in the individuals exposed to pesticides (Paudyal 2008).

As health is the most important thing to survive so, compromise in health can not be afforded. Protective measures can help farm workers to avoid get exposed from agrochemicals. Some preventive measures are such as goggles, mask, gloves, shielding clothes and boots may reduce the chances of exposure. It was also reported that lack of knowledge and skills about the application is the main source of indiscriminating usage of agrochemicals and becoming the possible health hazard foundations. The application of agrochemicals must be at suggested rate and use of bio-pesticides instead of synthetic pesticides can lessen the public health issues related to exposure of these agrochemicals (Elahi et al. 2019). For this purpose, improved monitering programmes should be planed, prefer application of bio-pesticides rather than synthetic chemical pesticides, use of resistant varieties and other alternative and preventive measures should be promoted.

8.8 Conclusion

Though, pesticide chemiclas were initiated to rescue human-life by increasing agricultural productions and regualting harmful insects, weeds and diseases but their hostile properties have created great concerns about paybacks related to their uses. Above mentioned facts clearly highlights the negative costs of undiscriminating use of pesticides. Some side-effects are appeared as increased resistance in pest populations, reduced beneficial organisms for example pollinators, endangered soil microbial diversity, water and air contamination in natural ecosystem. The persistency of pesticides has affected our ecosystem to a greater extent and now pesticides have gone into countless food-chains. They are biomagnified into the higher food trophic levels including humans and other mammals. Many human illnesses, acute and chronic, have now become a big health concern because of pesticides contaminated living resources.

Now, it is need of hour, we should focus on the proper and safer pesticide applications to shield our environmental resources and prevent health hazards. Alternate pest management approaches like integrated pest management, a combination of various controling tools such as, growing resistant genotype, cultural, physical and mechanical pest control, must be deployed with rational pesticide use to decrease the frequency and pesticide applications with least amounts. Moreover, progressive cropping approaches including bio and nano-echnology can also simplify the ways for developing pest resistant crop genotypes. Pesticides with low side-effects must be explored and replaced the existing pesticdes with higher health and environmental risks. Community awareness extension programs can be utilized and promoted to teach and encourage the farmers to implement the advanced and innovative integrated pest management approaches as key tool to decrease the harmful effects of pesticides to our agro-ecosystem and natural environment.

References

Abd-Alrahman SH, Elhalwagy ME, Kotb GA, Farid H, Farag AA, Draz HM, Isa AM, Sabico S (2014) Exposure to difenoconazole, diclofop-methyl alone and combination alters oxidative stress and biochemical parameters in albino rats. Int J Clin Exp Med 7(10):3637–3646. https://www.ncbi.nlm.nih.gov/pmc/articles/PMC4238557/

Alaux C, Brunet JL, Dussaubat C, Mondet F, Tchamitchan S (2010) Interactions between Nosema microspores and a neonicotinoid weaken honeybees (*Apis mellifera*). Environ Microbiol 12(3):774–782. https://doi.org/10.1111/j.1462-2920.2009.02123.x

Alavanja MC (2009) Pesticides use and exposure extensive worldwide. Rev Environ Health 24(4):303–309. https://doi.org/10.1515/reveh.2009.24.4.303

Alavanja MC, Sprince NE, Oliver E, Whitten P, Lynch C, Blair A (2001) Nested case-control analysis of high pesticide exposure events from the agricultural health study. Am J Ind Med 39:557–563. https://doi.org/10.1002/ajim.1054

Alavanja MC, Hoppin JA, Kamel F (2004) Health effects of chronic pesticide exposure: cancer and neurotoxicity. Annu Rev Public Health 25:155–197. https://doi.org/10.1146/annurev.publhealth.25.101802.123020

Ames BN, Profet M, Gold LS (1990) Dietary pesticides (99.99% all natural). Proc Natl Acad Sci USA 87(19):7777–7781. https://doi.org/10.1073/pnas.87.19.7777

Baron GL, Jansen VAA, Brown MJF, Raine NE (2017) Pesticide reduces bumblebee colony initiation and increases probability of population extinction. Nat Ecol Evol 1:1308–1316. https://doi.org/10.1038/s41559-017-0260-1

Bassil KL, Vakil C, Sanborn M, Cole DC, Kaur JS, Kerr KJ (2007) Cancer health effects of pesticides: systematic review. Can Fam Physician 53(10):1704–1711. https://www.ncbi.nlm.nih.gov/pubmed/17934034

Beketov MA, Kefford BJ, Schäfer RB, Liess M (2013) Pesticides reduce regional biodiversity of stream invertebrates. Proc Natl Acad Sci (PNAS) 110(27):11039–11043. https://doi.org/10.1073/pnas.1305618110

Berenbaum MR (2016) Does honey bee "risk cup" runneth over? Estimating aggregate exposures for assessing pesticide risks to honey bees in agroecosystems. J Agric Food Chem 64:13–20. https://doi.org/10.1021/acs.jafc.5b01067

Boldt TS, Jacobsen CS (1998) Different toxic effects of the sulphonylurea herbicides metsulfuron methyl, chlorsulfuron and thifensulfuron methyl on fluorescent pseudomonads isolated from an agricultural soil. FEMS Microbiol Lett 161(1):29–35. https://doi.org/10.1111/j.1574-6968.1998.tb12925.x

Brakes CR, Smith RH (2005) Exposure of non-target small mammals to rodenticides: short-term effects, recovery and implications for secondary poisoning. J Appl Ecol 42:18–128. https://doi.org/10.1111/j.1365-2664.2005.00997.x

Brandt A, Gorenflo A, Siede R, Meixner M, Büchler R (2016) The neonicotinoids thiacloprid, imidacloprid, and clothianidin affect the immunocompetence of honey bees (*Apis Mellifera* L.). J Insect Physiol 86:40–47. https://doi.org/10.1016/j.jinsphys.2016.01.001

Bretveld RW, Thomas CM, Scheepers PT, Zielhuis GA, Roeleveld N (2006) Pesticide exposure: the hormonal function of the female reproductive system disrupted? Reprod Biol Endocrinol 31(4):30. https://doi.org/10.1186/1477-7827-4-30

Brown JA, Johansen PH, Colgan PW, Mathers RA (2009) Changes in the predator-avoidance behaviour of juvenile guppies (*Poecilia Reticulata*) exposed to pentachlorophenol. Can J Zool 63(9):2001–2005. https://doi.org/10.1139/z85-294

Brühl CA, Schmidt T, Pieper S, Alscher A (2013) Terrestrial pesticide exposure of amphibians: an underestimated cause of global decline? Sci Rep 3:1135. https://doi.org/10.1038/srep01135

Buralli RJ, Ribeiro H, Leão RS, Marques RC, Guimarães JRD (2019) Data on pesticide exposure and mental health screening of family farmers in Brazil. Data in brief. In press. 103993. https://doi.org/10.1016/j.dib.2019.103993.

Cai DW (2008) Understand the role of chemical pesticides and prevent misuses of pesticides. Bullet Agric Sci Technol 1:36–38. http://www.academicjournals.org/sre

Calle EE, Frumkin H, Henley SJ, Savitz DA, Thun MJ (2002) Organochlorines and breast cancer risk. CA Cancer J Clin 52:301–309. https://doi.org/10.3322/canjclin.52.5.301

Carvalho FP (2006) Agriculture, pesticides, food security and food safety. Environ Sci Pol 9(7–8):685–692. https://doi.org/10.1016/j.envsci.2006.08.002

Carvalho FP (2017) Pesticides, environment, and food safety. Food Energy Secur 6(2):48–60. https://doi.org/10.1002/fes3.108

Clarkson TW (1995) Environmental contaminants in the food chain. Am J Clin Nutr 61(3):682–686. https://doi.org/10.1093/ajcn/61.3.682S

Connolly CN (2013) The risk of insecticides to pollinating insects. Commun Integr Biol 6(5): e25074. https://doi.org/10.4161/cib.25074

Culbreth ME, Harrill JA, Freudenrich TM, Mundy WR, Shafer TJ (2012) Comparison of chemical-induced changes in proliferation and apoptosis in human and mouse neuroprogenitor cells. Neurotoxicology 33:1499–1510. https://doi.org/10.1016/j.neuro.2012.05.012

Damalas CA, Eleftherohorinos IG (2011) Pesticide exposure, safety issues, and risk assessment indicators. Int J Environ Res Public Health 11:1402–1408. https://doi.org/10.3390/ijerph8051402

De A, Bose R, Kumar A, Mozumdar S (2014) Worldwide pesticide use. In: Targeted delivery of pesticides using biodegradable polymeric nanoparticles, SpringerBriefs in molecular science. Springer, New Delhi. https://doi.org/10.1007/978-81-322-1689-6_2

Démares FJ, Pirk CW, Nicolson SW, Human H (2018) Neonicotinoids decrease sucrose responsiveness of honey bees at first contact. J Insect Physiol 108:25–30. https://doi.org/10.1016/j.jinsphys.2018.05.004

Deribe E, Rosseland BO, Borgstrom R, Salbu B, Gebremariam Z, Dadebo E, Skipperud L, Eklo OM (2013) Biomagnification of DDT and its metabolites in four fish species of a tropical lake. Ecotoxicol Environ Saf 95:10–18. https://doi.org/10.1016/j.ecoenv.2013.03.020

Dos Anjos J, de Andrade J (2014) Determination of nineteen pesticides residues (organophosphates, organochlorine, pyrethroids, carbamate, thiocarbamate and strobilurin) in coconut water by SDME/GC-MS. Microchem J 112:119–126. https://doi.org/10.1016/j.microc.2013.10.001

Eddleston M (2000) Patterns and problems of deliberate self-poisoning in the developing world. Q J Med 93:715–731. https://doi.org/10.1093/qjmed/93.11.715

Eilers EJ, Kremen C, Greenleaf SS, Garber AK, Klein AM (2011) Contribution of pollinator-mediated crops to nutrients in the human food supply. PLoS ONE 6:e21363. https://doi.org/10.1371/journal.pone.0021363

Elahi E, Weijun C, Zhang H, Nazeer M (2019) Agricultural intensification and damages to human health in relation to agrochemicals: application of artificial intelligence. Land Use Policy 83:461–474. https://doi.org/10.1016/j.landusepol.2019.02.023

FAO (2010) Pesticide residues in food – 2010 evaluations. Part I. residues. FAO plant production and protection paper 206:2011. http://www.fao.org/fileadmin/templates/agphome/documents/Pests_Pesticides/JMPR/Evaluation10/2010_Evaluation.pdf

Favari L, Lopez E, Martinez-Tabche L, Dıaz-Pardo E (2002) Effect of insecticides on plankton and fish of Ignacio Ramirez reservoir (Mexico): a biochemical and biomagnification study. Ecotoxicol Environ Saf 51(3):177–186. https://doi.org/10.1006/eesa.2002.2142

Feltham H, Park K, Goulson D (2014) Field realistic doses of pesticide imidacloprid reduce bumblebee pollen foraging efficiency. Ecotoxicol 23:317–323. https://doi.org/10.1007/s10646-014-1189-7

Fimrite P (2011) Suit says EPA fails to shield species from poisons. The San Francisco Chronicle. http://www.sfgate.com/cgi-bin/article.cgi?f=/c/a/2011/01/21/BAQ21HC5CB.DTL

Food and Environment Research Agency UK (Fera) (2009) Pesticide usage statistics: tables (select year and chemical group). http://pusstats.csl.gov.uk/index.cfm

Gasnier C, Dumont C, Benachour N, Clair E, Chagnon MC, Séralini GE (2009) Glyphosate-based herbicides are toxic and endocrine disruptors in human cell lines. Toxicology 262:184–191. https://doi.org/10.1016/j.tox.2009.06.006

Gervais JA, Rosenberg DK, Fry DM, Trulio L, Sturm KK (2000) Burrowing owls and agricultural pesticides: evaluation of residues and risks for three populations in california, USA. Environ Toxicol Chem 19(2):337–343. https://doi.org/10.1002/etc.5620190213

Gibbs KE, MacKey RL, Currie DJ (2009) Human land use, agriculture, pesticides and losses of imperiled species. Divers Distrib 15(2):242–253. https://doi.org/10.1111/j.1472-4642.2008.00543.x

Gillion RJ, Barbash JE, Crawford GG, Hamilton PA, Martin, JD, Nakagaki N, Nowell LH, Scott JC, Stackelberg PE, Thelin GP, Wolock DM (2006) 1. Overview of findings and implications, pesticides in the nation's streams and ground water, 1992–2001. Report. https://pubs.usgs.gov/circ/2005/1291/

González-Alzaga B, Hernández AF, Gómez I, Aguilar-Garduño C, Parrón T, Lacasaña M (2015) Pre- and postnatal exposures to pesticides and neurodevelopmental effects in children living in agricultural communities from South-Eastern Spain. Environ Int 85:229–237. https://doi.org/10.1016/j.envint.2015.09.019

Grasman KA, Scanlon PF, Fox GA (1998) Reproductive and physiological effects of environmental contaminants in fish-eating birds of the Great Lakes: a review of historical trends. Environ Monit Assess 53:117. https://doi.org/10.1023/A:1005915514437

Green H, Broun P, Cakmak I, Condon L, Fedoroff N, Gonzalez-Valero J, Graham I, Lewis J, Moloney M, Oniango RK (2016) Planting seeds for the future of food. J Sci Food Agric 96(5):1409–1414. https://doi.org/10.1002/jsfa.7554

Hernández F, Marín JM, Pozo OJ, Sancho JV, López FJ, Morell I (2008) Pesticide residues and transformation products in groundwater from a Spanish agricultural region on the Mediterranean Coast. Int J Environ Ana Chem 88(6):409–424. https://doi.org/10.1080/03067310701724772

Heusinkveld HJ, Thomas GO, Lamot I, van den Berg M, Kroese ABA, Westerink RH (2010) Dual actions of lindane (γ-hexachlorocyclohexane) on calcium homeostasis and exocytosis in rat PC12 cells. Toxicol Appl Pharmacol 248:12–19. https://doi.org/10.1016/j.taap.2010.06.013

Hole DG, Perkins AJ, Wilson JD, Alexander IH, Grice PV, Evans AD (2005) Does organic farming benefit biodiversity? Biol Cons 122(1):113–130. https://doi.org/10.1016/j.biocon.2004.07.018

Hussain S, Siddique T, Saleem M, Arshad M, Khalid A (2009) Impact of pesticides on soil microbial diversity, enzymes, and biochemical reactions. Adv Agron 102:159–200. https://doi.org/10.1016/S0065-2113(09)01005-0

Jamieson AJ, Malkocs T, Piertney SB, Fujii T, Zhang Z (2017) Bioaccumulation of persistent organic pollutants in the deepest ocean fauna. Nat Ecol Evol 1(3):51. https://doi.org/10.1038/s41559-016-0051

Karunakaran CO (1958) The Kerala food poisoning. J Indian Med Assoc 31:204. https://www.ncbi.nlm.nih.gov/pubmed/13588025

Katagi T, Tanaka H (2016) Metabolism, bioaccumulation, and toxicity of pesticides in aquatic insect larvae. J Pestic Sci 41(2):25–37. https://doi.org/10.1584/jpestics.D15-064

Köhler HR, Triebskorn R (2013) Wildlife ecotoxicology of pesticides: can we track effects to the population level and beyond? Science 341(6147):759–765. https://doi.org/10.1126/science.1237591

Kole RK, Banerjee H, Bhattacharyya A (2001) Monitoring of market fish samples for endosulfan and hexachlorocyclohexane residues in and around Calcutta. Bull Environ Contam Toxicol 67:554–559. https://doi.org/10.1007/s001280159

Koutros S, Alavanja MCR, Lubin JH (2010) An update of cancer incidence in the agricultural health study. J Occup Environ Med 52:1098–1105. https://doi.org/10.1097/JOM.0b013e3181f72b7c

Lanfranchi AL, Menone ML, Miglioranza KSB, Janiot LJ, Aizpún JE, Moreno VJ (2006) Striped weakfish (*Cynoscion Guatucupa*): a biomonitor of organochlorine pesticides in Estuarine and near-Coastal Zones. Mar Pollut Bull 52(1):74–80. https://doi.org/10.1016/j.marpolbul.2005.08.008

Law RJ, Bersuder P, Barry J, Barber J, Deaville R, Barnett J, Jepson PD (2013) Organochlorine pesticides and chlorobiphenyls in the blubber of bycaught female common dolphins from England and Wales from 1992–2006. Mar Pollut Bull 69(1–2):238–242. https://doi.org/10.1016/j.marpolbul.2012.12.026

Lemaire G, Terouanne B, Mauvais P, Michel S, Rahman R (2004) Effect of organochlorine pesticides on human androgen receptor activation in vitro. Toxicol Appl Pharmacol 196:235–246. https://doi.org/10.1016/j.taap.2003.12.011

Leonel J, Sericano JL, Fillmann G, Secchi E, Montone RC (2010) Long-term trends of polychlorinated biphenyls and chlorinated pesticides in franciscana dolphin (*Pontoporia Blainvillei*) from Southern Brazil. Mar Pollut Bull 60(3):412–418. https://doi.org/10.1016/j.marpolbul.2009.10.011

Lu XM, Lu PZ (2018) Response of microbial communities to pesticide residues in soil restored with *Azolla imbricata*. Appl Microbiol Biotechnol 102(1):475–484. https://doi.org/10.1007/s00253-017-8596-7

Mansoor MM, Afzal M, Raza AB, Akram Z, Waqar A, Afzal MB (2015) Post-exposure temperature influence on the toxicity of conventional and new chemistry insecticides to green lacewing chrysoperla carnea (stephens) (neuroptera: chrysopidae). Saudi J Biol Sci 22(3):317–321. https://doi.org/10.1016/j.sjbs.2014.10.008

Mitra A, Chatterjee C, Mandal FB (2011) Synthetic chemical pesticides and their effects on birds. Res J Environ Toxicol 5(2):81–96. https://doi.org/10.3923/rjet.2011.81.96

Moscardini VF, Gontijo P, Carvalho GA, de Oliveira RL, Maia JB, De Silva FF (2013) Toxicity and sublethal effects of seven insecticides to eggs of the flower bug *Orius Insidiosus* (say) (hemiptera: anthocoridae). Chemosphere 92(5):490–496. https://doi.org/10.1016/j.chemosphere.2013.01.111

Mostafalou S, Abdollahi M (2012) Concerns of environmental persistence of pesticides and human chronic diseases. Clin Exp Pharmacol S5:e002. https://doi.org/10.4172/2161-1459.S5-e002

Mostafalou S, Abdollahi M (2013) Pesticides and human chronic diseases: evidences, mechanisms, and perspectives. Toxicol Appl Pharmacol 268(2):157–177. https://doi.org/10.1016/j.taap.2013.01.025

Mrema EJ, Rubino FM, Brambilla G, Moretto A, Tsatsakis AM, Colosio C (2013) Persistent organochlorinated pesticides and mechanisms of their toxicity. Toxicology 307:74–88. https://doi.org/10.1016/j.tox.2012.11.015

Newton (1976) Population limitation in diurnal raptors. Can Field Nat 90(3):274–300

Noori AW, Salom K, Al-Ghamdi A, Ansari MA (2012) Antibiotic, pesticide, and microbial contaminants of honey: human health hazards. Sci World J 2012(9). https://doi.org/10.1100/2012/930849

Ollerton J (2017) Pollinator diversity: distribution, ecological function, and conservation. Annu Rev Ecol Evol Syst 48:353–376. https://doi.org/10.1146/annurev-ecolsys-110316-022919

Parrón T, Requena M, Hernández AF, Alarcón R (2014) Environmental exposure to pesticides and cancer risk in multiple human organ systems. Toxicol Lett 230(2):157–165. https://doi.org/10.1016/j.toxlet.2013.11.009

Paudyal BP (2008) Organophosphorus poisoning. J Nepal Med Assoc 47(172):251–258. https://www.ncbi.nlm.nih.gov/pubmed/19079407

Pereira L, Fernandes MN, Martinez CB (2013) Hematological and biochemical alterations in the fish *Prochilodus lineatus* caused by the herbicide clomazone. Environ Toxicol Pharmacol 36(1):1–8. https://doi.org/10.1016/j.etap.2013.02.019

Perveen N, Ahmad M (2017) Toxicity of some insecticides to the haemocytes of giant honeybee, *Apis Dorsata* F. under laboratory conditions. Saudi J Biol Sci 24(5):1016–1022. https://doi.org/10.1016/j.sjbs.2016.12.011

Reeves WR, McGuire MK, Stokes M, Vicini JL (2019) Assessing the safety of pesticides in food: how current regulations protect human health. Adv Nutr 10(1):80–88. https://doi.org/10.1093/advances/nmy061

Robinson RA (2016) Post-war changes in arable farming and biodiversity in Great Britain. J Appl Ecol 39:157–176. https://doi.org/10.1046/j.1365-2664.2002.00695.x

Ross JH, Driver JH, Cochran RC, Thongsinthusak T, Krieger RI (2001) Could pesticide toxicology studies be more relevant to occupational risk assessment? Ann Occup Hyg 45(1):5–17. https://doi.org/10.1016/S0003-4878(00)00095-8

Sabarwal A, Kumar K, Singh RP (2018) Hazardous effects of chemical pesticides on human health–cancer and other associated disorders. Environ Toxicol Pharmacol 63:103–114. https://doi.org/10.1016/j.etap.2018.08.018

Sanborn M, Kerr KJ, Sanin LH, Cole DC, Bassil KL, Vakil C (2007) Non-cancer health effects of pesticides: systematic review and implications for family doctors. Can Fam Physician 53(10):1712–1720. https://www.ncbi.nlm.nih.gov/pubmed/17934035

Sanchez-Bayo F, Goka K (2014) Pesticide residues and bees – a risk assessment. PLoS ONE 9(4). https://doi.org/10.1371/journal.pone.0094482

Schneider CW, Tautz J, Grünewald B, Fuchs S (2012) RFID tracking of sublethal effects of two neonicotinoid insecticides on the foraging behavior of *Apis mellifera*. PloS One 7:e30023. https://doi.org/10.1371/journal.pone.0030023

Sebiomo A, Ogundero VW, Bankole SA (2011) Effect of four herbicides on microbial population, soil organic matter and dehydrogenase activity. Afr J of Biotechnol 10(5):770–778. https://doi.org/10.5897/AJB10.989

Serrano-Medina A, Ugalde-Lizárraga A, Bojorquez-Cuevas MS, Garnica-Ruiz J, González-Corral MA, García-Ledezma A, Pineda-García G, Cornejo-Bravo JM (2019) Neuropsychiatric disorders in farmers associated with organophosphorus pesticide exposure in a rural village of Northwest México. Int J Environ Res Public Health 16(5):689. https://doi.org/10.3390/ijerph16050689

Slotkin TA, Seidler FJ (2007) Comparative developmental neurotoxicity of organophosphates in vivo: transcriptional responses of pathways for brain cell development, cell signaling, cytotoxicity and neurotransmitter systems. Brain Res Bull 72:232–274. https://doi.org/10.1016/j.brainresbull.2007.01.005

Sponsler DB, Grozinger CM, Hitaj C, Rundlöf M, Botías C, Code A, Douglas MR (2019) Pesticides and pollinators: a socioecological synthesis. Sci Tot Environ 662:1012–1027. https://doi.org/10.1016/j.scitotenv.2019.01.016

Srivastava AK, Kesavachandran C (2019) Health effects of pesticides. CRC Press/Taylor & Francis Group, London. https://doi.org/10.1201/9780429058219

Stinson ER, Hayes LE, Bush PB, White DH (1994) Carbofuran affects wildlife on Virginia corn fields. Wild Soc Bull 22(4):566–575. https://www.jstor.org/stable/3783081

Storkey J, Meyer S, Still KS, Leuschner C (2012) The impact of agricultural intensification and land-use change on the European arable flora. Proc Biol Sci 279(1732):1421–1429. https://doi.org/10.1098/rspb.2011.1686

Thongprakaisang S, Thiantanawat A, Rangkadilok N, Suriyo T, Satayavivad J (2013) Glyphosate induces human breast cancer cells growth via estrogen receptors. Food Chem Toxicol 59:129–136. https://doi.org/10.1016/j.fct.2013.05.057

Tien CJ, Chen CS (2012) Assessing the toxicity of organophosphorous pesticides to indigenous algae with implication for their ecotoxicological impact to aquatic ecosystems. J Environ Sci Health Part B 47:901–912. https://doi.org/10.1080/03601234.2012.693870

Turusov V, Rakitsky V, Tomatis L (2002) Dichlorodiphenyltrichloroethane (DDT): ubiquity, persistence, and risks. Environ Health Perspect 110:125–128. https://doi.org/10.1289/ehp.02110125

Venter O, Brodeur NN, Nemiroff L, Belland B, Dolinsek IJ, Grant JWA (2006) Threats to endangered species in Canada. Bioscience 56(11):903–910. https://doi.org/10.1641/0006-3568(2006)56[903:TTESIC]2.0.CO;2

Vinson F, Merhi M, Baldi I, Raynal H, Gamet-Payrastre L (2011) Exposure to pesticides and risk of childhood cancer: a meta-analysis of recent epidemiological studies. Occup Environ Med 68(9):694–702. http://www.ncbi.nlm.nih.gov/pubmed/21606468

Walczak M, Reichert M (2016) Characteristics of selected bioaccumulative substances and their impact on fish health. J Vet Res 60(4):473–480. https://doi.org/10.1515/jvetres-2016-0070

Wasim MD, Sengupta D, Chowdhry A (2009) Impact of pesticides use in agriculture: their benefits and hazards. Interdiscip Toxicol 2(1):1–12. https://doi.org/10.2478/v10102-009-0001-7

Waskom R (1995) Best management practices for private well protection. Colorado State University Cooperative Extension. http://waterscape.org/home/vanduo/dw_lit/bmp_xcm179_private_well.pdf

Weber JB, Wilkerson GG, Reinhardt CF (2004) Calculating pesticide sorption coefficients (Kd) using selected soil properties. Chemosphere 55(2):157–166. https://doi.org/10.1016/j.chemosphere.2003.10.049

Weichenthal S, Moase C, Chan P (2010) A review of pesticide exposure and cancer incidence in the agricultural health study cohort. Environ Health Perspect 118(8):1117–1125. http://www.ncbi.nlm.nih.gov/pubmed/20444670

Whitehorn PR, O'Connor S, Wackers FL (2012) Neonicotinoid pesticide reduces bumble bee colony growth and queen production. Science 336(6079):351–352. https://doi.org/10.1126/science.1215025

Witczak A, Abdel-Gawad H (2014) Assessment of health risk from organochlorine pesticides residues in high-fat spreadable foods produced in Poland. J Environ Sci Health B 49:917–928. https://doi.org/10.1080/03601234.2014.951574

Wu WJ, Qin N, Zhu Y, He QS, Ouyang HL, He W, Liu WX, Xu FL (2013) The residual levels and Health Risks of Hexachlorocyclohexanes (HCHs) and Dichloro-Diphenyl-Trichloroethanes (DDTs) in the fish from Lake Baiyangdian, North China. Environ Sci Poll Res 20(9):5950–5962. https://doi.org/10.1007/s11356-013-1607-z

Xie B, Hu Y, Liang Z, Liu B, Zheng X, Xie L (2016) Association between pesticide exposure and risk of kidney cancer: a meta-analysis. Onco Targets Ther 28(9):3893–3900. https://doi.org/10.2147/OTT.S104334

Yadav SK (2017) Pesticide applications-threat to ecosystems. J Human Ecol 32(1):37–45. https://doi.org/10.1080/09709274.2010.11906319

Yokoyama K (2007) Our recent experiences with sarin poisoning cases in Japan and pesticide users with references to some selected chemicals. Neurotoxicology 28(2):364–373. https://doi.org/10.1016/j.neuro.2006.04.006

Chapter 9
Pesticides Pollution and Analysis in Water

P. Senthil Kumar and G. Janet Joshiba

Abstract Nowadays, individuals have been exposed to various kinds of substances with a wide range because of the evolution of industries, transport, science, and technology. One of the major categories of toxic compounds is pesticides which serve as a fundamental compound in elevating food production in agriculture by shielding the crops from harmful insects and pests. In addition, it is also helpful in food preservation and inhibition of vector-borne diseases. In spite, of all the advantages of the pesticide, it is not recommended in many developed countries because of its potential risk at human beings and the ecosystem. Due to various environmental conditions like agricultural runoff, precipitation, leaching and bioaccumulation the pesticide affects the human beings and animals in a wide spectrum by damaging the organs and causing various metabolic disorders. With a view, to combat all the deleterious effects of the pesticides on the environment and living beings, it is mandatory to monitor its concentration in the ecosystem. Various conventional techniques such as chromatography, electrophoresis, spectrometry and biological assay are used in the analysis of pesticides. In recent times, the biosensors and molecular imprinted polymers have outstripped all other conventional technologies. This chapter depicts the impacts of water pollution caused by pesticides and it also explains the various technologies used in the analysis of pesticide residues.

Keywords Pesticides · Agriculture · Analysis · Chromatography · Biosensors

9.1 Introduction

Water is an integral constituent of our mother nature and it is subsided from the clouds in the form of rain, streams, lakes, seas, etc., The progressive development in the industrialization and population explosion has led to the destruction in the quality

P. Senthil Kumar (✉) · G. Janet Joshiba
Department of Chemical Engineering, Sri Sivasubramaniya Nadar College of Engineering, Chennai, Tamil Nadu, India

Inamuddin et al. (eds.), *Sustainable Agriculture Reviews 48*, Sustainable Agriculture Reviews 48, https://doi.org/10.1007/978-3-030-54719-6_9

of water which is generally utilized in various purposes such as irrigation, industries, domestic and commercial activities (Dunca 2018). Water pollution is defined as the addition of various toxic contaminants to water which modifies and damages the quality of the water. Furthermore, water pollution also leads to the enumeration of unwanted harmful and toxic contaminants into the marine sources which affect the wellness of the living beings in the ecosystem. The polluted water causes a deficit of drinking water, in another way it also contaminates the groundwater and makes it unfit for drinking and domestic purposes (Dunca 2018).

In the current scenario, there are various anthropogenic and non-anthropogenic water pollution sources present in our environment, out of that water pollution caused due to pesticides is one of the major threats growing in our society. This can happen when pesticides are utilized in ashore zones to develop crops and are flushed away by wind and rain into water bodies. Accordingly, they modify the condition of such water bodies by changing to it's physical, compound or natural conditions, subsequently making it lethal, sullied and unsatisfactory for use (Permaculture news). Around 80% of the total population of the earth is majorly concerned about water pollution, water deficiency, water depreciation, and Weather change. Out of the various outrageous consequences, water pollution ranks first in affecting the livelihood of the ecosystem and around two million people are affected every year because of the various contagious disease caused by the polluted water (Chen et al. 2019). The outrageous growth of global population has led to intensification in the global food production sector and in another way, around 45% of the food crops are wasted every year because of the damage caused due to pests and insects. Further, the pest management using pesticides and insect repellents are considered to be an effective way of controlling the pests and its negative impacts on the environment (Abhilash and Singh 2009).

After the devastation and negative impacts of the Bhopal gas leak tragedy which happened on 1984 December, various changes have been incorporated in the legislation and new rules have been enunciated regarding the permissible limits of industrial effluents, vehicular emissions, toxic wastes clearance, etc., Water acts are one of the important pollution control law which restricts and regulates the elevation of water pollution and it also helps in controlling the water crisis (Dunca 2018). The addition of pesticides into the water bodies effects some of the harmful consequences such as destruction of aquatic organisms, groundwater contamination, Biomagnification, foodchain disruption, deleterious diseases, etc., (Permaculture news).In order to avoid pesticide water pollution and its harmful impacts, the periodical monitoring of the water quality should be made mandatory in all regions. For the sake of assessing the quality of the water some water quality indices such as Water pollution index, Water quality index and River habitat survey are used in the water and wastewater treatment sector (Dunca 2018). This chapter depicts the pesticide water pollution and its impact on the wellness of the ecosystem; also, it explains the various methods used for the analysis of pesticide contamination in water.

9.2 Pesticides as Water Contaminants

The pesticide is one of the ubiquitous synthetic compounds which is used in various sectors of our livelihood such as agriculture, industries, and medicines. These harmful synthetic pesticides initially came into existence by the year 1940; by then the demand for the pesticides begin to elevate which resulted in the consumption of pesticides by 25% of the total world population. Also by the year 2001, around 2.26 million active ingredients were incorporated in the manufacture of pesticides (World Health Organization). Pesticides are harmful chemicals which are generally composed of oxygen, nitrogen, sulphur, phosphorus, bromine, and nitrogen, Further it also consists of some of the toxic heavy metals such as arsenic, lead, mercury and sulphates which are highly harmful to the living organisms. The pesticides are majorly utilized in the food preservation, vector born disease prevention and to control the harmful pest such as weeds, rodents, and insects which damage crop cultivation. In spite of its harmful chemical nature, the pesticides are applied in agriculture to increase the crop productiveness and revenue (Permaculture news, World Health Organization 1988). The Pesticides play an important role in enhancing a country's economy by increasing the yield in the food production sector along with the support of several agricultural types of equipment and innovative techniques (Cycoń et al. 2016). From the reports of Pesticide Action Network delivered in 2008, it is clear that around 350 various pesticides are recognized in the edible foods consumed by everyone living in the European Union. As per the reports of the European Union, every year on an average of 140,000 tons of pesticides is utilized in agriculture for pest management. Some fruits like grapes, potatoes, citrus fruits, etc., are prone to pesticides and uptake pesticides at a higher dosage level (Fenik et al. 2011). Nearly 1 million children are getting affected every year due to the usage of pesticides in the prevention of certain diseases such as malaria, dengue, leishmaniasis, and Japanese encephalitis. The pesticide stability, persistence, toxicity, and environmental fate are some of the reasons behind the hazardous nature of the pesticides (World Health Organization).

A principal supporter of the Green Revolution has been the advancement and use of pesticides for the control of a wide assortment of insectivorous and herbaceous bugs that would damage the crop production and yield. The utilization of pesticides harmonizes with the "compound age" which has changed society since the 1950s (www.fao.org).There are in excess of 1000 pesticides utilized the world over to guarantee sustenance isn't harmed or demolished by insects and pests. Several primitive and cheaper pesticides get deposited in the soil for a considerable length of time. They have been prohibited in many developed nations, whereas in some developing nations they are still utilized for agricultural purposes. Pesticides are highly toxic to humans and other living organisms and it is capable of causing acute, chronic and genetic disorders based on the quantity and mode through which it is consumed (Safe drinking water foundation). The organochlorine pesticides such as hexachlorocyclohexane, dichlorodiphenyltrichloroethane, dieldrin and aldrinare the predominant pesticides used in several developing countries because of their feasibility and versatility, Though, these pesticides are banned in many developed countries because of their capacity to get accumulated and complexed degradation process (Abhilash and Singh 2009). As per the reports of Food and Agriculture Organization of the United Nations, It is inferred that five major countries such as Costa Rica, Maldives, Bahamas, Trinidad, and Tobago, and Barbados have utilized pesticides above a higher rate of about 20 kg/ha yr. in the year 2016, Then the develop nation like the United States of Asia has utilized one fifth of the global pesticide utilization level of about 372 and 389 million kg by the year 2006 and 2007 (Wang et al. 2019).

9.3 Classification of Pesticides

Generally, Pesticides are classified based on the five important principles such as chemical structure, the purpose of utilization, toxicity, degradation ability and mode of action on target organisms. The pesticides are classified into organic and inorganic pesticides based on their chemical structure. The compounds such as inorganic herbicides, arsenic insecticides, fluoride insecticides, and inorganic fungicides are categorized as inorganic pesticides, whereas the pesticide compounds such as organochlorine pesticides, organonitrogen and organophosphorus pesticides fall under the category of inorganic pesticides (Fenik et al. 2011). The pesticides are of various types, every variant of pesticide works specifically against some pest such as insects, rodents, moles, etc., Some of the pesticide types used commercially in the agricultural, industrial and household sectors are:

- Antimicrobials
- Algaecides
- Biopesticides
- Desiccants
- Disinfectants
- Defoliants
- Illegal and counterfeit pesticides
- Insecticides
- Fungicides
- Herbicides
- Insect growth regulators
- Mothballs
- Miticides
- Molluscicides
- Ovicides
- Natural and biological pesticides
- Pheromones
- Repellents
- Plant growth regulators
- Synergists
- Wood preservative

Based on the chemical constituents present in the pesticides it is divided into five major classes such as organochlorines, organophosphorus, carbamates, pyrethrins, pyrethroids, neonicotinoids and miscellaneous pesticides (Pundir et al. 2019; Schmuck et al. 1996).

9.4 Source, Fate and Transport of Pesticides

The pesticides after entering into the environment undergo various physical and chemical processes which include degradation, metamorphosis, adsorption, desorption, volatilization, accumulation and runoff into nearby water sources. Among these, Transformation or metamorphosis stands first in managing the outcome and conveyance of a pesticide into the environment. It undergoes some degradation processes such as oxidation, hydrolysis, photolysis, and biodegradation (Cycoń et al. 2016).

9.5 Environmental Effects of Pesticide Pollution

Due to their toxic nature and persistence pesticides exert various negative impacts on the ecosystem. Various classifications of pesticides have different kinds of consequences for living life forms, in this way speculation is troublesome. The two main primary terrestrial impacts of pesticides due to soil erosion and runoff are Bioconcentration and Biomaganification. Bioconcentration is the evolution of the harmful chemical compounds into the living cells of the organism. Generally, the pesticides get concentrated in the fatty tissues, In this regard, the pesticide like Dichlorodiphenyltrichloroethane which is lipophilic gets easily accumulated in the fatty tissues of humans and aquatic organisms. In addition, the glyphosate pesticide easily gets metabolized and degraded. Biomagnification is caused due to the deposition of harmful chemical components in the living organisms through the food cycle, various types of pesticides and different synthetic substances are progressively amplified in tissue and different organs (www.fao.org).The presence of highly toxic chemicals in the pesticides which is used in agriculture results in damaging the biology of the soil by interrupting the enzyme activities, nitrification, legume nodulation and symbiosis relationships of the soil (Cycoń et al. 2016). The utilization of high-level toxic pesticides without proper knowledge damages the nutrient conditions of the soil and it also affects the livelihood of natural Biocontrol agents which are known as the friends of cultivation. The Bio-control agents present in the soil environment generally enhances the nutritional source of the soil and it also helps in controlling pests to an extent (Abhilash & Singh 2009).

9.6 Health Effects Caused Due to Pesticides

Pesticides are classified among one of the dangerous synthetic components which is capable of causing deleterious and genetic disorders. The toxic chemical components of pesticides are ingested into the human body through skin contact, inspiration and consumption of pesticide-contaminated food items (www.fao.org). The human body when it comes in contact with the toxic pesticides it results in several acute and chronic health disorders. The pesticide accumulation in the human body can cause kidney dysfunctions, sterility issues, immune system suppressions, cancer, neurological dysfunctions, endocrine disorders and pesticide poisoning among kids. In the case of pesticides, the vigorousness of the health disorders varies based on the level of dosage of pesticides to the human body. Some of the mild health effects caused due to pesticides are skin rashes, headaches, flu, and vision problems, whereas the severe health effects of pesticides are blindness, paralysis, and death (Abhilash and Singh 2009; Kim et al. 2016). The pesticides easily get deposited with the human body tissues such as in blood, breast, and adipose tissues, further, they don't undergo any metabolic activity and get deposited within the living cells and tissues for a prolonged time interval. Results of various epidemiological studies have concluded that the pesticide is capable of causing Parkinson's disorder. The chlordane pesticide compounds cause anxiety, depressions, immune suppression, migraines, and diabetes. According to United States Environmental Protection Agency, pesticides such as aldrin and dieldrin are categorized as powerful carcinogens and prolonged exposure to these compounds causes vomiting, headaches, convulsions, irritations, and dizziness. Uptake of heavy doses of aldrin and dieldrin damages the excretory system and nervous system of the human body (Taiwo 2019). The organochlorine pesticide groups are highly hazardous and most complicated for degradation. This pesticide group is major cancer-causing, estrogenic and highly accumulating kind of pesticides. The organophosphorus and carbamate groups are also called enzyme inhibitors, most commonly they inhibit the enzyme activity of cholinesterase which leads to serious damage in the central nervous system (Samsidar et al. 2018). The toxicity of pesticides causes many epidemiology associated diseases such as diabetes, alzheimer disease, parkinson disease, sclerosis, birth defects and neurological disorders (Sabarwal et al. 2018).

9.7 Analysis of Pesticides

Monitoring of pesticide concentration in the water and land sources is essential to control its harmful impacts and to implement remediation techniques. In addition, the concentration of pesticides has also been monitored in aquatic biota and several other species to study their bioaccumulation and magnification rate in the living cells (Miller et al. 2019). The analysis of pesticides using any kind of conventional or modern technique involves four important phases such as sample preparation, sample separation, sample analysis, and a detection phase. During the examination of pesticide using any analytical technique, the sample preparation is known to be the most important and sensitive step. The precise monitoring of pesticides mainly depends on two factors such as effective separation and extent of purity of the pesticide sample extracted from the environmental sources. The pesticide sample is extracted using techniques such as solid-phase extraction, liquid-liquid extraction, solid-phase microextraction, and Matrix solid-phase dispersion are followed in the separation of pesticide samples to be quantified from the natural or man-made sources (Samsidar et al. 2018). Owing to their toxicity and complexity nature towards degradation, the pesticide compounds are monitored periodically using various effective and promising analytic techniques which includes thin layer chromatography, high pressure liquid chromatography, gas-liquid chromatography, colorimetry, mass spectrometry, nuclear magnetic resonance Spectroscopy, capillary electrophoresis, enzyme-linked immunosorbent assay and biosensors (Pundir et al. 2019). In a research work conducted by Millet et al. for analyzing the concentration of various contaminants in the marine environment using Gammaruspulex collected from 15 different sites. The results concluded that along with various other contaminants some pesticides which are banned and it also showed the presence of fenuron in a maximum amount of about 16.1 ng g^{-1} (Miller et al. 2019).

9.8 Thin-Layer Chromatography

Generally, the Thin Layer Chromatography and HighPerformance Thin Layer Chromatography are simple, rapid, feasible and effective techniques in the detection of pesticides. In the thin layer Chromatography, the various stationary phases used are aluminum, glass and plastic support, whereas in the mobile phase various solvents are used depending on the type of material to be eluted. Thin Layer Chromatography is highly preferred in the analysis of pesticides because of the factor that it does not require samples with high purity, easy separation, the economically feasible and wide decision of mobile phases. Over the period, several advancements have been made in the thin layer chromatography such as automated sample application; computer-aided scanning equipment and imaging software which helps in enhancing the accuracy and precision in the analysis of pesticides. In recent times, the high-performance thin-layer chromatography coupled with mass spectrometry (High performance thin layer chromatography –mass spectrometry) is utilized in the monitoring of various complicated compounds like pesticides (Tuzimski 2018).

9.9 Liquid Chromatography

Liquid chromatography is usually used in the analysis of non-volatile, high polar and thermally unstable pesticide compounds. High Pressure Liquid Chromatography is highly preferred in the monitoring of pesticides because of their effectiveness and feasibility. High Pressure Liquid Chromatography is made up of four important units such as stationary phase column, pump, mobile phase, and a detector. The C-18 column is the most preferred column for the detection of the pesticide compound. The type of column, mobile phase, and detector are chosen based on the type of compound to be eluted. Some of the factors which affect the retention time seem to be the strength of the stationary phase, solvent phase composition, flow rate and type of detector used. For detecting the pesticide compounds the detectors such as Ultra Violet and Fluorescence detectors are preferred, Furthermore some of the Mass spectrometry detectors such as tandem-Mass Spectrometry, ion trap, single quadrupole and time-of-flight for advanced detection of pesticide residues. The Time-of-flight –Mass Spectrometre detectors coupled with Liquid Chromatography are known to be the most effective analytical technique in determining the pesticide residues with high specificity and sensitivity (Tuzimski 2018).

9.10 Gas Chromatography

In the Gas chromatography technique, the selection of the appropriate column for the detection of pesticide compounds is more important. There are several varieties of columns available for the detection of pesticides. Based on the polarity and type of chemical present in every pesticide, the type of column is chosen. Basically, two-dimensional gas chromatography is preferably used in the detection of pesticides. Various stationary phases used in the detection of pesticides include:

- Silica column integrated with 95% dimethyl-5% diphenylpolysiloxane Silica with 1,4-bis(dimethylsiloxy) phenylenedimethylpolysiloxane
- Silica with 14% cyanopropylphenyl & 86% dimethylpolysiloxane Silica with 35% diphenyl & 65% dimethylpolysiloxane 8Silica with 95% dimethyl-5% diphenylpolysiloxane In the advanced analysis process, the Gas chromatography is merged with various other analytical equipment such as Electrolytic conductivity detection, Electron capture detection and Mass spectrometry depending on the type of pesticide to be detected. Using the gas chromatography, the pesticides can be detected with high resolution and specificity; also it can be used to detect all classes of the pesticide compounds (Tuzimski 2018).

9.11 Miscellaneous Chromatography

Capillary Electro chromatography is the consolidation of two technologies such as capillary electrophoresis and high-performance liquid chromatography. In this Capillary Electro chromatography, electroosmosis acts as the driving force through which the mobile phase is circulated into the bed of the chromatograph. In this Capillary Electro chromatography, the monolithic columns encompassing magnetic nanoparticles are used for the monitoring of organophosphorus pesticides. Supercritical fluid chromatography is a type of chromatography in which the supercritical fluid such as carbon dioxide is utilized as the mobile phase. It is well known that carbon dioxide has the tendency to get easily converted into a supercritical state with a critical temperature and pressure of about 31.1 °C and 7.38 MPa. This Supercritical fluid chromatography is highly preferred in pesticide analysis because of its unique properties such as diverse separation modes, high flow rate, and high sensitivity. For more effective and précised pesticide analysis the Supercritical fluid chromatography are coupled with Mass spectrometry detector. Micellar electrokinetic chromatography is a form of technology where the analytes of interest are segregated by differential partitioning amid mobile phase and pseudo stationary phase, In addition, it is the combination of chromatography and electrophoretic separation techniques. The micelles are present in the pseudo-stationary phase and the buffer solution serves as the mobile phase. Micellar electrokinetic chromatography is highly preferred in the analysis of pesticides such as carbamate and organophosphorus (Tuzimski 2018; Samsidar et al. 2018).

9.12 Spectrometry Techniques

Spectrometry is one of the most important analytical techniques used for the determination of pesticides. Nuclear magnetic resonance is one of the highly sensible types of spectrometers used in the analysis of pesticides. In addition, it can be also used in the determination of some parent pesticide compound. This method is coupled with other analytical techniques such as Liquid Chromatography and Gas Chromatography and used as the detectors (Tuzimski 2018). Tandem mass spectrometry is one of the sensitive analytical techniques applied in the monitoring of pesticides. Surface-enhanced Raman spectroscopy is the integration of two techniques such as nanotechnology and Raman spectroscopy. This Surface-enhanced Raman spectroscopy detects the pesticide residue effectively at a higher speed and efficiency, also, it is sensitive and detects the pesticide even at lower concentrations (Pang et al. 2016).

9.13 Electrophoresis Techniques

The capillary electrophoresis is the analytical method in which the separation is based upon the ratio of charge to mass of the target/ analyte of interest. The neutral and both charged targets are separated through the optimization of electrolyte parameters. Various factors such as pH, type of charge and concentration influence the efficiency of capillary electrophoresis by affecting the velocity of the charge, shape of peak and partition efficiency. The effective separation of charges at a faster rate is achieved by implementing narrow-bore capillary of about 25 to 100 mm. The detectors such as fluorescence detector, Ultra Violet detector, and Mass spectrometry detector are utilized in the Capillary electrophoresis method. The pesticides such as organophosphorus and trichlorfon are monitored using the capillary electrophoresis technique. For higher resolution, sensitivity, and specificity the capillary electrophoresis is coupled with Mass spectrometry detector.

9.14 BioAssays

The immunoassays are one type of sensitive and highly specific assay for determining the pesticide residue from the environment and it is completely based on the synergy between the antigen and its specific antibody. Basically, the pesticide compounds are not antigenic so it is required to affix a carrier molecule with the pesticide to bring out the immune reaction. The pesticide detection is performed using various methods such as chemiluminescence, fluorescence, and radiometric methods. In recent times, these immunoassays are ported to be an effective method for the analysis of pesticides because of its rapidness, feasibility, and specificity. Enzyme-linked immunosorbent assay is one of the well-known pesticide analysis techniques among the various other assays. It offers rapid, highly specific and précised examination of complex variants of pesticides. For the examination of cyprodinil compound in fruits, the monoclonal antibody-based immunoassay is performed (Tuzimski 2018).

9.15 Biosensors

Biosensors are a simple analytical instrument that identifies any targeted analyte compound of interest with the support of a biorecognition element and a signal transducer. The biorecognition element observes the physical, chemical and biological reverberations exerted by the analyte compound, Whereas, the signal transducers, in turn, transform all the responses into some form of quantifiable signals. The Biosensors have been applied in various applications such as environmental concerns, agriculture, Biomedical sector, Medical diagnostics, and analytical purposes. The biosensors are categorized into five major classes such as electrochemical, thermal, piezoelectric, optical and enzyme-based biosensors. They are known to be a simple, effective, specific and more promising method in monitoring pesticide contamination. In the enzyme-based sensors the BChE, tyrosinase, alkaline phosphate, etc., are conventionally used as the biorecognition elements (Pundir et al. 2019). Electrochemical sensors are a category of sensors which is based on the specific interaction between the analyte of interest and biorecognition element. These sensors are mainly based upon the type of enzyme used. For effective and advanced analysis of pesticides, these sensors are doped with various nanomaterials on the surface. The Acetylcholine esterase pesticide analysis is utilized in various applications such as monitoring of contaminants, food safety, and quality control. These biosensors have the capability to replace the existing conventional technologies by simplifying the sample preparation, simpler field testing and feasible cost for analysis. The integration of nanomaterials with biomolecules is seemed to be one of the effective approaches in the biosensing sector. The receptor recognizes the biomolecule and the transducer transmits the signals as electrical signals which are collected by the detectors such as amperometry, potentiometer, and voltammeter. The Molecular imprinted polymers are one of the advancements of the biosensors which are highly specific and versatile which can detect an analyte of various ranges of molecular weight. These MIP are similar as the biomolecules and it mocks the function of biorecognition elements. The carbamate pesticides are detected using the electrochemical sensor coupled with Boron-doped diamond electrode coupled with various detectors such as differential cyclic voltammetry, pulse voltammetry, and square wave voltammetry. Furthermore, this electrode is preferred in the detection of methomyl in surface waters and other water bodies. In the analysis of the metobromuron pesticide, the sensors with ultra-trace graphite electrode coated using graphene nanoplatelets and carbon nanotubes are used. The optical biosensors are used in wide applications such as food safety, protection, detection of contaminants and medicine. The optical transducers receive the signals from the analyte of interest and respond by altering their optical properties such as fluorescence, reflection, and adsorption. In the detection of organophosphorus pesticide, the luminescent europium-o-(4-methoxy benzoyl) benzoic acid [o-(4-anisoyl)] complex in ethanol-water solution is used (Tuzimski 2018; Samsidar et al. 2018). The nanomaterials integration on the biosensors helps in enhancing the uniqueness of the biosensors in the detection of pesticides. The type of biosensor used in the analysis of pesticides varies according to the complexity of the pesticides. Basically, the amperometric sensors are used in the fast, feasible and specific monitoring of pesticides, whereas, the fluorescent biosensors are utilized in the primary and secondary quantification of toxic pesticides (Zhao et al. 2018; Samsidar et al. 2018).

9.16 Conclusion & Future Perspectives

Pesticide monitoring has become one of the greatest concerns to various developing and developed nations. As the environmental and health effects of pesticides are vigorous, it is claimed to be one of the persistent organic pollutants by various organizations and highly toxic variants have been banned in various developed nations to control the harmful impacts of pesticides. Various analytical techniques such as Thin Layer Chromatography, Liquid Chromatography, Gas Chromatography, High-Pressure Liquid Chromatography, Capillary electrophoresis, Mass spectrometry, Enzyme Linked Immuno Sorbent Assay and some Bio-immunoassays have been followed conventionally for analysis of pesticides. For more advanced analysis many different detectors such as Mass Spectrometry, Fluorescent Detector, Nitrogen Phosphorus Detector, Flame Photometric Detector, Electron Capture Detector and Diode Array Detector are coupled with the analytical techniques. The above techniques are highly selective, highly sensitive and efficient in analyzing the pesticide residues, despite all the unique characteristics of these techniques; they exert two main drawbacks such as higher time consumption and high cost. The evolution of biosensors has eased the pesticide monitoring process, also it was found to be the most favorable analytical technique for analysis of pesticides. The enzyme immune sensors are has dominated all other conventional technologies and it is found to be sensitive towards the composition of pesticides. As the consequences of the pesticides on the environment are getting increased much more attention must be given to new innovative analytical approaches in the monitoring of pesticides.

References

Abhilash PC, Singh N (2009) Pesticide use and application: an Indian scenario. J Hazard Mat 165:1–12. https://doi.org/10.1016/j.jhazmat.2008.10.061

Chen B, Wang M, Duan M, Ma X, Hong J, Xie F, Zhang R, Li X (2019) In search of key: protecting human health and the ecosystem from water pollution in China. J Clean Prod 228:101–111. https://doi.org/10.1016/j.jclepro.2019.04.228

Copplestone JF (1988) The development of the WHO recommended classification of pesticides by hazard. Bull World Health Organ 66:545–551

Cycoń M, Mrozik A, Piotrowska-Seget Z (2016) Bioaugmentation as a strategy for the remediation of pesticide-polluted soil: a review. Chemosphere 172:52–71. https://doi.org/10.1016/j.chemosphere.2016.12.129

Dunca AM (2018) Water pollution and water quality assessment of major transboundary Rivers from Banat (Romania). J Chem. https://doi.org/10.1155/2018/9073763

Fenik J, Tankiewicz M, Biziuk M (2011) Properties and determinationof pesticides in fruits and vegetables. Trends Anal Chem 30:814–826. https://doi.org/10.1016/j.trac.2011.02.008

http://www.fao.org/3/w2598e/w2598e07.htm

https://permaculturenews.org/2017/05/24/pesticides-water-pollution/

https://www.safewater.org/fact-sheets-1/2017/1/23/pesticides

https://www.who.int/ceh/capacity/Pesticides.pdf

Kim KH, Kabir E, Jahan SA (2016) Exposure to pesticides and the associated human health effect. Sci Total Environ 575:525–535. https://doi.org/10.1016/j.scitotenv.2016.09.009

Miller T, Ng KT, Bury ST, Bury SE, Bury NR, Barron LP (2019) Biomonitoring of pesticides, pharmaceuticals and illcit drugs in a fresh water invertebrate to estimate toxic or effect pressure. Environ Int 129:595–606. https://doi.org/10.1016/j.envint.2019.04.038

Pang S, Yang T, He L (2016) Review of surface enhanced Raman spectroscopic (SERS) detection of synthetic chemical pesticides. Trends Anal Chem 85:73–82. https://doi.org/10.1016/j.trac.2016.06.017

Pundir CS, Malik A, Preety (2019) Bio-sensing of organophosphorus pesticides: a review. Biosens Bioelectron 140:111348. https://doi.org/10.1016/j.bios.2019.111348

Sabarwal A, Kumar K, Singh RP (2018) Hazardous effects of chemical pesticides on human health – cancer and other associated disorders. Environ Toxicol Pharmacol 63:103–114. https://doi.org/10.1016/j.scitotenv

Samsidar A, Siddiquee S, Shaarani SM (2018) A review of extraction, analytical and advanced methods for determination of pesticides in environment and foodstuffs. Trends Food Sci Technol 71:188–201. https://doi.org/10.1016/j.tifs.2017.11.011

Schmuck R, Mager H, Kunast C, Bock KD, Weyhermuller SS (1996) Variability in the reproductive performance of beneficial insects in standard laboratory toxicity assays – implications for hazard classification of pesticides. Ann Appl Biol 128:437–451. https://doi.org/10.1111/j.1744-7348.1996.tb07105.x

Taiwo AM (2019) A review of environmental and health effects of organochlorine pesticide residues in Africa. Chemosphere 220:1126–1140. https://doi.org/10.1016/j.chemosphere.2019.01.001

Tuzimski T (2018) Herbicides and Pesticides. Medical University of Lublin, Lublin

Wang R, Yuan Y, Yen H, Grieneisen M, Arnold J, Wang D, Wang C, Zhang M (2019) A review of pesticide fate and transport simulation at watershed level using SWAT: current status and research concerns. Sci Total Environ 669:512–526. https://doi.org/10.1016/j.scitotenv.2019.03.141

Zhao F, Wu J, Ying Y, She Y, Wang J, Ping J (2018) Carbon nanomaterial-enabled pesticide biosensors: design strategy, biosensing mechanism, and practical application. Trends Anal Chem 106:62–83. https://doi.org/10.1016/j.trac.2018.06.017

Index

Inamuddin et al. (eds.), *Sustainable Agriculture Reviews 48*, Sustainable Agriculture Reviews 48, https://doi.org/10.1007/978-3-030-54719-6

The manufacturer's authorised representative in the EU is Springer
Nature Customer Service Centre GmbH, Europaplatz 3, 69115 Heidelberg,
Germany. If you have any concerns regarding our products, please
contact ProductSafety@springernature.com

Printed and bound by CPI Group (UK) Ltd, Croydon, CR0 4YY
30/07/2026
02182022-0001